ASCE/SEI 31-03

American Society of Civil Engineers

Seismic Evaluation of Existing Buildings

This document uses both the International System of Units (SI) and customary units.

Published by the American Society of Civil Engineers

Library of Congress Cataloging-in-Publication Data

Seismic evaluation of existing buildings / American Society of Civil Engineers.
    p. cm. -- (ASCE standard)
  Includes bibliographical references and index.
  "SEI/ASCE 31/02."
  ISBN 0-7844-0670-7
 1. Buildings--Earthquake effects. I. Structural Engineering Institute II. American Society of Civil Engineers.

TH1095.S3842 2003
693.8'52--dc21

                              2003041921

Published by the American Society of Civil Engineers
1801 Alexander Bell Drive
Reston, Virginia 20191
www.asce pubs.asce.org

Any statements expressed in these materials are those of the individual authors and do not necessarily represent the views of ASCE, which takes no responsibility for any statement made herein. No reference made in this publication to any specific method, product, process or service constitutes or implies an endorsement, recommendation, or warranty thereof by ASCE. The materials are for general information only and do not represent a standard of ASCE, nor are they intended as a reference in purchase specifications, contracts, regulations, statutes, or any other legal document. ASCE makes no representation or warranty of any kind, whether express or implied, concerning the accuracy, completeness, suitability, or utility of any information, apparatus, product, or process discussed in this publication, and assumes no liability therefore. This information should not be used without first securing competent advice with respect to its suitability for any general or specific application. Anyone utilizing this information assumes all liability arising from such use, including but not limited to infringement of any patent or patents.

ASCE and American Society of Civil Engineers—Registered in U.S. Patent and Trademark Office.

Photocopies: Authorization to photocopy material for internal or personal use under circumstances not falling within the fair use provisions of the Copyright Act is granted by ASCE to libraries and other users registered with the Copyright Clearance Center (CCC) Transactional Reporting Service, provided that the base fee of $18.00 per article is paid directly to CCC, 222 Rosewood Drive, Danvers, MA 01923. The identification for ASCE Books is 0-7844-0670-7/03/ $18.00. Requests for special permission or bulk copying should be addressed to Permissions & Copyright Dept., ASCE.

Copyright © 2003 by the American Society of Civil Engineers.
All Rights Reserved.
Library of Congress Catalog Card No: 2003041921
ISBN 0-7844-0670-7
Manufactured in the United States of America.

STANDARDS

In April 1980, the Board of Direction approved ASCE Rules for Standards Committees to govern the writing and maintenance of standards developed by the Society. All such standards are developed by a consensus standards process managed by the Codes and Standards Activities Committee. The consensus process includes balloting by the balanced standards committee, which is composed of Society members and nonmembers, balloting by the membership of ASCE as a whole, and balloting by the public. All standards are updated or reaffirmed by the same process at intervals not exceeding 5 years.

The following standards have been issued:

ANSI/ASCE 1-82 N-725	Guideline for Design and Analysis of Nuclear Safety Related Earth Structures
ANSI/ASCE 2-91	Measurement of Oxygen Transfer in Clean Water
ANSI/ASCE 3-91	Standard for the Structural Design of Composite Slabs and ANSI/ASCE 9-91 Standard Practice for the Construction and Inspection of Composite Slabs
ASCE 4-98	Seismic Analysis of Safety-Related Nuclear Structures
Building Code Requirements for Masonry Structures (ACI 530-02/ASCE 5-02/TMS 402-02) and Specifications for Masonry Structures (ACI 530.1-02/ASCE 6-02/TMS 602-02)	
SEI/ASCE 7-02	Minimum Design Loads for Buildings and Other Structures
ASCE 8-02	Standard Specification for the Design of Cold-Formed Stainless Steel Structural Members
ANSI/ASCE 9-91	Listed with ASCE 3-91
ASCE 10-97	Design of Latticed Steel Transmission Structures
SEI/ASCE 11-99	Guideline for Structural Condition Assessment of Existing Buildings
ANSI/ASCE 12-91	Guideline for the Design of Urban Subsurface Drainage
ASCE 13-93	Standard Guidelines for Installation of Urban Subsurface Drainage
ASCE 14-93	Standard Guidelines for Operation and Maintenance of Urban Subsurface Drainage
ASCE 15-98	Standard Practice for Direct Design of Buried Precast Concrete Pipe Using Standard Installations (SIDD)
ASCE 16-95	Standard for Load and Resistance Factor Design (LRFD) of Engineered Wood Construction
ASCE 17-96	Air-Supported Structures
ASCE 18-96	Standard Guidelines for In-Process Oxygen Transfer Testing
ASCE 19-96	Structural Applications of Steel Cables for Buildings
ASCE 20-96	Standard Guidelines for the Design and Installation of Pile Foundations
ASCE 21-96	Automated People Mover Standards—Part 1
ASCE 21-98	Automated People Moves Standards—Part 2
ASCE 21-00	Automated People Mover Standards—Part 3
SEI/ASCE 23-97	Specification for Structural Steel Beams with Web Openings
SEI/ASCE 24-98	Flood Resistant Design and Construction

ASCE 25-97	Earthquake-Actuated Automatic Gas Shut-Off Devices
ASCE 26-97	Standard Practice for Design of Buried Precast Concrete Box Sections
ASCE 27-00	Standard Practice for Direct Design of Precast Concrete Pipe for Jacking in Trenchless Construction
ASCE 28-00	Standard Practice for Direct Design of Precast Concrete Box Sections for Jacking in Trenchless Construction
SEI/ASCE/SFPE 29-99	Standard Calculation Methods for Structural Fire Protection
SEI/ASCE 30-00	Guideline for Condition Assessment of the Building Envelope
ASCE/SEI 31-03	Seismic Evaluation of Existing Buildings
SEI/ASCE 32-01	Design and Construction of Frost-Protected Shallow Foundations
EWRI/ASCE 33-01	Comprehensive Transboundary International Water Quality Management Agreement
EWRI/ASCE 34-01	Standard Guidelines for Artificial Recharge of Ground Water
EWRI/ASCE 35-01	Guidelines for Quality Assurance of Installed Fine-Pore Aeration Equipment
CI/ASCE 36-01	Standard Construction Guidelines for Microtunneling
SEI/ASCE 37-02	Design Loads on Structures During Construction
CI/ASCE 38-02	Standard Guideline for the Collection and Depiction of Existing Subsurface Utility Data
EWRI/ASCE 39-03	Standard Practice for the Design and Operation of Hail Suppression Projects

FOREWORD

The material presented in this publication has been prepared in accordance with recognized engineering principles. This Standard and Commentary should not be used without first securing competent advice with respect to their suitability for any given application. The publication of the material contained herein is not intended as a representation or warranty on the part of the American Society of Civil Engineers, or of any other person named herein, that this information is suitable for any general or particular use or promises freedom from infringement of any patent or patents. Anyone making use of this information assumes all liability from such use.

ACKNOWLEDGEMENTS

The American Society of Civil Engineers (ASCE) acknowledges the work of the Seismic Rehabilitation of Existing Buildings Standards Committee of the Codes and Standards Activities Division of the Structural Engineering Institute. This group comprises individuals from many backgrounds including: consulting engineering, research, construction industry, education, government, design, and private practice. This Standard process began in 1998 and incorporates information as described in the commentary.

This Standard was prepared through the consensus standards process by balloting in compliance with procedures of ASCE's Codes and Standards Activities Committee. Those individuals who serve on the Standards Committee are:

Bechara E. Abboud
Aziz Alfi
David Allen
Prodyot K. Basu
Michael D. Blakely
Allen R. Bone
David C. Breiholz
James Brown
Thomas M. Bykonen
James R. Cagley
Hashu H. Chandwaney
Fu-Lien Chang
Chang Chen
Kevin C. K. Chueng
James H. Collins
W. G. Corley
Majed A. Dabdoub
Michael D. Davister
Steven L. Dickson
Max Falamaki
Richard B. Fallgren
Mark W. Fantozzi
Hans Gesund
Stephen H. Getz
Nader Ghafoori
Stayendra K. Ghosh
Sergio Gonzalez-Karg
Phillip Gould
Melvyn Green
Max A. Gregersen
Michael R. Hagerty
Harold S. Hamada
D. Kirk Harman
David B. Hattis
John R. Hayes

Richard L. Hess
James A. Hill
William T. Holmes
Darrick Hom, Secretary
Charles J. Hookham
J. Kent Hsiao
Tom C. Hui
Roy J. Hunt
Mohammad Iqbal
Robert C. Jackson
Wen-Chen Jau
Martin W. Johnson
John C. Kariotis
Brian E. Kehoe
Peter H. Lam
Patrick J. Lama
Jim E. Lapping
Darrell J. Lawver
Feng-Bao Lin
Phillip Line
David E. Linton
Rene W. Luft
Terry R. Lundeen
Charles R. Magadini
Ayaz H. Malik
Lincoln E. Malik
Rusk Masih
Vicki V. May
Frank E. McClure
Bruce H. McCracken
James B. McDermott
Richard McConnell
Mike Mehrain
Stanley H. Mendes
Martha Merriam

Thomas H. Miller
Andy H. Milligan
Andrew P. Misovec
Jack Moll
Myles A. Murray
Joseph F. Muessendorfer
Joseph P. Nicoletti
Glen J. Pappas
James C. Parkert
Mandakumaran Paruvakat
Celina U. Penalba
Mark A. Pickett
Jose A. Pincheira
Chris D. Poland, Chair
Daniel E. Pradel
Denis C. Pu
R. C. Richardson
Timothy E. Roecker
Charles W. Roeder
Abdulreza A. Sadjadi
Ali M. Sadre
Arthur B. Savery

Ashvin A. Shah
Daniel Shapiro
Richard L. Silva
Thomas D. Skaggs
Glenn R. Smith, Jr.
Charles A. Spitz
William W. Stewart
Eric C. Stovner
Donald R. Strand
Peter Tian
Eugene Trahern
Frederick M. Turner
Michael T. Valley
Ivan P. Vamos
Gara Varum
Thomas G. Williamson
Lyle L. Wilson
Lisa A. Wipplinger
Tom C. Xia
Wen-Huei Yen
Wade W. Younie

CONTENTS

1.0	**General Provisions**	1-1
1.1	Scope	1-1
1.2	Basic Requirements	1-3
1.3	Definitions	1-8
1.4	Notation	1-12
1.5	References	1-16
2.0	**Evaluation Requirements**	2-1
2.1	General	2-1
2.2	Level of Investigation Required	2-1
2.3	Site Visit	2-2
2.4	Level of Performance	2-3
2.5	Level of Seismicity	2-4
2.6	Building Type	2-5
3.0	**Screening Phase (Tier 1)**	3-1
3.1	General	3-1
3.2	Benchmark Buildings	3-3
3.3	Selection and Use of Checklists	3-5
3.4	Further Evaluation Requirements	3-7
3.5	Tier 1 Analysis	3-9
3.6	Level of Low Seismicity Checklist	3-20
3.7	Structural Checklists	3-21
3.8	Geologic Site Hazards And Foundations Checklist	3-119
3.9	Nonstructural Checklists	3-121
4.0	**Evaluation Phase (Tier 2)**	4-1
4.1	General	4-1
4.2	Tier 2 Analysis	4-1
4.3	Procedures for Building Systems	4-33
4.4	Procedures for Lateral-Force-Resisting Systems	4-50
4.5	Procedures for Diaphragms	4-94
4.6	Procedures for Connections	4-106
4.7	Procedures for Geologic Site Hazards and Foundations	4-118
4.8	Procedures for Nonstructural Components	4-124
5.0	**Detailed Evaluation Phase (Tier 3)**	5-1
5.1	General	5-1
5.2	Available Procedures	5-1
5.3	Selection of Detailed Procedures	5-3

Appendix A - Examples **A-1**
A1.0 Example 1: Building Type W1: Wood Light Frame A-2
A2.0 Example 2: Building Type S1A: Steel Moment Frame with Flexible Diaphragms A-14
A3.0 Example 3: Building Type C3:
 Concrete Frame with Infill Masonry Shear Walls and Stiff Diaphragms A-34
A4.0 Example 4: Building Type RM2:
 Reinforced Masonry Bearing Wall Building with Stiff Diaphragms A-44
A5.0 Example 5: Building Type W2: Wood Frame, Commercial and Industrial A-59
A6.0 Example 6: Building Type S2: Steel Braced Frame with Stiff Diaphragms A-73
A7.0 Example 7: Building Type URM:
 Unreinforced Masonry Bearing Wall Building with Flexible Diaphragms A-88

Appendix B - Summary Data Sheet **B-1**

Index **I-1**

1.0 General Provisions

1.1 Scope

This standard provides a three-tiered process for seismic evaluation of existing buildings in any level of seismicity (Section 2.5). Buildings are evaluated to either the Life Safety or Immediate Occupancy Performance Level (Section 2.4). The design of mitigation measures is not addressed in this standard.

This standard does not preclude a building from being evaluated by other well-established procedures based on rational methods of analysis in accordance with principles of mechanics and approved by the authority having jurisdiction (if any).

C1.1 Scope

This standard provides a process for seismic evaluation of existing buildings. It is intended to serve as a nationally applicable tool for design professionals, code officials, and building owners looking to seismically evaluate existing buildings. This standard may be used on a voluntary basis or may be required by the authority having jurisdiction. A major portion is dedicated to instructing the evaluating design professional on how to determine if a building is adequately designed and constructed to resist seismic forces. All aspects of building performance are considered and defined in terms of structural, nonstructural, and foundation/ geologic hazard issues. Lifelines such as water, electrical, natural gas supply lines, and waste disposal lines beyond the perimeter of the building, which may be necessary for buildings to be occupied, are not considered in this document.

The evaluation procedures include a consideration of ground shaking and to a limited extent other seismic hazards such as liquefaction, slope failure, surface fault rupture, and effects of neighboring structures. Other phenomena such as tsunami, lateral spreading, and local topological effects are not considered.

The need for evaluation using this standard may have been indicated by rapid visual screening using FEMA 154, *Rapid Visual Screening of Buildings for Potential Seismic Hazards: A Handbook* (FEMA, 1988a).

Mitigation strategies for rehabilitating buildings found to be deficient are not included in this standard; additional resources should be consulted for information regarding mitigation strategies.

Standard Basis

This standard has evolved from and is intended to replace FEMA 310, *Handbook for Seismic Evaluation of Buildings—A Prestandard* (FEMA, 1998). This standard was written to:

- Reflect advancements in technology,
- Incorporate the experience of design professionals,
- Incorporate lessons learned during recent earthquakes,
- Be compatible with FEMA 356, *Prestandard and Commentary for the Seismic Rehabilitation of Buildings* (FEMA, 2000c),
- Be suitable for adoption in building codes and contracts,
- Be nationally applicable, and
- Provide evaluation techniques for varying levels of building performance.

Since the development and publication of FEMA 178, *NEHRP Handbook for the Seismic Evaluation of Existing Buildings* (BSSC, 1992a), numerous significant earthquakes have occurred: the 1985 Michoacan earthquakes that affected the Mexico City area, the 1989 Loma Prieta earthquake in the San Francisco Bay Area, the 1994 Northridge earthquake in the Los Angeles area, and the 1995 Hyokogen-Nanbu earthquake in the Kobe area. While each earthquake validated the fundamental assumptions underlying the procedures presented in FEMA 178, each also offered new insights into the potential weaknesses in certain systems that should be evaluated. This knowledge was incorporated into FEMA 310.

Extent of Application

No buildings are automatically exempt from the evaluation provisions of this standard; exemptions should be defined by public policy. However, based on the exemption contained in the codes for new buildings, jurisdictions may choose to exempt the following classes of construction:

- Detached one- and two-family dwellings located where the design short-period spectral response acceleration parameter, S_{DS}, is less than 0.4g
- Detached one- and two-family wood frame dwellings located where the design short-period response acceleration parameter, S_{DS}, is equal to or greater than 0.4g that satisfy the light-frame construction requirements of FEMA 368 and 369, *2000 NEHRP Recommended Provisions for Seismic Regulations for New Buildings and Other Structures* (BSSC, 2000)
- Agricultural storage structures that are intended only for incidental human occupancy

Application to Historic Buildings

Although the principles for evaluating historic structures are similar to those for other buildings, the design professional should be aware of special conditions and considerations that may exist.

Historic structures often include archaic materials, systems, and details. It may be necessary to look at handbooks and building codes from the year of construction to determine details and material properties.

Although the expected performance of architectural elements and finishes must be considered for all types of buildings, the interaction of architectural and structural elements in historic buildings often plays a more important role in the overall seismic performance of the structural system. Disturbance of historic architectural elements and finishes to allow testing during evaluation and to implement the resulting rehabilitation measures may also be unacceptable. It is often necessary to evaluate historic buildings on a case-by-case basis and using general performance, rather than prescriptive, criteria.

There are national and often state and municipal registers of historic places, buildings, and districts (neighborhoods). Additionally for some programs, "eligibility" for the register is sufficient cause for special treatment. All U.S. states and territories have a designated State Historic Preservation Officer who should be consulted regarding these registers.

In addition, an appropriate level of performance for historic structures needs to be chosen that is acceptable to the local jurisdiction. Some feel that historic buildings should meet the safety levels of other buildings since they are a subset of the general seismic safety needs. Others feel that historic structures, because of their value to society, should meet a higher level of performance. In some cases, a reduced level of performance has been allowed to avoid damaging historic fabric during rehabilitation. In other cases, a higher performance objective has been used to enhance post-earthquake reparability of historic features.

General Provisions

> The following resources may be useful where evaluating historic structures:
>
> - *Standards for the Treatment of Historic Properties* (Secretary of the Interior, 1995)
> - *National Park Service Catalog of Historic Buildings Preservation Briefs* (Secretary of the Interior, 1975–2001) *California Historical Building Code* (CBSC, 1995)
> - *1998 Proceedings on Disaster Management Programs for Historic Sites* (Secretary of the Interior, 1998)
>
> **Alternative Methods**
>
> Alternative documents that may be used to evaluate existing buildings include, but are not limited to:
> - *Guidelines for the Seismic Retrofit of Existing Buildings* (ICBO, 2001)
> - *Los Angeles Division 91, Los Angeles Division 95* (LADBS, 1997)
> - ATC-40, *Seismic Evaluation and Retrofit of Concrete Buildings* (ATC, 1996)
>
> Some users have based the seismic evaluation of buildings on the provisions of new buildings. While this may seem appropriate, it must be done with full knowledge of the inherent assumptions. Codes for new buildings contain requirements that govern building configuration, strength, stiffness, detailing, and special inspection and testing. The strength and stiffness requirements are easily transferred to existing buildings; the other provisions are not. If the lateral-force-resisting elements of an existing building do not have details of construction similar to those required for new construction, the basic assumptions of ductility will not be met. Lateral-force-resisting elements that are not properly detailed should be omitted during an evaluation using a code for new buildings, unless the interaction of such elements may result in less desirable seismic performance.
>
> **Mitigation Strategies**
>
> Potential seismic deficiencies in existing buildings may be identified using this standard. If the evaluation is voluntary, the owner may choose to accept the risk of damage from future earthquakes rather than upgrade, or to demolish the building. If the evaluation is required by a local ordinance for a hazard-reduction program, the owner may have to choose among rehabilitation, demolition, or other options.
>
> The following documents may be useful in determining appropriate rehabilitation or mitigation strategies:
> - FEMA 172, *NEHRP Handbook of Techniques for the Seismic Rehabilitation of Existing Buildings* (BSSC, 1992b)
> - FEMA 227 and 228, *NEHRP Benefit-Cost Model for the Seismic Rehabilitation of Buildings*
> - FEMA 156 and 157, *NEHRP Typical Costs for Seismic Rehabilitation of Existing Buildings*
> - FEMA 356, *Prestandard and Commentary for the Seismic Rehabilitation of Buildings* (FEMA, 2000c)

1.2 Basic Requirements

Prior to completing a seismic evaluation, the requirements of Section 2 shall be met.

The three-tiered process for seismic evaluation of buildings is depicted in Figure 1-1.

> **C1.2 Basic Requirements**
>
> Prior to conducting the seismic evaluation based on this standard, the design professional should understand the evaluation process and the basic requirements specified in this section.

> The evaluation process consists of the following three tiers, as shown in Figure 1-1: Screening Phase (Tier 1), Evaluation Phase (Tier 2), and Detailed Evaluation Phase (Tier 3). As indicated in Figure 1-1, the design professional may choose to (1) report deficiencies and recommend mitigation or (2) conduct further evaluation, after any tier of the evaluation process.
>
> **Judgment by the Design Professional**
>
> While this standard provides very prescriptive direction for the evaluation of existing buildings, it is not to be taken as the only direction. This standard provides direction for common details, deficiencies, and behavior observed in past earthquakes that are found in common building types. However, every structure is unique and may contain features and details that are not covered by this standard. It is important that the design professional use judgment where applying the provisions of this standard. The design professional should always look for uncommon details and behavior about the structure that may have the potential for damage or collapse or that may improve the performance of the building relative to buildings of the same building type.

1.2.1 Tier 1 – Screening Phase

A Tier 1 Evaluation shall be conducted for all buildings in accordance with the requirements of Section 3. Benchmark building criteria shall be checked. Checklists, as applicable, of compliant/non-compliant statements related to structural, nonstructural, and foundation conditions, shall be selected and completed in accordance with the requirements of Section 3.3 (Table 3-2) for a Tier 1 Evaluation. It shall be permitted to substitute Tier 2 calculations for the Tier 1 Quick Checks. Potential deficiencies shall be summarized upon completion of the Tier 1 Evaluation.

> **C1.2.1 Tier 1 – Screening Phase**
>
> The screening phase, Tier 1, consists of three sets of checklists that allow a rapid evaluation of the structural, nonstructural, and foundation/geologic hazard elements of the building and site conditions. It shall be completed for all building evaluations conducted in accordance with this standard. The purpose of a Tier 1 Evaluation is to screen out buildings that comply with the provisions of this standard or quickly identify potential deficiencies. In some cases, "Quick Checks" may be required during a Tier 1 Evaluation; however, the level of analysis necessary is minimal. If deficiencies are identified for a building using the checklists, the design professional may proceed to Tier 2 and conduct a more detailed evaluation of the building or conclude the evaluation and state that potential deficiencies were identified. In some cases, a Tier 2 or Tier 3 Evaluation may be required, even if no deficiencies are noted in Tier 1.

1.2.2 Tier 2 – Evaluation Phase

For those buildings identified in Section 3.4 (Table 3-3), a Full-Building Tier 2 Evaluation or a Tier 3 Evaluation shall be performed. For those buildings not identified as requiring a Full Building Tier 2 Evaluation or a Tier 3 Evaluation, but for which potential deficiencies were identified in Tier 1, a Deficiency-Only Tier 2 Evaluation may be performed. For a Deficiency-Only Tier 2 Evaluation, only the procedures associated with non-compliant checklist statements need be completed. Potential deficiencies shall be summarized upon completion of the Tier 2 Evaluation. Alternatively, the design professional may choose to end the investigation and report the deficiencies in accordance with Section 1.

C1.2.2 Tier 2 – Evaluation Phase

Based on the ABK research (ABK, 1984), unreinforced masonry buildings with flexible diaphragms were shown to behave in a unique manner. The special analysis procedures provided in Section 4.2.6 were developed to predict the behavior of such buildings. For Tier 2, a complete analysis of the building that addresses all of the deficiencies identified in Tier 1 shall be performed. Analysis in Tier 2 is limited to simplified linear analysis methods. As in Tier 1, evaluation in Tier 2 is intended to identify buildings not requiring rehabilitation. If deficiencies are identified during a Tier 2 Evaluation, the design professional may choose to either conclude the evaluation and report the deficiencies or proceed to Tier 3 and conduct a detailed seismic evaluation.

1.2.3 Tier 3 – Detailed Evaluation Phase

A Tier 3 Evaluation shall be performed in accordance with the requirements of Section 5 for buildings identified in Section 3.4 (Table 3-3) or where the design professional chooses to further evaluate buildings for which potential deficiencies were identified in Tier 1 or Tier 2. Potential deficiencies shall be summarized upon completion of the Tier 3 Evaluation.

C1.2.3 Tier 3 – Detailed Evaluation Phase

References that describe methods for conducting a Tier 3 Detailed Evaluation are provided in Section 5 commentary of this standard. Recent research has shown that certain types of complex structures can be shown to be adequate using nonlinear analysis procedures even though other common procedures do not. While these procedures are complex and expensive to carry out, they often result in construction savings equal to many times their cost. The use of Tier 3 procedures must be limited to appropriate cases.

1.2.4 Final Report

After a seismic evaluation has been performed, a final report shall be prepared. As a minimum, the report shall include the following items:

1. Scope and Intent: The purpose for the evaluation including jurisdiction requirements (if any), a list of the tier(s) followed, and level of investigation conducted
2. Site and Building Data:
   - General building description (number of stories and dimensions)
   - Structural system description (framing, lateral-force-resisting system, floor and roof diaphragm construction, basement, and foundation system)
   - Nonstructural systems description (all nonstructural elements that affect seismic performance)
   - Building type
   - Performance level
   - Level of seismicity
   - Soil type
3. List of Assumptions: Material properties, site soil conditions
4. Findings: A prioritized list of deficiencies

C1.2.4 Final Report

The final report serves to communicate the results to the owner and record the process and assumptions used to complete the evaluation. Each section should be carefully written in a manner that is understandable to its intended audience. The extent of the final report may range from a letter to a detailed document. Depending on the availability of information and the scope of the evaluation effort, the final report may include the following items (in addition to the required items):

1. Site and Building Data:
   - Building occupancy and use
   - Level of inspections and testing conducted
   - Availability of original design and construction documents
   - Historical significance
   - Past performance of the building type in earthquakes

2. Recommendations: Mitigation schemes or further evaluation

3. Appendix: References, preliminary calculations, photographs, material test results, all necessary checklists, summary data sheet, and analysis procedure

General Provisions

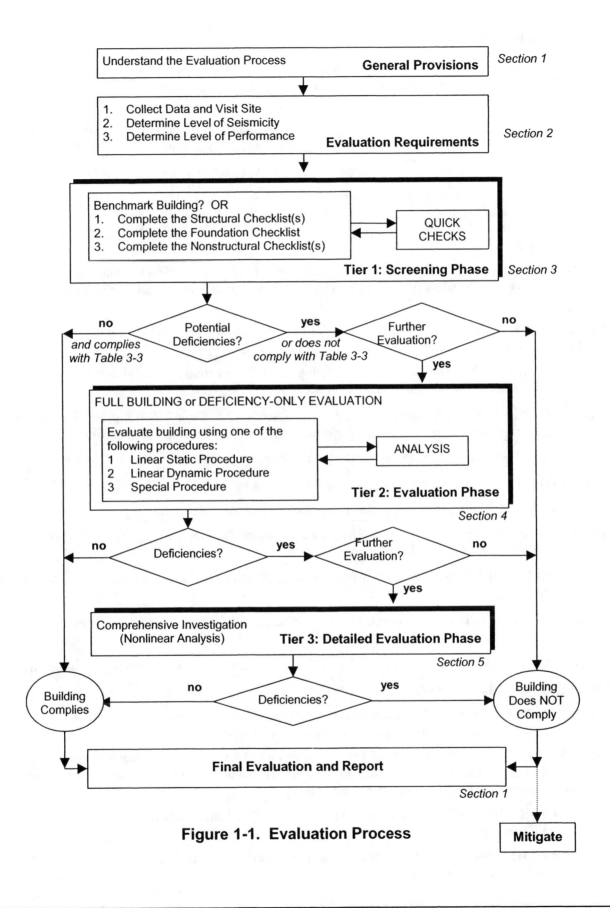

Figure 1-1. Evaluation Process

1.3 Definitions

ACTION: An internal moment, shear, torque, axial load, deformation, displacement, or rotation corresponding to a displacement due to a structural degree of freedom; designated as force- or deformation-controlled.

ASPECT RATIO: Ratio of full height to length for concrete and masonry shear walls; ratio of story height to length for wood shear walls; ratio of span to depth for horizontal diaphragms.

BASE: The level at which the horizontal seismic ground motions are considered to be imparted to the structure.

BASIC NONSTRUCTURAL CHECKLIST: Set of evaluation statements that shall be completed as part of the Tier 1 Evaluation. Each statement represents a potential nonstructural deficiency based on performance in past earthquakes.

BASIC STRUCTURAL CHECKLIST: Set of evaluation statements that shall be completed as part of the Tier 1 Evaluation. Each statement represents a potential structural deficiency based on performance in past earthquakes.

BENCHMARK BUILDING: A building designed and constructed or evaluated to a specific performance level using an acceptable code or standard listed in Table 3-1.

BRACED FRAME: A vertical lateral-force-resisting element consisting of vertical, horizontal, and diagonal components joined by concentric or eccentric connections.

BUILDING TYPE: A building classification defined in Section 2.6 (Table 2-2) that groups buildings with common lateral-force-resisting systems and performance characteristics in past earthquakes.

CAPACITY: The permissible strength or deformation for a component action.

CAVITY WALL: A masonry wall with an air space between wythes.

COLLAR JOINT: The vertical, longitudinal, mortared or grouted joint between wythes.

COLLECTOR: A member that transfers lateral forces from the diaphragm of the structure to vertical elements of the lateral-force-resisting system.

COMPONENT, FLEXIBLE: A component, including attachments, having a fundamental period greater than 0.06 seconds.

COMPONENT, RIGID: A component, including attachments, having a fundamental period less than or equal to 0.06 seconds.

CROSS WALL: A wood-framed wall sheathed with lumber, structural panels, or gypsum wallboard.

DEFAULT VALUE: Strengths as defined in Section 2.2.

DEFICIENCY-ONLY TIER 2 EVALUATION: An evaluation, beyond the Tier 1 Evaluation, that investigates only the non-compliant checklist evaluation statements.

DEFORMATION-CONTROLLED ACTION: An action that has an associated deformation that is allowed to exceed the yield value of the element being evaluated. The extent of permissible deformation beyond yield is based on component modification factors (m-factors).

DESIGN EARTHQUAKE: A percentage of the Maximum Considered Earthquake. *See* Maximum Considered Earthquake.

DIAPHRAGM: A floor or roof system that serves to interconnect the building and acts to transmit lateral forces to the vertical resisting elements.

DIAPHRAGM EDGE: The intersection of the floor or roof diaphragm and a shear wall, frame, or collector element.

EXPECTED STRENGTH: The probable strength of a material, as determined by testing, historical records, or other means. For the purposes of an evaluation using this standard, it shall be permitted to use the originally specified strength times 1.25 in the absence of more conclusive data.

FLEXIBLE DIAPHRAGM: A diaphragm with a maximum lateral deformation of twice or more the average story drift.

FORCE-CONTROLLED ACTION: An action that is not allowed to exceed the nominal strength of the element being evaluated.

FULL-BUILDING TIER 2 EVALUATION: An evaluation beyond a Tier 1 Evaluation that involves a complete analysis of the entire lateral-force-resisting system of the building using the Tier 2 analysis procedures defined in Section 4.2. While special attention should be given to the potential deficiencies identified in the Tier 1 Evaluation, all lateral-force-resisting elements must be evaluated. This evaluation is required where triggered by Table 3-3.

GEOLOGIC SITE HAZARDS AND FOUNDATIONS CHECKLIST: Set of evaluation statements that shall be completed as part of the Tier 1 Evaluation. Each statement represents a potential foundation or site deficiency based on the performance of buildings in past earthquakes.

IMMEDIATE OCCUPANCY PERFORMANCE LEVEL: Building performance that includes damage to both structural and nonstructural components during a design earthquake, such that: (a) the damage is not life-threatening, so as to permit immediate occupancy of the building after a design earthquake, and (b) the damage is repairable while the building is occupied.

INFILL: A panel of masonry placed within a steel or concrete frame. Panels separated from the surrounding frame by a gap are termed "isolated infills." Panels that are in full contact with a frame around its full perimeter are termed "shear infills."

LATERAL-FORCE-RESISTING SYSTEM: The collection of frames, shear walls, bearing walls, braced frames, and interconnecting roof and floor diaphragms that provides earthquake resistance to a building.

LEVEL OF LOW SEISMICITY CHECKLIST: Set of evaluation statements that are completed as part of the Tier 1 Evaluation for buildings in levels of low seismicity being evaluated to the Life Safety Performance Level.

LEVEL OF SEISMICITY: A degree of expected earthquake hazard. For this standard, levels are categorized as low, moderate, or high, based on mapped acceleration values and site amplification factors as defined in Section 2.5 (Table 2-1).

LIFE SAFETY PERFORMANCE LEVEL: Building performance that includes damage to both structural and nonstructural components during a design earthquake, such that: (a) partial or total structural collapse does not occur, and (b) damage to nonstructural components is non–life-threatening.

LINEAR DYNAMIC PROCEDURE (LDP): A Tier 2 response-spectrum-based, modal analysis procedure, the use of which is required where the distribution of lateral forces is expected to depart from that assumed for the Linear Static Procedure.

LINEAR STATIC PROCEDURE (LSP): A Tier 2 lateral force analysis procedure using a pseudo lateral force. This procedure is used for buildings for which the Linear Dynamic Procedure or the Special Procedure is not required.

LOAD PATH: A route or course along which seismic inertia forces are transferred from the superstructure to the foundation.

MAXIMUM CONSIDERED EARTHQUAKE (MCE): An earthquake based on the lesser of probabilistic values with a 2-percent probability of exceedence in 50 years and 150 percent of the median deterministic values at a given site. The response spectra associated with this earthquake shall be considered to have 5-percent damping.

MEANS OF EGRESS: A path for exiting a building, including, but not limited to, doors, corridors, ramps, and stairways.

MOMENT-RESISTING FRAME (MRF): A frame capable of resisting horizontal forces due to the members (beams and columns) and joints resisting forces primarily by flexure.

NORMAL WALL: A wall perpendicular to the direction of seismic forces.

OPEN FRONT: An exterior building wall plane on one side only, without vertical elements of the lateral-force-resisting system in one or more stories.

PIER: Vertical portion of a wall between two horizontally adjacent openings. Piers resist axial stresses from gravity forces and bending moments from combined gravity and lateral forces.

POINTING: The partial reconstruction of the bed joints of a masonry wall by removing unsound mortar and replacing it with new mortar.

PRIMARY COMPONENT: An element that is required to resist the seismic forces in order for the structure to achieve the selected performance level.

PSEUDO LATERAL FORCE (V): The calculated lateral force used for the Tier 1 Quick Checks and for the Tier 2 Linear Static Procedure. The pseudo lateral force represents the force required, in a linear analysis, to impose the expected actual deformation of the structure in its yielded state where subjected to the design earthquake motions.

QUICK CHECK: Analysis procedure used in Tier 1 Evaluations to determine if the lateral-force-resisting system has sufficient strength and/or stiffness.

REINFORCED MASONRY: Masonry having both vertical and horizontal reinforcement as follows: Vertical reinforcement of at least 0.20 in.^{2} in cross section at each corner or end, at each side of each opening, and at a maximum spacing of 4 feet throughout. Horizontal reinforcement of at least 0.20 in.^{2} in cross section at the top of the wall, at the top and bottom of wall openings, at structurally connected roof and floor openings, and at a maximum spacing of 10 feet throughout. The sum of the areas of horizontal and vertical reinforcement shall be at least 0.0005 times the gross cross-sectional area of the element, and the minimum area of reinforcement in either direction shall not be less than 0.000175 times the gross cross-sectional area of the element.

RIGID DIAPHRAGM: A diaphragm with a maximum lateral deformation of less than half the average story drift.

SECONDARY COMPONENT: An element that need not resist the seismic forces it may attract in order for the structure to achieve the selected performance level.

SHEAR WALL: A wall that resists lateral forces applied parallel with its plane. Also known as an in-plane wall.

SITE CLASS: A classification assigned to a site based on the types of soils present and their engineering properties as defined in Section 3.5.2.3.1.

SPECIAL PROCEDURE: Analysis procedure, used for unreinforced masonry bearing wall buildings with flexible diaphragms, that recognizes the diaphragm motion, strength, and damping as the predominant response parameters.

General Provisions

SPECIAL PROCEDURE TIER 2 EVALUATION: An evaluation procedure for unreinforced masonry bearing wall buildings with flexible diaphragms.

STIFF DIAPHRAGM: A diaphragm with a maximum lateral deformation equal to half or more than half but less than twice the average story drift.

STORY SHEAR FORCE: Portion of the pseudo lateral force carried by each story of the building.

SUPPLEMENTAL NONSTRUCTURAL CHECKLIST: Set of nonstructural evaluation statements that shall be completed as part of the Tier 1 Evaluation for buildings in levels of moderate or high seismicity being evaluated to the Immediate Occupancy Performance Level.

SUPPLEMENTAL STRUCTURAL CHECKLIST: Set of evaluation statements that shall be completed as part of the Tier 1 Evaluation for buildings in levels of moderate seismicity being evaluated to the Immediate Occupancy Performance Level, and for buildings in levels of high seismicity.

TIER 1 EVALUATION: Completion of checklists of evaluation statements that identifies potential deficiencies in a building based on performance of similar buildings in past earthquakes.

TIER 2 EVALUATION: The specific evaluation of potential deficiencies to determine if they represent actual deficiencies that may require mitigation. Depending on the building type, this evaluation may be a Full-Building Tier 2 Evaluation, Deficiency-Only Tier 2 Evaluation, or a Special Procedure Tier 2 Evaluation.

TIER 3 EVALUATION: A comprehensive building evaluation implicitly or explicitly recognizing nonlinear response.

UNREINFORCED MASONRY: Masonry construction that does not satisfy the definition of reinforced masonry.

UNREINFORCED MASONRY BEARING WALL: An unreinforced masonry wall that provides vertical support for a floor or roof for which the total superimposed vertical load exceeds 100 pounds per lineal foot of wall.

YIELD STORY DRIFT: The lateral displacement of one level relative to the level above or below at which yield stress is first developed in a frame member.

C1.3 Definitions

IMMEDIATE OCCUPANCY PERFORMANCE LEVEL: The definition of Immediate Occupancy Performance Level contains two performance criteria that require judgment to be exercised by the owner or the owner's agent and the building official (if any). The following guidance may be used to incorporate the two criteria in the design evaluation: (a) after a design earthquake, the basic vertical- and lateral-force-resisting systems retain nearly all of their pre-earthquake strength, and (b) very limited damage to both structural and nonstructural components is anticipated during the design earthquake that will require some minor repairs, but the critical parts of the building are habitable.

LIFE SAFETY PERFORMANCE LEVEL: The definition of Life Safety Performance Level contains two performance criteria that require judgment to be exercised by the owner or the owner's agent and the building official. The following guidance may be used to incorporate the two criteria in the design evaluation: (a) at least some margin against either partial or total structural collapse remains, and (b) injuries may occur, but the overall risk of life-threatening injury as a result of structural damage is expected to be low.

General Provisions

> It should be noted that the probability of achieving these performance levels for an existing building is less than that for a new or rehabilitation design. This is due to various factors, including construction type, age, materials used, ground motions used in the evaluation, and design factors of safety. A building meeting the provisions of this standard may or may not meet the requirements for a current code or rehabilitation design.

1.4 Notation

a_n	Diameter of core multiplied by its length or area of the side of a square prism
a_p	Component amplification factor
A_{br}	Average cross-sectional area of the diagonal brace
A_b	Total area of the bed joints above and below the test specimen for an in-place shear test
A_c	Summation of the cross-sectional area of all columns in the story under consideration
A_n	Area of net mortared/grouted section
A_p	Gross area of prestressed concrete elements
A_w	Summation of the horizontal cross-sectional area of all shear walls in the direction of loading
A_x	Amplification factor to account for accidental torsion
C	Modification factor to relate expected maximum inelastic displacements calculated for linear elastic response
C	Compliant
C_p	Horizontal force factor
C_t	Modification factor, based on earthquake records, used to adjust the building period to account for the characteristics of the building system
C_{vx}	Vertical distribution factor, based on story weights and heights
C1	Concrete Moment Frames building type, as defined in Table 2-2
C2	Concrete Shear Wall with Stiff Diaphragms building type, as defined in Table 2-2
C2A	Concrete Shear Wall with Flexible Diaphragms building type, as defined in Table 2-2
C3	Concrete Frames with Infill Masonry Shear Walls with Stiff Diaphragms building type, as defined in Table 2-2
C3A	Concrete Frames with Infill Masonry Shear Walls with Flexible Diaphragms building type, as defined in Table 2-2
d	Depth
d_b	Diameter of reinforcing steel bar
D	In-plane width dimension of masonry or depth of diaphragm
DCR	Demand-capacity ratio
D_p	Relative displacement
D_r	Drift ratio
E	Modulus of elasticity
f_j^{avg}	Average axial stress in diagonal bracing elements at level j

f'_c	Compressive strength of concrete
f'_m	Compressive strength of masonry
f_{sp}	Tensile splitting strength of masonry
f_y	Yield stress of reinforcing steel
F_a	Site coefficient, defined in Table 3-6
F_i	Lateral force applied at floor level i
F_{pe}	Effective prestressing force of a prestressing tendon
F_{px}	Total diaphragm force at level x
F_v	Site coefficient, defined in Table 3-5
F_{wx}	Force applied to a wall at level x
F_x	Total story force at level x
F_y	Yield stress
h	Story height
h_i, h_x	Height from the base to floor level i or x
h_n	Height above the base to the roof level
H	Least clear height of opening on either side of pier
I	Moment of inertia
IO	Immediate Occupancy Performance Level
I_p	Nonstructural component modification factor
j	Number of story level under consideration
J	Force-delivery reduction factor
k	Exponent related to the building period, used to define the vertical distribution of lateral forces
k_b	Stiffness of a representative beam
k_c	Stiffness of a representative column
l_b	Clear length of beam
L	Length
L_{br}	Average length of the diagonal brace
L_c	Length of cross wall
L_i	Effective span for an open-front building
LS	Life Safety Performance Level
m	Component modification factor
MCE	Maximum Considered Earthquake
M_{gj}	Moment in girder at level j
n	Number of stories above ground
n_c	Total number of columns

General Provisions

n_f	Total number of frames
n_p	Number of prestressed strands
$\overline{N}$	Average field standard penetration test for the top 100 feet
N/A	Not applicable
N_{br}	Number of diagonal braces in tension and compression if the braces are designed for compression, number of diagonal braces in tension if the braces are designed for tension only
NC	Non-compliant
NL	No limit
p_{D+L}	Stress resulting from actual dead plus live loads in place at the time of testing (psi)
P_{CE}	Expected gravity compressive force applied to a wall or pier component stress
P_D	Superimposed dead load at the top of the pier under consideration
P_{test}	Splitting test load
P_W	Weight of wall
PC1	Precast/Tilt-up Concrete Shear Walls with Flexible Diaphragms building type, as defined in Table 2-2
PC1A	Precast/Tilt-up Concrete Shear Walls with Stiff Diaphragms building type, as defined in Table 2-2
PC2	Precast Concrete Frames with Shear Walls building type, as defined in Table 2-2
PC2A	Precast Concrete Frames without Shear Walls building type, as defined in Table 2-2
Q_{CE}	Expected strength
Q_D	Actions due to dead load
Q_E	Actions due to earthquake loads
Q_G	Actions due to gravity load
Q_L	Actions due to effective live load
Q_S	Actions due to effective snow load
Q_{UD}	Deformation-controlled design actions
Q_{UF}	Force-controlled design actions
R_p	Nonstructural component response modification factor
RM1	Reinforced Masonry Bearing Walls with Flexible Diaphragms building type, as defined in Table 2-2
RM2	Reinforced Masonry Bearing Walls with Stiff Diaphragms building type, as defined in Table 2-2
s	Average span length of braced spans
$\overline{s_u}$	Average undrained shear strength in the top 100 feet
S_a	Response spectral acceleration parameter
S_{DS}	Design short-period spectral response acceleration parameter

S_{D1}	Design spectral response acceleration parameter at a one-second period
SRSS	Square root sum of squares
S_S	Short-period spectral response acceleration parameter
S_1	Spectral response acceleration parameter at a one-second period
S1	Steel Moment Frames with Stiff Diaphragms building type, as defined in Table 2-2
S1A	Steel Moment Frames with Flexible Diaphragms building type, as defined in Table 2-2
S2	Steel Braced Frames with Stiff Diaphragms building type, as defined in Table 2-2
S2A	Steel Braced Frames with Flexible Diaphragms building type, as defined in Table 2-2
S3	Steel Light Frames building type, as defined in Table 2-2
S4	Steel Frames with Concrete Shear Walls building type, as defined in Table 2-2
S5	Steel Frames with Infill Masonry Shear Walls and Stiff Diaphragms building type, as defined in Table 2-2
S5A	Steel Frames with Infill Masonry Shear Walls and Flexible Diaphragms building type, as defined in Table 2-2
t	Thickness of wall
T	Fundamental period of vibration
T1	Tier 1 Evaluation
T2	Tier 2 Evaluation
T3	Tier 3 Evaluation
URM	Unreinforced Masonry Bearing Walls with Flexible Diaphragms building type as defined in Table 2-2
URMA	Unreinforced Masonry Bearing Walls with Stiff Diaphragms building type as defined in Table 2-2
v	Shear stress
v_a	Shear stress for unreinforced masonry
v_j^{avg}	Average shear stress at level j
v_c	Unit shear strength for a cross wall
v_{me}	Expected masonry shear strength
v_{te}	Mortar shear strength
v_{to}	Mortar shear test value
v_u	Unit shear capacity for a diaphragm
V	Pseudo lateral force
V_a	Shear strength of an unreinforced masonry pier
V_c	Column shear force
V_{ca}	Total shear capacity of cross walls in the direction of analysis immediately above the diaphragm level being investigated
V_{cb}	Total shear capacity of cross walls in the direction of analysis immediately below the diaphragm level being investigated

V_d	Diaphragm shear
V_g	Shear due to gravity loads, in accordance with Section 4.2.4.2
V_j	Story shear force
V_o	Punching shear capacity
V_p	Shear force on an unreinforced masonry wall pier
V_r	Pier rocking shear capacity of an unreinforced masonry wall or wall pier
V_{test}	Load at first observed movement for an in-place masonry shear test
V_{wx}	Total shear force resisted by a shear wall at the level under consideration
w_i, w_x	Portion of the total building weight assigned to floor level i or x
W	Total seismic weight
W_d	Total dead load tributary to a diaphragm
W_j	Total seismic weight of all stories above level j
W_p	Component operating weight
W_w	Total dead load of an unreinforced masonry wall above the level under consideration or above an open front of a building
W_{wx}	Dead load of an unreinforced masonry wall assigned to level x, taken from mid-story below level x to mid-story above level x
W1	Wood Light Frames building type, as defined in Table 2-2
W1A	Multi-Story, Multi-Unit Residential Wood Frames building type, as defined in Table 2-2
W2	Wood Frames, Commercial and Industrial building type, as defined in Table 2-2
x	Height in structure of highest point of attachment of component
X, Y	Height of lower support attachment at level x or y as measured from grade
Δ_d	Diaphragm displacement
Δ_w	In-plane wall displacement
δ_{avg}	The average of displacements at the extreme points of the diaphragm at level x
δ_{max}	The maximum displacement at any point of the diaphragm at level x
δ_{xA}, δ_{yA}	Deflection at building level x or y of building A
δ_B	Deflection at building level x of building B
$\bar{v}_s$	Average shear wave velocity in the top 100 feet
ρ''	Volumetric ratio of horizontal confinement reinforcement in a joint

1.5 References

ACI, 1999. *ACI 318-99. Building Code Requirements for Reinforced Concrete.* Detroit, Mich.: American Concrete Institute.

ACI, 1999. *ACI 530-99. Building Code Requirements for Masonry Structures.* Detroit, Mich.: American Concrete Institute.

AISC, 1989. *Allowable Stress Design Specification for Structural Steel Buildings.* Chicago, Ill.: American Institute of Steel Construction, Inc.

AISC, 1997. *Seismic Provisions for Structural Steel Buildings.* Chicago, Ill.: American Institute of Steel Construction, Inc.

AISC, 1999. *Load and Resistance Factor Design Specification for Structural Steel Buildings.* Chicago, Ill.: American Institute of Steel Construction, Inc.

ASCE, 2003. ASCE 7-02, *Minimum Design Loads for Buildings and Other Structures.* Reston, Va.: American Society of Civil Engineers.

ASME, 2000. *A17.1: Safety Code for Elevators and Escalators.* New York, N.Y.: American Society of Mechanical Engineers.

ASTM, 1993. *E519-81: Test Method Diagonal Tension (Shear) for Masonry Assemblages.* West Conshohocken, Pa.: American Society for Testing and Materials.

ASTM, 1996. C496-96: Test Method for Splitting Tensile Strength of Cylindrical Concrete Specimens. West Conshohocken, Pa.: American Society for Testing and Materials.

BOCA, 1993. *National Building Code.* Country Club Hill, Ill.: Building Officials and Code Administrators International.

CBSC, 1995. *California Building Code (Title 24).* Sacramento, Calif.: California Building Standards Commission.

ICBO, 1994a. *Uniform Building Code.* Whittier, Calif.: International Conference of Building Officials.

ICBO, 1994b. *Emergency Provisions to the Uniform Building Code.* Whittier, Calif.: International Conference of Building Officials.

ICBO, 1997. *Uniform Building Code.* Whittier, Calif.: International Conference of Building Officials.

ICC, 2000. *International Building Code.* Falls Church, Va.: International Code Council.

MSS, 1993. SP-58, *Pipe Hangers and Supports: Materials, Design and Manufacture.* Vienna, Va.: Manufacturers Standardization Society of the Valve and Fitting Industry.

NFPA, 1996. NFPA-13, *Standard for the Installation of Sprinkler Systems.* Quincy, Mass.: National Fire Protection Association.

SBCC, 1994. *Standard Building Code.* Birmingham, Ala.: Southern Building Code Congress International.

General Provisions

C1.5 References

Agbabian, Barnes, Kariotis (ABK), 1981. Topical Report 03, *Methodology for Mitigation of Seismic Hazards in Existing Unreinforced Masonry Buildings: Diaphragm Testing*. Washington, D.C.: National Science Foundation.

Agbabian, Barnes, Kariotis (ABK), 1984. Topical Report 08, *Methodology for Mitigation of Seismic Hazards in Existing Unreinforced Masonry Buildings: The Methodology*. Washington, D.C.: National Science Foundation.

ANSI, 1997. *Z21.15: Manually Operated Gas Valves for Appliances, Appliance Connector Valves, and Hose end Valves*. New York, N.Y.: American National Standards Institute.

Army, Navy, and Air Force, 1992. *Tri-Services Manual TM 5-809-10/NAVFAC P-355/AFM 88-3: Seismic Design for Buildings*. Washington D.C.: Departments of the Army, the Navy, and the Air Force.

ASCE, 1997. ASCE 25-97: *Earthquake-Actuated Automatic Gas Shutoff Devices*. Reston, Va.: American Society of Civil Engineers.

ASME, 2002. *B16.33: Manually Operated Metallic Gas Valves for Use in Gas Piping Systems up to 125 psig*. New York, New York: American Society of Mechanical Engineers.

ATC, 1987. *ATC-14: Evaluation the Seismic Resistance of Existing Buildings*. Redwood City, Calif.: Prepared by H.J. Degenkolb Associates for the Applied Technology Council.

ATC, 1996. *ATC-40: Seismic Evaluation and Retrofit of Concrete Buildings*. Redwood City, Calif.: Prepared by the Applied Technology Council for the California Seismic Safety Commission.

ANSI, 1997. *Z21.15: Manually Operated Gas Valves for Appliances, Appliance Connector Valves and Hose End Valves*. New York, New York: American National Standards Institute.

ASHRAE, 1995. *Heat* ANSI, 1997. *Z21.15: Manually Operated Gas Valves for Appliances, Appliance Connector Valves and Hose End Valves*. New York, New York: American National Standards Institute.
ing, Ventilating, and Air-Conditioning Applications. Atlanta, Ga.: American Society of Heating, Refrigeration and Air-Conditioning Engineers.

BSSC, 1988. *NEHRP Recommended Provisions for Seismic Regulations for New Buildings and Other Structures*. Washington, D.C.: Prepared by the Building Seismic Safety Council for the Federal Emergency Management Agency.

BSSC, 1992a. FEMA 178, *NEHRP Handbook for the Seismic Evaluation of Existing Buildings*. Washington, D.C.: Building Seismic Safety Council for the Federal Emergency Management Agency.

BSSC, 1992b. FEMA 172, *NEHRP Handbook of Techniques for the Seismic Rehabilitation of Existing Buildings*. Washington, D.C.: Building Seismic Safety Council for the Federal Emergency Management Agency.

BSSC, 2000. FEMA 368 and FEMA 369, *NEHRP Recommended Provisions for Seismic Regulations for New Buildings and Other Structures, Part 1, Provisions and Part 2, Commentary*. Washington, D.C.: Building Seismic Safety Council for the Federal Emergency Management Agency.

CBSC, 1998. *California State Historical Building Code (Title 24, Part 8)*. Sacramento, Calif.: California Building Standards Commission.

FEMA, 1988a. FEMA 154, *Rapid Visual Screening of Buildings for Potential Seismic Hazards: A Handbook*. Washington, D.C.: Applied Technology Council for the Federal Emergency Management Agency.

FEMA, 1988b. FEMA 155, *Rapid Visual Screening of Buildings for Potential Seismic Hazards: Supporting Documentation.* Washington, D.C.: Applied Technology Council for the Federal Emergency Management Agency.

FEMA, 1992. *NEHRP Benefit-Cost Model for the Seismic Rehabilitation of Buildings*, Federal Emergency Management Agency, Washington, D.C. (FEMA Publication No. 227 and 228).

FEMA, 1994. *NEHRP Typical Costs for Seismic Rehabilitation of Existing Buildings*, Federal Emergency Management Agency, Washington, D.C. (FEMA Publication No. 156 and 157).

FEMA, 1998. FEMA 310, *Handbook for the Seismic Evaluation of Buildings—A Prestandard.* Washington, D.C.: American Society of Civil Engineers for the Federal Emergency Management Agency.

FEMA, 1999a. FEMA 306, *Evaluation of Earthquake Damaged Concrete and Masonry Wall Buildings: Basic Procedures Manual.* Washington, D.C.: Applied Technology Council (ATC-43 Project) for The Partnership for Response and Recovery, published by the Federal Emergency Management Agency.

FEMA, 1999b. FEMA 307, *Evaluation of Earthquake Damaged Concrete and Masonry Wall Buildings: Technical Resources.* Washington, D.C.: Applied Technology Council (ATC-43 Project) for The Partnership for Response and Recovery, published by the Federal Emergency Management Agency.

FEMA, 2000a. FEMA 350, *Recommended Seismic Design Criteria for New Steel Moment-Frame Buildings.* Washington, D.C.: SAC Joint Venture for the Federal Emergency Management Agency.

FEMA, 2000b. FEMA 351, *Recommended Seismic Evaluation and Upgrade Criteria for Existing Steel Moment-Frame Buildings.* Washington, D.C.: SAC Joint Venture for the Federal Emergency Management Agency.

FEMA, 2000c. FEMA 356, *Prestandard and Commentary for the Seismic Rehabilitation of Buildings.* Washington, D.C.: American Society of Civil Engineers for the Federal Emergency Management Agency.

FEMA, 2000d. FEMA 357, *Global Topics Report on the Prestandard and Commentary for the Seismic Rehabilitation of Buildings.* Washington, D.C.: American Society of Civil Engineers for the Federal Emergency Management Agency.

ICBO, 1997. UBC 21-6, *In-Place Masonry Shear Tests.* Whittier, Calif.: International Conference of Building Officials.

ICBO, 2001. *Guidelines for Seismic Retrofit of Existing Buildings (GSREB).* Whittier, Calif.: International Conference of Building Officials.

LADBS, 1997a. *Chapter 91: Earthquake Hazard Reduction in Existing Tilt-Up Concrete Wall Buildings (Division 91).* Los Angeles, Calif.: Los Angeles Department of Building and Safety.

LADBS, 1997b. *Chapter 95: Voluntary Earthquake Hazard Reduction in Existing Reinforced Concrete Buildings and Concrete Frame Buildings with Masonry Infill (Division 95).* Los Angeles, Calif.: Los Angeles Department of Building and Safety.

SEAOC, 1995. *Vision 2000: Performance Based Seismic Engineering of Buildings.* Sacramento, Calif.: Prepared by the Structural Engineers Association of California for the California Office of Emergency Services.

SEAOC, 1996. *Recommended Lateral Force Requirements and Commentary,* 6th ed. Sacramento, Calif.: Structural Engineers Association of California.

Secretary of the Interior, 1990. *Standards for Rehabilitation with Guidelines for Rehabilitating Historic Buildings.* Washington, D.C.: National Park Service.

Secretary of the Interior, 1975–2001 *Catalog of Historic Buildings Preservation Briefs*, National Park Service, Washington, D.C.

Secretary of the Interior, 1995. *Standards for the Treatment of Historic Properties.* Washington, D.C.: National Park Service.

Secretary of the Interior, 1998. *Disaster Management Programs for Historic Sites.* Washington, D.C.: National Park Service.

2.0 Evaluation Requirements

2.1 General

Prior to completing a seismic evaluation, the requirements of Section 2 shall be met.

2.2 Level of Investigation Required

The information collected during the investigation shall be sufficient to define the level of seismicity in accordance with Section 2.5, and the building type in accordance with Section 2.6. In addition, the level of investigation shall be sufficient to complete Tier 1 Checklists. Where conducting a Tier 1 Evaluation, a review shall be conducted of readily available documents pertaining to the original design and construction and subsequent service life of the building. If construction documents are available, the review shall include verification that the building was constructed in general conformance with the documents. Significant alterations and deviations that can be identified by visual observation shall be noted. Destructive examination shall be conducted as required to complete the checklists for buildings being evaluated to the Immediate Occupancy Performance Level; judgment shall be used regarding the need for destructive evaluation for buildings being evaluated to the Life Safety Performance Level. Limited non-destructive examination of a representative sample of connections and conditions shall be performed for all Tier 1 Evaluations. The use of default values is permitted for material properties for Tier 1 and Tier 2 Evaluations. The following default values are to be assumed unless otherwise indicated by the available documents:

- f'_c = 2,000 psi for concrete
- f_y = 33 ksi for reinforcing steel
- F_y = 33 ksi for structural steel
- f'_m = 1,000 psi
- v_{te} = 20 psi for concrete masonry units
- v_{te} = 10 psi for clay masonry units
- F_{pe} = 25 kips for effective force of a prestressing tendon

In addition to the information required for a Tier 1 Evaluation, sufficient information shall be collected for a Tier 2 Evaluation to complete the required Tier 2 Procedures. Destructive examination shall be conducted as required to complete the procedures for buildings being evaluated to the Immediate Occupancy Performance Level. Non-destructive examination of connections and conditions shall be performed for all Tier 2 Evaluations. Although the use of default material properties is permitted for Tier 2, it is recommended that material property data be obtained from building codes from the year of construction of the building being evaluated, from as-built plans, or from physical tests.

> **Exception:** Unreinforced masonry bearing wall buildings with flexible diaphragms using the Tier 2 Special Procedure of Section 4.2.6 shall have destructive tests conducted to determine the mortar shear strength, v_{te}, and the strength of the anchors.

Detailed information about the building is required for a Tier 3 Evaluation. If no documents are available, a set of drawings shall be created indicating critical components and connections of the existing lateral-force-resisting system. Non-destructive and destructive examination and testing shall be conducted for a Tier 3 Evaluation in accordance with the document used for the Tier 3 Evaluation.

> **C2.2 Level of Investigation Required**
>
> Building evaluation involves many substantial difficulties. One is the matter of uncovering the structure since plans and calculations often are not available. In many buildings, the structure is concealed by architectural finishes, and the design professional will have to get into attics, crawl spaces, and plenums to investigate. Some destructive testing may be necessary to determine material quality and allowable stresses. If reinforcing plans are available, some exposure of critical reinforcement may be necessary to verify conformance with the plans. The extent of investigation required depends on the level of evaluation because the conservatism inherent in both the Tier 1 and Tier 2 analyses covers the lack of detailed information in most cases. The evaluating design professional is encouraged to balance the investigation with the sophistication of the evaluation technique.
>
> The design professional in responsible charge during the original construction should be consulted if possible. In addition, the evaluating design professional may find it helpful to do some research on historical building systems, consult old handbooks and building codes, and perhaps consult with older engineers who have knowledge of early structural work in the community or region.
>
> The evaluation should be based on facts, as opposed to assumptions, to the greatest extent possible. If assumptions are made, the evaluating engineer should consider a range of values to assess the implications of the assumptions on the results. If the results are sensitive to the assumptions, more detailed information should be obtained.
>
> Two of the more important factors in any evaluation are the material properties and strengths. For a Tier 1 Evaluation, the default values may be used. For a Tier 2 Evaluation, it is recommended that the material strengths be determined by use of existing documentation or material testing. For a Tier 3 Evaluation, material testing is required to verify the existing documentation or establish the strengths if existing documentation is not available.
>
> Where evaluating a building using this standard, the design professional should:
>
> - Look for an existing geotechnical report on site soil conditions;
> - Establish site and soil parameters;
> - Assemble building design data, including contract drawings, specifications, and calculations;
> - Look for other data, such as assessments of the building performance during past earthquakes; and
> - Select and review the appropriate sets of evaluation statements included in Section 3.

2.3 Site Visit

A site visit shall be conducted by the evaluating design professional(s) to verify existing data or collect additional data, determine the general condition of the building, and verify or assess the site conditions. Relevant building data which shall be determined or confirmed during a site visit includes the following:

1. General building description: Number of stories, year(s) of construction, and dimensions
2. Structural system description: Framing, lateral-force-resisting system(s), floor and roof diaphragm construction, basement, and foundation system
3. Nonstructural element description: Nonstructural elements that could affect seismic performance
4. Nonstructural component connections: Anchorage conditions, location of connections, or support
5. Building type(s)
6. Site class
7. Building use
8. Special architectural features: Finishes, registered historic features
9. Adjacent buildings: Pounding concerns, falling hazards

10. Building condition: Dry-rot, fire, insect, corrosion, water, chemical, settlement, past-earthquake, wind, and other damage, and related repairs, alterations, and additions that could affect seismic performance

---

C2.3　Site Visit

Many of the Tier 1 Checklist items can be completed during the initial site visit. Subsequent assessment of the evaluation statements may indicate a need for more information about the building. The design professional may need to revisit the site to:

- Verify existing data;
- Develop other required data;
- Verify the vertical and lateral-force-resisting systems;
- Verify the condition of the building;
- Look for special conditions and anomalies;
- Address the evaluation statements again while in the field; and
- Perform non-destructive and destructive material tests, as necessary.

2.4　Level of Performance

A desired level of performance shall be defined prior to conducting a seismic evaluation using this standard. The level of performance shall be determined by the owner in consultation with the design professional and by the authority having jurisdiction (if any). Two performance levels for both structural and nonstructural components are defined in Section 1.3 of this standard: (1) Life Safety (LS) and (2) Immediate Occupancy (IO). For both performance levels, the seismic demand is based on a fraction of the Maximum Considered Earthquake (MCE) spectral response acceleration values. Buildings complying with the criteria of this standard shall be deemed to meet the specified performance level.

C2.4　Level of Performance

FEMA 178 addresses only the Life Safety Performance Level for buildings, whereas FEMA 310 addresses both the Life Safety and Immediate Occupancy Performance Levels. This standard addresses both the Life Safety and Immediate Occupancy Performance Levels.

The seismic analysis and design of buildings has traditionally focused on one performance level: reducing the risk to life loss in the design earthquake. Building codes for new buildings and the wide variety of evaluation guidelines developed in the last 30 years have based their provisions on the historic performance of buildings and the deficiencies that caused life safety concerns to develop. Beginning with the damage to hospitals in the 1971 San Fernando earthquake, there has been a growing desire to design and construct certain "essential facilities" that are needed immediately after an earthquake. In addition, there has been a growing recognition that new buildings can be designed for some measure of damage protection with only a small increase in construction cost over a design that achieves a lower level of performance; whereas the cost of rehabilitating existing buildings to a level of performance beyond the minimum safety standard may be considered onerous to stakeholders and policy-makers. In recent years, a new style of design guidelines began appearing that have been tailored to a variety of performance levels.

Evaluation Requirements

> The extensive and expensive, non–life-threatening damage that occurred in the 1994 Northridge, California, earthquake brought these various performance levels to the point of formalization. "Performance Based Engineering" was described by the Structural Engineers Association of California (SEAOC) in their Vision 2000 document. At the same time, the Earthquake Engineering Research Center (EERC) published a research and development plan for the development of "Performance Based Engineering Guidelines and Standards." The first formal application in published guidelines occurred in FEMA 273, where the range of possible performance levels and hazard levels was combined to define specific rehabilitation objectives to be used to rehabilitate buildings. FEMA 273 was subsequently revised and republished as FEMA 356.
>
> This standard defines and uses performance levels in a manner consistent with FEMA 356. The Life Safety and Immediate Occupancy Performance Levels are the same as defined in FEMA 356. The hazard level used is the third in a series of four levels as defined in FEMA 356. The level chosen is consistent with the hazard traditionally used for seismic analysis and the same as that used in the 1998 FEMA 310 handbook. Other performance levels are available and can be defined, but the standard is not applicable to other performance levels. For other performance levels and/or hazard levels, the user should perform a Tier 3 analysis; but consideration of the checklist items found in this standard is still recommended because they are based on the observed performance of buildings in past earthquakes.
>
> The process for defining the appropriate level of performance is the responsibility of the owner, design professional, and, where required, the authority having jurisdiction. Considerations in choosing an appropriate level of performance should include achieving basic safety, a cost-benefit analysis, the building occupancy type, and economic constraints.
>
> In general, buildings classified as essential facilities should be evaluated to the Immediate Occupancy Performance Level. The FEMA 368/369, *2000 NEHRP Recommended Provisions for Seismic Regulations for New Buildings and Other Structures* (BSSC, 2000) categorizes the following buildings as essential facilities " . . . required for post-earthquake recovery":
>
> - Fire or rescue and police stations
> - Hospitals
> - Designated other medical facilities having surgery or emergency treatment facilities
> - Designated emergency preparedness centers including the equipment therein
> - Power generating stations or other utilities required to provide emergency back-up service for other facilities listed here
> - Emergency vehicle garages
> - Designated communication centers
> - Structures containing sufficient quantities of toxic or explosive substances deemed to be dangerous to the public if released
> - Other facilities may be deemed "essential" by local jurisdictions.

2.5 Level of Seismicity

The level of seismicity of the building shall be defined as low, moderate, or high in accordance with Table 2-1. Levels of seismicity are defined in terms of mapped response acceleration values and site amplification factors.

Evaluation Requirements

Table 2-1. Levels of Seismicity Definitions

Level of Seismicity[1]	S_{DS}	S_{D1}
Low	<0.167g	<0.067g
Moderate	≥0.167g <0.500g	≥0.067g <0.200g
High	≥0.500g	≥0.200g

[1]Sites with S_{DS} and S_{D1} values in different levels of seismicity shall be classified as moderate.

where:

S_{DS} = Design short-period spectral response acceleration parameter (Sec. 3.5.2.3.1)

S_{D1} = Design spectral response acceleration parameter at a one-second period (Sec. 3.5.2.3.1)

C2.5 Level of Seismicity

The successful performance of buildings in levels of high seismicity depends on a combination of strength, ductility (manifested in the details of construction), and the presence of a fully interconnected, balanced, and complete lateral-force-resisting system. As these fundamentals are applied in levels of lower seismicity, the need for strength and ductility reduces substantially; in fact, strength can substitute for a lack of ductility. ATC-14, the first-generation version of FEMA 178, recognized this fact and defined separate provisions for levels of low and high seismicity. Based in part on work sponsored by the National Center for Earthquake Engineering Research (NCEER), FEMA 178 eliminated the separate provisions and elected to permit the lateral force calculations to determine where there was sufficient strength to make up for a lack of detailing and ductility.

The collective experience of the engineers using FEMA 178 is that the requirements too often require calculations for deficiencies that are never a problem because of the low lateral forces. Therefore, this standard includes three separate Tier 1 procedures for the three basic levels of seismicity. The levels are defined in terms of the expected spectral response for the site under consideration. Thus, the criteria for an area depends both on the expected MCE accelerations and on the site adjustment factors. Soil conditions can vary dramatically throughout local areas and result in multiple levels of seismicity within a single geographical region. For example, homes located on soft soils in the Marina District of San Francisco experienced more severe shaking and damage than homes located on rock sites a few blocks south.

2.6 Building Type

The building type shall be classified as one or more of the building types listed in Table 2-2 based on the lateral-force-resisting system(s) and the diaphragm type. Separate building types shall be used for buildings with different lateral-force-resisting systems in different directions, areas, or levels. Where the subject building does not match a Common Building Type, the General Structural Checklists shall be used.

C2.6 Building Type

Fundamental to the Tier 1 analysis of buildings is the grouping of buildings into sets that have similar behavioral characteristics. These groups of "building types" were first defined in ATC-14 and have since been used in most of the FEMA guideline documents. During the development of FEMA 356, it was determined that a number of additional building types were needed to cover all common styles of construction. These were fully developed and presented in that document. The added building styles included type W1A apartment buildings that performed poorly in the Northridge and Loma Prieta earthquakes, and a number of variations on diaphragm type for the basic building systems. The new types are included as subtypes to the original fifteen, so there remain fifteen model building types.

The common building types are defined in Table 2-2. Because most structures are unique in some fashion, judgment should be used where selecting the building type, with the focus on the lateral-force-resisting system and elements.

Separate checklists for each of the Common Building Types are included in this standard as well as General Structural Checklists for buildings that do not fit the descriptions of the Common Building Types. Procedures for using the general checklists are provided in Section 3.3.

Evaluation Requirements

Table 2-2. Common Building Types

Building Type 1: WOOD LIGHT FRAMES	
W1	These buildings are single- or multiple-family dwellings of one or more stories in height. Building loads are light and the framing spans are short. Floor and roof framing consists of wood joists or rafters on wood studs spaced no more than 24 inches apart. The first-floor framing is supported directly on the foundation, or is raised up on cripple studs and post-and-beam supports. The foundation consists of spread footings constructed on concrete, concrete masonry block, or brick masonry or wood in older construction. Chimneys, where present, consist of solid brick masonry, masonry veneer, or wood frame with internal metal flues. Lateral forces are resisted by wood frame diaphragms and shear walls. Floor and roof diaphragms consist of straight or diagonal lumber sheathing, tongue-and-groove planks, oriented strand board, or plywood. Shear walls consist of straight or lumber sheathing, plank siding, oriented strand board, plywood, stucco, gypsum board, particle board, or fiberboard. Interior partitions are sheathed with plaster or gypsum board.
W1A (Multi-Story, Multi-Unit, Residential)	These buildings are multi-story, similar in construction to W1 buildings, but have plan areas on each floor of greater than 3000 square feet. Older construction often has open front garages at the lowest story.
Building Type 2: WOOD FRAMES, COMMERCIAL AND INDUSTRIAL	
W2	These buildings are commercial or industrial buildings with a floor area of 5,000 square feet or more. There are few, if any, interior walls. The floor and roof framing consists of wood or steel trusses, glulam or steel beams, and wood posts or steel columns. Lateral forces are resisted by wood diaphragms and exterior stud walls sheathed with plywood, oriented strand board, stucco, plaster, straight or diagonal wood sheathing, or braced with rod bracing. Wall openings for storefronts and garages, where present, are framed by post-and-beam framing.
Building Type 3: STEEL MOMENT FRAMES	
S1 (With Stiff Diaphragms)	These buildings consist of a frame assembly of steel beams and steel columns. Floor and roof framing consists of cast-in-place concrete slabs or metal deck with concrete fill supported on steel beams, open web joists, or steel trusses. Lateral forces are resisted by steel moment frames that develop their stiffness through rigid or semi-rigid beam-column connections. Where all connections are moment-resisting connections, the entire frame participates in lateral force resistance. Where only selected connections are moment-resisting connections, resistance is provided along discrete frame lines. Columns are oriented so that each principal direction of the building has columns resisting forces in strong axis bending. Diaphragms consist of concrete or metal deck with concrete fill and are stiff relative to the frames. Where the exterior of the structure is concealed, walls consist of metal panel curtain walls, glazing, brick masonry, or precast concrete panels. Where the interior of the structure is finished, frames are concealed by ceilings, partition walls, and architectural column furring. Foundations consist of concrete spread footings or deep pile foundations.

Building Type 4: STEEL BRACED FRAMES	
S1A (With Flexible Diaphragms)	These buildings are similar to S1 buildings, except that diaphragms consist of wood framing; untopped metal deck; or metal deck with lightweight insulating concrete, poured gypsum, or similar nonstructural topping and are flexible relative to the frames.
S2 (With Stiff Diaphragms)	These buildings have a frame of steel columns, beams, and braces. Braced frames develop resistance to lateral forces by the bracing action of the diagonal members. The braces induce forces in the associated beams and columns such that all elements work together in a manner similar to a truss with all element stresses being primarily axial. Where the braces do not completely triangulate the panel, some of the members are subjected to shear and flexural stresses; eccentrically braced frames are one such case (refer to Sec. 4.4.3.3). Diaphragms transfer lateral loads to braced frames. The diaphragms consist of concrete or metal deck with concrete fill and are stiff relative to the frames.
S2A (With Flexible Diaphragms)	These buildings are similar to S2 buildings, except that diaphragms consist of wood framing; untopped metal deck; or metal deck with lightweight insulating concrete, poured gypsum, or similar nonstructural topping and are flexible relative to the frames.
Building Type 5: STEEL LIGHT FRAMES	
S3	These buildings are pre-engineered and prefabricated with transverse rigid steel frames. They are onestory in height. The roof and walls consist of lightweight metal, fiberglass, or cementitious panels. The frames are designed for maximum efficiency, and the beams and columns consist of tapered, built-up sections with thin plates. The frames are built in segments and assembled in the field with bolted or welded joints. Lateral forces in the transverse direction are resisted by the rigid frames. Lateral forces in the longitudinal direction are resisted by wall panel shear elements or rod bracing. Diaphragm forces are resisted by untopped metal deck, roof panel shear elements, or a system of tension-only rod bracing.
Building Type 6: STEEL FRAMES WITH CONCRETE SHEAR WALLS	
S4	These buildings consist of a frame assembly of steel beams and steel columns. The floor and roof diaphragms consist of cast-in-place concrete slabs or metal deck with or without concrete fill. Framing consists of steel beams, open web joists, or steel trusses. Lateral forces are resisted by cast-in-place concrete shear walls. These walls are bearing walls where the steel frame does not provide a complete vertical support system. In older construction the steel frame is designed for vertical loads only. In modern dual systems, the steel moment frames are designed to work together with the concrete shear walls in proportion to their relative rigidity. In the case of a dual system, the walls shall be evaluated under this building type and the frames shall be evaluated under S1 or S1A, Steel Moment Frames. The steel frame may provide a secondary lateral-force-resisting system depending on the stiffness of the frame and the moment capacity of the beam-column connections.

Evaluation Requirements

Building Type 7: STEEL FRAMES WITH INFILL MASONRY SHEAR WALLS	
S5 **(With Stiff Diaphragms)**	This is an older type of building construction that consists of a frame assembly of steel beams and steel columns. The floor and roof diaphragms consist of cast-in-place concrete slabs or metal deck with concrete fill and are stiff relative to the walls. Framing consists of steel beams, open web joists, or steel trusses. Walls consist of infill panels constructed of solid clay brick, concrete block, or hollow clay tile masonry. Infill walls may completely encase the frame members and present a smooth masonry exterior with no indication of the frame. The seismic performance of this type of construction depends on the interaction between the frame and infill panels. The combined behavior is more like a shear wall structure than a frame structure. Solidly infilled masonry panels form diagonal compression struts between the intersections of the frame members. If the walls are offset from the frame and do not fully engage the frame members, the diagonal compression struts will not develop. The strength of the infill panel is limited by the shear capacity of the masonry bed joint or the compression capacity of the strut. The post-cracking strength is determined by an analysis of a moment frame that is partially restrained by the cracked infill.
S5A **(With Flexible Diaphragms)**	These buildings are similar to S5 buildings, except that diaphragms consist of wood sheathing or untopped metal deck, or have large aspect ratios and are flexible relative to the walls.
Building Type 8: CONCRETE MOMENT FRAMES	
C1	These buildings consist of a frame assembly of cast-in-place concrete beams and columns. Floor and roof framing consists of cast-in-place concrete slabs, concrete beams, one-way joists, two-way waffle joists, or flat slabs. Lateral forces are resisted by concrete moment frames that develop their stiffness through monolithic beam-column connections. In older construction, or in levels of low seismicity, the moment frames may consist of the column strips of two-way flat slab systems. Modern frames in levels of high seismicity have joint reinforcing, closely spaced ties, and special detailing to provide ductile performance. This detailing is not present in older construction. Foundations consist of concrete spread footings, mat foundations, or deep foundations.
Building Type 9: CONCRETE SHEAR WALLS	
C2 **(With Stiff Diaphragms)**	These buildings have floor and roof framing that consists of cast-in-place concrete slabs, concrete beams, one-way joists, two-way waffle joists, or flat slabs. Floors are supported on concrete columns or bearing walls. Lateral forces are resisted by cast-in-place concrete shear walls. In older construction, shear walls are lightly reinforced but often extend throughout the building. In more recent construction, shear walls occur in isolated locations, are more heavily reinforced with concrete slabs, and are stiff relative to the walls. Foundations consist of concrete spread footings, mat foundations, or deep foundations.
C2A **(With Flexible Diaphragms)**	These buildings are similar to C2 buildings, except that diaphragms consist of wood sheathing, or have large aspect ratios, and are flexible relative to the walls.

Evaluation Requirements

	Building Type 10: CONCRETE FRAMES WITH INFILL MASONRY SHEAR WALLS
C3 (With Stiff Diaphragms)	This is an older type of building construction that consists of a frame assembly of cast-in-place concrete beams and columns. The floor and roof diaphragms consist of cast-in-place concrete slabs and are stiff relative to the walls. Walls consist of infill panels constructed of solid clay brick, concrete block, or hollow clay tile masonry. The seismic performance of this type of construction depends on the interaction between the frame and the infill panels. The combined behavior is more like a shear wall structure than a frame structure. Solidly infilled masonry panels form diagonal compression struts between the intersections of the frame members. If the walls are offset from the frame and do not fully engage the frame members, the diagonal compression struts will not develop. The strength of the infill panel is limited by the shear capacity of the masonry bed joint or the compression capacity of the strut. The post-cracking strength is determined by an analysis of a moment frame that is partially restrained by the cracked infill. The shear strength of the concrete columns, after racking of the infill, may limit the semiductile behavior of the system.
C3A (With Flexible Diaphragms)	These buildings are similar to C3 buildings, except that diaphragms consist of wood sheathing or untopped metal deck, or have large aspect ratios and are flexible relative to the walls.
	Building Type 11: PRECAST/TILT-UP CONCRETE SHEAR WALLS
PC1 (With Flexible Diaphragms)	These buildings have precast concrete perimeter wall panels that are cast on-site and tilted into place. Floor and roof framing consists of wood joists, glulam beams, steel beams, or open web joists. Framing is supported on interior steel columns and perimeter concrete bearing walls. The floors and roof consist of wood sheathing or untopped metal deck. Lateral forces are resisted by the precast concrete perimeter wall panels. Wall panels may be solid or have large window and door openings that cause the panels to behave more as frames than as shear walls. In older construction, wood framing is attached to the walls with wood ledgers. Foundations consist of concrete spread footings or deep pile foundations.
PC1A (With Stiff Diaphragms)	These buildings are similar to PC1 buildings, except that diaphragms consist of precast elements, cast-in-place concrete, or metal deck with concrete fill, and are stiff relative to the walls.
	Building Type 12: PRECAST CONCRETE FRAMES
PC2 (With Shear Walls)	These buildings consist of a frame assembly of precast concrete girders and columns with the presence of shear walls. Floor and roof framing consists of precast concrete planks, tees, or double-tees supported on precast concrete girders and columns. Lateral forces are resisted by precast or cast-in-place concrete shear walls. Diaphragms consist of precast elements interconnected with welded inserts, cast-in-place closure strips, or reinforced concrete topping slabs.

PC2A **(Without Shear Walls)**	These buildings are similar to PC2 buildings, except that concrete shear walls are not present. Lateral forces are resisted by precast concrete moment frames that develop their stiffness through beam-column joints rigidly connected by welded inserts or cast-in-place concrete closures. Diaphragms consist of precast elements interconnected with welded inserts, cast-in-place closure strips, or reinforced concrete topping slabs.
Building Type 13: REINFORCED MASONRY BEARING WALLS WITH FLEXIBLE DIAPHRAGMS	
RM1	These buildings have bearing walls that consist of reinforced brick or concrete block masonry. The floor and roof framing consists of steel or wood beams and girders or open web joists, and are supported by steel, wood, or masonry columns. Lateral forces are resisted by the reinforced brick or concrete block masonry shear walls. Diaphragms consist of straight or diagonal wood sheathing, plywood, or untopped metal deck, and are flexible relative to the walls. Foundations consist of brick or concrete spread footings or deep foundations.
Building Type 14: REINFORCED MASONRY BEARING WALLS WITH STIFF DIAPHRAGMS	
RM2	These building are similar to RM1 buildings, except that the diaphragms consist of metal deck with concrete fill, precast concrete planks, tees, or double-tees, with or without a cast-in-place concrete topping slab, and are stiff relative to the walls. The floor and roof framing is supported on interior steel or concrete frames or interior reinforced masonry walls.
Building Type 15: UNREINFORCED MASONRY BEARING WALLS	
URM **(With Flexible Diaphragms)**	These buildings have perimeter bearing walls that consist of unreinforced clay brick, stone, or concrete masonry. Interior bearing walls, where present, also consist of unreinforced clay brick, stone, or concrete masonry. In older construction, floor and roof framing consists of straight or diagonal lumber sheathing supported by wood joists, which, in turn, are supported on posts and timbers. In more recent construction, floors consist of structural panel or plywood sheathing rather than lumber sheathing. The diaphragms are flexible relative to the walls. Where they exist, ties between the walls and diaphragms consist of anchors or bent steel plates embedded in the mortar joints and attached to framing. Foundations consist of brick or concrete spread footings, or deep foundations.
URMA **(With Stiff Diaphragms)**	These buildings are similar to URM buildings, except that the diaphragms are stiff relative to the unreinforced masonry walls and interior framing. In older construction or large, multi-story buildings, diaphragms consist of cast-in-place concrete. In levels of low seismicity, more recent construction consists of metal deck and concrete fill supported on steel framing.

3.0 Screening Phase (Tier 1)

3.1 General

Prior to conducting a Tier 1 Evaluation, the requirements of Chapter 2 shall be met. Tier 1 of the evaluation process is shown schematically in Figure 3-1.

Initially, the design professional shall determine whether the building meets the benchmark building criteria of Section 3.2. If the building meets the benchmark building criteria, it is deemed to meet the structural requirements of this standard for the specified level of performance; however, a Tier 1 Evaluation for foundations and nonstructural elements is still required.

If the building is not a benchmark building, the design professional shall select and complete the appropriate checklists in accordance with Section 3.3.

A list of potential deficiencies identified by evaluation statements for which the building was found to be non-compliant shall be compiled upon completion of the Tier 1 Checklists.

Further evaluation requirements shall be determined in accordance with Section 3.4 once the checklists have been completed.

C3.1 General

The purpose of the screening phase of the evaluation process is to quickly identify buildings that comply with the provisions of this standard. A Tier 1 Evaluation also familiarizes the design professional with the building, its potential deficiencies, and its potential behavior.

A Tier 1 Evaluation is required for all buildings so that potential deficiencies may be quickly identified. Further evaluation using a Tier 2 or Tier 3 Evaluation will then focus, at a minimum, on the potential deficiencies identified in Tier 1.

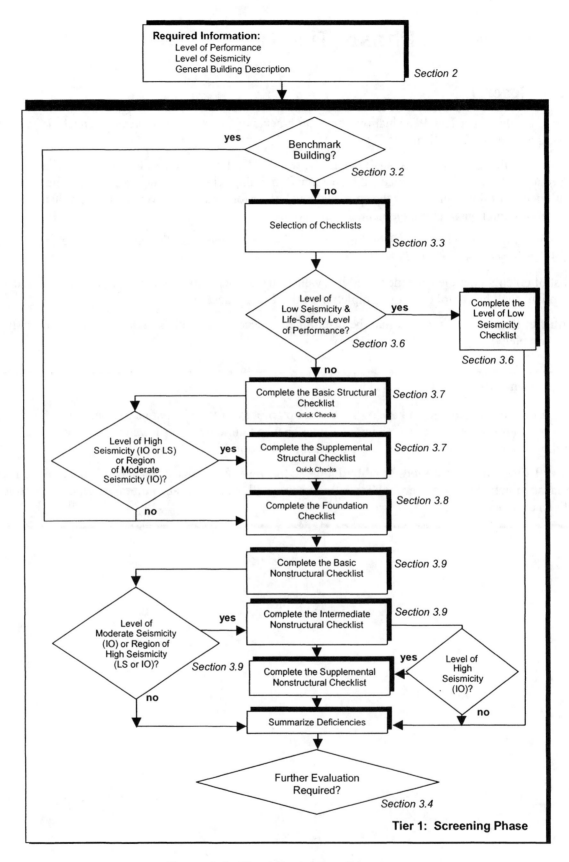

Figure 3-1. Tier 1 Evaluation Process

3.2 Benchmark Buildings

A structural seismic evaluation using this standard need not be performed for buildings designed and constructed or evaluated in accordance with the benchmark provisions listed in Table 3-1; however, an evaluation for foundations and nonstructural elements is still required. Only the provisions under which the structure was originally designed or evaluated may be permitted for the application of the provisions of this section and Table 3-1. Table 3-1 identifies the first year of publication of provisions whose seismic criteria are acceptable for certain building types so that further evaluation is not required. If knowledge of the level of seismicity has changed since the benchmark dates listed in Table 3-1, a building must have been designed and constructed or evaluated in accordance with the current or greater level of seismicity to be compliant with this section. The design professional shall document in the final report the evidence used to determine that the building is designed and constructed or evaluated in accordance with the provisions listed in Table 3-1 and current level of seismicity.

The applicable level of performance is indicated in Table 3-1 for each provision as a superscript.

C3.2 Benchmark Buildings

While benchmark buildings need not proceed with further evaluation, it should be noted that they are not simply exempt from the criteria of this standard. The design professional must determine that the building is compliant with the benchmark provisions. Knowledge that a code was in effect at the time of construction is not sufficient. A statement on the drawings simply stating that it was designed to the benchmark provisions will not suffice. Sometimes, details in the existing building will not correspond to the construction documents. Sometimes, the building is not properly detailed to meet the benchmark provisions. This may occur due to renovations or poor construction. Only through a site visit, an examination of existing documentation, and other requirements of Chapter 2 will the design professional be able to determine whether the structure being evaluated complies with this section.

The methodology in this standard is substantially compatible with building code provisions; however, the nature of methodology is such that complete compatibility may not be achievable. From observed earthquake damage, it may be inferred that certain building types built to a building code after a certain year will provide basic life safety. However, without the benchmark building provisions in this standard, all buildings would need to be evaluated, regardless of the year built. While many of the buildings would likely pass the Tier 1 methodology, there could be buildings that would be flagged even though damage observed in past earthquakes would indicate that the building may be adequate for life safety. The design professional would then have to resolve the non-compliant items by recommending mitigation or by performing Tier 2 calculations, which may not have been anticipated. The intent of this section is to cover this incompatibility between the methodologies and accept buildings thought to produce life safety performance in past earthquakes.

Note that the benchmark building provisions need not be followed. A design professional may choose to proceed with a structural Tier 1 Evaluation, even if the building meets the requirements of Section 3.2. Also note that benchmark building provisions only apply to the structural aspects of the evaluation. Nonstructural and foundation elements still require a Tier 1 Evaluation.

Screening Phase (Tier 1)

Table 3-1. Benchmark Buildings

Building Type[1,2]	Model Building Seismic Design Provisions					FEMA 178[ls]	FEMA 310[ls, io]	CBC[io]
	NBC[ls]	SBC[ls]	UBC[ls]	IBC[ls]	NEHRP[ls]			
Wood Frame, Wood Shear Panels (Type W1 & W2)	1993	1994	1976	2000	1985	*	1998	1973
Wood Frame, Wood Shear Panels (Type W1A)	*	*	1997	2000	1997	*	1998	1973
Steel Moment-Resisting Frame (Type S1 & S1A)	*	*	1994[4]	2000	**	*	1998	1995
Steel Braced Frame (Type S2 & S2A)	1993	1994	1988	2000	1991	1992	1998	1973
Light Metal Frame (Type S3)	*	*	*	2000	*	1992	1998	1973
Steel Frame w/ Concrete Shear Walls (Type S4)	1993	1994	1976	2000	1985	1992	1998	1973
Reinforced Concrete Moment-Resisting Frame (Type C1)[3]	1993	1994	1976	2000	1985	*	1998	1973
Reinforced Concrete Shear Walls (Type C2 & C2A)	1993	1994	1976	2000	1985	*	1998	1973
Steel Frame with URM Infill (Type S5, S5A)	*	*	*	2000	*	*	1998	*
Concrete Frame with URM Infill (Type C3 & C3A)	*	*	*	2000	*	*	1998	*
Tilt-up Concrete (Type PC1 & PC1A)	*	*	1997	2000	*	*	1998	*
Precast Concrete Frame (Type PC2 & PC2A)	*	*	*	2000	*	1992	1998	1973
Reinforced Masonry (Type RM1)	*	*	1997	2000	*	*	1998	*
Reinforced Masonry (Type RM2)	1993	1994	1976	2000	1985	*	1998	*
Unreinforced Masonry (Type URM)[5]	*	*	1991[6]	2000	*	1992	*	*
Unreinforced Masonry (Type URMA)	*	*	*	2000	*	*	1998	*

[1] "Building Type" refers to one of the Common Building Types defined in Table 2-2.
[2] Buildings on hillside sites shall not be considered Benchmark Buildings.
[3] Flat Slab Buildings shall not be considered Benchmark Buildings.
[4] Steel Moment-Resisting Frames shall comply with the 1994 UBC Emergency Provisions, published September/October 1994, or subsequent requirements.
[5] URM buildings evaluated using the ABK Methodology (ABK, 1984) may be considered benchmark buildings.
[6] Refers to the GSREB or its predecessor, the Uniform Code of Building Conservation (UCBC).

[ls] Only buildings designed and constructed or evaluated in accordance with these documents and being evaluated to the Life Safety (LS) Performance Level may be considered Benchmark Buildings.
[io] Buildings designed and constructed or evaluated in accordance with these documents and being evaluated to either the Life Safety or Immediate Occupancy (IO) Performance Level may be considered Benchmark Buildings.

* No benchmark year; buildings shall be evaluated using this standard.
** Local provisions shall be compared with the UBC.

NBC = *National Building Code* (BOCA, 1993).
SBC = *Standard Building Code* (SBCC, 1994).
UBC = *Uniform Building Code* (ICBO, 1997)
GSREB = *Guidelines for Seismic Retrofit of Existing Buildings* (ICBO, 2001).
IBC = *International Building Code* (ICC, 2000).
NEHRP = FEMA 368 and 369, *NEHRP Recommended Provisions for the Development of Seismic Regulations for New Buildings* (BSSC, 2000)
FEMA 178 (*See* BSSC, 1992a)
FEMA 310 (*See* FEMA, 1998)
CBC = *California Building Code, California Code of Regulations, Title 24* (CBSC, 1995).

3.3 Selection and Use of Checklists

Required checklists, as a function of level of seismicity and level of performance, are listed in Table 3-2. Each of the required checklists designated in Table 3-2 shall be completed for a Tier 1 Evaluation. Each of the evaluation statements on the checklists shall be marked "Compliant" (C), "Non-compliant" (NC), or "Not Applicable" (N/A). Compliant statements identify issues that are acceptable according to the criteria of this standard, while non-compliant statements identify issues that require further investigation. Certain statements may not apply to the buildings being evaluated.

Quick Checks for Tier 1 shall be performed in accordance with Section 3.5 where necessary to complete an evaluation statement.

The Level of Low Seismicity Checklist, located in Section 3.6, shall be completed for buildings in levels of low seismicity being evaluated to the Life Safety Performance Level. For buildings in levels of low seismicity being evaluated to the Immediate Occupancy Performance Level and buildings in levels of moderate or high seismicity, the appropriate Structural, Geologic Site Hazards, and Nonstructural Checklists shall be completed in accordance with Table 3-2.

The appropriate Structural Checklists shall be selected based on the Common Building Types defined in Table 2-2. The General Structural Checklists shall be used for buildings that cannot be classified as one of the Common Building Types defined in Table 2-2.

A building with a different lateral-force-resisting system in each principal direction shall use two sets of structural checklists, one for each direction. A building with more than one type of lateral-force-resisting system along a single axis of the building shall be classified as a "mixed" system. The General Structural Checklists shall be used for this type of building.

Two separate Structural Checklists are provided for each building type: a Basic Structural Checklist and a Supplemental Structural Checklist. As shown in Table 3-2, the Basic Structural Checklist shall be completed for buildings in levels of low seismicity being evaluated to the Immediate Occupancy Performance Level and for buildings in levels of moderate and high seismicity. The Supplemental Structural Checklist shall be completed in addition to the Basic Structural Checklist for buildings in levels of moderate seismicity being evaluated to the Immediate Occupancy Performance Level and for buildings in levels of high seismicity.

The Geologic Site Hazards and Foundations Checklist shall be completed for all buildings except those in levels of low seismicity being evaluated to the Life Safety Performance Level.

Three separate Nonstructural Checklists also are provided: Basic, Intermediate, and Supplemental. As shown in Table 3-2, the Basic Nonstructural Component Checklist shall be completed for all buildings except those in levels of low seismicity being evaluated to the Life Safety Performance Level. The Intermediate Nonstructural Component Checklist, in addition to the Basic Nonstructural Component Checklist, shall be completed for buildings in levels of moderate seismicity being evaluated for the Immediate Occupancy Performance Level or for buildings in levels of high seismicity. The Supplemental Nonstructural Component Checklist, in addition to the Basic and Intermediate Nonstructural Component Checklists, shall be completed for buildings in levels of high seismicity being evaluated to the Immediate Occupancy Performance Level.

Screening Phase (Tier 1)

> **C3.3 Selection and Use of Checklists**
>
> The evaluation statements provided in the checklists form the core of the Tier 1 Evaluation methodology. These evaluation statements are based on observed earthquake structural damage during actual earthquakes. The checklists do not necessarily identify the response of the structure to ground motion; rather, the design professional obtains a general sense of the structure's deficiencies and potential behavior during an earthquake. By quickly identifying the potential deficiencies in the structure, the design professional has a better idea of what to examine and analyze in a Tier 2 or Tier 3 Evaluation.
>
> The General Structural Checklists are a complete listing of all evaluation statements used in the Common Building Type checklists. They should be used for buildings with structural systems that do not match the Common Building Types. While the general purpose of the Tier 1 Checklists is to identify potential weak links associated with structures of a specific type that have been observed in past significant earthquakes, the General Checklists, by virtue of their design, do not accomplish this. They only represent a listing of possible deficiencies. The design professional must consider first the applicability of the potential deficiency to the building system being considered. Generally, only the deficiencies applicable to the primary lateral-force-resisting elements of the building need be considered.
>
> While the section numbers in parentheses following each evaluation statement correspond to Tier 2 Evaluation procedures, they also correspond to commentary in Chapter 4 regarding the statement's purpose. If additional information on the evaluation statement is required, please refer to the commentary in the Tier 2 procedure for that evaluation statement.

Table 3-2. Checklists Required for a Tier 1 Evaluation

Level of Seismicity[3]	Level of Performance[2]	Required Checklists[1]						
		Level of Low Seismicity (Sec. 3.6)	Basic Structural (Sec. 3.7)	Supplemental Structural (Sec. 3.7)	Geologic Site Hazard and Foundation (Sec. 3.8)	Basic Nonstructural (Sec. 3.9.1)	Intermediate Nonstructural (Sec. 3.9.2)	Supplemental Nonstructural (Sec. 3.9.3)
Low	LS	►						
	IO		►		►	►		
Moderate	LS		►		►	►		
	IO		►	►	►	►	►	
High	LS		►	►	►	►	►	
	IO		►	►	►	►	►	►

[1] A checkmark (►) designates the checklist that must be completed for a Tier 1 Evaluation as a function of the level of seismicity and level of performance.
[2] LS = Life Safety; IO = Immediate Occupancy (defined in Section 2.4).
[3] Defined in Section 2.5.

3.4 Further Evaluation Requirements

Upon completion of the Tier 1 Evaluation, further evaluation shall be conducted in accordance with Table 3-3.

A Full-Building Tier 2 Evaluation shall be completed for buildings with more than the number of stories listed in Table 3-3. "NL" designates No Limit on the number of stories.

A Full-Building Tier 2 Evaluation also is required for buildings designated in Table 3-3 by 'T2'. A Tier 3 Evaluation shall be required for buildings designated by 'T3' in Table 3-3.

For buildings not requiring a Full-Building Tier 2 Evaluation or a Tier 3 Evaluation, a Deficiency-Only Tier 2 Evaluation may be conducted to assess deficiencies identified by the Tier 1 Evaluation. Alternatively, the design professional may choose to end the investigation and report the deficiencies in accordance with Chapter 1.

C3.4 Further Evaluation Requirements

The purpose of Table 3-3 is to identify buildings where the Tier 1 Checklist methodology alone may not be adequate to come to the correct conclusion about the building. If the number of stories is exceeded in Table 3-3, further evaluation is required to adequately assess the building.

In most cases, the Tier 1 identification of potential deficiencies leads to further evaluation of only these deficiencies. As defined in Chapter 4, the required analysis may be localized to the specific deficiencies, or it may involve a global analysis to evaluate the specific deficiency. Each checklist evaluation statement concludes with a reference to the applicable section in Chapter 4 with the Tier 2 procedures, as well as commentary on the statement's purpose.

The "NL" designation for most buildings being evaluated to the Life Safety Performance Level is consistent with FEMA 178, which has no restriction on the use of the checklists.

The "T2," "T3," and number of story designations in the Immediate Occupancy Performance Level category indicates that the building cannot be deemed to meet the requirements of this standard without a full evaluation of the building. Based on past performance of these types of buildings in earthquakes, the behavior of the structure must be examined and understood. However, the Tier 1 Checklists will provide insight and information about the structure prior to a Tier 2 or Tier 3 Evaluation.

Screening Phase (Tier 1)

Table 3-3. Further Evaluation Requirements[1]

Model Building Type	Number of Stories[2] beyond which a Full-Building Tier 2 Evaluation is Required					
	Levels of Seismicity					
	Low		Moderate		High	
	LS	IO	LS	IO	LS	IO
Wood Frames						
Light (W1)	NL	2	4	2	2	2
Multi-Story, Multi-Unit Residential (W1A)	NL	3	4	2	2	2
Commercial and Industrial (W2)	NL	2	4	2	2	2
Steel Moment Frames						
Rigid Diaphragm (S1)	NL	3	6	T2	3	T2
Flexible Diaphragm (S1A)	NL	3	6	T2	3	T2
Steel Braced Frames						
Rigid Diaphragm (S2)	NL	3	6	2	6	2
Flexible Diaphragm (S2A)	NL	3	6	2	6	2
Steel Light Frames (S3)	NL	1	2	1	2	1
Steel Frames with Concrete Shear Walls (S4)	NL	4	6	4	6	3
Steel Frames with Infill Masonry Shear Walls						
Rigid Diaphragm (S5)	NL	2	6	T2	2	T2
Flexible Diaphragm (S5A)	NL	2	6	T2	2	T2
Concrete Moment Frames (C1)	NL	2	6	T2	6	T2
Concrete Shear Walls						
Rigid Diaphragm (C2)	NL	4	6	4	6	3
Flexible Diaphragm (C2A)	NL	4	6	4	6	3
Concrete Frame with Infill Masonry Shear Walls						
Rigid Diaphragm (C3)	NL	2	6	T2	2	T2
Flexible Diaphragm (C3A)	NL	2	6	T2	2	T2
Precast/Tilt-up Concrete Shear Walls						
Flexible Diaphragm (PC1)	NL	1	2	T2	2	T2
Rigid Diaphragm (PC1A)	NL	1	2	T2	2	T2
Precast Concrete Frames						
With Shear Walls (PC2)	NL	4	6	4	4	T2
Without Shear Walls (PC2A)	NL	T2	6	T2	3	T2
Reinforced Masonry Bearing Walls						
Flexible Diaphragm (RM1)	NL	3	6	T2	3	T2
Rigid Diaphragm (RM2)	NL	3	6	3	3	2
Unreinforced Masonry Bearing Walls						
Flexible Diaphragm (URM)	NL	1	NL	T3	NL	T3
Rigid Diaphragm (URMA)	NL	1	6	T3	3	T3
Mixed Systems	NL	2	6	T2	6	T2

[1] A Full-Building Tier 2 or Tier 3 Evaluation shall be completed for buildings with more than the number of stories listed herein.
[2] Number of stories shall be considered as the number of stories above lowest adjacent grade.

NL = No Limit (No limit on the number of stories).
T2 = Tier 2 (A Full-Building Tier 2 Evaluation is required; proceed to Chapter 4).
T3 = Tier 3 (A Tier 3 Evaluation is required; proceed to Chapter 5).

3.5 Tier 1 Analysis

3.5.1 Overview

Analyses performed as part of the Tier 1 Evaluation process are limited to Quick Checks. Quick Checks shall be used to calculate the stiffness and strength of certain building components to determine whether the building complies with certain evaluation criteria. Quick Checks shall be performed in accordance with Section 3.5.3 where triggered by evaluation statements from the checklists of Section 3.7. Seismic forces for use in the Quick Checks shall be computed in accordance with Section 3.5.2.

3.5.2 Seismic Shear Forces

3.5.2.1 Pseudo Lateral Force

The pseudo lateral force, in a given horizontal direction of a building, shall be calculated in accordance with Equations (3-1) and (3-2).

$$V = CS_a W \qquad (3-1)$$

where:
  V = Pseudo lateral force.
  C = Modification factor to relate expected maximum inelastic displacements to displacements
calculated for linear elastic response; C shall be taken from Table 3-4.
  S_a = Response spectral acceleration at the fundamental period of the building in the direction under
    consideration. The value of S_a shall be calculated in accordance with the procedures in
    Section 3.5.2.3.
  W = Effective seismic weight of the building including the total dead load and applicable portions of
    other gravity loads listed below:
    1. In areas used for storage, a minimum of 25 percent of the floor live load shall be applicable. The live load shall be permitted to be reduced for tributary area as approved by the code official. Floor live load in public garages and open parking structures need not be considered.
    2. Where an allowance for partition load is included in the floor load design, the actual partition weight or a minimum weight of 10 psf of floor area, whichever is greater, shall be applied.
    3. Total operating weight of permanent equipment.
    4. Where the design flat roof snow load calculated in accordance with ASCE 7-02 exceeds 30 psf, the effective snow load shall be taken as 20 percent of the design snow load. Where the design flat roof snow load is 30 psf or less, the effective snow load shall be permitted to be zero.

Alternatively, for buildings in which the bottom of the foundation is less than 3 feet below exterior grade with a slab or tie beams to connect interior footings and, being evaluated for the Life Safety Performance Level, Equation (3-2) shall be permitted to be used to compute the pseudo lateral force:

$$V = 0.75W \qquad (3-2)$$

If Equation (3-2) is used, an m-factor of 1.0 shall be used to compute the component forces and stresses for the Quick Checks of Section 3.5.3 and acceptance criteria of Section 4.2.4.

Table 3-4. Modification Factor, C

Building Type[1]	Number of Stories			
	1	2	3	=4
Wood (W1, W1A, W2) Moment Frame (S1, S3, C1, PC2A)	1.3	1.1	1.0	1.0
Shear Wall (S4, S5, C2, C3, PC1A, PC2, RM2, URMA) Braced Frame (S2)	1.4	1.2	1.1	1.0
Unreinforced Masonry (URM) Flexible Diaphragms (S1A, S2A, S5A, C2A, C3A, PC1, RM1)	1.0	1.0	1.0	1.0

[1] Defined in Table 2-2.

C3.5.2.1 Pseudo Lateral Force

The seismic evaluation procedure of this standard, as well as the FEMA 368/369, *NEHRP Recommended Provisions for Seismic Regulations for New Buildings and Other Structures* (BSSC, 2000) and the *Uniform Building Code* (ICBO, 1997), is based on a widely accepted philosophy that permits nonlinear response of a building where subjected to a ground motion that is representative of the design earthquake. The FEMA 368/369, *Uniform Building Code,* and FEMA 178 account for nonlinear seismic response in a linear static analysis procedure by including a response modification factor, R, in calculating a reduced equivalent base shear to produce a rough approximation of the internal forces during a design earthquake. In other words, the base shear is representative of the force that the building is expected to resist, but the building displacements are significantly less than the actual displacements of the building during a design earthquake. Thus, in this R-factor approach, displacements calculated from the reduced base shear need to be increased by another factor (C_d or R) where checking drift or ductility requirements. In summary, this procedure is based on equivalent lateral forces and pseudo displacements.

The linear static analysis procedure in this standard, as well as in FEMA 356, takes a different approach to account for the nonlinear seismic response. Pseudo static lateral forces are applied to the structure to obtain "actual" displacements during a design earthquake. The pseudo lateral force of Equation (3-1) represents the force required, in a linear static analysis, to impose the expected actual deformation of the structure in its yielded state where subjected to the design earthquake motions. The modification factor C in Equation (3-1) is intended to replace the product of modification factors C_1, C_2, and C_3 in FEMA 356. The factor C increases the pseudo lateral force where the period of the structure is low. The effect of the period of the structure is replaced by the number of stories in Table 3-4. Furthermore, the factor C is larger where a higher level of ductility in the building is relied upon. Thus, unreinforced masonry buildings have a lower factor as compared to concrete shear wall or moment frame structures. In assigning values for coefficient C, representative average values (instead of using most conservative values) for coefficients C_1 C_2 C_3 were considered.

The pseudo lateral force does not represent an actual lateral force that the building must resist in traditional design codes or FEMA 178. In summary, this procedure is based on equivalent displacements and pseudo lateral forces. For additional commentary regarding this linear static analysis approach, please refer to the commentary for Section 4.2.2.1 and FEMA 356.

Screening Phase (Tier 1)

> Instead of applying a ductility related response reduction factor, R, to the applied loads, this standard uses ductility related m-factors in the acceptability checks of each component. These m-factors are conceptually similar to the m-factors in FEMA 356 but are approximated for the simplified procedures of this standard. Thus, instead of using a single R-value for the entire structure, different m-factors are used depending on the ductility of the component being evaluated. The m-factors specified for each tier of analysis shall not be used for other tiers of analysis (i.e., Tier 2 values of m may not be used where a Tier 1 analysis is performed).
>
> For short and stiff buildings with low ductility located in levels of high seismicity, the required building strength in accordance with Equation (3-1) may exceed the force required to cause sliding at the foundation level. The strength of the structure, however, does not need to exceed the sliding resistance at the foundation soil interface. It is assumed that this sliding resistance is equal to $0.75W$. Thus, where Equation (3-2) is applied to these buildings, the required strength of structural components need not exceed $0.75W$.

3.5.2.2 Story Shear Forces

The pseudo lateral force calculated in accordance with Section 3.5.2.1 shall be distributed vertically in accordance with Equations (3-3a and 3-3b).

$$F_x = \frac{w_x h_x^k}{\sum_{i=1}^{n} w_i h_i^k} V \tag{3-3a}$$

$$V_j = \sum_{x=j}^{n} F_x \tag{3-3b}$$

where:
- V_j = Story shear at story level j
- n = Total number of stories above ground level
- j = Number of story level under consideration
- W = Total seismic weight per Section 3.5.2.1
- V = Pseudo lateral force from Equation (3-1) or (3-2)
- w_i = Portion of total building weight W located on or assigned to floor level i
- w_x = Portion of total building weight W located on or assigned to floor level x
- h_i = Height (ft) from the base to floor level i
- h_x = Height (ft) from the base to floor level x
- k = 1.0 for T = 0.5 second
- = 2.0 for T > 2.5 seconds; linear interpolation shall be used for intermediate values of k

For buildings with stiff or rigid diaphragms, the story shear forces shall be distributed to the lateral-force-resisting elements based on their relative rigidities. For buildings with flexible diaphragms (Types S1A, S2A, S5A, C2A, C3A, PC1, RM1, URM), story shear shall be calculated separately for each line of lateral resistance.

3.5.2.3 Spectral Acceleration

Spectral acceleration for use in computing the pseudo lateral force shall be computed in accordance with this section. Spectral acceleration shall be based on mapped spectral accelerations, defined in Section 3.5.2.3.1, for the site of the building being evaluated. Alternatively, a site-specific response spectrum shall be permitted to be developed according to Section 3.5.2.3.2.

3.5.2.3.1 Mapped Spectral Acceleration

The spectral acceleration, S_a, shall be computed in accordance with Equation (3-4).

$$S_a = \frac{S_{D1}}{T}, \text{ but} \tag{3-4}$$

S_a shall not exceed S_{DS}

where:

$$S_{D1} = \frac{2}{3} F_v S_1 \tag{3-5}$$

$$S_{DS} = \frac{2}{3} F_a S_s \tag{3-6}$$

T = Fundamental period of vibration of the building, calculated in accordance with Section 3.5.2.4

S_s and S_1 = Short period response acceleration and spectral response acceleration at a one-second period, respectively, for the Maximum Considered Earthquake (MCE) obtained (ASCE 7-02)

F_v and F_a = Site coefficients determined from Tables 3-5 and 3-6, respectively, based on the site class and the values of the response acceleration parameters S_s and S_1. The site class of the building shall be defined as one of the following:

- **Class A:** Hard rock with measured shear wave velocity, $\overline{v_s} > 5{,}000$ ft/sec
- **Class B:** Rock with $2{,}500$ ft/sec $< \overline{v_s} < 5{,}000$ ft/sec
- **Class C:** Very dense soil and soft rock with $1{,}200$ ft/sec $< \overline{v_s} < 2{,}500$ ft/sec or with either standard blow count $\overline{N} > 50$ or undrained shear strength $\overline{s_u} > 2{,}000$ psf
- **Class D:** Stiff soil with 600 ft/sec $< \overline{v_s} < 1{,}200$ ft/sec or with $15 < \overline{N} < 50$ or $1{,}000$ psf $< \overline{s_u} < 2{,}000$ psf
- **Class E:** Any profile with more than 10 feet of soft clay defined as soil with plasticity index $PI > 20$, or water content $w > 40$ percent, and $\overline{s_u} < 500$ psf or a soil profile with $\overline{v_s} < 600$ ft/sec
- **Class F:** Soils requiring a site-specific geotechnical investigation and dynamic site response analyses:
  - Soils vulnerable to potential failure or collapse under seismic loading, such as liquefiable soils; quick, highly sensitive clays; collapsible, weakly cemented soils
  - Peats and/or highly organic clays ($H > 10$ feet of peat and/or highly organic clay; where H = thickness of soil)
  - Very high plasticity clays ($H > 25$ feet with $PI > 75$ percent)
  - Very thick soft/medium stiff clays ($H > 120$ feet)

The parameters $\overline{v_s}$, $\overline{N}$, and $\overline{s_u}$ are, respectively, the average values of the shear wave velocity, Standard Penetration Test (SPT) blow count, and undrained shear strength of the upper 100 feet of soils at the site. These values shall be calculated from Equation (3-7), below:

Screening Phase (Tier 1)

$$\overline{v_s}, \overline{N}, \overline{s_u} = \frac{\sum_{i=1}^{n} d_i}{\sum_{i=1}^{n} \frac{d_i}{v_{si}}, \frac{d_i}{N_i}, \frac{d_i}{s_{ui}}} \quad (3\text{-}7)$$

where:
- N_i = SPT blow count in soil layer i
- n = Number of layers of similar soil materials for which data is available
- d_i = Depth of layer i
- s_{ui} = Undrained shear strength in layer i
- v_{si} = Shear wave velocity of the soil in layer i

Except where explicitly defined otherwise, site classification shall be based on the soil properties averaged over the top 100 feet of soil. For a soil profile classified as Class F, a Class E soil profile shall be permitted for a Tier 1 Evaluation. If sufficient data are not available to classify a soil profile, a Class D profile shall be assumed, except where site-specific geotechnical investigation is required by the authority having jurisdiction to determine the presence of site Class E or F.

Table 3-5. Values of F_v as a Function of Site Class and Mapped Spectral Acceleration at a One Second Period, S_1

Site Class	Mapped Spectral Acceleration at One-Second Period[1]				
	$S_1 < 0.1$	$S_1 = 0.2$	$S_1 = 0.3$	$S_1 = 0.4$	$S_1 > 0.5$
A	0.8	0.8	0.8	0.8	0.8
B	1.0	1.0	1.0	1.0	1.0
C	1.7	1.6	1.5	1.4	1.3
D	2.4	2.0	1.8	1.6	1.5
E	3.5	3.2	2.8	2.4	2.4
F	*	*	*	*	*

[1]Note: Use straight-line interpolation for intermediate values of S_1.

* Site-specific geotechnical investigation and dynamic site response analyses required.

Table 3-6. Values of F_a as a Function of Site Class and Mapped Short-Period Spectral Acceleration, S_s

Site Class	Mapped Spectral Acceleration at Short Periods[1]				
	$S_s < 0.25$	$S_s = 0.5$	$S_s = 0.75$	$S_s = 1.00$	$S_s > 1.25$
A	0.8	0.8	0.8	0.8	0.8
B	1.0	1.0	1.0	1.0	1.0
C	1.2	1.2	1.1	1.0	1.0
D	1.6	1.4	1.2	1.1	1.0
E	2.5	1.7	1.2	0.9	0.9
F	*	*	*	*	*

[1]Note: Use straight-line interpolation for intermediate values of S_s.

* Site-specific geotechnical investigation and dynamic site response analyses required.

C3.5.2.3.1 Mapped Spectral Acceleration

The short-period response acceleration and spectral response acceleration at a one-second period parameters, S_s and S_1, are provided in ASCE 7-02. The values of S_s and S_1 represent an earthquake with a 2 percent probability of exceedence in 50 years with deterministic-based maximum values near known fault sources.

The 2/3 factor in the calculation of S_{D1} and S_{DS} is the same as used for seismic rehabilitation in FEMA 356 where further information about this factor is provided. It is intended to provide a 50 percent margin of safety between the loss of the first primary element and collapse. However, because the seismic acceleration values in the building codes for new construction prior to the *2000 International Building Code* have been roughly based on 10 percent exceedence in 50-year probabilities, this margin may not exist for all structures being evaluated to this standard. On the other hand, the use of 2/3 MCE accelerations will result in significantly higher seismic design values in some areas of the country. The result is that many relatively new buildings, while being considered life safe in a 10 percent in 50 year (500-year return period) event, may not be adequately protected against collapse in an extreme rare earthquake (2,500 return period). When using this document, the design professional, the building owner, and the regulatory authority may wish to consider the use of the current 10 percent in 50-year values in lieu of 2/3 of the MCE values, where appropriate.

3.5.2.3.2 Site-Specific Spectral Acceleration

Development of site-specific response spectra shall be based on the geologic, seismological, and soil characteristics associated with the specific site of the building being evaluated. Site-specific response spectra shall be mean spectra based on input ground motions at the 2-percent/50-year probability of exceedence. The site-specific response spectra need not exceed 150 percent of the median deterministic spectra for the characteristic event on the controlling fault. Spectral amplitudes of the 5-percent damped site response spectrum, in the period range of greatest significance to the structural response, shall not be less than 70 percent of the mapped spectral accelerations defined in Section 3.5.2.3.1 unless an independent third party review is performed to confirm its acceptability for use.

3.5.2.4 Period

The fundamental period of a building, in the direction under consideration, shall be calculated in accordance with Equation (3-8).

$$T = C_t h_n^\beta \tag{3-8}$$

where:
- T = Fundamental period (in seconds) in the direction under consideration
- C_t = 0.060 for wood buildings (Building Types W1, W1A, and W2)
  - = 0.035 for moment-resisting frame systems of steel (Building Types S1 and S1A)
  - = 0.030 for moment-resisting frames of reinforced concrete (Building Type C1)
  - = 0.030 for eccentrically braced steel frames (Building Types S2 and S2A)
  - = 0.020 for all other framing systems
- h_n = height (in feet) above the base to the roof level
- β = 0.80 for moment-resisting frame systems of steel (Building Types S1 and S1A)
  - = 0.90 for moment-resisting frame systems of reinforced concrete (Building Type C1)
  - = 0.75 for all other framing systems

Alternatively, for steel or reinforced-concrete moment frames of 12 stories or less, the fundamental period of the building may be calculated as follows:

$$T = 0.10n \tag{3-9}$$

where:
- n = number of stories above the base

Screening Phase (Tier 1)

> **C3.5.2.4 Period**
>
> The value of $C_t = 0.06$ for wood buildings is imported directly from FEMA 356. This value is based on engineering judgment and is not based on measured or calculated period for this type of structure. With the exception of the C_t value for wood buildings, the values of C_t given in this standard are intended to be reasonable lower bound (not mean) values for structures including the contribution of nonstructural elements. The value of T used in the evaluation should be as close as possible to but less than the true period of the structure.

3.5.3 Quick Checks for Strength and Stiffness

Quick Checks shall be used to compute the stiffness and strength of building components. Quick Checks are triggered by evaluation statements in the checklists of Section 3.7 and are required to determine the compliance of certain building components. The seismic shear forces used in the Quick Checks shall be calculated in accordance with Section 3.5.2.

> **C3.5.3 Quick Checks for Strength and Stiffness**
>
> The Quick Check equations used here are essentially the same as those used in FEMA 178, modified for use with the pseudo lateral forces and the appropriate material m-factors.

3.5.3.1 Story Drift for Moment Frames

Equation (3-10) shall be used to calculate the drift ratios of regular, multi-story, multi-bay moment frames with columns continuous above and below the story under consideration. The drift ratio is based on the deflection due to flexural displacement of a representative column, including the effect of end rotation due to bending of the representative beam.

$$D_r = \left(\frac{k_b + k_c}{k_b k_c}\right)\left(\frac{h}{12E}\right) V_c \qquad (3\text{-}10)$$

where:
- D_r = Drift ratio: Interstory displacement divided by story height
- k_b = I/L for the representative beam
- k_c = I/h for the representative column
- h = Story height (in.)
- I = Moment of inertia (in.^{4})
- L = Beam length from center-to-center of adjacent columns (in.)
- E = Modulus of elasticity (ksi)
- V_c = Shear in the column (kips)

The column shear forces are calculated using the story shear forces in accordance with Section 3.5.2.2. For reinforced concrete frames, an effective cracked section moment of inertia equal to one-half of gross value shall be used.

Equation (3-10) also may be used for the first floor of the frame if columns are fixed against rotation at the bottom. However, if columns are pinned at the bottom, the drift ratio shall be multiplied by two.

Screening Phase (Tier 1)

For other configurations of frames, the Quick Check need not be performed; however, a Full-Building Tier 2 Evaluation, including calculation of the drift ratio, shall be completed based on principles of structural mechanics.

> **C3.5.3.1 Story Drifts for Moment Frames**
>
> Equation (3-10) assumes that all of the columns in the frame have similar stiffness.

3.5.3.2 Shear Stress in Concrete Frame Columns

The average shear stress, v_j^{avg}, in the columns of concrete frames shall be computed in accordance with Equation (3-11).

$$v_j^{avg} = \frac{1}{m}\left(\frac{n_c}{n_c - n_f}\right)\left(\frac{V_j}{A_c}\right) \qquad (3\text{-}11)$$

where:
- n_c = Total number of columns
- n_f = Total number of frames in the direction of loading
- A_c = Summation of the cross-sectional area of all columns in the story under consideration
- V_j = Story shear computed in accordance with Section 3.5.2.2
- m = Component modification factor; m shall be taken equal to 2.0 for buildings being evaluated to the Life Safety Performance Level and equal-to-1.3 for buildings being evaluated to the Immediate Occupancy Performance Level

> **C3.5.3.2 Shear Stress in Concrete Frame Columns**
>
> Equation (3-11) assumes that all of the columns in the frame have similar stiffness.
>
> The inclusion of the term $[n_c/(n_c-n_f)]$ in Equation 3-11 is based on the assumption that the end column carries half the load of a typical interior column. This equation is not theoretically correct for a one-bay frame and yields shear forces that are twice the correct force; but due to the lack of redundancy in one-bay frame, this level of conservatism is considered appropriate.

3.5.3.3 Shear Stress in Shear Walls

The average shear stress in shear walls, v_j^{avg}, shall be calculated in accordance with Equation (3-12).

$$v_j^{avg} = \frac{1}{m}\left(\frac{V_j}{A_w}\right) \qquad (3\text{-}12)$$

where:
- V_j = Story shear at level j computed in accordance with Section 3.5.2.2.
- A_w = Summation of the horizontal cross-sectional area of all shear walls in the direction of loading. Openings shall be taken into consideration where computing A_w. For masonry walls, the net area shall be used. For wood-framed walls, the length shall be used rather than the area.
- m = Component modification factor; m shall be taken from Table 3-7.

Screening Phase (Tier 1)

Table 3-7. *m*-factors for Shear Walls

Wall Type	Level of Performance[1]	
	LS	IO
Reinforced Concrete, Precast Concrete, Wood, and Reinforced Masonry	4.0	2.0
Unreinforced Masonry	1.5	N/A

[1] Defined in Section 2.4.

3.5.3.4 Diagonal Bracing

The average axial stress in diagonal bracing elements, f_j^{avg}, shall be calculated in accordance with Equation (3-13).

$$f_j^{avg} = \frac{1}{m}\left(\frac{V_j}{sN_{br}}\right)\left(\frac{L_{br}}{A_{br}}\right) \qquad (3\text{-}13)$$

where:
- L_{br} = Average length of the braces (ft)
- N_{br} = Number of braces in tension and compression if the braces are designed for compression; number of diagonal braces in tension if the braces are designed for tension only
- s = Average span length of braced spans (ft)
- A_{br} = Average area of a diagonal brace (in.²)
- V_j = Maximum story shear at each level (kips)
- m = Component modification factor; m shall be taken from Table 3-8

Table 3-8. *m*-factors for Diagonal Braces

Brace Type	d/t [2]	Level of Performance[1]	
		LS	IO
Tube[3]	< $90/(F_{ye})^{1/2}$	6.0	2.5
	> $190/(F_{ye})^{1/2}$	3.0	1.5
Pipe[3]	< $1500/F_{ye}$	6.0	2.5
	> $6000/F_{ye}$	3.0	1.5
Tension-only		3.0	1.5
All others		6.0	2.5

[1] Defined in Section 2.4.
[2] Depth-to-thickness ratio.
[3] Interpolation to be used for tubes and pipes.
$F_{ye} = 1.25 F_y$; expected yield stress as defined by Section 4.2.4.4.

3.5.3.5 Precast Connections

The strength of the connection in precast concrete moment frames shall be greater than the moment in the girder, M_{gj}, calculated in accordance with Equation (3-14).

$$M_{gj} = \frac{V_j}{m}\left(\frac{1}{n_c - n_f}\right)\left(\frac{h}{2}\right) \qquad (3\text{-}14)$$

where:
- n_c = Total number of columns
- n_f = Total number of frames in the direction of loading
- V_j = Story shear at the level directly below the connection under consideration
- h = Typical column story height
- m = Component modification factor taken equal to 2.0 for buildings being evaluated to the Life Safety Performance Level and equal-to-1.3 for buildings being evaluated to the Immediate Occupancy Performance Level

---

C3.5.3.5 Precast Connections

The term $[1/(n_c-n_f)]$ in Equation (3-14) is based on the assumption that the end column carries half the load of a typical interior column.

3.5.3.6 Axial Stress Due to Overturning

The axial stress of columns in moment frames at the base subjected to overturning forces, p_{ot}, shall be calculated in accordance with Equation (3-15).

$$p_{ot} = \frac{1}{m}\left(\frac{2}{3}\right)\left(\frac{Vh_n}{Ln_f}\right)\left(\frac{1}{A_{col}}\right) \qquad (3\text{-}15)$$

where:
- n_f = Total number of frames in the direction of loading
- V = Pseudo lateral force
- h_n = Height (in feet) above the base to the roof level
- L = Total length of the frame (in feet)
- m = Component modification factor taken equal to 2.0 for buildings being evaluated to the Life Safety Performance Level and equal-to-1.3 for buildings being evaluated to the Immediate Occupancy Performance Level
- A_{col} = Area of the end column of the frame

C3.5.3.6 Axial Stress Due to Overturning

The 2/3 factor in Equation (3-15) assumes a triangular force distribution with the resultant applied at 2/3 the height of the building.

3.5.3.7 Flexible Diaphragm Connection Forces

The horizontal seismic forces associated with the connection of a flexible diaphragm to either concrete or masonry walls, T_c, shall be calculated in accordance with Equation (3-16).

$$T_c = ?\, S_{DS}\, w_p\, A_p \qquad (3\text{-}16)$$

where:
- w_p = unit weight of the wall
- A_p = area of wall tributary to the connection
- $?$ = 0.9 for Life Safety and 1.4 for Immediate Occupancy
- S_{DS} = Value calculated from Equation (3-6)

C3.5.3.7 Flexible Diaphragm Connection Forces

Equation (3-16) is based on FEMA 356 out-of-plane forces. The factor ? assumes the flexible diaphragm has a 0.75 factor applied for existing buildings.

3.5.3.8 Prestressed Elements

The average prestress in prestressed or post-tensioned elements, f_p, shall be calculated in accordance with Equation (3-17).

$$f_p = \frac{f_{pe} n_p}{A_p} \qquad (3\text{-}17)$$

where:
- f_{pe} = Effective force of a prestressed strand
- n_p = Number of prestressed strands
- A_p = Gross area of prestressed concrete elements

C3.5.3.8 Prestressed Elements

The average prestress is simply calculated as the effective force of a prestressed strand times the number of strands divided by the gross concrete area. In many cases half-inch strands are used, which corresponds to an effective force of 25 kips per strand.

Screening Phase (Tier 1)

3.6 Level of Low Seismicity Checklist

This Level of Low Seismicity Checklist shall be completed where required by Table 3-2.

Each of the evaluation statements on this checklist shall be marked Compliant (C), Non-compliant (NC), or Not Applicable (N/A) for a Tier 1 Evaluation. Compliant statements identify issues that are acceptable according to the criteria of this standard, while non-compliant statements identify issues that require further investigation. Certain statements may not apply to the buildings being evaluated. For non-compliant evaluation statements, the design professional may choose to conduct further investigation using the corresponding Tier 2 Evaluation procedure; corresponding section numbers are in parentheses following each evaluation statement.

Structural Components

C   NC   N/A   LOAD PATH: The structure shall contain a minimum of one complete load path for Life Safety and Immediate Occupancy for seismic force effects from any horizontal direction that serves to transfer the inertial forces from the mass to the foundation. (Tier 2: Sec. 4.3.1.1)

C   NC   N/A   WALL ANCHORAGE: Exterior concrete or masonry walls that are dependent on the diaphragm for lateral support shall be anchored for out-of-plane forces at each diaphragm level with steel anchors, reinforcing dowels, or straps that are developed into the diaphragm. Connections shall have adequate strength to resist the connection force calculated in the Quick Check procedure of Section 3.5.3.7. (Tier 2: Sec. 4.6.1.1)

Geologic Site and Foundation Components

C   NC   N/A   FOUNDATION PERFORMANCE: There shall be no evidence of excessive foundation movement such as settlement or heave that would affect the integrity or strength of the structure. (Tier 2: Sec. 4.7.2.1)

Nonstructural Components

C   NC   N/A   EMERGENCY LIGHTING: Emergency lighting equipment shall be anchored or braced to prevent falling during an earthquake. (Tier 2: Sec. 4.8.3.1)

C   NC   N/A   CLADDING ANCHORS: Cladding components weighing more than 10 psf shall be mechanically anchored to the exterior wall framing at a spacing equal to or less than 6 feet. (Tier 2: Sec. 4.8.4.1)

C   NC   N/A   CLADDING DETERIORATION: There shall be no evidence of deterioration, damage, or corrosion in any of the connection elements. (Tier 2: Sec. 4.8.4.2)

C   NC   N/A   PARAPETS: There shall be no laterally unsupported unreinforced masonry parapets or cornices with height-to-thickness ratios greater than 2.5. (Tier 2: Sec. 4.8.8.1)

C   NC   N/A   CANOPIES: Canopies located at building exits shall be anchored to the structural framing at a spacing of 10 feet or less. (Tier 2: Sec. 4.8.8.2)

C   NC   N/A   EMERGENCY POWER: Equipment used as part of an emergency power system shall be mounted to maintain continued operation after an earthquake. (Tier 2: Sec. 4.8.12.1)

C   NC   N/A   HAZARDOUS MATERIAL EQUIPMENT: HVAC or other equipment containing hazardous material shall not have damaged supply lines or unbraced isolation supports. (Tier 2: Sec. 4.8.12.2)

Screening Phase (Tier 1)

3.7 Structural Checklists

This section provides Basic and Supplemental Structural Checklists for the following building types:

BUILDING TYPE		CHECKLIST LOCATION
W1	Wood Light Frames	3.7.1, 3.7.1S
W1A	Multi-Story, Multi-Unit Residential Wood Frames	3.7.1A, 3.7.1AS
W2	Wood Frames, Commercial and Industrial	3.7.2
S1	Steel Moment Frames with Stiff Diaphragms	3.7.3, 3.7.3S
S1A	Steel Moment Frames with Flexible Diaphragms	3.7.3A, 3.7.3AS
S2	Steel Braced Frames with Stiff Diaphragms	3.7.4, 3.7.4S
S2A	Steel Braced Frames with Flexible Diaphragms	3.7.4A, 3.7.4AS
S3	Steel Light Frames	3.7.5, 3.7.5S
S4	Steel Frames with Concrete Shear Walls	3.7.6, 3.7.6S
S5	Steel Frames with Infill Masonry Shear Walls and Stiff Diaphragms	3.7.7, 3.7.7S
S5A	Steel Frames with Infill Masonry Shear Walls and Flexible Diaphragms	3.7.7A, 3.7.7.AS
C1	Concrete Moment Frames	3.7.8, 3.7.8S
C2	Concrete Shear Walls with Stiff Diaphragms	3.7.9, 3.7.9S
C2A	Concrete Shear Walls with Flexible Diaphragms	3.7.9A, 3.7.9AS
C3	Concrete Frames with Infill Masonry Shear Walls and Stiff Diaphragms	3.7.10
C3A	Concrete Frames with Infill Masonry Shear Walls and Flexible Diaphragms	3.7.10A, 3.7.10AS
PC1	Precast/Tilt-up Concrete Shear Walls with Flexible Diaphragms	3.7.11, 3.7.11S
PC1A	Precast/Tilt-up Concrete Shear Walls with Stiff Diaphragms	3.7.11A, 3.7.11AS
PC2	Precast Concrete Frames with Shear Walls	3.7.12, 3.7.12S
PC2A	Precast Concrete Frames without Shear Walls	3.7.12A, 3.7.12AS
RM1	Reinforced Masonry Bearing Walls with Flexible Diaphragms	3.7.13, 3.7.13S
RM2	Reinforced Masonry Bearing Walls with Stiff Diaphragms	3.7.14, 3.7.14S
URM	Unreinforced Masonry Bearing Walls with Flexible Diaphragms	3.7.15, 3.7.15S
URMA	Unreinforced Masonry Bearing Walls with Stiff Diaphragms	3.7.15A, 3.7.15AS
	General Basic Structural Checklist	3.7.16
	General Supplemental Structural Checklist	3.7.16S

For a description of the specific building types listed above, refer to Table 2-2.

The appropriate Basic Structural Checklist shall be completed where required by Table 3-2.

The appropriate Supplemental Structural Checklist shall be completed where required by Table 3-2. The appropriate Basic Structural Checklist shall be completed prior to completing the appropriate Supplemental Structural Checklist.

3.7.1 Basic Structural Checklist for Building Type W1: Wood Light Frames

This Basic Structural Checklist shall be completed where required by Table 3-2.

Each of the evaluation statements on this checklist shall be marked Compliant (C), Non-compliant (NC), or Not Applicable (N/A) for a Tier 1 Evaluation. Compliant statements identify issues that are acceptable according to the criteria of this standard, while non-compliant statements identify issues that require further investigation. Certain statements may not apply to the buildings being evaluated. For non-compliant evaluation statements, the design professional may choose to conduct further investigation using the corresponding Tier 2 Evaluation procedure; corresponding section numbers are in parentheses following each evaluation statement.

> **C3.7.1 Basic Structural Checklist for Building Type W1**
>
> These buildings are single- or multiple-family dwellings of one or more stories in height. Building loads are light and the framing spans are short. Floor and roof framing consists of wood joists or rafters on wood studs spaced no more than 24 inches apart. The first floor framing is supported directly on the foundation, or is raised up on cripple studs and post-and-beam supports. The foundation consists of spread footings constructed on concrete, concrete masonry block, or brick masonry or even wood in older construction. Chimneys, where present, consist of solid brick masonry, masonry veneer, or wood frame with internal metal flues. Lateral forces are resisted by wood frame diaphragms and shear walls. Floor and roof diaphragms consist of straight or diagonal lumber sheathing, tongue-and-groove planks, oriented strand board, or plywood. Shear walls consist of straight or diagonal lumber sheathing, plank siding, plywood, oriented strand board, stucco, gypsum board, particle board, or fiberboard. Interior partitions are sheathed with plaster or gypsum board.

Building System

C   NC   N/A   LOAD PATH: The structure shall contain a minimum of one complete load path for Life Safety and Immediate Occupancy for seismic force effects from any horizontal direction that serves to transfer the inertial forces from the mass to the foundation. (Tier 2: Sec. 4.3.1.1)

C   NC   N/A   VERTICAL DISCONTINUITIES: All vertical elements in the lateral-force-resisting system shall be continuous to the foundation. (Tier 2: Sec. 4.3.2.4)

C   NC   N/A   DETERIORATION OF WOOD: There shall be no signs of decay, shrinkage, splitting, fire damage, or sagging in any of the wood members, and none of the metal connection hardware shall be deteriorated, broken, or loose. (Tier 2: Sec. 4.3.3.1)

C   NC   N/A   WOOD STRUCTURAL PANEL SHEAR WALL FASTENERS: There shall be no more than 15 percent of inadequate fastening such as overdriven fasteners, omitted blocking, excessive fastening spacing, or inadequate edge distance. This statement shall apply to the Immediate Occupancy Performance Level only. (Tier 2: Sec. 4.3.3.2)

Lateral-Force-Resisting System

C   NC   N/A   REDUNDANCY: The number of lines of shear walls in each principal direction shall be greater than or equal to 2 for Life Safety and Immediate Occupancy. (Tier 2: Sec. 4.4.2.1.1)

Screening Phase (Tier 1)

C NC N/A SHEAR STRESS CHECK: The shear stress in the shear walls, calculated using the Quick Check procedure of Section 3.5.3.3, shall be less than the following values for Life Safety and Immediate Occupancy (Tier 2: Sec. 4.4.2.7.1):

Structural panel sheathing	1,000 plf
Diagonal sheathing	700 plf
Straight sheathing	100 plf
All other conditions	100 plf

C NC N/A STUCCO (EXTERIOR PLASTER) SHEAR WALLS: Multi-story buildings shall not rely on exterior stucco walls as the primary lateral-force-resisting system. (Tier 2: Sec. 4.4.2.7.2)

C NC N/A GYPSUM WALLBOARD OR PLASTER SHEAR WALLS: Interior plaster or gypsum wallboard shall not be used as shear walls on buildings over one story in height with the exception of the uppermost level of a multi-story building. (Tier 2: Sec. 4.4.2.7.3)

C NC N/A NARROW WOOD SHEAR WALLS: Narrow wood shear walls with an aspect ratio greater than 2-to-1 for Life Safety and 1.5-to-1 for Immediate Occupancy shall not be used to resist lateral forces developed in the building in levels of moderate and high seismicity. Narrow wood shear walls with an aspect ratio greater than 2-to-1 for Immediate Occupancy shall not be used to resist lateral forces developed in the building in levels of low seismicity. (Tier 2: Sec. 4.4.2.7.4)

C NC N/A WALLS CONNECTED THROUGH FLOORS: Shear walls shall have interconnection between stories to transfer overturning and shear forces through the floor. (Tier 2: Sec. 4.4.2.7.5)

C NC N/A HILLSIDE SITE: For structures that are taller on at least one side by more than one-half story due to a sloping site, all shear walls on the downhill slope shall have an aspect ratio less than 1-to-1 for Life Safety and 1 to 2 for Immediate Occupancy. (Tier 2: Sec. 4.4.2.7.6)

C NC N/A CRIPPLE WALLS: Cripple walls below first-floor-level shear walls shall be braced to the foundation with wood structural panels. (Tier 2: Sec. 4.4.2.7.7)

C NC N/A OPENINGS: Walls with openings greater than 80 percent of the length shall be braced with wood structural panel shear walls with aspect ratios of not more than 1.5-to-1 or shall be supported by adjacent construction through positive ties capable of transferring the lateral forces. (Tier 2: Sec. 4.4.2.7.8)

Connections

C NC N/A WOOD POSTS: There shall be a positive connection of wood posts to the foundation. (Tier 2: Sec. 4.6.3.3)

C NC N/A WOOD SILLS: All wood sills shall be bolted to the foundation. (Tier 2: Sec. 4.6.3.4)

C NC N/A GIRDER/COLUMN CONNECTION: There shall be a positive connection utilizing plates, connection hardware, or straps between the girder and the column support. (Tier 2: Sec. 4.6.4.1)

3.7.1S Supplemental Structural Checklist for Building Type W1: Wood Light Frames

This Supplemental Structural Checklist shall be completed where required by Table 3-2. The Basic Structural Checklist shall be completed prior to completing this Supplemental Structural Checklist.

Lateral-Force-Resisting System

C NC N/A HOLD-DOWN ANCHORS: All shear walls shall have hold-down anchors constructed per acceptable construction practices, attached to the end studs. This statement shall apply to the Immediate Occupancy Performance Level only. (Tier 2: Sec. 4.4.2.7.9)

Diaphragms

C NC N/A DIAPHRAGM CONTINUITY: The diaphragms shall not be composed of split-level floors and shall not have expansion joints. (Tier 2: Sec. 4.5.1.1)

C NC N/A ROOF CHORD CONTINUITY: All chord elements shall be continuous, regardless of changes in roof elevation. (Tier 2: Sec. 4.5.1.3)

C NC N/A PLAN IRREGULARITIES: There shall be tensile capacity to develop the strength of the diaphragm at re-entrant corners or other locations of plan irregularities. This statement shall apply to the Immediate Occupancy Performance Level only. (Tier 2: Sec. 4.5.1.7)

C NC N/A DIAPHRAGM REINFORCEMENT AT OPENINGS: There shall be reinforcing around all diaphragm openings larger than 50 percent of the building width in either major plan dimension. This statement shall apply to the Immediate Occupancy Performance Level only. (Tier 2: Sec. 4.5.1.8)

C NC N/A STRAIGHT SHEATHING: All straight sheathed diaphragms shall have aspect ratios less than 2-to-1 for Life Safety and 1-to-1 for Immediate Occupancy in the direction being considered. (Tier 2: Sec. 4.5.2.1)

C NC N/A SPANS: All wood diaphragms with spans greater than 24 feet for Life Safety and 12 feet for Immediate Occupancy shall consist of wood structural panels or diagonal sheathing. (Tier 2: Sec. 4.5.2.2)

C NC N/A UNBLOCKED DIAPHRAGMS: All diagonally sheathed or unblocked wood structural panel diaphragms shall have horizontal spans less than 40 feet for Life Safety and 30 feet for Immediate Occupancy and shall have aspect ratios less than or equal to 4-to-1 for Life Safety and 3-to-1 for Immediate Occupancy. (Tier 2: Sec. 4.5.2.3)

C NC N/A OTHER DIAPHRAGMS: The diaphragm shall not consist of a system other than wood, metal deck, concrete, or horizontal bracing. (Tier 2: Sec. 4.5.7.1)

Connections

C NC N/A WOOD SILL BOLTS: Sill bolts shall be spaced at 6 feet or less for Life Safety and 4 feet or less for Immediate Occupancy, with proper edge and end distance provided for wood and concrete. (Tier 2: Sec. 4.6.3.9)

Screening Phase (Tier 1)

3.7.1A Basic Structural Checklist for Building Type W1A: Multi-Story, Multi-Unit Residential Wood Frames

This Basic Structural Checklist shall be completed where required by Table 3-2.

Each of the evaluation statements on this checklist shall be marked Compliant (C), Non-compliant (NC), or Not Applicable (N/A) for a Tier 1 Evaluation. Compliant statements identify issues that are acceptable according to the criteria of this standard, while non-compliant statements identify issues that require further investigation. Certain statements may not apply to the buildings being evaluated. For non-compliant evaluation statements, the design professional may choose to conduct further investigation using the corresponding Tier 2 Evaluation procedure; corresponding section numbers are in parentheses following each evaluation statement.

> **C3.7.1A Basic Structural Checklist for Building Type W1A**
>
> These buildings are multi-story, similar in construction to W1 buildings, but have plan areas on each floor typically greater than 3,000 square feet. Older construction often has open front garages at the lowest story. The foundation consists of spread footings constructed on concrete, concrete masonry block, or brick masonry in older construction. Chimneys, where present, consist of solid brick masonry, masonry veneer, or wood frame with internal metal flues. Lateral forces are resisted by wood frame diaphragms and shear walls. Floor and roof diaphragms consist of straight or diagonal lumber sheathing, tongue-and-groove planks, oriented strand board, or plywood. Shear walls consist of straight or diagonal lumber sheathing, plank siding, oriented strand board, plywood, stucco, gypsum board, particle board, or fiberboard. Interior partitions are sheathed with plaster or gypsum board.

Building System

C NC N/A   **LOAD PATH:** The structure shall contain a minimum of one complete load path for Life Safety and Immediate Occupancy for seismic force effects from any horizontal direction that serves to transfer the inertial forces from the mass to the foundation. (Tier 2: Sec. 4.3.1.1)

C NC N/A   **WEAK STORY:** The strength of the lateral-force-resisting system in any story shall not be less than 80 percent of the strength in an adjacent story, above or below, for Life Safety and Immediate Occupancy. (Tier 2: Sec. 4.3.2.1)

C NC N/A   **SOFT STORY:** The stiffness of the lateral-force-resisting system in any story shall not be less than 70 percent of the lateral-force-resisting system stiffness in an adjacent story above or below, or less than 80 percent of the average lateral-force-resisting system stiffness of the three stories above or below for Life Safety and Immediate Occupancy. (Tier 2: Sec. 4.3.2.2)

C NC N/A   **VERTICAL DISCONTINUITIES:** All vertical elements in the lateral-force-resisting system shall be continuous to the foundation. (Tier 2: Sec. 4.3.2.4)

C NC N/A   **DETERIORATION OF WOOD:** There shall be no signs of decay, shrinkage, splitting, fire damage, or sagging in any of the wood members, and none of the metal connection hardware shall be deteriorated, broken, or loose. (Tier 2: Sec. 4.3.3.1)

C NC N/A   **WOOD STRUCTURAL PANEL SHEAR WALL FASTENERS:** There shall be no more than 15 percent of inadequate fastening such as overdriven fasteners, omitted blocking, excessive fastening spacing, or inadequate edge distance. This statement shall apply to the Immediate Occupancy Performance Level only. (Tier 2: Sec. 4.3.3.2)

Screening Phase (Tier 1)

Lateral-Force-Resisting System

C NC N/A REDUNDANCY: The number of lines of shear walls in each principal direction shall be greater than or equal to 2 for Life Safety and Immediate Occupancy. (Tier 2: Sec. 4.4.2.1.1)

C NC N/A SHEAR STRESS CHECK: The shear stress in the shear walls, calculated using the Quick Check procedure of Section 3.5.3.3, shall be less than the following values for Life Safety and Immediate Occupancy (Tier 2: Sec. 4.4.2.7.1):

Structural panel sheathing	1,000 plf
Diagonal sheathing	700 plf
Straight sheathing	100 plf
All other conditions	100 plf

C NC N/A STUCCO (EXTERIOR PLASTER) SHEAR WALLS: Multi-story buildings shall not rely on exterior stucco walls as the primary lateral-force-resisting system. (Tier 2: Sec. 4.4.2.7.2)

C NC N/A GYPSUM WALLBOARD OR PLASTER SHEAR WALLS: Interior plaster or gypsum wallboard shall not be used as shear walls on buildings over one story in height with the exception of the uppermost level of a multi-story building. (Tier 2: Sec. 4.4.2.7.3)

C NC N/A NARROW WOOD SHEAR WALLS: Narrow wood shear walls with an aspect ratio greater than 2-to-1 for Life Safety and 1.5-to-1 for Immediate Occupancy shall not be used to resist lateral forces developed in the building in levels of moderate and high seismicity. Narrow wood shear walls with an aspect ratio greater than 2-to-1 for Immediate Occupancy shall not be used to resist lateral forces developed in the building in levels of low seismicity. (Tier 2: Sec. 4.4.2.7.4)

C NC N/A WALLS CONNECTED THROUGH FLOORS: Shear walls shall have interconnection between stories to transfer overturning and shear forces through the floor. (Tier 2: Sec. 4.4.2.7.5)

C NC N/A HILLSIDE SITE: For structures that are taller on at least one side by more than one-half story due to a sloping site, all shear walls on the downhill slope shall have an aspect ratio less than 1-to-1 for Life Safety and 1-to-2 for Immediate Occupancy. (Tier 2: Sec. 4.4.2.7.6)

C NC N/A CRIPPLE WALLS: Cripple walls below first-floor-level shear walls shall be braced to the foundation with wood structural panels. (Tier 2: Sec. 4.4.2.7.7)

C NC N/A OPENINGS: Walls with openings greater than 80 percent of the length shall be braced with wood structural panel shear walls with aspect ratios of not more than 1.5-to-1 or shall be supported by adjacent construction through positive ties capable of transferring the lateral forces. (Tier 2: Sec. 4.4.2.7.8)

Connections

C NC N/A WOOD POSTS: There shall be a positive connection of wood posts to the foundation. (Tier 2: Sec. 4.6.3.3)

C NC N/A WOOD SILLS: All wood sills shall be bolted to the foundation. (Tier 2: Sec. 4.6.3.4)

C NC N/A GIRDER/COLUMN CONNECTION: There shall be a positive connection utilizing plates, connection hardware, or straps between the girder and the column support. (Tier 2: Sec. 4.6.4.1)

3.7.1AS Supplemental Structural Checklist for Building Type W1A: Multi-Story, Multi-Unit Residential Wood Frame

This Supplemental Structural Checklist shall be completed where required by Table 3-2. The Basic Structural Checklist shall be completed prior to completing this Supplemental Structural Checklist.

Lateral-Force-Resisting System

C   NC   N/A   HOLD-DOWN ANCHORS: All shear walls shall have hold-down anchors constructed per acceptable construction practices, attached to the end studs. This statement shall apply to the Immediate Occupancy Performance Level only. (Tier 2: Sec. 4.4.2.7.9)

Diaphragms

C   NC   N/A   DIAPHRAGM CONTINUITY: The diaphragms shall not be composed of split-level floors and shall not have expansion joints. (Tier 2: Sec. 4.5.1.1)

C   NC   N/A   ROOF CHORD CONTINUITY: All chord elements shall be continuous, regardless of changes in roof elevation. (Tier 2: Sec. 4.5.1.3)

C   NC   N/A   PLAN IRREGULARITIES: There shall be tensile capacity to develop the strength of the diaphragm at re-entrant corners or other locations of plan irregularities. This statement shall apply to the Immediate Occupancy Performance Level only. (Tier 2: Sec. 4.5.1.7)

C   NC   N/A   DIAPHRAGM REINFORCEMENT AT OPENINGS: There shall be reinforcing around all diaphragm openings larger than 50 percent of the building width in either major plan dimension. This statement shall apply to the Immediate Occupancy Performance Level only. (Tier 2: Sec. 4.5.1.8)

C   NC   N/A   STRAIGHT SHEATHING: All straight sheathed diaphragms shall have aspect ratios less than 2-to-1 for Life Safety and 1-to-1 for Immediate Occupancy in the direction being considered. (Tier 2: Sec. 4.5.2.1)

C   NC   N/A   SPANS: All wood diaphragms with spans greater than 24 feet for Life Safety and 12 feet for Immediate Occupancy shall consist of wood structural panels or diagonal sheathing. (Tier 2: Sec. 4.5.2.2)

C   NC   N/A   UNBLOCKED DIAPHRAGMS: All diagonally sheathed or unblocked wood structural panel diaphragms shall have horizontal spans less than 40 feet for Life Safety and 30 feet for Immediate Occupancy and shall have aspect ratios less than or equal to 4-to-1 for Life Safety and 3-to-1 for Immediate Occupancy. (Tier 2: Sec. 4.5.2.3)

C   NC   N/A   OTHER DIAPHRAGMS: The diaphragm shall not consist of a system other than wood, metal deck, concrete, or horizontal bracing. (Tier 2: Sec. 4.5.7.1)

Connections

C   NC   N/A   WOOD SILL BOLTS: Sill bolts shall be spaced at 6 feet or less for Life Safety and 4 feet or less for Immediate Occupancy, with proper edge and end distance provided for wood and concrete. (Tier 2: Sec. 4.6.3.9)

Screening Phase (Tier 1)

3.7.2 Basic Structural Checklist for Building Type W2: Wood Frames, Commercial and Industrial

This Basic Structural Checklist shall be completed where required by Table 3-2.

Each of the evaluation statements on this checklist shall be marked Compliant (C), Non-compliant (NC), or Not Applicable (N/A) for a Tier 1 Evaluation. Compliant statements identify issues that are acceptable according to the criteria of this standard, while non-compliant statements identify issues that require further investigation. Certain statements may not apply to the buildings being evaluated. For non-compliant evaluation statements, the design professional may choose to conduct further investigation using the corresponding Tier 2 Evaluation procedure; corresponding section numbers are in parentheses following each evaluation statement.

C3.7.2 Basic Structural Checklist for Building Type W2

These buildings are commercial or industrial buildings with a floor area of 5,000 square feet or more. There are few, if any, interior walls. The floor and roof framing consists of wood or steel trusses, glulam or steel beams, and wood posts or steel columns. Lateral forces are resisted by wood diaphragms and exterior stud walls sheathed with plywood, oriented strand board, stucco, plaster, straight or diagonal wood sheathing, or braced with rod bracing. Wall openings for storefronts and garages, where present, are framed by post-and-beam framing.

Building System

C NC N/A   **LOAD PATH:** The structure shall contain a minimum of one complete load path for Life Safety and Immediate Occupancy for seismic force effects from any horizontal direction that serves to transfer the inertial forces from the mass to the foundation. (Tier 2: Sec. 4.3.1.1)

C NC N/A   **MEZZANINES:** Interior mezzanine levels shall be braced independently from the main structure, or shall be anchored to the lateral-force-resisting elements of the main structure. (Tier 2: Sec. 4.3.1.3)

C NC N/A   **WEAK STORY:** The strength of the lateral-force-resisting system in any story shall not be less than 80 percent of the strength in an adjacent story, above or below, for Life Safety and Immediate Occupancy. (Tier 2: Sec. 4.3.2.1)

C NC N/A   **SOFT STORY:** The stiffness of the lateral-force-resisting system in any story shall not be less than 70 percent of the lateral-force-resisting system stiffness in an adjacent story above or below, or less than 80 percent of the average lateral-force-resisting system stiffness of the three stories above or below for Life Safety and Immediate Occupancy. (Tier 2: Sec. 4.3.2.2)

C NC N/A   **GEOMETRY:** There shall be no changes in horizontal dimension of the lateral-force-resisting system of more than 30 percent in a story relative to adjacent stories for Life Safety and Immediate Occupancy, excluding one-story penthouses and mezzanines. (Tier 2: Sec. 4.3.2.3)

C NC N/A   **VERTICAL DISCONTINUITIES:** All vertical elements in the lateral-force-resisting system shall be continuous to the foundation. (Tier 2: Sec. 4.3.2.4)

C NC N/A   **MASS:** There shall be no change in effective mass more than 50 percent from one story to the next for Life Safety and Immediate Occupancy. Light roofs, penthouses, and mezzanines need not be considered. (Tier 2: Sec. 4.3.2.5)

Screening Phase (Tier 1)

C NC N/A DETERIORATION OF WOOD: There shall be no signs of decay, shrinkage, splitting, fire damage, or sagging in any of the wood members, and none of the metal connection hardware shall be deteriorated, broken, or loose. (Tier 2: Sec. 4.3.3.1)

C NC N/A WOOD STRUCTURAL PANEL SHEAR WALL FASTENERS: There shall be no more than 15 percent of inadequate fastening such as overdriven fasteners, omitted blocking, excessive fastening spacing, or inadequate edge distance. This statement shall apply to the Immediate Occupancy Performance Level only. (Tier 2: Sec. 4.3.3.2)

Lateral-Force-Resisting System

C NC N/A REDUNDANCY: The number of lines of shear walls in each principal direction shall be greater than or equal to 2 for Life Safety and Immediate Occupancy. (Tier 2: Sec. 4.4.2.1.1)

C NC N/A SHEAR STRESS CHECK: The shear stress in the shear walls, calculated using the Quick Check procedure of Section 3.5.3.3, shall be less than the following values for Life Safety and Immediate Occupancy (Tier 2: Sec. 4.4.2.7.1):

Structural panel sheathing	1,000 plf
Diagonal sheathing	700 plf
Straight sheathing	100 plf
All other conditions	100 plf

C NC N/A STUCCO (EXTERIOR PLASTER) SHEAR WALLS: Multi-story buildings shall not rely on exterior stucco walls as the primary lateral-force-resisting system. (Tier 2: Sec. 4.4.2.7.2)

C NC N/A GYPSUM WALLBOARD OR PLASTER SHEAR WALLS: Interior plaster or gypsum wallboard shall not be used as shear walls on buildings over one story in height with the exception of the uppermost level of a multi-story building. (Tier 2: Sec. 4.4.2.7.3)

C NC N/A NARROW WOOD SHEAR WALLS: Narrow wood shear walls with an aspect ratio greater than 2-to-1 for Life Safety and 1.5-to-1 for Immediate Occupancy shall not be used to resist lateral forces developed in the building in levels of moderate and high seismicity. Narrow wood shear walls with an aspect ratio greater than 2-to-1 for Immediate Occupancy shall not be used to resist lateral forces developed in the building in levels of low seismicity. (Tier 2: Sec. 4.4.2.7.4)

C NC N/A WALLS CONNECTED THROUGH FLOORS: Shear walls shall have interconnection between stories to transfer overturning and shear forces through the floor. (Tier 2: Sec. 4.4.2.7.5)

C NC N/A HILLSIDE SITE: For structures that are taller on at least one side by more than one-half story due to a sloping site, all shear walls on the downhill slope shall have an aspect ratio less than 1-to-1 for Life Safety and 1-to-2 for Immediate Occupancy. (Tier 2: Sec. 4.4.2.7.6)

C NC N/A CRIPPLE WALLS: Cripple walls below first-floor-level shear walls shall be braced to the foundation with wood structural panels. (Tier 2: Sec. 4.4.2.7.7)

C NC N/A OPENINGS: Walls with openings greater than 80 percent of the length shall be braced with wood structural panel shear walls with aspect ratios of not more than 1.5-to-1 or shall be supported by adjacent construction through positive ties capable of transferring the lateral forces. (Tier 2: Sec. 4.4.2.7.8)

Connections

C NC N/A WOOD POSTS: There shall be a positive connection of wood posts to the foundation. (Tier 2: Sec. 4.6.3.3)

C NC N/A WOOD SILLS: All wood sills shall be bolted to the foundation. (Tier 2: Sec. 4.6.3.4)

C NC N/A GIRDER/COLUMN CONNECTION: There shall be a positive connection utilizing plates, connection hardware, or straps between the girder and the column support. (Tier 2: Sec. 4.6.4.1)

Screening Phase (Tier 1)

3.7.2S Supplemental Structural Checklist for Building Type W2: Wood Frames, Commercial and Industrial

This Supplemental Structural Checklist shall be completed where required by Table 3-2. The Basic Structural Checklist shall be completed prior to completing this Supplemental Structural Checklist.

Lateral-Force-Resisting System

C   NC   N/A   HOLD-DOWN ANCHORS: All shear walls shall have hold-down anchors constructed per acceptable construction practices, attached to the end studs. This statement shall apply to the Immediate Occupancy Performance Level only. (Tier 2: Sec. 4.4.2.7.9)

Diaphragms

C   NC   N/A   DIAPHRAGM CONTINUITY: The diaphragms shall not be composed of split-level floors and shall not have expansion joints. (Tier 2: Sec. 4.5.1.1)

C   NC   N/A   ROOF CHORD CONTINUITY: All chord elements shall be continuous, regardless of changes in roof elevation. (Tier 2: Sec. 4.5.1.3)

C   NC   N/A   PLAN IRREGULARITIES: There shall be tensile capacity to develop the strength of the diaphragm at re-entrant corners or other locations of plan irregularities. This statement shall apply to the Immediate Occupancy Performance Level only. (Tier 2: Sec. 4.5.1.7)

C   NC   N/A   DIAPHRAGM REINFORCEMENT AT OPENINGS: There shall be reinforcing around all diaphragm openings larger than 50 percent of the building width in either major plan dimension. This statement shall apply to the Immediate Occupancy Performance Level only. (Tier 2: Sec. 4.5.1.8)

C   NC   N/A   STRAIGHT SHEATHING: All straight sheathed diaphragms shall have aspect ratios less than 2-to-1 for Life Safety and 1-to-1 for Immediate Occupancy in the direction being considered. (Tier 2: Sec. 4.5.2.1)

C   NC   N/A   SPANS: All wood diaphragms with spans greater than 24 feet for Life Safety and 12 feet for Immediate Occupancy shall consist of wood structural panels or diagonal sheathing. Wood commercial and industrial buildings may have rod-braced systems. (Tier 2: Sec. 4.5.2.2)

C   NC   N/A   UNBLOCKED DIAPHRAGMS: All diagonally sheathed or unblocked wood structural panel diaphragms shall have horizontal spans less than 40 feet for Life Safety and 30 feet for Immediate Occupancy and shall have aspect ratios less than or equal to 4-to-1 for Life Safety and 3-to-1 for Immediate Occupancy. (Tier 2: Sec. 4.5.2.3)

C   NC   N/A   OTHER DIAPHRAGMS: The diaphragm shall not consist of a system other than wood, metal deck, concrete, or horizontal bracing. (Tier 2: Sec. 4.5.7.1)

Connections

C   NC   N/A   WOOD SILL BOLTS: Sill bolts shall be spaced at 6 feet or less for Life Safety and 4 feet or less for Immediate Occupancy, with proper edge and end distance provided for wood and concrete. (Tier 2: Sec. 4.6.3.9)

Screening Phase (Tier 1)

3.7.3 Basic Structural Checklist for Building Type S1: Steel Moment Frames with Stiff Diaphragms

This Basic Structural Checklist shall be completed where required by Table 3-2.

Each of the evaluation statements on this checklist shall be marked Compliant (C), Non-compliant (NC), or Not Applicable (N/A) for a Tier 1 Evaluation. Compliant statements identify issues that are acceptable according to the criteria of this standard, while non-compliant statements identify issues that require further investigation. Certain statements may not apply to the buildings being evaluated. For non-compliant evaluation statements, the design professional may choose to conduct further investigation using the corresponding Tier 2 Evaluation procedure; corresponding section numbers are in parentheses following each evaluation statement.

> **C3.7.3 Basic Structural Checklist for Building Type S1**
>
> These buildings consist of a frame assembly of steel beams and steel columns. Floor and roof framing consists of cast-in-place concrete slabs or metal deck with concrete fill supported on steel beams, open web joists, or steel trusses. Lateral forces are resisted by steel moment frames that develop their stiffness through rigid or semi-rigid beam-column connections. Where all connections are moment-resisting connections, the entire frame participates in lateral force resistance. Where only selected connections are moment-resisting connections, resistance is provided along discrete frame lines. Columns are oriented so that each principal direction of the building has columns resisting forces in strong axis bending. Diaphragms consist of concrete or metal deck with concrete fill and are stiff relative to the frames. Where the exterior of the structure is concealed, walls consist of metal panel curtain walls, glazing, brick masonry, or precast concrete panels. Where the interior of the structure is finished, frames are concealed by ceilings, partition walls, and architectural column furring. Foundations consist of concrete spread footings or deep pile foundations.

Building System

C NC N/A **LOAD PATH:** The structure shall contain a minimum of one complete load path for Life Safety and Immediate Occupancy for seismic force effects from any horizontal direction that serves to transfer the inertial forces from the mass to the foundation. (Tier 2: Sec. 4.3.1.1)

C NC N/A **ADJACENT BUILDINGS:** The clear distance between the building being evaluated and any adjacent building shall be greater than 4 percent of the height of the shorter building for Life Safety and Immediate Occupancy. (Tier 2: Sec. 4.3.1.2)

C NC N/A **MEZZANINES:** Interior mezzanine levels shall be braced independently from the main structure, or shall be anchored to the lateral-force-resisting elements of the main structure. (Tier 2: Sec. 4.3.1.3)

C NC N/A **WEAK STORY:** The strength of the lateral-force-resisting system in any story shall not be less than 80 percent of the strength in an adjacent story, above or below, for Life Safety and Immediate Occupancy. (Tier 2: Sec. 4.3.2.1)

C NC N/A **SOFT STORY:** The stiffness of the lateral-force-resisting system in any story shall not be less than 70 percent of the lateral-force-resisting system stiffness in an adjacent story above or below, or less than 80 percent of the average lateral-force-resisting system stiffness of the three stories above or below for Life Safety and Immediate Occupancy. (Tier 2: Sec. 4.3.2.2)

Screening Phase (Tier 1)

C NC N/A GEOMETRY: There shall be no changes in horizontal dimension of the lateral-force-resisting system of more than 30 percent in a story relative to adjacent stories for Life Safety and Immediate Occupancy, excluding one-story penthouses and mezzanines. (Tier 2: Sec. 4.3.2.3)

C NC N/A VERTICAL DISCONTINUITIES: All vertical elements in the lateral-force-resisting system shall be continuous to the foundation. (Tier 2: Sec. 4.3.2.4)

C NC N/A MASS: There shall be no change in effective mass more than 50 percent from one story to the next for Life Safety and Immediate Occupancy. Light roofs, penthouses, and mezzanines need not be considered. (Tier 2: Sec. 4.3.2.5)

C NC N/A TORSION: The estimated distance between the story center of mass and the story center of rigidity shall be less than 20 percent of the building width in either plan dimension for Life Safety and Immediate Occupancy. (Tier 2: Sec. 4.3.2.6)

C NC N/A DETERIORATION OF STEEL: There shall be no visible rusting, corrosion, cracking, or other deterioration in any of the steel elements or connections in the vertical- or lateral-force-resisting systems. (Tier 2: Sec. 4.3.3.3)

C NC N/A DETERIORATION OF CONCRETE: There shall be no visible deterioration of concrete or reinforcing steel in any of the vertical- or lateral-force-resisting elements. (Tier 2: Sec. 4.3.3.4)

Lateral-Force-Resisting System

C NC N/A REDUNDANCY: The number of lines of moment frames in each principal direction shall be greater than or equal to 2 for Life Safety and Immediate Occupancy. The number of bays of moment frames in each line shall be greater than or equal to 2 for Life Safety and 3 for Immediate Occupancy. (Tier 2: Sec. 4.4.1.1.1)

C NC N/A INTERFERING WALLS: All concrete and masonry infill walls placed in moment frames shall be isolated from structural elements. (Tier 2: Sec. 4.4.1.2.1)

C NC N/A DRIFT CHECK: The drift ratio of the steel moment frames, calculated using the Quick Check procedure of Section 3.5.3.1, shall be less than 0.025 for Life Safety and 0.015 for Immediate Occupancy. (Tier 2: Sec. 4.4.1.3.1)

C NC N/A AXIAL STRESS CHECK: The axial stress due to gravity loads in columns subjected to overturning forces shall be less than $0.10F_y$ for Life Safety and Immediate Occupancy. Alternatively, the axial stress due to overturning forces alone, calculated using the Quick Check procedure of Section 3.5.3.6, shall be less than $0.30F_y$ for Life Safety and Immediate Occupancy. (Tier 2: Sec. 4.4.1.3.2)

Connections

C NC N/A TRANSFER TO STEEL FRAMES: Diaphragms shall be connected for transfer of loads to the steel frames for Life Safety, and the connections shall be able to develop the lesser of the strength of the frames or the diaphragms for Immediate Occupancy. (Tier 2: Sec. 4.6.2.2)

C NC N/A STEEL COLUMNS: The columns in lateral-force-resisting frames shall be anchored to the building foundation for Life Safety, and the anchorage shall be able to develop the lesser of the tensile capacity of the column, the tensile capacity of the lowest level column splice (if any), or the uplift capacity of the foundation, for Immediate Occupancy. (Tier 2: Sec. 4.6.3.1)

Screening Phase (Tier 1)

3.7.3S Supplemental Structural Checklist for Building Type S1: Steel Moment Frames with Stiff Diaphragms

This Supplemental Structural Checklist shall be completed where required by Table 3-2. The Basic Structural Checklist shall be completed prior to completing this Supplemental Structural Checklist.

Lateral-Force-Resisting System

C   NC   N/A   **MOMENT-RESISTING CONNECTIONS:** All moment connections shall be able to develop the strength of the adjoining members or panel zones. (Tier 2: Sec. 4.4.1.3.3)

C   NC   N/A   **PANEL ZONES:** All panel zones shall have the shear capacity to resist the shear demand required to develop 0.8 times the sum of the flexural strengths of the girders framing in at the face of the column. (Tier 2: Sec. 4.4.1.3.4)

C   NC   N/A   **COLUMN SPLICES:** All column splice details located in moment-resisting frames shall include connection of both flanges and the web for Life Safety, and the splice shall develop the strength of the column for Immediate Occupancy. (Tier 2: Sec. 4.4.1.3.5)

C   NC   N/A   **STRONG COLUMN/WEAK BEAM:** The percentage of strong column/weak beam joints in each story of each line of moment-resisting frames shall be greater than 50 percent for Life Safety and Immediate Occupancy. (Tier 2: Sec. 4.4.1.3.6)

C   NC   N/A   **COMPACT MEMBERS:** All frame elements shall meet section requirements set forth by *Seismic Provisions for Structural Steel Buildings* Table I-9-1 (AISC, 1997). (Tier 2: Sec. 4.4.1.3.7)

C   NC   N/A   **BEAM PENETRATIONS:** All openings in frame-beam webs shall be less than ¼ of the beam depth and shall be located in the center half of the beams. This statement shall apply to the Immediate Occupancy Performance Level only. (Tier 2: Sec. 4.4.1.3.8)

C   NC   N/A   **GIRDER FLANGE CONTINUITY PLATES:** There shall be girder flange continuity plates at all moment-resisting frame joints. This statement shall apply to the Immediate Occupancy Performance Level only. (Tier 2: Sec. 4.4.1.3.9)

C   NC   N/A   **OUT-OF-PLANE BRACING:** Beam-column joints shall be braced out-of-plane. This statement shall apply to the Immediate Occupancy Performance Level only. (Tier 2: Sec. 4.4.1.3.10)

C   NC   N/A   **BOTTOM FLANGE BRACING:** The bottom flanges of beams shall be braced out-of-plane. This statement shall apply to the Immediate Occupancy Performance Level only. (Tier 2: Sec. 4.4.1.3.11)

Diaphragms

C   NC   N/A   **PLAN IRREGULARITIES:** There shall be tensile capacity to develop the strength of the diaphragm at re-entrant corners or other locations of plan irregularities. This statement shall apply to the Immediate Occupancy Performance Level only. (Tier 2: Sec. 4.5.1.7)

C   NC   N/A   **DIAPHRAGM REINFORCEMENT AT OPENINGS:** There shall be reinforcing around all diaphragm openings larger than 50 percent of the building width in either major plan dimension. This statement shall apply to the Immediate Occupancy Performance Level only. (Tier 2: Sec. 4.5.1.8)

Connections

C   NC   N/A   **UPLIFT AT PILE CAPS:** Pile caps shall have top reinforcement and piles shall be anchored to the pile caps for Life Safety, and the pile cap reinforcement and pile anchorage shall be able to develop the tensile capacity of the piles for Immediate Occupancy. (Tier 2: Sec. 4.6.3.10)

Screening Phase (Tier 1)

3.7.3A Basic Structural Checklist for Building Type S1A: Steel Moment Frames with Flexible Diaphragms

This Basic Structural Checklist shall be completed where required by Table 3-2.

Each of the evaluation statements on this checklist shall be marked Compliant (C), Non-compliant (NC), or Not Applicable (N/A) for a Tier 1 Evaluation. Compliant statements identify issues that are acceptable according to the criteria of this standard, while non-compliant statements identify issues that require further investigation. Certain statements may not apply to the buildings being evaluated. For non-compliant evaluation statements, the design professional may choose to conduct further investigation using the corresponding Tier 2 Evaluation procedure; corresponding section numbers are in parentheses following each evaluation statement.

C3.7.3A Basic Structural Checklist for Building Type S1A

These buildings consist of a frame assembly of steel beams and steel columns. Floor and roof framing consists of cast-in-place concrete slabs or metal deck with concrete fill supported on steel beams, open web joists, or steel trusses. Lateral forces are resisted by steel moment frames that develop their stiffness through rigid or semi-rigid beam-column connections. Where all connections are moment-resisting connections the entire frame participates in lateral force resistance. Where only selected connections are moment-resisting connections, resistance is provided along discrete frame lines. Columns are oriented so that each principal direction of the building has columns resisting forces in strong axis bending. Diaphragms consist of wood framing; untopped metal deck; or metal deck with lightweight insulating concrete, poured gypsum, or similar nonstructural topping and are flexible relative to the frames. Where the exterior of the structure is concealed, walls consist of metal panel curtain walls, glazing, brick masonry, or precast concrete panels. Where the interior of the structure is finished, frames are concealed by ceilings, partition walls, and architectural column furring. Foundations consist of concrete spread footings or deep pile foundations.

Building System

C   NC   N/A   LOAD PATH: The structure shall contain a minimum of one complete load path for Life Safety and Immediate Occupancy for seismic force effects from any horizontal direction that serves to transfer the inertial forces from the mass to the foundation. (Tier 2: Sec. 4.3.1.1)

C   NC   N/A   ADJACENT BUILDINGS: The clear distance between the building being evaluated and any adjacent building shall be greater than 4 percent of the height of the shorter building for Life Safety and Immediate Occupancy. (Tier 2: Sec. 4.3.1.2)

C   NC   N/A   MEZZANINES: Interior mezzanine levels shall be braced independently from the main structure, or shall be anchored to the lateral-force-resisting elements of the main structure. (Tier 2: Sec. 4.3.1.3)

C   NC   N/A   WEAK STORY: The strength of the lateral-force-resisting system in any story shall not be less than 80 percent of the strength in an adjacent story, above or below, for Life Safety and Immediate Occupancy. (Tier 2: Sec. 4.3.2.1)

C   NC   N/A   SOFT STORY: The stiffness of the lateral-force-resisting system in any story shall not be less than 70 percent of the lateral-force-resisting system stiffness in an adjacent story above or below, or less than 80 percent of the average lateral-force-resisting system stiffness of the three stories above or below for Life Safety and Immediate Occupancy. (Tier 2: Sec. 4.3.2.2)

Screening Phase (Tier 1)

C　NC　N/A　　GEOMETRY: There shall be no changes in horizontal dimension of the lateral-force-resisting system of more than 30 percent in a story relative to adjacent stories for Life Safety and Immediate Occupancy, excluding one-story penthouses and mezzanines. (Tier 2: Sec. 4.3.2.3)

C　NC　N/A　　VERTICAL DISCONTINUITIES: All vertical elements in the lateral-force-resisting system shall be continuous to the foundation. (Tier 2: Sec. 4.3.2.4)

C　NC　N/A　　MASS: There shall be no change in effective mass more than 50 percent from one story to the next for Life Safety and Immediate Occupancy. Light roofs, penthouses, and mezzanines need not be considered. (Tier 2: Sec. 4.3.2.5)

C　NC　N/A　　DETERIORATION OF WOOD: There shall be no signs of decay, shrinkage, splitting, fire damage, or sagging in any of the wood members, and none of the metal connection hardware shall be deteriorated, broken, or loose. (Tier 2: Sec. 4.3.3.1)

C　NC　N/A　　DETERIORATION OF STEEL: There shall be no visible rusting, corrosion, cracking, or other deterioration in any of the steel elements or connections in the vertical- or lateral-force-resisting systems. (Tier 2: Sec. 4.3.3.3)

Lateral-Force-Resisting System

C　NC　N/A　　REDUNDANCY: The number of lines of moment frames in each principal direction shall be greater than or equal to 2 for Life Safety and Immediate Occupancy. The number of bays of moment frames in each line shall be greater than or equal to 2 for Life Safety and 3 for Immediate Occupancy. (Tier 2: Sec. 4.4.1.1.1)

C　NC　N/A　　INTERFERING WALLS: All concrete and masonry infill walls placed in moment frames shall be isolated from structural elements. (Tier 2: Sec. 4.4.1.2.1)

C　NC　N/A　　DRIFT CHECK: The drift ratio of the steel moment frames, calculated using the Quick Check procedure of Section 3.5.3.1, shall be less than 0.025 for Life Safety and 0.015 for Immediate Occupancy. (Tier 2: Sec. 4.4.1.3.1)

C　NC　N/A　　AXIAL STRESS CHECK: The axial stress due to gravity loads in columns subjected to overturning forces shall be less than $0.10F_y$ for Life Safety and Immediate Occupancy. Alternatively, the axial stress due to overturning forces alone, calculated using the Quick Check procedure of Section 3.5.3.6, shall be less than $0.30F_y$ for Life Safety and Immediate Occupancy. (Tier 2: Sec. 4.4.1.3.2)

Connections

C　NC　N/A　　TRANSFER TO STEEL FRAMES: Diaphragms shall be connected for transfer of loads to the steel frames for Life Safety, and the connections shall be able to develop the lesser of the strength of the frames or the diaphragms for Immediate Occupancy. (Tier 2: Sec. 4.6.2.2)

C　NC　N/A　　STEEL COLUMNS: The columns in lateral-force-resisting frames shall be anchored to the building foundation for Life Safety, and the anchorage shall be able to develop the lesser of the tensile capacity of the column, the tensile capacity of the lowest level column splice (if any), or the uplift capacity of the foundation, for Immediate Occupancy. (Tier 2: Sec. 4.6.3.1)

Screening Phase (Tier 1)

3.7.3AS Supplemental Structural Checklist for Building Type S1A: Steel Moment Frames with Flexible Diaphragms

This Supplemental Structural Checklist shall be completed where required by Table 3-2. The Basic Structural Checklist shall be completed prior to completing this Supplemental Structural Checklist.

Lateral-Force-Resisting System

C NC N/A MOMENT-RESISTING CONNECTIONS: All moment connections shall be able to develop the strength of the adjoining members or panel zones. (Tier 2: Sec. 4.4.1.3.3)

C NC N/A PANEL ZONES: All panel zones shall have the shear capacity to resist the shear demand required to develop 0.8 times the sum of the flexural strengths of the girders framing in at the face of the column. (Tier 2: Sec. 4.4.1.3.4)

C NC N/A COLUMN SPLICES: All column splice details located in moment-resisting frames shall include connection of both flanges and the web for Life Safety,, and the splice shall develop the strength of the column for Immediate Occupancy. (Tier 2: Sec. 4.4.1.3.5)

C NC N/A STRONG COLUMN/WEAK BEAM: The percentage of strong column/weak beam joints in each story of each line of moment-resisting frames shall be greater than 50 percent for Life Safety and Immediate Occupancy. (Tier 2: Sec. 4.4.1.3.6)

C NC N/A COMPACT MEMBERS: All frame elements shall meet section requirements set forth by *Seismic Provisions for Structural Steel Buildings* Table I-9-1 (AISC, 1997). (Tier 2: Sec. 4.4.1.3.7)

C NC N/A BEAM PENETRATIONS: All openings in frame-beam webs shall be less than ¼ of the beam depth and shall be located in the center half of the beams. This statement shall apply to the Immediate Occupancy Performance Level only. (Tier 2: Sec. 4.4.1.3.8)

C NC N/A GIRDER FLANGE CONTINUITY PLATES: There shall be girder flange continuity plates at all moment-resisting frame joints. This statement shall apply to the Immediate Occupancy Performance Level only. (Tier 2: Sec. 4.4.1.3.9)

C NC N/A OUT-OF-PLANE BRACING: Beam-column joints shall be braced out-of-plane. This statement shall apply to the Immediate Occupancy Performance Level only. (Tier 2: Sec. 4.4.1.3.10)

C NC N/A BOTTOM FLANGE BRACING: The bottom flanges of beams shall be braced out-of-plane. This statement shall apply to the Immediate Occupancy Performance Level only. (Tier 2: Sec. 4.4.1.3.11)

Diaphragms

C NC N/A CROSS TIES: There shall be continuous cross ties between diaphragm chords. (Tier 2: Sec. 4.5.1.2)

C NC N/A PLAN IRREGULARITIES: There shall be tensile capacity to develop the strength of the diaphragm at re-entrant corners or other locations of plan irregularities. This statement shall apply to the Immediate Occupancy Performance Level only. (Tier 2: Sec. 4.5.1.7)

C NC N/A DIAPHRAGM REINFORCEMENT AT OPENINGS: There shall be reinforcing around all diaphragm openings larger than 50 percent of the building width in either major plan dimension. This statement shall apply to the Immediate Occupancy Performance Level only. (Tier 2: Sec. 4.5.1.8)

Screening Phase (Tier 1)

C   NC   N/A   **STRAIGHT SHEATHING:** All straight sheathed diaphragms shall have aspect ratios less than 2-to-1 for Life Safety and 1-to-1 for Immediate Occupancy in the direction being considered. (Tier 2: Sec. 4.5.2.1)

C   NC   N/A   **SPANS:** All wood diaphragms with spans greater than 24 feet for Life Safety and 12 feet for Immediate Occupancy shall consist of wood structural panels or diagonal sheathing. (Tier 2: Sec. 4.5.2.2)

C   NC   N/A   **UNBLOCKED DIAPHRAGMS:** All diagonally sheathed or unblocked wood structural panel diaphragms shall have horizontal spans less than 40 feet for Life Safety and 30 feet for Immediate Occupancy and shall have aspect ratios less than or equal to 4-to-1 for Life Safety and 3-to-1 for Immediate Occupancy. (Tier 2: Sec. 4.5.2.3)

C   NC   N/A   **NON-CONCRETE FILLED DIAPHRAGMS:** Untopped metal deck diaphragms or metal deck diaphragms with fill other than concrete shall consist of horizontal spans of less than 40 feet and shall have span/depth ratios less than 4-to-1. This statement shall apply to the Immediate Occupancy Performance Level only. (Tier 2: Sec. 4.5.3.1)

C   NC   N/A   **OTHER DIAPHRAGMS:** The diaphragm shall not consist of a system other than wood, metal deck, concrete, or horizontal bracing. (Tier 2: Sec. 4.5.7.1)

Connections

C   NC   N/A   **UPLIFT AT PILE CAPS:** Pile caps shall have top reinforcement and piles shall be anchored to the pile caps for Life Safety, and the pile cap reinforcement and pile anchorage shall be able to develop the tensile capacity of the piles for Immediate Occupancy. (Tier 2: Sec. 4.6.3.10)

Screening Phase (Tier 1)

3.7.4 Basic Structural Checklist for Building Type S2: Steel Braced Frames with Stiff Diaphragms

This Basic Structural Checklist shall be completed where required by Table 3-2.

Each of the evaluation statements on this checklist shall be marked Compliant (C), Non-compliant (NC), or Not Applicable (N/A) for a Tier 1 Evaluation. Compliant statements identify issues that are acceptable according to the criteria of this standard, while non-compliant statements identify issues that require further investigation. Certain statements may not apply to the buildings being evaluated. For non-compliant evaluation statements, the design professional may choose to conduct further investigation using the corresponding Tier 2 Evaluation procedure; corresponding section numbers are in parentheses following each evaluation statement.

C3.7.4 Basic Structural Checklist for Building Type S2

These buildings have a frame of steel columns, beams, and braces. Braced frames develop resistance to lateral forces by the bracing action of the diagonal members. The braces induce forces in the associated beams and columns such that all elements work together in a manner similar to a truss with all element stresses being primarily axial. Where the braces do not completely triangulate the panel, some of the members are subjected to shear and flexural stresses; eccentrically braced frames are one such case (refer to Sec. 4.4.3.3). Diaphragms transfer lateral loads to braced frames. The diaphragms consist of concrete or metal deck with concrete fill and are stiff relative to the frames.

Building System

C   NC   N/A   LOAD PATH: The structure shall contain a minimum of one complete load path for Life Safety and Immediate Occupancy for seismic force effects from any horizontal direction that serves to transfer the inertial forces from the mass to the foundation. (Tier 2: Sec. 4.3.1.1)

C   NC   N/A   ADJACENT BUILDINGS: The clear distance between the building being evaluated and any adjacent building shall be greater than 4 percent of the height of the shorter building for Life Safety and Immediate Occupancy. (Tier 2: Sec. 4.3.1.2)

C   NC   N/A   MEZZANINES: Interior mezzanine levels shall be braced independently from the main structure, or shall be anchored to the lateral-force-resisting elements of the main structure. (Tier 2: Sec. 4.3.1.3)

C   NC   N/A   WEAK STORY: The strength of the lateral-force-resisting system in any story shall not be less than 80 percent of the strength in an adjacent story, above or below, for Life Safety and Immediate Occupancy. (Tier 2: Sec. 4.3.2.1)

C   NC   N/A   SOFT STORY: The stiffness of the lateral-force-resisting system in any story shall not be less than 70 percent of the lateral-force-resisting system stiffness in an adjacent story above or below, or less than 80 percent of the average lateral-force-resisting system stiffness of the three stories above or below for Life Safety and Immediate Occupancy. (Tier 2: Sec. 4.3.2.2)

C   NC   N/A   GEOMETRY: There shall be no changes in horizontal dimension of the lateral-force-resisting system of more than 30 percent in a story relative to adjacent stories for Life Safety and Immediate Occupancy, excluding one-story penthouses and mezzanines. (Tier 2: Sec. 4.3.2.3)

C   NC   N/A   VERTICAL DISCONTINUITIES: All vertical elements in the lateral-force-resisting system shall be continuous to the foundation. (Tier 2: Sec. 4.3.2.4)

Screening Phase (Tier 1)

C   NC   N/A   **MASS:** There shall be no change in effective mass more than 50 percent from one story to the next for Life Safety and Immediate Occupancy. Light roofs, penthouses, and mezzanines need not be considered. (Tier 2: Sec. 4.3.2.5)

C   NC   N/A   **TORSION:** The estimated distance between the story center of mass and the story center of rigidity shall be less than 20 percent of the building width in either plan dimension for Life Safety and Immediate Occupancy. (Tier 2: Sec. 4.3.2.6)

C   NC   N/A   **DETERIORATION OF STEEL:** There shall be no visible rusting, corrosion, cracking, or other deterioration in any of the steel elements or connections in the vertical- or lateral-force-resisting systems. (Tier 2: Sec. 4.3.3.3)

C   NC   N/A   **DETERIORATION OF CONCRETE:** There shall be no visible deterioration of concrete or reinforcing steel in any of the vertical- or lateral-force-resisting elements. (Tier 2: Sec. 4.3.3.4)

Lateral-Force-Resisting System

C   NC   N/A   **AXIAL STRESS CHECK:** The axial stress due to gravity loads in columns subjected to overturning forces shall be less than $0.10F_y$ for Life Safety and Immediate Occupancy. Alternatively, the axial stress due to overturning forces alone, calculated using the Quick Check procedure of Section 3.5.3.6, shall be less than $0.30F_y$ for Life Safety and Immediate Occupancy. (Tier 2: Sec. 4.4.1.3.2)

C   NC   N/A   **REDUNDANCY:** The number of lines of braced frames in each principal direction shall be greater than or equal to 2 for Life Safety and Immediate Occupancy. The number of braced bays in each line shall be greater than 2 for Life Safety and 3 for Immediate Occupancy. (Tier 2: Sec. 4.4.3.1.1)

C   NC   N/A   **AXIAL STRESS CHECK:** The axial stress in the diagonals, calculated using the Quick Check procedure of Section 3.5.3.4, shall be less than $0.50F_y$ for Life Safety and for Immediate Occupancy. (Tier 2: Sec. 4.4.3.1.2)

C   NC   N/A   **COLUMN SPLICES:** All column splice details located in braced frames shall develop the tensile strength of the column. This statement shall apply to the Immediate Occupancy Performance Level only. (Tier 2: Sec. 4.4.3.1.3)

Connections

C   NC   N/A   **TRANSFER TO STEEL FRAMES:** Diaphragms shall be connected for transfer of loads to the steel frames for Life Safety, and the connections shall be able to develop the lesser of the strength of the frames or the diaphragms for Immediate Occupancy. (Tier 2: Sec. 4.6.2.2)

C   NC   N/A   **STEEL COLUMNS:** The columns in lateral-force-resisting frames shall be anchored to the building foundation for Life Safety, and the anchorage shall be able to develop the lesser of the tensile capacity of the column, the tensile capacity of the lowest level column splice (if any), or the uplift capacity of the foundation, for Immediate Occupancy. (Tier 2: Sec. 4.6.3.1)

Screening Phase (Tier 1)

3.7.4S Supplemental Structural Checklist for Building Type S2: Steel Braced Frames with Stiff Diaphragms

This Supplemental Structural Checklist shall be completed where required by Table 3-2. The Basic Structural Checklist shall be completed prior to completing this Supplemental Structural Checklist.

Lateral-Force-Resisting System

C NC N/A   **COMPACT MEMBERS:** All frame elements shall meet section requirements set forth by *Seismic Provisions for Structural Steel Buildings* Table I-9-1 (AISC, 1997). (Tier 2: Sec. 4.4.1.3.7)

C NC N/A   **SLENDERNESS OF DIAGONALS:** All diagonal elements required to carry compression shall have Kl/r ratios less than 120. (Tier 2: Sec. 4.4.3.1.4)

C NC N/A   **CONNECTION STRENGTH:** All the brace connections shall develop the yield capacity of the diagonals. (Tier 2: Sec. 4.4.3.1.5)

C NC N/A   **OUT-OF-PLANE BRACING:** Braced frame connections attached to beam bottom flanges located away from beam-column joints shall be braced out-of-plane at the bottom flange of the beams. This statement shall apply to the Immediate Occupancy Performance Level only. (Tier 2: Sec. 4.4.3.1.6)

C NC N/A   **K-BRACING:** The bracing system shall not include K-braced bays. (Tier 2: Sec. 4.4.3.2.1)

C NC N/A   **TENSION-ONLY BRACES:** Tension-only braces shall not comprise more than 70 percent of the total lateral-force-resisting capacity in structures over two stories in height. This statement shall apply to the Immediate Occupancy Performance Level only. (Tier 2: Sec. 4.4.3.2.2)

C NC N/A   **CHEVRON BRACING:** The bracing system shall not include chevron, or V-braced, bays. This statement shall apply to the Immediate Occupancy Performance Level only. (Tier 2: Sec. 4.4.3.2.3)

C NC N/A   **CONCENTRICALLY BRACED FRAME JOINTS:** All the diagonal braces shall frame into the beam-column joints concentrically. This statement shall apply to the Immediate Occupancy Performance Level only. (Tier 2: Sec. 4.4.3.2.4)

Diaphragms

C NC N/A   **OPENINGS AT BRACED FRAMES:** Diaphragm openings immediately adjacent to the braced frames shall extend less than 25 percent of the frame length for Life Safety and 15 percent of the frame length for Immediate Occupancy. (Tier 2: Sec. 4.5.1.5)

C NC N/A   **PLAN IRREGULARITIES:** There shall be tensile capacity to develop the strength of the diaphragm at re-entrant corners or other locations of plan irregularities. This statement shall apply to the Immediate Occupancy Performance Level only. (Tier 2: Sec. 4.5.1.7)

C NC N/A   **DIAPHRAGM REINFORCEMENT AT OPENINGS:** There shall be reinforcing around all diaphragm openings larger than 50 percent of the building width in either major plan dimension. This statement shall apply to the Immediate Occupancy Performance Level only. (Tier 2: Sec. 4.5.1.8)

Connections

C NC N/A   **UPLIFT AT PILE CAPS:** Pile caps shall have top reinforcement and piles shall be anchored to the pile caps for Life Safety, and the pile cap reinforcement and pile anchorage shall be able to develop the tensile capacity of the piles for Immediate Occupancy. (Tier 2: Sec. 4.6.3.10)

3.7.4A Basic Structural Checklist for Building Type S2A: Steel Braced Frames with Flexible Diaphragms

This Basic Structural Checklist shall be completed where required by Table 3-2.

Each of the evaluation statements on this checklist shall be marked Compliant (C), Non-compliant (NC), or Not Applicable (N/A) for a Tier 1 Evaluation. Compliant statements identify issues that are acceptable according to the criteria of this standard, while non-compliant statements identify issues that require further investigation. Certain statements may not apply to the buildings being evaluated. For non-compliant evaluation statements, the design professional may choose to conduct further investigation using the corresponding Tier 2 Evaluation procedure; corresponding section numbers are in parentheses following each evaluation statement.

> **C3.7.4A Basic Structural Checklist for Building Type S2A**
>
> These buildings consist of a frame assembly of steel beams and steel columns. Floor and roof framing consists of wood framing or untopped metal deck supported on steel beams, open web joists, or steel trusses. Lateral forces are resisted by tension and compression forces in diagonal steel members. As such, this selection is not intended to explicitly address eccentrically braced frame systems. See Section 4.4.3.3 for discussion of eccentrically braced frames. Where diagonal brace connections are concentric to beam column joints, all member stresses are primarily axial. Where diagonal brace connections are eccentric to the joints, members are subjected to bending and axial stresses. Diaphragms consist of wood framing; untopped metal deck; or metal deck with lightweight insulating concrete, poured gypsum, or similar nonstructural topping and are flexible relative to the frames. Where the exterior of the structure is concealed, walls consist of metal panel curtain walls, glazing, brick masonry, or precast concrete panels. Where the interior of the structure is finished, frames are concealed by ceilings, partition walls, and architectural furring. Foundations consist of concrete spread footings or deep pile foundations.

Building System

C NC N/A   **LOAD PATH:** The structure shall contain a minimum of one complete load path for Life Safety and Immediate Occupancy for seismic force effects from any horizontal direction that serves to transfer the inertial forces from the mass to the foundation. (Tier 2: Sec. 4.3.1.1)

C NC N/A   **ADJACENT BUILDINGS:** The clear distance between the building being evaluated and any adjacent building shall be greater than 4 percent of the height of the shorter building for Life Safety and Immediate Occupancy. (Tier 2: Sec. 4.3.1.2)

C NC N/A   **MEZZANINES:** Interior mezzanine levels shall be braced independently from the main structure, or shall be anchored to the lateral-force-resisting elements of the main structure. (Tier 2: Sec. 4.3.1.3)

C NC N/A   **WEAK STORY:** The strength of the lateral-force-resisting system in any story shall not be less than 80 percent of the strength in an adjacent story, above or below, for Life Safety and Immediate Occupancy. (Tier 2: Sec. 4.3.2.1)

C NC N/A   **SOFT STORY:** The stiffness of the lateral-force-resisting system in any story shall not be less than 70 percent of the lateral-force-resisting system stiffness in an adjacent story above or below, or less than 80 percent of the average lateral-force-resisting system stiffness of the three stories above or below for Life Safety and Immediate Occupancy. (Tier 2: Sec. 4.3.2.2)

Screening Phase (Tier 1)

C   NC   N/A   **GEOMETRY:** There shall be no changes in horizontal dimension of the lateral-force-resisting system of more than 30 percent in a story relative to adjacent stories for Life Safety and Immediate Occupancy, excluding one-story penthouses and mezzanines. (Tier 2: Sec. 4.3.2.3)

C   NC   N/A   **VERTICAL DISCONTINUITIES:** All vertical elements in the lateral-force-resisting system shall be continuous to the foundation. (Tier 2: Sec. 4.3.2.4)

C   NC   N/A   **MASS:** There shall be no change in effective mass more than 50 percent from one story to the next for Life Safety and Immediate Occupancy. Light roofs, penthouses, and mezzanines need not be considered. (Tier 2: Sec. 4.3.2.5)

C   NC   N/A   **DETERIORATION OF WOOD:** There shall be no signs of decay, shrinkage, splitting, fire damage, or sagging in any of the wood members, and none of the metal connection hardware shall be deteriorated, broken, or loose. (Tier 2: Sec. 4.3.3.1)

C   NC   N/A   **DETERIORATION OF STEEL:** There shall be no visible rusting, corrosion, cracking, or other deterioration in any of the steel elements or connections in the vertical- or lateral-force-resisting systems. (Tier 2: Sec. 4.3.3.3)

Lateral-Force-Resisting System

C   NC   N/A   **AXIAL STRESS CHECK:** The axial stress due to gravity loads in columns subjected to overturning forces shall be less than $0.10F_y$ for Life Safety and Immediate Occupancy. Alternatively, the axial stress due to overturning forces alone, calculated using the Quick Check procedure of Section 3.5.3.6, shall be less than $0.30F_y$ for Life Safety and Immediate Occupancy. (Tier 2: Sec. 4.4.1.3.2)

C   NC   N/A   **REDUNDANCY:** The number of lines of braced frames in each principal direction shall be greater than or equal to 2 for Life Safety and Immediate Occupancy. The number of braced bays in each line shall be greater than 2 for Life Safety and 3 for Immediate Occupancy. (Tier 2: Sec. 4.4.3.1.1)

C   NC   N/A   **AXIAL STRESS CHECK:** The axial stress in the diagonals, calculated using the Quick Check procedure of Section 3.5.3.4, shall be less than $0.50F_y$ for Life Safety and for Immediate Occupancy. (Tier 2: Sec. 4.4.3.1.2)

C   NC   N/A   **COLUMN SPLICES:** All column splice details located in braced frames shall develop the tensile strength of the column. This statement shall apply to the Immediate Occupancy Performance Level only. (Tier 2: Sec. 4.4.3.1.3)

Connections

C   NC   N/A   **TRANSFER TO STEEL FRAMES:** Diaphragms shall be connected for transfer of loads to the steel frames for Life Safety, and the connections shall be able to develop the lesser of the strength of the frames or the diaphragms for Immediate Occupancy. (Tier 2: Sec. 4.6.2.2)

C   NC   N/A   **STEEL COLUMNS:** The columns in lateral-force-resisting frames shall be anchored to the building foundation for Life Safety, and the anchorage shall be able to develop the lesser of the tensile capacity of the column, the tensile capacity of the lowest level column splice (if any), or the uplift capacity of the foundation, for Immediate Occupancy. (Tier 2: Sec. 4.6.3.1)

Screening Phase (Tier 1)

3.7.4AS Supplemental Structural Checklist for Building Type S2A: Steel Braced Frames with Flexible Diaphragms

This Supplemental Structural Checklist shall be completed where required by Table 3-2. The Basic Structural Checklist shall be completed prior to completing this Supplemental Structural Checklist.

Lateral-Force-Resisting System

C NC N/A   COMPACT MEMBERS: All frame elements shall meet section requirements set forth by Seismic Provisions for Structural Steel Buildings Table I-9-1 (AISC, 1997). (Tier 2: Sec. 4.4.1.3.7)

C NC N/A   SLENDERNESS OF DIAGONALS: All diagonal elements required to carry compression shall have Kl/r ratios less than 120. (Tier 2: Sec. 4.4.3.1.4)

C NC N/A   CONNECTION STRENGTH: All the brace connections shall develop the yield capacity of the diagonals. (Tier 2: Sec. 4.4.3.1.5)

C NC N/A   OUT-OF-PLANE BRACING: Braced frame connections attached to beam bottom flanges located away from beam-column joints shall be braced out-of-plane at the bottom flange of the beams. This statement shall apply to the Immediate Occupancy Performance Level only. (Tier 2: Sec. 4.4.3.1.6)

C NC N/A   K-BRACING: The bracing system shall not include K-braced bays. (Tier 2: Sec. 4.4.3.2.1)

C NC N/A   TENSION-ONLY BRACES: Tension-only braces shall not comprise more than 70 percent of the total lateral-force-resisting capacity in structures over two stories in height. This statement shall apply to the Immediate Occupancy Performance Level only. (Tier 2: Sec. 4.4.3.2.2)

C NC N/A   CHEVRON BRACING: The bracing system shall not include chevron, or V-braced, bays. This statement shall apply to the Immediate Occupancy Performance Level only. (Tier 2: Sec. 4.4.3.2.3)

C NC N/A   CONCENTRICALLY BRACED FRAME JOINTS: All the diagonal braces shall frame into the beam-column joints concentrically. This statement shall apply to the Immediate Occupancy Performance Level only. (Tier 2: Sec. 4.4.3.2.4)

Diaphragms

C NC N/A   CROSS TIES: There shall be continuous cross ties between diaphragm chords. (Tier 2: Sec. 4.5.1.2)

C NC N/A   OPENINGS AT BRACED FRAMES: Diaphragm openings immediately adjacent to the braced frames shall extend less than 25 percent of the frame length for Life Safety and 15 percent of the frame length for Immediate Occupancy. (Tier 2: Sec. 4.5.1.5)

C NC N/A   PLAN IRREGULARITIES: There shall be tensile capacity to develop the strength of the diaphragm at re-entrant corners or other locations of plan irregularities. This statement shall apply to the Immediate Occupancy Performance Level only. (Tier 2: Sec. 4.5.1.7)

C NC N/A   DIAPHRAGM REINFORCEMENT AT OPENINGS: There shall be reinforcing around all diaphragm openings larger than 50 percent of the building width in either major plan dimension. This statement shall apply to the Immediate Occupancy Performance Level only. (Tier 2: Sec. 4.5.1.8)

C NC N/A   STRAIGHT SHEATHING: All straight sheathed diaphragms shall have aspect ratios less than 2-to-1 for Life Safety and 1-to-1 for Immediate Occupancy in the direction being considered. (Tier 2: Sec. 4.5.2.1)

Screening Phase (Tier 1)

C NC N/A   SPANS: All wood diaphragms with spans greater than 24 feet for Life Safety and 12 feet for Immediate Occupancy shall consist of wood structural panels or diagonal sheathing. (Tier 2: Sec. 4.5.2.2)

C NC N/A   UNBLOCKED DIAPHRAGMS: All diagonally sheathed or unblocked wood structural panel diaphragms shall have horizontal spans less than 40 feet for Life Safety and 30 feet for Immediate Occupancy and shall have aspect ratios less than or equal to 4-to-1 for Life Safety and 3-to-1 for Immediate Occupancy. (Tier 2: Sec. 4.5.2.3)

C NC N/A   NON-CONCRETE FILLED DIAPHRAGMS: Untopped metal deck diaphragms or metal deck diaphragms with fill other than concrete shall consist of horizontal spans of less than 40 feet and shall have span/depth ratios less than 4-to-1. This statement shall apply to the Immediate Occupancy Performance Level only. (Tier 2: Sec. 4.5.3.1)

Connections

C NC N/A   UPLIFT AT PILE CAPS: Pile caps shall have top reinforcement and piles shall be anchored to the pile caps for Life Safety, and the pile cap reinforcement and pile anchorage shall be able to develop the tensile capacity of the piles for Immediate Occupancy. (Tier 2: Sec. 4.6.3.10)

Screening Phase (Tier 1)

3.7.5 Basic Structural Checklist for Building Type S3: Steel Light Frames

This Basic Structural Checklist shall be completed where required by Table 3-2. This Basic Structural Checklist shall not be used for a structure with a roof dead load greater than 25 psf or a building area greater than 20,000 feet. Where either limit is exceeded, a Steel Moment Frame Basic Structural Checklist (Type S1 or S1A) shall be used.

Each of the evaluation statements on this checklist shall be marked Compliant (C), Non-compliant (NC), or Not Applicable (N/A) for a Tier 1 Evaluation. Compliant statements identify issues that are acceptable according to the criteria of this standard, while non-compliant statements identify issues that require further investigation. Certain statements may not apply to the buildings being evaluated. For non-compliant evaluation statements, the design professional may choose to conduct further investigation using the corresponding Tier 2 Evaluation procedure; corresponding section numbers are in parentheses following each evaluation statement.

C3.7.5 Basic Structural Checklist for Building Type S3

These buildings are pre-engineered and prefabricated with transverse rigid steel frames. They are one story in height. The roof and walls consist of lightweight metal, fiberglass, or cementitious panels. The frames are designed for maximum efficiency, and the beams and columns consist of tapered, built-up sections with thin plates. The frames are built in segments and assembled in the field with bolted or welded joints. Lateral forces in the transverse direction are resisted by the rigid frames. Lateral forces in the longitudinal direction are resisted by wall panel shear elements or rod bracing. Diaphragm forces are resisted by untopped metal deck, roof panel shear elements, or a system of tension-only rod bracing.

Building System

C NC N/A   **LOAD PATH:** The structure shall contain a minimum of one complete load path for Life Safety and Immediate Occupancy for seismic force effects from any horizontal direction that serves to transfer the inertial forces from the mass to the foundation. (Tier 2: Sec. 4.3.1.1)

C NC N/A   **MEZZANINES:** Interior mezzanine levels shall be braced independently from the main structure, or shall be anchored to the lateral-force-resisting elements of the main structure. (Tier 2: Sec. 4.3.1.3)

C NC N/A   **VERTICAL DISCONTINUITIES:** All vertical elements in the lateral-force-resisting system shall be continuous to the foundation. (Tier 2: Sec. 4.3.2.4)

C NC N/A   **TORSION:** The estimated distance between the story center of mass and the story center of rigidity shall be less than 20 percent of the building width in either plan dimension for Life Safety and Immediate Occupancy. (Tier 2: Sec. 4.3.2.6)

C NC N/A   **DETERIORATION OF STEEL:** There shall be no visible rusting, corrosion, cracking, or other deterioration in any of the steel elements or connections in the vertical- or lateral-force-resisting systems. (Tier 2: Sec. 4.3.3.3)

Lateral-Force-Resisting System

C NC N/A   **AXIAL STRESS CHECK:** The axial stress in the diagonals, calculated using the Quick Check procedure of Section 3.5.3.4, shall be less than $0.50 F_y$ for Life Safety and for Immediate Occupancy. (Tier 2: Sec. 4.4.3.1.2)

Connections

C NC N/A TRANSFER TO STEEL FRAMES: Diaphragms shall be connected for transfer of loads to the steel frames for Life Safety, and the connections shall be able to develop the lesser of the strength of the frames or the diaphragms for Immediate Occupancy. (Tier 2: Sec. 4.6.2.2)

C NC N/A STEEL COLUMNS: The columns in lateral-force-resisting frames shall be anchored to the building foundation for Life Safety, and the anchorage shall be able to develop the lesser of the tensile capacity of the column, the tensile capacity of the lowest level column splice (if any), or the uplift capacity of the foundation, for Immediate Occupancy. (Tier 2: Sec. 4.6.3.1)

C NC N/A WALL PANELS: Metal, fiberglass, or cementitious wall panels shall be positively attached to the foundation for Life Safety and Immediate Occupancy. (Tier 2: Sec. 4.6.3.8)

C NC N/A ROOF PANELS: Metal, plastic, or cementitious roof panels shall be positively attached to the roof framing to resist seismic forces for Life Safety and Immediate Occupancy. (Tier 2: Sec. 4.6.5.1)

C NC N/A WALL PANELS: Metal, fiberglass, or cementitious wall panels shall be positively attached to the framing to resist seismic forces for Life Safety and Immediate Occupancy. (Tier 2: Sec. 4.6.5.2)

3.7.5S Supplemental Structural Checklist for Building Type S3: Steel Light Frames

This Supplemental Structural Checklist shall be completed where required by Table 3-2. The Basic Structural Checklist shall be completed prior to completing this Supplemental Structural Checklist. This Supplemental Structural Checklist shall not be used for a structure with a roof dead load greater than 25 psf or a building area greater than 20,000 feet. Where either limit is exceeded, a Steel Moment Frame Supplemental Structural Checklist (Type S1 or S1A) shall be used.

Lateral-Force-Resisting System

C NC N/A **MOMENT-RESISTING CONNECTIONS:** All moment connections shall be able to develop the strength of the adjoining members or panel zones. (Tier 2: Sec. 4.4.1.3.3)

C NC N/A **BEAM PENETRATIONS:** All openings in frame-beam webs shall be less than ¼ of the beam depth and shall be located in the center half of the beams. This statement shall apply to the Immediate Occupancy Performance Level only. (Tier 2: Sec. 4.4.1.3.8)

C NC N/A **COMPACT MEMBERS:** All frame elements shall meet section requirements set forth by Seismic Provisions for Structural Steel Buildings Table I-9-1 (AISC, 1997). (Tier 2: Sec. 4.4.1.3.7)

C NC N/A **OUT-OF-PLANE BRACING:** Beam-column joints shall be braced out-of-plane. This statement shall apply to the Immediate Occupancy Performance Level only. (Tier 2: Sec. 4.4.1.3.10)

C NC N/A **BOTTOM FLANGE BRACING:** The bottom flanges of beams shall be braced out-of-plane. This statement shall apply to the Immediate Occupancy Performance Level only. (Tier 2: Sec. 4.4.1.3.11)

Diaphragms

C NC N/A **PLAN IRREGULARITIES:** There shall be tensile capacity to develop the strength of the diaphragm at re-entrant corners or other locations of plan irregularities. This statement shall apply to the Immediate Occupancy Performance Level only. (Tier 2: Sec. 4.5.1.7)

C NC N/A **DIAPHRAGM REINFORCEMENT AT OPENINGS:** There shall be reinforcing around all diaphragm openings larger than 50 percent of the building width in either major plan dimension. This statement shall apply to the Immediate Occupancy Performance Level only. (Tier 2: Sec. 4.5.1.8)

C NC N/A **OTHER DIAPHRAGMS:** The diaphragm shall not consist of a system other than wood, metal deck, concrete, or horizontal bracing. (Tier 2: Sec. 4.5.7.1)

Connections

C NC N/A **UPLIFT AT PILE CAPS:** Pile caps shall have top reinforcement and piles shall be anchored to the pile caps for Life Safety, and the pile cap reinforcement and pile anchorage shall be able to develop the tensile capacity of the piles for Immediate Occupancy. (Tier 2: Sec. 4.6.3.10)

SCREENING PHASE (Tier 1)

3.7.6 Basic Structural Checklist for Building Type S4: Steel Frames with Concrete Shear Walls

This Basic Structural Checklist shall be completed where required by Table 3-2.

Each of the evaluation statements on this checklist shall be marked Compliant (C), Non-compliant (NC), or Not Applicable (N/A) for a Tier 1 Evaluation. Compliant statements identify issues that are acceptable according to the criteria of this standard, while non-compliant statements identify issues that require further investigation. Certain statements may not apply to the buildings being evaluated. For non-compliant evaluation statements, the design professional may choose to conduct further investigation using the corresponding Tier 2 Evaluation procedure; corresponding section numbers are in parentheses following each evaluation statement.

> **C3.7.6 Basic Structural Checklist for Building Type S4**
>
> These buildings consist of a frame assembly of steel beams and steel columns. The floor and roof diaphragms consist of cast-in-place concrete slabs or metal deck with or without concrete fill. Framing consists of steel beams, open web joists, or steel trusses. Lateral forces are resisted by cast-in-place concrete shear walls. These walls are bearing walls where the steel frame does not provide a complete vertical support system. In older construction the steel frame is designed for vertical loads only. In modern dual systems, the steel moment frames are designed to work together with the concrete shear walls in proportion to their relative rigidity. In the case of a dual system, the walls shall be evaluated under this building type and the frames shall be evaluated under S1 or S1A, Steel Moment Frames. The steel frame may provide a secondary lateral-force-resisting system depending on the stiffness of the frame and the moment capacity of the beam-column connections.

Building System

C NC N/A **LOAD PATH:** The structure shall contain a minimum of one complete load path for Life Safety and Immediate Occupancy for seismic force effects from any horizontal direction that serves to transfer the inertial forces from the mass to the foundation. (Tier 2: Sec. 4.3.1.1)

C NC N/A **MEZZANINES:** Interior mezzanine levels shall be braced independently from the main structure, or shall be anchored to the lateral-force-resisting elements of the main structure. (Tier 2: Sec. 4.3.1.3)

C NC N/A **WEAK STORY:** The strength of the lateral-force-resisting system in any story shall not be less than 80 percent of the strength in an adjacent story, above or below, for Life Safety and Immediate Occupancy. (Tier 2: Sec. 4.3.2.1)

C NC N/A **SOFT STORY:** The stiffness of the lateral-force-resisting system in any story shall not be less than 70 percent of the lateral-force-resisting system stiffness in an adjacent story above or below, or less than 80 percent of the average lateral-force-resisting system stiffness of the three stories above or below for Life Safety and Immediate Occupancy. (Tier 2: Sec. 4.3.2.2)

C NC N/A **GEOMETRY:** There shall be no changes in horizontal dimension of the lateral-force-resisting system of more than 30 percent in a story relative to adjacent stories for Life Safety and Immediate Occupancy; excluding one-story penthouses and mezzanines. (Tier 2: Sec. 4.3.2.3)

C NC N/A **VERTICAL DISCONTINUITIES:** All vertical elements in the lateral-force-resisting system shall be continuous to the foundation. (Tier 2: Sec. 4.3.2.4)

Screening Phase (Tier 1)

C　NC　N/A　MASS: There shall be no change in effective mass more than 50 percent from one story to the next for Life Safety and Immediate Occupancy. Light roofs, penthouses, and mezzanines need not be considered. (Tier 2: Sec. 4.3.2.5)

C　NC　N/A　TORSION: The estimated distance between the story center of mass and the story center of rigidity shall be less than 20 percent of the building width in either plan dimension for Life Safety and Immediate Occupancy. (Tier 2: Sec. 4.3.2.6)

C　NC　N/A　DETERIORATION OF STEEL: There shall be no visible rusting, corrosion, cracking, or other deterioration in any of the steel elements or connections in the vertical- or lateral-force-resisting systems. (Tier 2: Sec. 4.3.3.3)

C　NC　N/A　DETERIORATION OF CONCRETE: There shall be no visible deterioration of concrete or reinforcing steel in any of the vertical- or lateral-force-resisting elements. (Tier 2: Sec. 4.3.3.4)

C　NC　N/A　CONCRETE WALL CRACKS: All existing diagonal cracks in wall elements shall be less than 1/8 inch for Life Safety and 1/16 inch for Immediate Occupancy, shall not be concentrated in one location, and shall not form an X pattern. (Tier 2: Sec. 4.3.3.9)

Lateral-Force-Resisting System

C　NC　N/A　COMPLETE FRAMES: Steel or concrete frames classified as secondary components shall form a complete vertical-load-carrying system. (Tier 2: Sec. 4.4.1.6.1)

C　NC　N/A　REDUNDANCY: The number of lines of shear walls in each principal direction shall be greater than or equal to 2 for Life Safety and Immediate Occupancy. (Tier 2: Sec. 4.4.2.1.1)

C　NC　N/A　SHEAR STRESS CHECK: The shear stress in the concrete shear walls, calculated using the Quick Check procedure of Section 3.5.3.3, shall be less than the greater of 100 psi or $2\sqrt{f'c}$ for Life Safety and Immediate Occupancy. (Tier 2: Sec. 4.4.2.2.1)

C　NC　N/A　REINFORCING STEEL: The ratio of reinforcing steel area to gross concrete area shall be not less than 0.0015 in the vertical direction and 0.0025 in the horizontal direction for Life Safety and Immediate Occupancy. The spacing of reinforcing steel shall be equal to or less than 18 inches for Life Safety and Immediate Occupancy. (Tier 2: Sec. 4.4.2.2.2)

C　NC　N/A　COLUMN SPLICES: Steel columns encased in shear-wall-boundary elements shall have splices that develop the tensile strength of the column. This statement shall apply to the Immediate Occupancy Performance Level only. (Tier 2: Sec. 4.4.2.2.9)

Connections

C　NC　N/A　TRANSFER TO SHEAR WALLS: Diaphragms shall be connected for transfer of loads to the shear walls for Life Safety and the connections shall be able to develop the lesser of the shear strength of the walls or diaphragms for Immediate Occupancy. (Tier 2 Sec. 4.6.2.1)

C　NC　N/A　FOUNDATION DOWELS: Wall reinforcement shall be doweled into the foundation for Life Safety, and the dowels shall be able to develop the lesser of the strength of the walls or the uplift capacity of the foundation for Immediate Occupancy. (Tier 2: Sec. 4.6.3.5)

C　NC　N/A　SHEAR-WALL-BOUNDARY COLUMNS: The shear-wall-boundary columns shall be anchored to the building foundation for Life Safety, and the anchorage shall be able to develop the tensile capacity of the column for Immediate Occupancy. (Tier 2: Sec.4.6.3.6)

3.7.6S Supplemental Structural Checklist for Building Type S4: Steel Frames with Concrete Shear Walls

This Supplemental Structural Checklist shall be completed where required by Table 3-2. The Basic Structural Checklist shall be completed prior to completing this Supplemental Structural Checklist.

Lateral-Force-Resisting System

C   NC   N/A   **COUPLING BEAMS:** The stirrups in coupling beams over means of egress shall be spaced at or less than $d/2$ and shall be anchored into the confined core of the beam with hooks of 135° or more for Life Safety. All coupling beams shall comply with the requirements above and shall have the capacity in shear to develop the uplift capacity of the adjacent wall for Immediate Occupancy. (Tier 2: Sec. 4.4.2.2.3)

C   NC   N/A   **OVERTURNING:** All shear walls shall have aspect ratios less than 4-to-1. Wall piers need not be considered. This statement shall apply to the Immediate Occupancy Performance Level only. (Tier 2: Sec. 4.4.2.2.4)

C   NC   N/A   **CONFINEMENT REINFORCING:** For shear walls with aspect ratios greater than 2-to-1, the boundary elements shall be confined with spirals or ties with spacing less than $8d_b$. This statement shall apply to the Immediate Occupancy Performance Level only. (Tier 2: Sec. 4.4.2.2.5)

C   NC   N/A   **REINFORCING AT OPENINGS:** There shall be added trim reinforcement around all wall openings with a dimension greater than three times the thickness of the wall. This statement shall apply to the Immediate Occupancy Performance Level only. (Tier 2: Sec. 4.4.2.2.6)

C   NC   N/A   **WALL THICKNESS:** Thickness of bearing walls shall not be less than 1/25 the unsupported height or length, whichever is shorter, nor less than 4 inches. This statement shall apply to the Immediate Occupancy Performance Level only. (Tier 2: Sec. 4.4.2.2.7)

C   NC   N/A   **WALL CONNECTIONS:** There shall be a positive connection between the shear walls and the steel beams and columns for Life Safety and the connection shall be able to develop the strength of the walls for Immediate Occupancy. (Tier 2: Sec. 4.4.2.2.8)

Diaphragms

C   NC   N/A   **OPENINGS AT SHEAR WALLS:** Diaphragm openings immediately adjacent to the shear walls shall be less than 25 percent of the wall length for Life Safety and 15 percent of the wall length for Immediate Occupancy. (Tier 2: Sec. 4.5.1.4)

C   NC   N/A   **PLAN IRREGULARITIES:** There shall be tensile capacity to develop the strength of the diaphragm at re-entrant corners or other locations of plan irregularities. This statement shall apply to the Immediate Occupancy Performance Level only. (Tier 2: Sec. 4.5.1.7)

C   NC   N/A   **DIAPHRAGM REINFORCEMENT AT OPENINGS:** There shall be reinforcing around all diaphragm openings larger than 50 percent of the building width in either major plan dimension. This statement shall apply to the Immediate Occupancy Performance Level only. (Tier 2: Sec. 4.5.1.8)

Connections

C   NC   N/A   **UPLIFT AT PILE CAPS:** Pile caps shall have top reinforcement and piles shall be anchored to the pile caps for Life Safety, and the pile cap reinforcement and pile anchorage shall be able to develop the tensile capacity of the piles for Immediate Occupancy. (Tier 2: Sec. 4.6.3.10)

Screening Phase (Tier 1)

3.7.7 Basic Structural Checklist for Building Type S5: Steel Frames with Infill Masonry Shear Walls and Stiff Diaphragms

This Basic Structural Checklist shall be completed where required by Table 3-2.

Each of the evaluation statements on this checklist shall be marked Compliant (C), Non-compliant (NC), or Not Applicable (N/A) for a Tier 1 Evaluation. Compliant statements identify issues that are acceptable according to the criteria of this standard, while non-compliant statements identify issues that require further investigation. Certain statements may not apply to the buildings being evaluated. For non-compliant evaluation statements, the user may choose to conduct further investigation using the corresponding Tier 2 Evaluation procedure; corresponding section numbers are in parentheses following each evaluation statement.

C3.7.7 Basic Structural Checklist for Building Type S5

This is an older type of building construction that consists of a frame assembly of steel beams and steel columns. The floor and roof diaphragms consist of cast-in-place concrete slabs or metal deck with concrete fill and are stiff relative to the walls. Framing consists of steel beams, open web joists, or steel trusses. Walls consist of infill panels constructed of solid clay brick, concrete block, or hollow clay tile masonry. Infill walls may completely encase the frame members and present a smooth masonry exterior with no indication of the frame. The seismic performance of this type of construction depends on the interaction between the frame and infill panels. The combined behavior is more like a shear wall structure than a frame structure. Solidly infilled masonry panels form diagonal compression struts between the intersections of the frame members. If the walls are offset from the frame and do not fully engage the frame members, the diagonal compression struts will not develop. The strength of the infill panel is limited by the shear capacity of the masonry bed joint or the compression capacity of the strut. The post-cracking strength is determined by an analysis of a moment frame that is partially restrained by the cracked infill.

Building System

C   NC   N/A   **LOAD PATH:** The structure shall contain a minimum of one complete load path for Life Safety and Immediate Occupancy for seismic force effects from any horizontal direction that serves to transfer the inertial forces from the mass to the foundation. (Tier 2: Sec. 4.3.1.1)

C   NC   N/A   **MEZZANINES:** Interior mezzanine levels shall be braced independently from the main structure, or shall be anchored to the lateral-force-resisting elements of the main structure. (Tier 2: Sec. 4.3.1.3)

C   NC   N/A   **WEAK STORY:** The strength of the lateral-force-resisting system in any story shall not be less than 80 percent of the strength in an adjacent story, above or below, for Life Safety and Immediate Occupancy. (Tier 2: Sec. 4.3.2.1)

C   NC   N/A   **SOFT STORY:** The stiffness of the lateral-force-resisting system in any story shall not be less than 70 percent of the lateral-force-resisting system stiffness in an adjacent story above or below, or less than 80 percent of the average lateral-force-resisting system stiffness of the three stories above or below for Life Safety and Immediate Occupancy. (Tier 2: Sec. 4.3.2.2)

C   NC   N/A   **GEOMETRY:** There shall be no changes in horizontal dimension of the lateral-force-resisting system of more than 30 percent in a story relative to adjacent stories for Life Safety and Immediate Occupancy, excluding one-story penthouses and mezzanines. (Tier 2: Sec. 4.3.2.3)

Screening Phase (Tier 1)

C NC N/A **VERTICAL DISCONTINUITIES:** All vertical elements in the lateral-force-resisting system shall be continuous to the foundation. (Tier 2: Sec. 4.3.2.4)

C NC N/A **MASS:** There shall be no change in effective mass more than 50 percent from one story to the next for Life Safety and Immediate Occupancy. Light roofs, penthouses, and mezzanines need not be considered. (Tier 2: Sec. 4.3.2.5)

C NC N/A **TORSION:** The estimated distance between the story center of mass and the story center of rigidity shall be less than 20 percent of the building width in either plan dimension for Life Safety and Immediate Occupancy. (Tier 2: Sec. 4.3.2.6)

C NC N/A **DETERIORATION OF STEEL:** There shall be no visible rusting, corrosion, cracking, or other deterioration in any of the steel elements or connections in the vertical- or lateral-force-resisting systems. (Tier 2: Sec. 4.3.3.3)

C NC N/A **DETERIORATION OF CONCRETE:** There shall be no visible deterioration of concrete or reinforcing steel in any of the vertical- or lateral-force-resisting elements. (Tier 2: Sec. 4.3.3.4)

C NC N/A **MASONRY UNITS:** There shall be no visible deterioration of masonry units. (Tier 2: Sec. 4.3.3.7)

C NC N/A **MASONRY JOINTS:** The mortar shall not be easily scraped away from the joints by hand with a metal tool, and there shall be no areas of eroded mortar. (Tier 2: Sec.4.3.3.8)

C NC N/A **CRACKS IN INFILL WALLS:** There shall be no existing diagonal cracks in the infilled walls that extend throughout a panel greater than 1/8 inch for Life Safety and 1/16 inch for Immediate Occupancy, or out-of-plane offsets in the bed joint greater than 1/8 inch for Life Safety and 1/16 inch for Immediate Occupancy. (Tier 2: Sec. 4.3.3.12)

Lateral-Force-Resisting System

C NC N/A **REDUNDANCY:** The number of lines of shear walls in each principal direction shall be greater than or equal to 2 for Life Safety and Immediate Occupancy. (Tier 2: Sec. 4.4.2.1.1)

C NC N/A **SHEAR STRESS CHECK:** The shear stress in the reinforced masonry shear walls, calculated using the Quick Check procedure of Section 3.5.3.3, shall be less than 70 psi for Life Safety and Immediate Occupancy. (Tier 2: Sec. 4.4.2.4.1)

C NC N/A **SHEAR STRESS CHECK:** The shear stress in the unreinforced masonry shear walls, calculated using the Quick Check procedure of Section 3.5.3.3, shall be less than 30 psi for clay units and 70 psi for concrete units for Life Safety and Immediate Occupancy. (Tier 2: Sec. 4.4.2.5.1)

C NC N/A **WALL CONNECTIONS:** Masonry shall be in full contact with frame for Life Safety and Immediate Occupancy. (Tier 2: Sec. 4.4.2.6.1)

Connections

C NC N/A **TRANSFER TO SHEAR WALLS:** Diaphragms shall be connected for transfer of loads to the shear walls for Life Safety and the connections shall be able to develop the lesser of the shear strength of the walls or diaphragms for Immediate Occupancy. (Tier 2 Sec. 4.6.2.1)

C NC N/A **STEEL COLUMNS:** The columns in lateral-force-resisting frames shall be anchored to the building foundation for Life Safety, and the anchorage shall be able to develop the lesser of the tensile capacity of the column, the tensile capacity of the lowest level column splice (if any), or the uplift capacity of the foundation, for Immediate Occupancy. (Tier 2: Sec. 4.6.3.1)

Screening Phase (Tier 1)

3.7.7S Supplemental Structural Checklist for Building Type S5: Steel Frames with Infill Masonry Shear Walls and Stiff Diaphragms

This Supplemental Structural Checklist shall be completed where required by Table 3-2. The Basic Structural Checklist shall be completed prior to completing this Supplemental Structural Checklist.

Lateral-Force-Resisting System

C NC N/A REINFORCING AT OPENINGS: All wall openings that interrupt rebar shall have trim reinforcing on all sides. This statement shall apply to the Immediate Occupancy Performance Level only. (Tier 2: Sec. 4.4.2.4.3)

C NC N/A PROPORTIONS: The height-to-thickness ratio of the infill walls at each story shall be less than 9 for Life Safety in levels of high seismicity, 13 for Immediate Occupancy in levels of moderate seismicity, and 8 for Immediate Occupancy in levels of high seismicity. (Tier 2: Sec. 4.4.2.6.2)

C NC N/A SOLID WALLS: The infill walls shall not be of cavity construction. (Tier 2: Sec. 4.4.2.6.3)

Diaphragms

C NC N/A PLAN IRREGULARITIES: There shall be tensile capacity to develop the strength of the diaphragm at re-entrant corners or other locations of plan irregularities. This statement shall apply to the Immediate Occupancy Performance Level only. (Tier 2: Sec. 4.5.1.7)

C NC N/A DIAPHRAGM REINFORCEMENT AT OPENINGS: There shall be reinforcing around all diaphragm openings larger than 50 percent of the building width in either major plan dimension. This statement shall apply to the Immediate Occupancy Performance Level only. (Tier 2: Sec. 4.5.1.8)

Connections

C NC N/A UPLIFT AT PILE CAPS: Pile caps shall have top reinforcement and piles shall be anchored to the pile caps for Life Safety, and the pile cap reinforcement and pile anchorage shall be able to develop the tensile capacity of the piles for Immediate Occupancy. (Tier 2: Sec. 4.6.3.10)

Screening Phase (Tier 1)

3.7.7A Basic Structural Checklist for Building Type S5A: Steel Frames with Infill Masonry Shear Walls and Flexible Diaphragm

This Basic Structural Checklist shall be completed where required by Table 3-2.

Each of the evaluation statements on this checklist shall be marked Compliant (C), Non-compliant (NC), or Not Applicable (N/A) for a Tier 1 Evaluation. Compliant statements identify issues that are acceptable according to the criteria of this standard, while non-compliant statements identify issues that require further investigation. Certain statements may not apply to the buildings being evaluated. For non-compliant evaluation statements, the design professional may choose to conduct further investigation using the corresponding Tier 2 Evaluation procedure; corresponding section numbers are in parentheses following each evaluation statement.

C3.7.7A Basic Structural Checklist for Building Type S5A

This is an older type of building construction that consists of a frame assembly of steel beams and steel columns. The floors and roof consist of untopped metal deck or wood framing between the steel beams and are flexible relative to the walls. Framing consists of steel beams, open web joists or steel trusses. Walls consist of infill panels constructed of solid clay brick, concrete block, or hollow clay tile masonry. Infill walls may completely encase the frame members, and present a smooth masonry exterior with no indication of the frame. The seismic performance of this type of construction depends on the interaction between the frame and infill panels. The combined behavior is more like a shear wall structure than a frame structure. Solidly infilled masonry panels form diagonal compression struts between the intersections of the frame members. If the walls are offset from the frame and do not fully engage the frame members, the diagonal compression struts will not develop. The strength of the infill panel is limited by the shear capacity of the masonry bed joint or the compression capacity of the strut. The post-cracking strength is determined by an analysis of a moment frame that is partially restrained by the cracked infill.

Building System

C   NC   N/A   **LOAD PATH:** The structure shall contain a minimum of one complete load path for Life Safety and Immediate Occupancy for seismic force effects from any horizontal direction that serves to transfer the inertial forces from the mass to the foundation. (Tier 2: Sec. 4.3.1.1)

C   NC   N/A   **MEZZANINES:** Interior mezzanine levels shall be braced independently from the main structure, or shall be anchored to the lateral-force-resisting elements of the main structure. (Tier 2: Sec. 4.3.1.3)

C   NC   N/A   **ADJACENT BUILDINGS:** The clear distance between the building being evaluated and any adjacent building shall be greater than 4 percent of the height of the shorter building for Life Safety and Immediate Occupancy. (Tier 2: Sec. 4.3.1.2)

C   NC   N/A   **WEAK STORY:** The strength of the lateral-force-resisting system in any story shall not be less than 80 percent of the strength in an adjacent story, above or below, for Life Safety and Immediate Occupancy. (Tier 2: Sec. 4.3.2.1)

C   NC   N/A   **SOFT STORY:** The stiffness of the lateral-force-resisting system in any story shall not be less than 70 percent of the lateral-force-resisting system stiffness in an adjacent story above or below, or less than 80 percent of the average lateral-force-resisting system stiffness of the three stories above or below for Life Safety and Immediate Occupancy. (Tier 2: Sec. 4.3.2.2)

Screening Phase (Tier 1)

C   NC   N/A   **GEOMETRY:** There shall be no changes in horizontal dimension of the lateral-force-resisting system of more than 30 percent in a story relative to adjacent stories for Life Safety and Immediate Occupancy, excluding one-story penthouses and mezzanines. (Tier 2: Sec. 4.3.2.3)

C   NC   N/A   **VERTICAL DISCONTINUITIES:** All vertical elements in the lateral-force-resisting system shall be continuous to the foundation. (Tier 2: Sec. 4.3.2.4)

C   NC   N/A   **MASS:** There shall be no change in effective mass more than 50 percent from one story to the next for Life Safety and Immediate Occupancy. Light roofs, penthouses, and mezzanines need not be considered. (Tier 2: Sec. 4.3.2.5)

C   NC   N/A   **DETERIORATION OF WOOD:** There shall be no signs of decay, shrinkage, splitting, fire damage, or sagging in any of the wood members, and none of the metal connection hardware shall be deteriorated, broken, or loose. (Tier 2: Sec. 4.3.3.1)

C   NC   N/A   **DETERIORATION OF STEEL:** There shall be no visible rusting, corrosion, cracking, or other deterioration in any of the steel elements or connections in the vertical- or lateral-force-resisting systems. (Tier 2: Sec. 4.3.3.3)

C   NC   N/A   **MASONRY UNITS:** There shall be no visible deterioration of masonry units. (Tier 2: Sec. 4.3.3.7)

C   NC   N/A   **MASONRY JOINTS:** The mortar shall not be easily scraped away from the joints by hand with a metal tool, and there shall be no areas of eroded mortar. (Tier 2: Sec. 4.3.3.8)

C   NC   N/A   **CRACKS IN INFILL WALLS:** There shall be no existing diagonal cracks in the infilled walls that extend throughout a panel greater than 1/8 inch for Life Safety and 1/16 inch for Immediate Occupancy, or out-of-plane offsets in the bed joint greater than 1/8 inch for Life Safety and 1/16 inch for Immediate Occupancy. (Tier 2: Sec. 4.3.3.12)

Lateral-Force-Resisting System

C   NC   N/A   **REDUNDANCY:** The number of lines of shear walls in each principal direction shall be greater than or equal to 2 for Life Safety and Immediate Occupancy. (Tier 2: Sec. 4.4.2.1.1)

C   NC   N/A   **SHEAR STRESS CHECK:** The shear stress in the reinforced masonry shear walls, calculated using the Quick Check procedure of Section 3.5.3.3, shall be less than 70 psi for Life Safety and Immediate Occupancy. (Tier 2: Sec. 4.4.2.4.1)

C   NC   N/A   **SHEAR STRESS CHECK:** The shear stress in the unreinforced masonry shear walls, calculated using the Quick Check procedure of Section 3.5.3.3, shall be less than 30 psi for clay units and 70 psi for concrete units for Life Safety and Immediate Occupancy. (Tier 2: Sec. 4.4.2.5.1)

C   NC   N/A   **WALL CONNECTIONS:** Masonry shall be in full contact with frame for Life Safety and Immediate Occupancy. (Tier 2: Sec. 4.4.2.6.1)

Connections

C   NC   N/A   **TRANSFER TO SHEAR WALLS:** Diaphragms shall be connected for transfer of loads to the shear walls for Life Safety and the connections shall be able to develop the lesser of the shear strength of the walls or diaphragms for Immediate Occupancy. (Tier 2 Sec. 4.6.2.1)

C   NC   N/A   **STEEL COLUMNS:** The columns in lateral-force-resisting frames shall be anchored to the building foundation for Life Safety, and the anchorage shall be able to develop the lesser of the tensile capacity of the column, the tensile capacity of the lowest level column splice (if any), or the uplift capacity of the foundation, for Immediate Occupancy. (Tier 2: Sec. 4.6.3.1)

3.7.7AS Supplemental Structural Checklist for Building Type S5A: Steel Frames with Infill Masonry Shear Walls and Flexible Diaphragms

This Supplemental Structural Checklist shall be completed where required by Table 3-2. The Basic Structural Checklist shall be completed prior to completing this Supplemental Structural Checklist.

Lateral-Force-Resisting System

C   NC   N/A   REINFORCING AT OPENINGS: All wall openings that interrupt rebar shall have trim reinforcing on all sides. This statement shall apply to the Immediate Occupancy Performance Level only. (Tier 2: Sec. 4.4.2.4.3)

C   NC   N/A   PROPORTIONS: The height-to-thickness ratio of the infill walls at each story shall be less than 9 for Life Safety in levels of high seismicity, 13 for Immediate Occupancy in levels of moderate seismicity, and 8 for Immediate Occupancy in levels of high seismicity. (Tier 2: Sec. 4.4.2.6.2)

C   NC   N/A   SOLID WALLS: The infill walls shall not be of cavity construction. (Tier 2: Sec. 4.4.2.6.3)

Diaphragms

C   NC   N/A   CROSS TIES: There shall be continuous cross ties between diaphragm chords. (Tier 2: Sec. 4.5.1.2)

C   NC   N/A   PLAN IRREGULARITIES: There shall be tensile capacity to develop the strength of the diaphragm at re-entrant corners or other locations of plan irregularities. This statement shall apply to the Immediate Occupancy Performance Level only. (Tier 2: Sec. 4.5.1.7)

C   NC   N/A   DIAPHRAGM REINFORCEMENT AT OPENINGS: There shall be reinforcing around all diaphragm openings larger than 50 percent of the building width in either major plan dimension. This statement shall apply to the Immediate Occupancy Performance Level only. (Tier 2: Sec. 4.5.1.8)

C   NC   N/A   STRAIGHT SHEATHING: All straight sheathed diaphragms shall have aspect ratios less than 2-to-1 for Life Safety and 1-to-1 for Immediate Occupancy in the direction being considered. (Tier 2: Sec. 4.5.2.1)

C   NC   N/A   SPANS: All wood diaphragms with spans greater than 24 feet for Life Safety and 12 feet for Immediate Occupancy shall consist of wood structural panels or diagonal sheathing. (Tier 2: Sec. 4.5.2.2)

C   NC   N/A   UNBLOCKED DIAPHRAGMS: All diagonally sheathed or unblocked wood structural panel diaphragms shall have horizontal spans less than 40 feet for Life Safety and 30 feet for Immediate Occupancy and shall have aspect ratios less than or equal to 4-to-1 for Life Safety and 3-to-1 for Immediate Occupancy. (Tier 2: Sec. 4.5.2.3)

C   NC   N/A   NON-CONCRETE FILLED DIAPHRAGMS: Untopped metal deck diaphragms or metal deck diaphragms with fill other than concrete shall consist of horizontal spans of less than 40 feet and shall have span/depth ratios less than 4-to-1. This statement shall apply to the Immediate Occupancy Performance Level only. (Tier 2: Sec. 4.5.3.1)

C   NC   N/A   OTHER DIAPHRAGMS: The diaphragm shall not consist of a system other than wood, metal deck, concrete, or horizontal bracing. (Tier 2: Sec. 4.5.7.1)

Screening Phase (Tier 1)

Connections

C NC N/A **STIFFNESS OF WALL ANCHORS:** Anchors of concrete or masonry walls to wood structural elements shall be installed taut and shall be stiff enough to limit the relative movement between the wall and the diaphragm to no greater than 1/8 inch prior to engagement of the anchors. (Tier 2: Sec. 4.6.1.4)

C NC N/A **UPLIFT AT PILE CAPS:** Pile caps shall have top reinforcement and piles shall be anchored to the pile caps for Life Safety, and the pile cap reinforcement and pile anchorage shall be able to develop the tensile capacity of the piles for Immediate Occupancy. (Tier 2: Sec. 4.6.3.10)

3.7.8 Basic Structural Checklist for Building Type C1: Concrete Moment Frames

This Basic Structural Checklist shall be completed where required by Table 3-2.

Each of the evaluation statements on this checklist shall be marked Compliant (C), Non-compliant (NC), or Not Applicable (N/A) for a Tier 1 Evaluation. Compliant statements identify issues that are acceptable according to the criteria of this standard, while non-compliant statements identify issues that require further investigation. Certain statements may not apply to the buildings being evaluated. For non-compliant evaluation statements, the design professional may choose to conduct further investigation using the corresponding Tier 2 Evaluation procedure; corresponding section numbers are in parentheses following each evaluation statement.

C3.7.8 Basic Structural Checklist for Building Type C1

These buildings consist of a frame assembly of cast-in-place concrete beams and columns. Floor and roof framing consists of cast-in-place concrete slabs, concrete beams, one-way joists, two-way waffle joists, or flat slabs. Lateral forces are resisted by concrete moment frames that develop their stiffness through monolithic beam-column connections. In older construction, or in levels of low seismicity, the moment frames may consist of the column strips of two-way flat slab systems. Modern frames in levels of high seismicity have joint reinforcing, closely spaced ties, and special detailing to provide ductile performance. This detailing is not present in older construction. Foundations consist of concrete spread footings, mat foundations, or deep foundations.

Building System

C NC N/A **LOAD PATH:** The structure shall contain a minimum of one complete load path for Life Safety and Immediate Occupancy for seismic force effects from any horizontal direction that serves to transfer the inertial forces from the mass to the foundation. (Tier 2: Sec. 4.3.1.1)

C NC N/A **ADJACENT BUILDINGS:** The clear distance between the building being evaluated and any adjacent building shall be greater than 4 percent of the height of the shorter building for Life Safety and Immediate Occupancy. (Tier 2: Sec. 4.3.1.2)

C NC N/A **MEZZANINES:** Interior mezzanine levels shall be braced independently from the main structure, or shall be anchored to the lateral-force-resisting elements of the main structure. (Tier 2: Sec. 4.3.1.3)

C NC N/A **WEAK STORY:** The strength of the lateral-force-resisting system in any story shall not be less than 80 percent of the strength in an adjacent story, above or below, for Life Safety and Immediate Occupancy. (Tier 2: Sec. 4.3.2.1)

C NC N/A **SOFT STORY:** The stiffness of the lateral-force-resisting system in any story shall not be less than 70 percent of the lateral-force-resisting system stiffness in an adjacent story above or below, or less than 80 percent of the average lateral-force-resisting system stiffness of the three stories above or below for Life Safety and Immediate Occupancy. (Tier 2: Sec. 4.3.2.2)

C NC N/A **GEOMETRY:** There shall be no changes in horizontal dimension of the lateral-force-resisting system of more than 30 percent in a story relative to adjacent stories for Life Safety and Immediate Occupancy, excluding one-story penthouses and mezzanines. (Tier 2: Sec. 4.3.2.3)

C NC N/A **VERTICAL DISCONTINUITIES:** All vertical elements in the lateral-force-resisting system shall be continuous to the foundation. (Tier 2: Sec. 4.3.2.4)

Screening Phase (Tier 1)

C NC N/A **MASS:** There shall be no change in effective mass more than 50 percent from one story to the next for Life Safety and Immediate Occupancy. Light roofs, penthouses, and mezzanines need not be considered. (Tier 2: Sec. 4.3.2.5)

C NC N/A **TORSION:** The estimated distance between the story center of mass and the story center of rigidity shall be less than 20 percent of the building width in either plan dimension for Life Safety and Immediate Occupancy. (Tier 2: Sec. 4.3.2.6)

C NC N/A **DETERIORATION OF CONCRETE:** There shall be no visible deterioration of concrete or reinforcing steel in any of the vertical- or lateral-force-resisting elements. (Tier 2: Sec. 4.3.3.4)

C NC N/A **POST-TENSIONING ANCHORS:** There shall be no evidence of corrosion or spalling in the vicinity of post-tensioning or end fittings. Coil anchors shall not have been used. (Tier 2: Sec. 4.3.3.5)

Lateral-Force-Resisting System

C NC N/A **REDUNDANCY:** The number of lines of moment frames in each principal direction shall be greater than or equal to 2 for Life Safety and Immediate Occupancy. The number of bays of moment frames in each line shall be greater than or equal to 2 for Life Safety and 3 for Immediate Occupancy. (Tier 2: Sec. 4.4.1.1.1)

C NC N/A **INTERFERING WALLS:** All concrete and masonry infill walls placed in moment frames shall be isolated from structural elements. (Tier 2: Sec. 4.4.1.2.1)

C NC N/A **SHEAR STRESS CHECK:** The shear stress in the concrete columns, calculated using the Quick Check procedure of Section 3.5.3.2, shall be less than the greater of 100 psi or $2\sqrt{f'_c}$ for Life Safety and Immediate Occupancy. (Tier 2: Sec. 4.4.1.4.1)

C NC N/A **AXIAL STRESS CHECK:** The axial stress due to gravity loads in columns subjected to overturning forces shall be less than $0.10f'_c$ for Life Safety and Immediate Occupancy. Alternatively, the axial stresses due to overturning forces alone, calculated using the Quick Check procedure of Section 3.5.3.6, shall be less than $0.30f'_c$ for Life Safety and Immediate Occupancy. (Tier 2: Sec. 4.4.1.4.2)

Connections

C NC N/A **CONCRETE COLUMNS:** All concrete columns shall be doweled into the foundation for Life Safety, and the dowels shall be able to develop the tensile capacity of reinforcement in columns of lateral-force-resisting system for Immediate Occupancy. (Tier 2: Sec. 4.6.3.2)

3.7.8S Supplemental Structural Checklist for Building Type C1: Concrete Moment Frames

This Supplemental Structural Checklist shall be completed where required by Table 3-2. The Basic Structural Checklist shall be completed prior to completing this Supplemental Structural Checklist.

Lateral-Force-Resisting System

C   NC   N/A   **FLAT SLAB FRAMES:** The lateral-force-resisting system shall not be a frame consisting of columns and a flat slab/plate without beams. (Tier 2: Sec. 4.4.1.4.3)

C   NC   N/A   **PRESTRESSED FRAME ELEMENTS:** The lateral-force-resisting frames shall not include any prestressed or post-tensioned elements where the average prestress exceeds the lesser of 700 psi or $f'_c/6$ at potential hinge locations. The average prestress shall be calculated in accordance with the Quick Check procedure of Section 3.5.3.8. (Tier 2: Sec. 4.4.1.4.4)

C   NC   N/A   **CAPTIVE COLUMNS:** There shall be no columns at a level with height/depth ratios less than 50 percent of the nominal height/depth ratio of the typical columns at that level for Life Safety and 75 percent for Immediate Occupancy. (Tier 2: Sec. 4.4.1.4.5)

C   NC   N/A   **NO SHEAR FAILURES:** The shear capacity of frame members shall be able to develop the moment capacity at the ends of the members. (Tier 2: Sec. 4.4.1.4.6)

C   NC   N/A   **STRONG COLUMN/WEAK BEAM:** The sum of the moment capacity of the columns shall be 20 percent greater than that of the beams at frame joints. (Tier 2: Sec. 4.4.1.4.7)

C   NC   N/A   **BEAM BARS:** At least two longitudinal top and two longitudinal bottom bars shall extend continuously throughout the length of each frame beam. At least 25 percent of the longitudinal bars provided at the joints for either positive or negative moment shall be continuous throughout the length of the members for Life Safety and Immediate Occupancy. (Tier 2: Sec. 4.4.1.4.8)

C   NC   N/A   **COLUMN-BAR SPLICES:** All column bar lap splice lengths shall be greater than $35d_b$ for Life Safety and $50d_b$ for Immediate Occupancy, and shall be enclosed by ties spaced at or less than $8d_b$ for Life Safety and Immediate Occupancy. Alternatively, column bars shall be spliced with mechanical couplers with a capacity of at least 1.25 times the nominal yield strength of the spliced bar. (Tier 2: Sec. 4.4.1.4.9)

C   NC   N/A   **BEAM-BAR SPLICES:** The lap splices or mechanical couplers for longitudinal beam reinforcing shall not be located within $l_b/4$ of the joints and shall not be located in the vicinity of potential plastic hinge locations. (Tier 2: Sec. 4.4.1.4.10)

C   NC   N/A   **COLUMN-TIE SPACING:** Frame columns shall have ties spaced at or less than $d/4$ for Life Safety and Immediate Occupancy throughout their length and at or less than $8d_b$ for Life Safety and Immediate Occupancy at all potential plastic hinge locations. (Tier 2: Sec. 4.4.1.4.11)

C   NC   N/A   **STIRRUP SPACING:** All beams shall have stirrups spaced at or less than $d/2$ for Life Safety and Immediate Occupancy throughout their length. At potential plastic hinge locations, stirrups shall be spaced at or less than the minimum of $8d_b$ or $d/4$ for Life Safety and Immediate Occupancy. (Tier 2: Sec. 4.4.1.4.12)

C   NC   N/A   **JOINT REINFORCING:** Beam-column joints shall have ties spaced at or less than $8d_b$ for Life Safety and Immediate Occupancy. (Tier 2: Sec. 4.4.1.4.13)

C   NC   N/A   **JOINT ECCENTRICITY:** There shall be no eccentricities larger than 20 percent of the smallest column plan dimension between girder and column centerlines. This statement shall apply to the Immediate Occupancy Performance Level only. (Tier 2: Sec. 4.4.1.4.14)

Screening Phase (Tier 1)

C　NC　N/A　　STIRRUP AND TIE HOOKS: The beam stirrups and column ties shall be anchored into the member cores with hooks of 135° or more. This statement shall apply to the Immediate Occupancy Performance Level only. (Tier 2: Sec. 4.4.1.4.15)

C　NC　N/A　　DEFLECTION COMPATIBILITY: Secondary components shall have the shear capacity to develop the flexural strength of the components for Life Safety and shall meet the requirements of Sections 4.4.1.4.9, 4.4.1.4.10, 4.4.1.4.11, 4.4.1.4.12 and 4.4.1.4.15 for Immediate Occupancy. (Tier 2: Sec. 4.4.1.6.2)

C　NC　N/A　　FLAT SLABS: Flat slabs/plates not part of lateral-force-resisting system shall have continuous bottom steel through the column joints for Life Safety and Immediate Occupancy. (Tier 2: Sec. 4.4.1.6.3)

Diaphragms

C　NC　N/A　　DIAPHRAGM CONTINUITY: The diaphragms shall not be composed of split-level floors and shall not have expansion joints. (Tier 2: Sec. 4.5.1.1)

C　NC　N/A　　PLAN IRREGULARITIES: There shall be tensile capacity to develop the strength of the diaphragm at re-entrant corners or other locations of plan irregularities. This statement shall apply to the Immediate Occupancy Performance Level only. (Tier 2: Sec. 4.5.1.7)

C　NC　N/A　　DIAPHRAGM REINFORCEMENT AT OPENINGS: There shall be reinforcing around all diaphragm openings larger than 50 percent of the building width in either major plan dimension. This statement shall apply to the Immediate Occupancy Performance Level only. (Tier 2: Sec. 4.5.1.8)

Connections

C　NC　N/A　　UPLIFT AT PILE CAPS: Pile caps shall have top reinforcement and piles shall be anchored to the pile caps for Life Safety, and the pile cap reinforcement and pile anchorage shall be able to develop the tensile capacity of the piles for Immediate Occupancy. (Tier 2: Sec. 4.6.3.10)

3.7.9 Basic Structural Checklist for Building Type C2: Concrete Shear Walls with Stiff Diaphragms

This Basic Structural Checklist shall be completed where required by Table 3-2.

Each of the evaluation statements on this checklist shall be marked Compliant (C), Non-compliant (NC), or Not Applicable (N/A) for a Tier 1 Evaluation. Compliant statements identify issues that are acceptable according to the criteria of this standard, while non-compliant statements identify issues that require further investigation. Certain statements may not apply to the buildings being evaluated. For non-compliant evaluation statements, the design professional may choose to conduct further investigation using the corresponding Tier 2 Evaluation procedure; corresponding section numbers are in parentheses following each evaluation statement.

> **C3.7.9 Basic Structural Checklist for Building Type C2**
>
> These buildings have floor and roof framing that consists of cast-in-place concrete slabs, concrete beams, one-way joists, two-way waffle joists, or flat slabs. Floors are supported on concrete columns or bearing walls. Lateral forces are resisted by cast-in-place concrete shear walls. In older construction, shear walls are lightly reinforced but often extend throughout the building. In more recent construction, shear walls occur in isolated locations and are more heavily reinforced with boundary elements and closely spaced ties to provide ductile performance. The diaphragms consist of concrete slabs and are stiff relative to the walls. Foundations consist of concrete spread footings, mat foundations, or deep foundations.

Building System

C   NC   N/A   **LOAD PATH:** The structure shall contain a minimum of one complete load path for Life Safety and Immediate Occupancy for seismic force effects from any horizontal direction that serves to transfer the inertial forces from the mass to the foundation. (Tier 2: Sec. 4.3.1.1)

C   NC   N/A   **MEZZANINES:** Interior mezzanine levels shall be braced independently from the main structure, or shall be anchored to the lateral-force-resisting elements of the main structure. (Tier 2: Sec. 4.3.1.3)

C   NC   N/A   **WEAK STORY:** The strength of the lateral-force-resisting system in any story shall not be less than 80 percent of the strength in an adjacent story, above or below, for Life Safety and Immediate Occupancy. (Tier 2: Sec. 4.3.2.1)

C   NC   N/A   **SOFT STORY:** The stiffness of the lateral-force-resisting system in any story shall not be less than 70 percent of the lateral-force-resisting system stiffness in an adjacent story above or below, or less than 80 percent of the average lateral-force-resisting system stiffness of the three stories above or below for Life Safety and Immediate Occupancy. (Tier 2: Sec. 4.3.2.2)

C   NC   N/A   **GEOMETRY:** There shall be no changes in horizontal dimension of the lateral-force-resisting system of more than 30 percent in a story relative to adjacent stories for Life Safety and Immediate Occupancy, excluding one-story penthouses and mezzanines. (Tier 2: Sec. 4.3.2.3)

C   NC   N/A   **VERTICAL DISCONTINUITIES:** All vertical elements in the lateral-force-resisting system shall be continuous to the foundation. (Tier 2: Sec. 4.3.2.4)

C   NC   N/A   **MASS:** There shall be no change in effective mass more than 50 percent from one story to the next for Life Safety and Immediate Occupancy. Light roofs, penthouses, and mezzanines need not be considered. (Tier 2: Sec. 4.3.2.5)

Screening Phase (Tier 1)

C NC N/A **TORSION:** The estimated distance between the story center of mass and the story center of rigidity shall be less than 20 percent of the building width in either plan dimension for Life Safety and Immediate Occupancy. (Tier 2: Sec. 4.3.2.6)

C NC N/A **DETERIORATION OF CONCRETE:** There shall be no visible deterioration of concrete or reinforcing steel in any of the vertical- or lateral-force-resisting elements. (Tier 2: Sec. 4.3.3.4)

C NC N/A **POST-TENSIONING ANCHORS:** There shall be no evidence of corrosion or spalling in the vicinity of post-tensioning or end fittings. Coil anchors shall not have been used. (Tier 2: Sec. 4.3.3.5)

C NC N/A **CONCRETE WALL CRACKS:** All existing diagonal cracks in wall elements shall be less than 1/8 inch for Life Safety and 1/16 inch for Immediate Occupancy, shall not be concentrated in one location, and shall not form an X pattern. (Tier 2: Sec. 4.3.3.9)

Lateral-Force-Resisting System

C NC N/A **COMPLETE FRAMES:** Steel or concrete frames classified as secondary components shall form a complete vertical-load-carrying system. (Tier 2: Sec. 4.4.1.6.1)

C NC N/A **REDUNDANCY:** The number of lines of shear walls in each principal direction shall be greater than or equal to 2 for Life Safety and Immediate Occupancy. (Tier 2: Sec. 4.4.2.1.1)

C NC N/A **SHEAR STRESS CHECK:** The shear stress in the concrete shear walls, calculated using the Quick Check procedure of Section 3.5.3.3, shall be less than the greater of 100 psi or $2\sqrt{f'c}$ for Life Safety and Immediate Occupancy. (Tier 2: Sec. 4.4.2.2.1)

C NC N/A **REINFORCING STEEL:** The ratio of reinforcing steel area to gross concrete area shall be not less than 0.0015 in the vertical direction and 0.0025 in the horizontal direction for Life Safety and Immediate Occupancy. The spacing of reinforcing steel shall be equal to or less than 18 inches for Life Safety and Immediate Occupancy. (Tier 2: Sec. 4.4.2.2.2)

Connections

C NC N/A **TRANSFER TO SHEAR WALLS:** Diaphragms shall be connected for transfer of loads to the shear walls for Life Safety and the connections shall be able to develop the lesser of the shear strength of the walls or diaphragms for Immediate Occupancy. (Tier 2: Sec. 4.6.2.1)

C NC N/A **FOUNDATION DOWELS:** Wall reinforcement shall be doweled into the foundation for Life Safety, and the dowels shall be able to develop the lesser of the strength of the walls or the uplift capacity of the foundation for Immediate Occupancy. (Tier 2: Sec. 4.6.3.5)

3.7.9S Supplemental Structural Checklist for Building Type C2: Concrete Shear Walls with Stiff Diaphragms

This Supplemental Structural Checklist shall be completed where required by Table 3-2. The Basic Structural Checklist shall be completed prior to completing this Supplemental Structural Checklist.

Lateral-Force-Resisting System

C NC N/A DEFLECTION COMPATIBILITY: Secondary components shall have the shear capacity to develop the flexural strength of the components for Life Safety and shall meet the requirements of Sections 4.4.1.4.9, 4.4.1.4.10, 4.4.1.4.11, 4.4.1.4.12 and 4.4.1.4.15 for Immediate Occupancy. (Tier 2: Sec. 4.4.1.6.2)

C NC N/A FLAT SLABS: Flat slabs/plates not part of lateral-force-resisting system shall have continuous bottom steel through the column joints for Life Safety and Immediate Occupancy. (Tier 2: Sec. 4.4.1.6.3)

C NC N/A COUPLING BEAMS: The stirrups in coupling beams over means of egress shall be spaced at or less than $d/2$ and shall be anchored into the confined core of the beam with hooks of 135° or more for Life Safety. All coupling beams shall comply with the requirements above and shall have the capacity in shear to develop the uplift capacity of the adjacent wall for Immediate Occupancy. (Tier 2: Sec. 4.4.2.2.3)

C NC N/A OVERTURNING: All shear walls shall have aspect ratios less than 4-to-1. Wall piers need not be considered. This statement shall apply to the Immediate Occupancy Performance Level only. (Tier 2: Sec. 4.4.2.2.4)

C NC N/A CONFINEMENT REINFORCING: For shear walls with aspect ratios greater than 2-to-1, the boundary elements shall be confined with spirals or ties with spacing less than $8d_b$. This statement shall apply to the Immediate Occupancy Performance Level only. (Tier 2: Sec. 4.4.2.2.5)

C NC N/A REINFORCING AT OPENINGS: There shall be added trim reinforcement around all wall openings with a dimension greater than three times the thickness of the wall. This statement shall apply to the Immediate Occupancy Performance Level only. (Tier 2: Sec. 4.4.2.2.6)

C NC N/A WALL THICKNESS: Thickness of bearing walls shall not be less than 1/25 the unsupported height or length, whichever is shorter, nor less than 4 inches. This statement shall apply to the Immediate Occupancy Performance Level only. (Tier 2: Sec. 4.4.2.2.7)

Diaphragms

C NC N/A DIAPHRAGM CONTINUITY: The diaphragms shall not be composed of split-level floors and shall not have expansion joints. (Tier 2: Sec. 4.5.1.1)

C NC N/A OPENINGS AT SHEAR WALLS: Diaphragm openings immediately adjacent to the shear walls shall be less than 25 percent of the wall length for Life Safety and 15 percent of the wall length for Immediate Occupancy. (Tier 2: Sec. 4.5.1.4)

C NC N/A PLAN IRREGULARITIES: There shall be tensile capacity to develop the strength of the diaphragm at re-entrant corners or other locations of plan irregularities. This statement shall apply to the Immediate Occupancy Performance Level only. (Tier 2: Sec. 4.5.1.7)

C NC N/A DIAPHRAGM REINFORCEMENT AT OPENINGS: There shall be reinforcing around all diaphragm openings larger than 50 percent of the building width in either major plan dimension. This statement shall apply to the Immediate Occupancy Performance Level only. (Tier 2: Sec. 4.5.1.8)

Screening Phase (Tier 1)

Connections

C   NC   N/A   UPLIFT AT PILE CAPS: Pile caps shall have top reinforcement and piles shall be anchored to the pile caps for Life Safety, and the pile cap reinforcement and pile anchorage shall be able to develop the tensile capacity of the piles for Immediate Occupancy. (Tier 2: Sec. 4.6.3.10)

Screening Phase (Tier 1)

3.7.9A Basic Structural Checklist for Building Type C2A: Concrete Shear Walls with Flexible Diaphragms

This Basic Structural Checklist shall be completed where required by Table 3-2.

Each of the evaluation statements on this checklist shall be marked Compliant (C), Non-compliant (NC), or Not Applicable (N/A) for a Tier 1 Evaluation. Compliant statements identify issues that are acceptable according to the criteria of this standard, while non-compliant statements identify issues that require further investigation. Certain statements may not apply to the buildings being evaluated. For non-compliant evaluation statements, the design professional may choose to conduct further investigation using the corresponding Tier 2 Evaluation procedure; corresponding section numbers are in parentheses following each evaluation statement.

C3.7.9A Basic Structural Checklist for Building Type C2A

These buildings have floor and roof framing that consists of wood sheathing on wood framing and concrete beams. Floors are supported on concrete columns or bearing walls. Lateral forces are resisted by cast-in-place concrete shear walls. In older construction, shear walls are lightly reinforced but often extend throughout the building. In more recent construction, shear walls occur in isolated locations and are more heavily reinforced with boundary elements and closely spaced ties to provide ductile performance. The diaphragms consist of wood sheathing or have large aspect ratios and are flexible relative to the walls. Foundations consist of concrete spread footings or deep pile foundations.

Building System

C　NC　N/A　　**LOAD PATH:** The structure shall contain a minimum of one complete load path for Life Safety and Immediate Occupancy for seismic force effects from any horizontal direction that serves to transfer the inertial forces from the mass to the foundation. (Tier 2: Sec. 4.3.1.1)

C　NC　N/A　　**ADJACENT BUILDINGS:** The clear distance between the building being evaluated and any adjacent building shall be greater than 4 percent of the height of the shorter building for Life Safety and Immediate Occupancy. (Tier 2: Sec. 4.3.1.2)

C　NC　N/A　　**MEZZANINES:** Interior mezzanine levels shall be braced independently from the main structure, or shall be anchored to the lateral-force-resisting elements of the main structure. (Tier 2: Sec. 4.3.1.3)

C　NC　N/A　　**WEAK STORY:** The strength of the lateral-force-resisting system in any story shall not be less than 80 percent of the strength in an adjacent story, above or below, for Life Safety and Immediate Occupancy. (Tier 2: Sec. 4.3.2.1)

C　NC　N/A　　**SOFT STORY:** The stiffness of the lateral-force-resisting system in any story shall not be less than 70 percent of the lateral-force-resisting system stiffness in an adjacent story above or below, or less than 80 percent of the average lateral-force-resisting system stiffness of the three stories above or below for Life Safety and Immediate Occupancy. (Tier 2: Sec. 4.3.2.2)

C　NC　N/A　　**GEOMETRY:** There shall be no changes in horizontal dimension of the lateral-force-resisting system of more than 30 percent in a story relative to adjacent stories for Life Safety and Immediate Occupancy, excluding one-story penthouses and mezzanines. (Tier 2: Sec. 4.3.2.3)

C　NC　N/A　　**VERTICAL DISCONTINUITIES:** All vertical elements in the lateral-force-resisting system shall be continuous to the foundation. (Tier 2: Sec. 4.3.2.4)

Screening Phase (Tier 1)

C NC N/A   **MASS:** There shall be no change in effective mass more than 50 percent from one story to the next for Life Safety and Immediate Occupancy. Light roofs, penthouses, and mezzanines need not be considered. (Tier 2: Sec. 4.3.2.5)

C NC N/A   **DETERIORATION OF WOOD:** There shall be no signs of decay, shrinkage, splitting, fire damage, or sagging in any of the wood members, and none of the metal connection hardware shall be deteriorated, broken, or loose. (Tier 2: Sec. 4.3.3.1)

C NC N/A   **DETERIORATION OF CONCRETE:** There shall be no visible deterioration of concrete or reinforcing steel in any of the vertical- or lateral-force-resisting elements. (Tier 2: Sec. 4.3.3.4)

C NC N/A   **POST-TENSIONING ANCHORS:** There shall be no evidence of corrosion or spalling in the vicinity of post-tensioning or end fittings. Coil anchors shall not have been used. (Tier 2: Sec. 4.3.3.5)

C NC N/A   **CONCRETE WALL CRACKS:** All existing diagonal cracks in wall elements shall be less than 1/8 inch for Life Safety and 1/16 inch for Immediate Occupancy, shall not be concentrated in one location, and shall not form an X pattern. (Tier 2: Sec. 4.3.3.9)

Lateral-Force-Resisting System

C NC N/A   **REDUNDANCY:** The number of lines of shear walls in each principal direction shall be greater than or equal to 2 for Life Safety and Immediate Occupancy. (Tier 2: Sec. 4.4.2.1.1)

C NC N/A   **SHEAR STRESS CHECK:** The shear stress in the concrete shear walls, calculated using the Quick Check procedure of Section 3.5.3.3, shall be less than the greater of 100 psi or $2\sqrt{f'c}$ for Life Safety and Immediate Occupancy. (Tier 2: Sec. 4.4.2.2.1)

C NC N/A   **REINFORCING STEEL:** The ratio of reinforcing steel area to gross concrete area shall be not less than 0.0015 in the vertical direction and 0.0025 in the horizontal direction for Life Safety and Immediate Occupancy. The spacing of reinforcing steel shall be equal to or less than 18 inches for Life Safety and Immediate Occupancy. (Tier 2: Sec. 4.4.2.2.2)

Connections

C NC N/A   **WALL ANCHORAGE:** Exterior concrete or masonry walls that are dependent on the diaphragm for lateral support shall be anchored for out-of-plane forces at each diaphragm level with steel anchors, reinforcing dowels, or straps that are developed into the diaphragm. Connections shall have adequate strength to resist the connection force calculated in the Quick Check procedure of Section 3.5.3.7. (Tier 2: Sec. 4.6.1.1)

C NC N/A   **TRANSFER TO SHEAR WALLS:** Diaphragms shall be connected for transfer of loads to the shear walls for Life Safety and the connections shall be able to develop the lesser of the shear strength of the walls or diaphragms for Immediate Occupancy. (Tier 2 Sec. 4.6.2.1)

C NC N/A   **FOUNDATION DOWELS:** Wall reinforcement shall be doweled into the foundation for Life Safety, and the dowels shall be able to develop the lesser of the strength of the walls or the uplift capacity of the foundation for Immediate Occupancy. (Tier 2: Sec. 4.6.3.5)

Screening Phase (Tier 1)

3.7.9AS Supplemental Structural Checklist for Building Type C2A: Concrete Shear Walls with Flexible Diaphragms

This Supplemental Structural Checklist shall be completed where required by Table 3-2. The Basic Structural Checklist shall be completed prior to completing this Supplemental Structural Checklist.

Lateral-Force-Resisting System

C NC N/A   COUPLING BEAMS: The stirrups in coupling beams over means of egress shall be spaced at or less than $d/2$ and shall be anchored into the confined core of the beam with hooks of 135° or more for Life Safety. All coupling beams shall comply with the requirements above and shall have the capacity in shear to develop the uplift capacity of the adjacent wall for Immediate Occupancy. (Tier 2: Sec. 4.4.2.2.3)

C NC N/A   OVERTURNING: All shear walls shall have aspect ratios less than 4-to-1. Wall piers need not be considered. This statement shall apply to the Immediate Occupancy Performance Level only. (Tier 2: Sec. 4.4.2.2.4)

C NC N/A   CONFINEMENT REINFORCING: For shear walls with aspect ratios greater than 2-to-1, the boundary elements shall be confined with spirals or ties with spacing less than $8d_b$. This statement shall apply to the Immediate Occupancy Performance Level only. (Tier 2: Sec. 4.4.2.2.5)

C NC N/A   REINFORCING AT OPENINGS: There shall be added trim reinforcement around all wall openings with a dimension greater than three times the thickness of the wall. This statement shall apply to the Immediate Occupancy Performance Level only. (Tier 2: Sec. 4.4.2.2.6)

C NC N/A   WALL THICKNESS: Thickness of bearing walls shall not be less than 1/25 the unsupported height or length, whichever is shorter, nor less than 4 inches. This statement shall apply to the Immediate Occupancy Performance Level only. (Tier 2: Sec. 4.4.2.2.7)

Diaphragms

C NC N/A   DIAPHRAGM CONTINUITY: The diaphragms shall not be composed of split-level floors and shall not have expansion joints. (Tier 2: Sec. 4.5.1.1)

C NC N/A   CROSS TIES: There shall be continuous cross ties between diaphragm chords. (Tier 2: Sec. 4.5.1.2)

C NC N/A   OPENINGS AT SHEAR WALLS: Diaphragm openings immediately adjacent to the shear walls shall be less than 25 percent of the wall length for Life Safety and 15 percent of the wall length for Immediate Occupancy. (Tier 2: Sec. 4.5.1.4)

C NC N/A   PLAN IRREGULARITIES: There shall be tensile capacity to develop the strength of the diaphragm at re-entrant corners or other locations of plan irregularities. This statement shall apply to the Immediate Occupancy Performance Level only. (Tier 2: Sec. 4.5.1.7)

C NC N/A   DIAPHRAGM REINFORCEMENT AT OPENINGS: There shall be reinforcing around all diaphragm openings larger than 50 percent of the building width in either major plan dimension. This statement shall apply to the Immediate Occupancy Performance Level only. (Tier 2: Sec. 4.5.1.8)

C NC N/A   STRAIGHT SHEATHING: All straight sheathed diaphragms shall have aspect ratios less than 2-to-1 for Life Safety and 1-to-1 for Immediate Occupancy in the direction being considered. (Tier 2: Sec. 4.5.2.1)

Screening Phase (Tier 1)

C NC N/A   SPANS: All wood diaphragms with spans greater than 24 feet for Life Safety and 12 feet for Immediate Occupancy shall consist of wood structural panels or diagonal sheathing. (Tier 2: Sec. 4.5.2.2)

C NC N/A   UNBLOCKED DIAPHRAGMS: All diagonally sheathed or unblocked wood structural panel diaphragms shall have horizontal spans less than 40 feet for Life Safety and 30 feet for Immediate Occupancy and shall have aspect ratios less than or equal to 4-to-1 for Life Safety and 3-to-1 for Immediate Occupancy. (Tier 2: Sec. 4.5.2.3)

C NC N/A   NON-CONCRETE FILLED DIAPHRAGMS: Untopped metal deck diaphragms or metal deck diaphragms with fill other than concrete shall consist of horizontal spans of less than 40 feet and shall have span/depth ratios less than 4-to-1. This statement shall apply to the Immediate Occupancy Performance Level only. (Tier 2: Sec. 4.5.3.1)

C NC N/A   OTHER DIAPHRAGMS: The diaphragm shall not consist of a system other than wood, metal deck, concrete, or horizontal bracing. (Tier 2: Sec. 4.5.7.1)

Connections

C NC N/A   UPLIFT AT PILE CAPS: Pile caps shall have top reinforcement and piles shall be anchored to the pile caps for Life Safety, and the pile cap reinforcement and pile anchorage shall be able to develop the tensile capacity of the piles for Immediate Occupancy. (Tier 2: Sec. 4.6.3.10)

Screening Phase (Tier 1)

3.7.10 Basic Structural Checklist for Building Type C3: Concrete Frames with Infill Masonry Shear Walls and Stiff Diaphragms

This Basic Structural Checklist shall be completed where required by Table 3-2.

Each of the evaluation statements on this checklist shall be marked Compliant (C), Non-compliant (NC), or Not Applicable (N/A) for a Tier 1 Evaluation. Compliant statements identify issues that are acceptable according to the criteria of this standard, while non-compliant statements identify issues that require further investigation. Certain statements may not apply to the buildings being evaluated. For non-compliant evaluation statements, the design professional may choose to conduct further investigation using the corresponding Tier 2 Evaluation procedure; corresponding section numbers are in parentheses following each evaluation statement.

C3.7.10 Basic Structural Checklist for Building Type C3

This is an older type of building construction that consists of a frame assembly of cast-in-place concrete beams and columns. The floor and roof diaphragms consist of cast-in-place concrete slabs and are stiff relative to the walls. Walls consist of infill panels constructed of solid clay brick, concrete block, or hollow clay tile masonry. The seismic performance of this type of construction depends on the interaction between the frame and infill panels. The combined behavior is more like a shear wall structure than a frame structure. Solidly infilled masonry panels form diagonal compression struts between the intersections of the frame members. If the walls are offset from the frame and do not fully engage the frame members, the diagonal compression struts will not develop. The strength of the infill panel is limited by the shear capacity of the masonry bed joint or the compression capacity of the strut. The post-cracking strength is determined by an analysis of a moment frame that is partially restrained by the cracked infill. The shear strength of the concrete columns, after cracking of the infill, may limit the semiductile behavior of the system.

Building System

C   NC   N/A   **LOAD PATH:** The structure shall contain a minimum of one complete load path for Life Safety and Immediate Occupancy for seismic force effects from any horizontal direction that serves to transfer the inertial forces from the mass to the foundation. (Tier 2: Sec. 4.3.1.1)

C   NC   N/A   **MEZZANINES:** Interior mezzanine levels shall be braced independently from the main structure, or shall be anchored to the lateral-force-resisting elements of the main structure. (Tier 2: Sec. 4.3.1.3)

C   NC   N/A   **WEAK STORY:** The strength of the lateral-force-resisting system in any story shall not be less than 80 percent of the strength in an adjacent story, above or below, for Life Safety and Immediate Occupancy. (Tier 2: Sec. 4.3.2.1)

C   NC   N/A   **SOFT STORY:** The stiffness of the lateral-force-resisting system in any story shall not be less than 70 percent of the lateral-force-resisting system stiffness in an adjacent story above or below, or less than 80 percent of the average lateral-force-resisting system stiffness of the three stories above or below for Life Safety and Immediate Occupancy. (Tier 2: Sec. 4.3.2.2)

C   NC   N/A   **GEOMETRY:** There shall be no changes in horizontal dimension of the lateral-force-resisting system of more than 30 percent in a story relative to adjacent stories for Life Safety and Immediate Occupancy, excluding one-story penthouses and mezzanines. (Tier 2: Sec. 4.3.2.3)

C   NC   N/A   **VERTICAL DISCONTINUITIES:** All vertical elements in the lateral-force-resisting system shall be continuous to the foundation. (Tier 2: Sec. 4.3.2.4)

Screening Phase (Tier 1)

C NC N/A **MASS:** There shall be no change in effective mass more than 50 percent from one story to the next for Life Safety and Immediate Occupancy. Light roofs, penthouses, and mezzanines need not be considered. (Tier 2: Sec. 4.3.2.5)

C NC N/A **TORSION:** The estimated distance between the story center of mass and the story center of rigidity shall be less than 20 percent of the building width in either plan dimension for Life Safety and Immediate Occupancy. (Tier 2: Sec. 4.3.2.6)

C NC N/A **DETERIORATION OF CONCRETE:** There shall be no visible deterioration of concrete or reinforcing steel in any of the vertical- or lateral-force-resisting elements. (Tier 2: Sec. 4.3.3.4)

C NC N/A **MASONRY UNITS:** There shall be no visible deterioration of masonry units. (Tier 2: Sec. 4.3.3.7)

C NC N/A **MASONRY JOINTS:** The mortar shall not be easily scraped away from the joints by hand with a metal tool, and there shall be no areas of eroded mortar. (Tier 2: Sec. 4.3.3.8)

C NC N/A **CRACKS IN INFILL WALLS:** There shall be no existing diagonal cracks in the infilled walls that extend throughout a panel greater than 1/8 inch for Life Safety and 1/16 inch for Immediate Occupancy, or out-of-plane offsets in the bed joint greater than 1/8 inch for Life Safety and 1/16 inch for Immediate Occupancy. (Tier 2: Sec. 4.3.3.12)

C NC N/A **CRACKS IN BOUNDARY COLUMNS:** There shall be no existing diagonal cracks wider than 1/8 inch for Life Safety and 1/16 inch for Immediate Occupancy in concrete columns that encase masonry infills. (Tier 2: Sec. 4.3.3.13)

Lateral-Force-Resisting System

C NC N/A **REDUNDANCY:** The number of lines of shear walls in each principal direction shall be greater than or equal to 2 for Life Safety and Immediate Occupancy. (Tier 2: Sec. 4.4.2.1.1)

C NC N/A **SHEAR STRESS CHECK:** The shear stress in the reinforced masonry shear walls, calculated using the Quick Check procedure of Section 3.5.3.3, shall be less than 70 psi for Life Safety and Immediate Occupancy. (Tier 2: Sec. 4.4.2.4.1)

C NC N/A **SHEAR STRESS CHECK:** The shear stress in the unreinforced masonry shear walls, calculated using the Quick Check procedure of Section 3.5.3.3, shall be less than 30 psi for clay units and 70 psi for concrete units for Life Safety and Immediate Occupancy. (Tier 2: Sec. 4.4.2.5.1)

C NC N/A **WALL CONNECTIONS:** Masonry shall be in full contact with frame for Life Safety and Immediate Occupancy. (Tier 2: Sec. 4.4.2.6.1)

Connections

C NC N/A **TRANSFER TO SHEAR WALLS:** Diaphragms shall be connected for transfer of loads to the shear walls for Life Safety and the connections shall be able to develop the lesser of the shear strength of the walls or diaphragms for Immediate Occupancy. (Tier 2: Sec. 4.6.2.1)

C NC N/A **CONCRETE COLUMNS:** All concrete columns shall be doweled into the foundation for Life Safety, and the dowels shall be able to develop the tensile capacity of reinforcement in columns of lateral-force-resisting system for Immediate Occupancy. (Tier 2: Sec. 4.6.3.2)

Screening Phase (Tier 1)

3.7.10S Supplemental Structural Checklist for Building Type C3: Concrete Frames with Infill Masonry Shear Walls and Stiff Diaphragms

This Supplemental Structural Checklist shall be completed where required by Table 3-2. The Basic Structural Checklist shall be completed prior to completing this Supplemental Structural Checklist.

Lateral-Force-Resisting System

C   NC   N/A   **DEFLECTION COMPATIBILITY:** Secondary components shall have the shear capacity to develop the flexural strength of the components for Life Safety and shall meet the requirements of Sections 4.4.1.4.9, 4.4.1.4.10, 4.4.1.4.11, 4.4.1.4.12 and 4.4.1.4.15 for Immediate Occupancy. (Tier 2: Sec. 4.4.1.6.2)

C   NC   N/A   **FLAT SLABS:** Flat slabs/plates not part of lateral-force-resisting system shall have continuous bottom steel through the column joints for Life Safety and Immediate Occupancy. (Tier 2: Sec. 4.4.1.6.3)

C   NC   N/A   **REINFORCING AT OPENINGS:** All wall openings that interrupt rebar shall have trim reinforcing on all sides. This statement shall apply to the Immediate Occupancy Performance Level only. (Tier 2: Sec. 4.4.2.4.3)

C   NC   N/A   **PROPORTIONS:** The height-to-thickness ratio of the infill walls at each story shall be less than 9 for Life Safety in levels of high seismicity, 13 for Immediate Occupancy in levels of moderate seismicity, and 8 for Immediate Occupancy in levels of high seismicity. (Tier 2: Sec. 4.4.2.6.2)

C   NC   N/A   **SOLID WALLS:** The infill walls shall not be of cavity construction. (Tier 2: Sec. 4.4.2.6.3)

C   NC   N/A   **INFILL WALLS:** The infill walls shall be continuous to the soffits of the frame beams and to the columns to either side. (Tier 2: Sec. 4.4.2.6.4)

Diaphragms

C   NC   N/A   **DIAPHRAGM CONTINUITY:** The diaphragms shall not be composed of split-level floors and shall not have expansion joints. (Tier 2: Sec. 4.5.1.1)

C   NC   N/A   **OPENINGS AT SHEAR WALLS:** Diaphragm openings immediately adjacent to the shear walls shall be less than 25 percent of the wall length for Life Safety and 15 percent of the wall length for Immediate Occupancy. (Tier 2: Sec. 4.5.1.4)

C   NC   N/A   **OPENINGS AT EXTERIOR MASONRY SHEAR WALLS:** Diaphragm openings immediately adjacent to exterior masonry shear walls shall not be greater than 8 feet long for Life Safety and 4 feet long for Immediate Occupancy. (Tier 2: Sec. 4.5.1.6)

C   NC   N/A   **PLAN IRREGULARITIES:** There shall be tensile capacity to develop the strength of the diaphragm at re-entrant corners or other locations of plan irregularities. This statement shall apply to the Immediate Occupancy Performance Level only. (Tier 2: Sec. 4.5.1.7)

C   NC   N/A   **DIAPHRAGM REINFORCEMENT AT OPENINGS:** There shall be reinforcing around all diaphragm openings larger than 50 percent of the building width in either major plan dimension. This statement shall apply to the Immediate Occupancy Performance Level only. (Tier 2: Sec. 4.5.1.8)

Connections

C   NC   N/A   **UPLIFT AT PILE CAPS:** Pile caps shall have top reinforcement and piles shall be anchored to the pile caps for Life Safety, and the pile cap reinforcement and pile anchorage shall be able to develop the tensile capacity of the piles for Immediate Occupancy. (Tier 2: Sec. 4.6.3.10)

Screening Phase (Tier 1)

3.7.10A Basic Structural Checklist for Building Type C3A: Concrete Frames with Infill Masonry Shear Walls and Flexible Diaphragms

This Basic Structural Checklist shall be completed where required by Table 3-2.

Each of the evaluation statements on this checklist shall be marked Compliant (C), Non-compliant (NC), or Not Applicable (N/A) for a Tier 1 Evaluation. Compliant statements identify issues that are acceptable according to the criteria of this standard, while non-compliant statements identify issues that require further investigation. Certain statements may not apply to the buildings being evaluated. For non-compliant evaluation statements, the design professional may choose to conduct further investigation using the corresponding Tier 2 Evaluation procedure; corresponding section numbers are in parentheses following each evaluation statement.

> **C3.7.10A Basic Structural Checklist for Building Type C3A**
>
> This is an older type of building construction that consists of a frame assembly of cast-in-place concrete beams and columns. The floors and roof consist of wood sheathing on wood framing between concrete beams. Walls consist of infill panels constructed of solid clay brick, concrete block, or hollow clay tile masonry. The seismic performance of this type of construction depends on the interaction between the frame and infill panels. The combined behavior is more like a shear wall structure than a frame structure. Solidly infilled masonry panels form diagonal compression struts between the intersections of the frame members. If the walls are offset from the frame and do not fully engage the frame members, the diagonal compression struts will not develop. The strength of the infill panel is limited by the shear capacity of the masonry bed joint or the compression capacity of the strut. The post-cracking strength is determined by an analysis of a moment frame that is partially restrained by the cracked infill. The shear strength of the concrete columns, after cracking of the infill, may limit the semiductile behavior of the system. Diaphragms consist of wood sheathing or have large aspect ratios and are flexible relative to the walls.

Building System

C   NC   N/A   **LOAD PATH:** The structure shall contain a minimum of one complete load path for Life Safety and Immediate Occupancy for seismic force effects from any horizontal direction that serves to transfer the inertial forces from the mass to the foundation. (Tier 2: Sec. 4.3.1.1)

C   NC   N/A   **ADJACENT BUILDINGS:** The clear distance between the building being evaluated and any adjacent building shall be greater than 4 percent of the height of the shorter building for Life Safety and Immediate Occupancy. (Tier 2: Sec. 4.3.1.2)

C   NC   N/A   **MEZZANINES:** Interior mezzanine levels shall be braced independently from the main structure, or shall be anchored to the lateral-force-resisting elements of the main structure. (Tier 2: Sec. 4.3.1.3)

C   NC   N/A   **WEAK STORY:** The strength of the lateral-force-resisting system in any story shall not be less than 80 percent of the strength in an adjacent story, above or below, for Life Safety and Immediate Occupancy. (Tier 2: Sec. 4.3.2.1)

C   NC   N/A   **SOFT STORY:** The stiffness of the lateral-force-resisting system in any story shall not be less than 70 percent of the lateral-force-resisting system stiffness in an adjacent story above or below, or less than 80 percent of the average lateral-force-resisting system stiffness of the three stories above or below for Life Safety and Immediate Occupancy. (Tier 2: Sec. 4.3.2.2)

Screening Phase (Tier 1)

C   NC   N/A   **GEOMETRY:** There shall be no changes in horizontal dimension of the lateral-force-resisting system of more than 30 percent in a story relative to adjacent stories for Life Safety and Immediate Occupancy, excluding one-story penthouses and mezzanines. (Tier 2: Sec. 4.3.2.3)

C   NC   N/A   **VERTICAL DISCONTINUITIES:** All vertical elements in the lateral-force-resisting system shall be continuous to the foundation. (Tier 2: Sec. 4.3.2.4)

C   NC   N/A   **MASS:** There shall be no change in effective mass more than 50 percent from one story to the next for Life Safety and Immediate Occupancy. Light roofs, penthouses, and mezzanines need not be considered. (Tier 2: Sec. 4.3.2.5)

C   NC   N/A   **DETERIORATION OF WOOD:** There shall be no signs of decay, shrinkage, splitting, fire damage, or sagging in any of the wood members, and none of the metal connection hardware shall be deteriorated, broken, or loose. (Tier 2: Sec. 4.3.3.1)

C   NC   N/A   **DETERIORATION OF CONCRETE:** There shall be no visible deterioration of concrete or reinforcing steel in any of the vertical- or lateral-force-resisting elements. (Tier 2: Sec. 4.3.3.4)

C   NC   N/A   **MASONRY UNITS:** There shall be no visible deterioration of masonry units. (Tier 2: Sec. 4.3.3.7)

C   NC   N/A   **MASONRY JOINTS:** The mortar shall not be easily scraped away from the joints by hand with a metal tool, and there shall be no areas of eroded mortar. (Tier 2: Sec. 4.3.3.8)

C   NC   N/A   **CRACKS IN INFILL WALLS:** There shall be no existing diagonal cracks in the infilled walls that extend throughout a panel greater than 1/8 inch for Life Safety and 1/16 inch for Immediate Occupancy, or out-of-plane offsets in the bed joint greater than 1/8 inch for Life Safety and 1/16 inch for Immediate Occupancy. (Tier 2: Sec. 4.3.3.12)

C   NC   N/A   **CRACKS IN BOUNDARY COLUMNS:** There shall be no existing diagonal cracks wider than 1/8 inch for Life Safety and 1/16 inch for Immediate Occupancy in concrete columns that encase masonry infills. (Tier 2: Sec. 4.3.3.13)

Lateral-Force-Resisting System

C   NC   N/A   **REDUNDANCY:** The number of lines of shear walls in each principal direction shall be greater than or equal to 2 for Life Safety and Immediate Occupancy. (Tier 2: Sec. 4.4.2.1.1)

C   NC   N/A   **SHEAR STRESS CHECK:** The shear stress in the reinforced masonry shear walls, calculated using the Quick Check procedure of Section 3.5.3.3, shall be less than 70 psi for Life Safety and Immediate Occupancy. (Tier 2: Sec. 4.4.2.4.1)

C   NC   N/A   **SHEAR STRESS CHECK:** The shear stress in the unreinforced masonry shear walls, calculated using the Quick Check procedure of Section 3.5.3.3, shall be less than 30 psi for clay units and 70 psi for concrete units for Life Safety and Immediate Occupancy. (Tier 2: Sec. 4.4.2.5.1)

C   NC   N/A   **WALL CONNECTIONS:** Masonry shall be in full contact with frame for Life Safety and Immediate Occupancy. (Tier 2: Sec. 4.4.2.6.1)

Connections

C   NC   N/A   **TRANSFER TO SHEAR WALLS:** Diaphragms shall be connected for transfer of loads to the shear walls for Life Safety and the connections shall be able to develop the lesser of the shear strength of the walls or diaphragms for Immediate Occupancy. (Tier 2: Sec. 4.6.2.1)

C   NC   N/A   **CONCRETE COLUMNS:** All concrete columns shall be doweled into the foundation for Life Safety, and the dowels shall be able to develop the tensile capacity of reinforcement in columns of lateral-force-resisting system for Immediate Occupancy. (Tier 2: Sec. 4.6.3.2)

Screening Phase (Tier 1)

3.7.10AS Supplemental Structural Checklist for Building Type C3A: Concrete Frames with Infill Masonry Shear Walls and Flexible Diaphragms

This Supplemental Structural Checklist shall be completed where required by Table 3-2. The Basic Structural Checklist shall be completed prior to completing this Supplemental Structural Checklist.

Lateral-Force-Resisting System

C NC N/A REINFORCING AT OPENINGS: All wall openings that interrupt rebar shall have trim reinforcing on all sides. This statement shall apply to the Immediate Occupancy Performance Level only. (Tier 2: Sec. 4.4.2.4.3)

C NC N/A PROPORTIONS: The height-to-thickness ratio of the infill walls at each story shall be less than 9 for Life Safety in levels of high seismicity, 13 for Immediate Occupancy in levels of moderate seismicity, and 8 for Immediate Occupancy in levels of high seismicity. (Tier 2: Sec. 4.4.2.6.2)

C NC N/A SOLID WALLS: The infill walls shall not be of cavity construction. (Tier 2: Sec. 4.4.2.6.3)

C NC N/A INFILL WALLS: The infill walls shall be continuous to the soffits of the frame beams and to the columns to either side. (Tier 2: Sec. 4.4.2.6.4)

Diaphragms

C NC N/A DIAPHRAGM CONTINUITY: The diaphragms shall not be composed of split-level floors and shall not have expansion joints. (Tier 2: Sec. 4.5.1.1)

C NC N/A CROSS TIES: There shall be continuous cross ties between diaphragm chords. (Tier 2: Sec. 4.5.1.2)

C NC N/A OPENINGS AT SHEAR WALLS: Diaphragm openings immediately adjacent to the shear walls shall be less than 25 percent of the wall length for Life Safety and 15 percent of the wall length for Immediate Occupancy. (Tier 2: Sec. 4.5.1.4)

C NC N/A OPENINGS AT EXTERIOR MASONRY SHEAR WALLS: Diaphragm openings immediately adjacent to exterior masonry shear walls shall not be greater than 8 feet long for Life Safety and 4 feet long for Immediate Occupancy. (Tier 2: Sec. 4.5.1.6)

C NC N/A PLAN IRREGULARITIES: There shall be tensile capacity to develop the strength of the diaphragm at re-entrant corners or other locations of plan irregularities. This statement shall apply to the Immediate Occupancy Performance Level only. (Tier 2: Sec. 4.5.1.7)

C NC N/A DIAPHRAGM REINFORCEMENT AT OPENINGS: There shall be reinforcing around all diaphragm openings larger than 50 percent of the building width in either major plan dimension. This statement shall apply to the Immediate Occupancy Performance Level only. (Tier 2: Sec. 4.5.1.8)

C NC N/A STRAIGHT SHEATHING: All straight sheathed diaphragms shall have aspect ratios less than 2-to-1 for Life Safety and 1-to-1 for Immediate Occupancy in the direction being considered. (Tier 2: Sec. 4.5.2.1)

C NC N/A SPANS: All wood diaphragms with spans greater than 24 feet for Life Safety and 12 feet for Immediate Occupancy shall consist of wood structural panels or diagonal sheathing. (Tier 2: Sec. 4.5.2.2)

Screening Phase (Tier 1)

C NC N/A   UNBLOCKED DIAPHRAGMS: All diagonally sheathed or unblocked wood structural panel diaphragms shall have horizontal spans less than 40 feet for Life Safety and 30 feet for Immediate Occupancy and shall have aspect ratios less than or equal to 4-to-1 for Life Safety and 3-to-1 for Immediate Occupancy. (Tier 2: Sec. 4.5.2.3)

C NC N/A   NON-CONCRETE FILLED DIAPHRAGMS: Untopped metal deck diaphragms or metal deck diaphragms with fill other than concrete shall consist of horizontal spans of less than 40 feet and shall have span/depth ratios less than 4-to-1. This statement shall apply to the Immediate Occupancy Performance Level only. (Tier 2: Sec. 4.5.3.1)

C NC N/A   OTHER DIAPHRAGMS: The diaphragm shall not consist of a system other than wood, metal deck, concrete, or horizontal bracing. (Tier 2: Sec. 4.5.7.1)

Connections

C NC N/A   STIFFNESS OF WALL ANCHORS: Anchors of concrete or masonry walls to wood structural elements shall be installed taut and shall be stiff enough to limit the relative movement between the wall and the diaphragm to no greater than 1/8 inch prior to engagement of the anchors. (Tier 2: Sec. 4.6.1.4)

C NC N/A   UPLIFT AT PILE CAPS: Pile caps shall have top reinforcement and piles shall be anchored to the pile caps for Life Safety, and the pile cap reinforcement and pile anchorage shall be able to develop the tensile capacity of the piles for Immediate Occupancy. (Tier 2: Sec. 4.6.3.10)

Screening Phase (Tier 1)

3.7.11 Basic Structural Checklist for Building Type PC1: Precast/Tilt-Up Concrete Shear Walls with Flexible Diaphragms

This Basic Structural Checklist shall be completed where required by Table 3-2.

Each of the evaluation statements on this checklist shall be marked Compliant (C), Non-compliant (NC), or Not Applicable (N/A) for a Tier 1 Evaluation. Compliant statements identify issues that are acceptable according to the criteria of this standard, while non-compliant statements identify issues that require further investigation. Certain statements may not apply to the buildings being evaluated. For non-compliant evaluation statements, the design professional may choose to conduct further investigation using the corresponding Tier 2 Evaluation procedure; corresponding section numbers are in parentheses following each evaluation statement.

C3.7.11 Basic Structural Checklist for Building Type PC1

These buildings have precast concrete perimeter wall panels that are cast on-site and tilted into place. Floor and roof framing consists of wood joists, glulam beams, steel beams, or open web joists. Framing is supported on interior steel or concrete columns and perimeter concrete bearing walls. The floors and roof consist of wood sheathing or untopped metal deck. Lateral forces are resisted by the precast concrete perimeter wall panels. Wall panels may be solid or have large window and door openings that cause the panels to behave more as frames than as shear walls. In older construction, wood framing is attached to the walls with wood ledgers. Foundations consist of concrete spread footings or deep pile foundations.

Building System

C NC N/A **LOAD PATH:** The structure shall contain a minimum of one complete load path for Life Safety and Immediate Occupancy for seismic force effects from any horizontal direction that serves to transfer the inertial forces from the mass to the foundation. (Tier 2: Sec. 4.3.1.1)

C NC N/A **ADJACENT BUILDINGS:** The clear distance between the building being evaluated and any adjacent building shall be greater than 4 percent of the height of the shorter building for Life Safety and Immediate Occupancy. (Tier 2: Sec. 4.3.1.2)

C NC N/A **MEZZANINES:** Interior mezzanine levels shall be braced independently from the main structure, or shall be anchored to the lateral-force-resisting elements of the main structure. (Tier 2: Sec. 4.3.1.3)

C NC N/A **WEAK STORY:** The strength of the lateral-force-resisting system in any story shall not be less than 80 percent of the strength in an adjacent story, above or below, for Life Safety and Immediate Occupancy. (Tier 2: Sec. 4.3.2.1)

C NC N/A **SOFT STORY:** The stiffness of the lateral-force-resisting system in any story shall not be less than 70 percent of the lateral-force-resisting system stiffness in an adjacent story above or below, or less than 80 percent of the average lateral-force-resisting system stiffness of the three stories above or below for Life Safety and Immediate Occupancy. (Tier 2: Sec. 4.3.2.2)

C NC N/A **GEOMETRY:** There shall be no changes in horizontal dimension of the lateral-force-resisting system of more than 30 percent in a story relative to adjacent stories for Life Safety and Immediate Occupancy, excluding one-story penthouses and mezzanines. (Tier 2: Sec. 4.3.2.3)

C NC N/A **VERTICAL DISCONTINUITIES:** All vertical elements in the lateral-force-resisting system shall be continuous to the foundation. (Tier 2: Sec. 4.3.2.4)

Screening Phase (Tier 1)

C   NC   N/A   MASS: There shall be no change in effective mass more than 50 percent from one story to the next for Life Safety and Immediate Occupancy. Light roofs, penthouses, and mezzanines need not be considered. (Tier 2: Sec. 4.3.2.5)

C   NC   N/A   DETERIORATION OF WOOD: There shall be no signs of decay, shrinkage, splitting, fire damage, or sagging in any of the wood members, and none of the metal connection hardware shall be deteriorated, broken, or loose. (Tier 2: Sec. 4.3.3.1)

C   NC   N/A   PRECAST CONCRETE WALLS: There shall be no visible deterioration of concrete or reinforcing steel or evidence of distress, especially at the connections. (Tier 2: Sec. 4.3.3.6)

Lateral-Force-Resisting System

C   NC   N/A   REDUNDANCY: The number of lines of shear walls in each principal direction shall be greater than or equal to 2 for Life Safety and Immediate Occupancy. (Tier 2: Sec. 4.4.2.1.1)

C   NC   N/A   SHEAR STRESS CHECK: The shear stress in the precast panels, calculated using the Quick Check procedure of Section 3.5.3.3, shall be less than the greater of 100 psi or $2\sqrt{f'c}$ for Life Safety and Immediate Occupancy. (Tier 2: Sec. 4.4.2.3.1)

C   NC   N/A   REINFORCING STEEL: The ratio of reinforcing steel area to gross concrete area shall be not less than 0.0015 in the vertical direction and 0.0025 in the horizontal direction for Life Safety and Immediate Occupancy. The spacing of reinforcing steel shall be equal to or less than 18 inches for Life Safety and Immediate Occupancy. (Tier 2: Sec. 4.4.2.3.2)

Connections

C   NC   N/A   WALL ANCHORAGE: Exterior concrete or masonry walls that are dependent on the diaphragm for lateral support shall be anchored for out-of-plane forces at each diaphragm level with steel anchors, reinforcing dowels, or straps that are developed into the diaphragm. Connections shall have adequate strength to resist the connection force calculated in the Quick Check procedure of Section 3.5.3.7. (Tier 2: Sec. 4.6.1.1)

C   NC   N/A   WOOD LEDGERS: The connection between the wall panels and the diaphragm shall not induce cross-grain bending or tension in the wood ledgers. (Tier 2: Sec. 4.6.1.2)

C   NC   N/A   TRANSFER TO SHEAR WALLS: Diaphragms shall be connected for transfer of loads to the shear walls for Life Safety and the connections shall be able to develop the lesser of the shear strength of the walls or diaphragms for Immediate Occupancy. (Tier 2: Sec. 4.6.2.1)

C   NC   N/A   PRECAST WALL PANELS: Precast wall panels shall be connected to the foundation for Life Safety and the connections shall be able to develop the strength of the walls for Immediate Occupancy. (Tier 2: Sec. 4.6.3.7)

C   NC   N/A   GIRDER/COLUMN CONNECTION: There shall be a positive connection utilizing plates, connection hardware, or straps between the girder and the column support. (Tier 2: Sec. 4.6.4.1)

Screening Phase (Tier 1)

3.7.11S Supplemental Structural Checklist for Building Type PC1: Precast/Tilt-Up Concrete Shear Walls with Flexible Diaphragms

This Supplemental Structural Checklist shall be completed where required by Table 3-2. The Basic Structural Checklist shall be completed prior to completing this Supplemental Structural Checklist.

Lateral-Force-Resisting System

C NC N/A **COUPLING BEAMS:** The stirrups in coupling beams over means of egress shall be spaced at or less than $d/2$ and shall be anchored into the confined core of the beam with hooks of 135° or more for Life Safety. All coupling beams shall comply with the requirements above and shall have the capacity in shear to develop the uplift capacity of the adjacent wall for Immediate Occupancy. (Tier 2: Sec. 4.4.2.2.3)

C NC N/A **WALL OPENINGS:** The total width of openings along any perimeter wall line shall constitute less than 75 percent of the length of any perimeter wall for Life Safety and 50 percent for Immediate Occupancy with the wall piers having aspect ratios of less than 2-to-1 for Life Safety and Immediate Occupancy. (Tier 2: Sec. 4.4.2.3.3)

C NC N/A **CORNER OPENINGS:** Walls with openings at a building corner larger than the width of a typical panel shall be connected to the remainder of the wall with collector reinforcing. (Tier 2: Sec. 4.4.2.3.4)

C NC N/A **PANEL-TO-PANEL CONNECTIONS:** Adjacent wall panels shall be interconnected to transfer overturning forces between panels by methods other than welded steel inserts. This statement shall apply to the Immediate Occupancy Performance Level only. (Tier 2: Sec. 4.4.2.3.5)

C NC N/A **WALL THICKNESS:** Thickness of bearing walls shall not be less than 1/25 the unsupported height or length, whichever is shorter, nor less than 4 inches. This statement shall apply to the Immediate Occupancy Performance Level only. (Tier 2: Sec. 4.4.2.3.6)

Diaphragms

C NC N/A **CROSS TIES:** There shall be continuous cross ties between diaphragm chords. (Tier 2: Sec. 4.5.1.2)

C NC N/A **PLAN IRREGULARITIES:** There shall be tensile capacity to develop the strength of the diaphragm at re-entrant corners or other locations of plan irregularities. This statement shall apply to the Immediate Occupancy Performance Level only. (Tier 2: Sec. 4.5.1.7)

C NC N/A **DIAPHRAGM REINFORCEMENT AT OPENINGS:** There shall be reinforcing around all diaphragm openings larger than 50 percent of the building width in either major plan dimension. This statement shall apply to the Immediate Occupancy Performance Level only. (Tier 2: Sec. 4.5.1.8)

C NC N/A **STRAIGHT SHEATHING:** All straight sheathed diaphragms shall have aspect ratios less than 2-to-1 for Life Safety and 1-to-1 for Immediate Occupancy in the direction being considered. (Tier 2: Sec. 4.5.2.1)

C NC N/A **SPANS:** All wood diaphragms with spans greater than 24 feet for Life Safety and 12 feet for Immediate Occupancy shall consist of wood structural panels or diagonal sheathing. (Tier 2: Sec. 4.5.2.2)

C NC N/A **UNBLOCKED DIAPHRAGMS:** All diagonally sheathed or unblocked wood structural panel diaphragms shall have horizontal spans less than 40 feet for Life Safety and 30 feet for Immediate Occupancy and shall have aspect ratios less than or equal to 4-to-1 for Life Safety and 3-to-1 for Immediate Occupancy. (Tier 2: Sec. 4.5.2.3)

Screening Phase (Tier 1)

C   NC   N/A   OTHER DIAPHRAGMS: The diaphragm shall not consist of a system other than wood, metal deck, concrete, or horizontal bracing. (Tier 2: Sec. 4.5.7.1)

Connections

C   NC   N/A   PRECAST PANEL CONNECTIONS: There shall be at least two anchors from each precast wall panel into the diaphragm elements for Life Safety and the anchors shall be able to develop the strength of the panels for Immediate Occupancy. (Tier 2: Sec. 4.6.1.3)

C   NC   N/A   UPLIFT AT PILE CAPS: Pile caps shall have top reinforcement and piles shall be anchored to the pile caps for Life Safety, and the pile cap reinforcement and pile anchorage shall be able to develop the tensile capacity of the piles for Immediate Occupancy. (Tier 2: Sec. 4.6.3.10)

C   NC   N/A   GIRDERS: Girders supported by walls or pilasters shall have at least two ties securing the anchor bolts for Life Safety and Immediate Occupancy. (Tier 2: Sec. 4.6.4.2)

Screening Phase (Tier 1)

3.7.11A Basic Structural Checklist for Building Type PC1A: Precast/Tilt-Up Concrete Shear Walls with Stiff Diaphragms

This Basic Structural Checklist shall be completed where required by Table 3-2.

Each of the evaluation statements on this checklist shall be marked Compliant (C), Non-compliant (NC), or Not Applicable (N/A) for a Tier 1 Evaluation. Compliant statements identify issues that are acceptable according to the criteria of this standard, while non-compliant statements identify issues that require further investigation. Certain statements may not apply to the buildings being evaluated. For non-compliant evaluation statements, the design professional may choose to conduct further investigation using the corresponding Tier 2 Evaluation procedure; corresponding section numbers are in parentheses following each evaluation statement.

> **C3.7.11A Basic Structural Checklist for Building Type PC1A**
>
> These buildings are one or more stories in height and have precast concrete perimeter wall panels that are cast on-site and tilted into place. The floors and roof consist of precast elements, cast-in-place concrete, or metal deck with concrete fill, and are stiff relative to the walls. Framing is supported on interior steel or concrete columns and perimeter concrete bearing walls. Lateral forces are resisted by the precast concrete perimeter wall panels. Wall panels may be solid or have large window and door openings that cause the panels to behave more as frames than as shear walls. Foundations consist of concrete spread footings or deep pile foundations.

Building System

C NC N/A   **LOAD PATH:** The structure shall contain a minimum of one complete load path for Life Safety and Immediate Occupancy for seismic force effects from any horizontal direction that serves to transfer the inertial forces from the mass to the foundation. (Tier 2: Sec. 4.3.1.1)

C NC N/A   **MEZZANINES:** Interior mezzanine levels shall be braced independently from the main structure, or shall be anchored to the lateral-force-resisting elements of the main structure. (Tier 2: Sec. 4.3.1.3)

C NC N/A   **WEAK STORY:** The strength of the lateral-force-resisting system in any story shall not be less than 80 percent of the strength in an adjacent story, above or below, for Life Safety and Immediate Occupancy. (Tier 2: Sec. 4.3.2.1)

C NC N/A   **SOFT STORY:** The stiffness of the lateral-force-resisting system in any story shall not be less than 70 percent of the lateral-force-resisting system stiffness in an adjacent story above or below, or less than 80 percent of the average lateral-force-resisting system stiffness of the three stories above or below for Life Safety and Immediate Occupancy. (Tier 2: Sec. 4.3.2.2)

C NC N/A   **GEOMETRY:** There shall be no changes in horizontal dimension of the lateral-force-resisting system of more than 30 percent in a story relative to adjacent stories for Life Safety and Immediate Occupancy, excluding one-story penthouses and mezzanines. (Tier 2: Sec. 4.3.2.3)

C NC N/A   **VERTICAL DISCONTINUITIES:** All vertical elements in the lateral-force-resisting system shall be continuous to the foundation. (Tier 2: Sec. 4.3.2.4)

C NC N/A   **MASS:** There shall be no change in effective mass more than 50 percent from one story to the next for Life Safety and Immediate Occupancy. Light roofs, penthouses, and mezzanines need not be considered. (Tier 2: Sec. 4.3.2.5)

Screening Phase (Tier 1)

C   NC   N/A   **TORSION:** The estimated distance between the story center of mass and the story center of rigidity shall be less than 20 percent of the building width in either plan dimension for Life Safety and Immediate Occupancy. (Tier 2: Sec. 4.3.2.6)

C   NC   N/A   **POST-TENSIONING ANCHORS:** There shall be no evidence of corrosion or spalling in the vicinity of post-tensioning or end fittings. Coil anchors shall not have been used. (Tier 2: Sec. 4.3.3.5)

C   NC   N/A   **PRECAST CONCRETE WALLS:** There shall be no visible deterioration of concrete or reinforcing steel or evidence of distress, especially at the connections. (Tier 2: Sec. 4.3.3.6)

Lateral-Force-Resisting System

C   NC   N/A   **REDUNDANCY:** The number of lines of shear walls in each principal direction shall be greater than or equal to 2 for Life Safety and Immediate Occupancy. (Tier 2: Sec. 4.4.2.1.1)

C   NC   N/A   **SHEAR STRESS CHECK:** The shear stress in the precast panels, calculated using the Quick Check procedure of Section 3.5.3.3, shall be less than the greater of 100 psi or $2\sqrt{f'c}$ for Life Safety and Immediate Occupancy. (Tier 2: Sec. 4.4.2.3.1)

C   NC   N/A   **REINFORCING STEEL:** The ratio of reinforcing steel area to gross concrete area shall be not less than 0.0015 in the vertical direction and 0.0025 in the horizontal direction for Life Safety and Immediate Occupancy. The spacing of reinforcing steel shall be equal to or less than 18 inches for Life Safety and Immediate Occupancy. (Tier 2: Sec. 4.4.2.3.2)

Diaphragms

C   NC   N/A   **TOPPING SLAB:** Precast concrete diaphragm elements shall be interconnected by a continuous reinforced concrete topping slab. (Tier 2: Sec. 4.5.5.1)

Connections

C   NC   N/A   **WALL ANCHORAGE:** Exterior concrete or masonry walls that are dependent on the diaphragm for lateral support shall be anchored for out-of-plane forces at each diaphragm level with steel anchors, reinforcing dowels, or straps that are developed into the diaphragm. Connections shall have adequate strength to resist the connection force calculated in the Quick Check procedure of Section 3.5.3.7. (Tier 2: Sec. 4.6.1.1)

C   NC   N/A   **TRANSFER TO SHEAR WALLS:** Diaphragms shall be connected for transfer of loads to the shear walls for Life Safety and the connections shall be able to develop the lesser of the shear strength of the walls or diaphragms for Immediate Occupancy. (Tier 2: Sec. 4.6.2.1)

C   NC   N/A   **TOPPING SLAB TO WALLS OR FRAMES:** Reinforced concrete topping slabs that interconnect the precast concrete diaphragm elements shall be doweled for transfer of forces into the shear wall or frame elements for Life Safety, and the dowels shall be able to develop the lesser of the shear strength of the walls, frames, or slabs for Immediate Occupancy. (Tier 2: Sec. 4.6.2.3)

C   NC   N/A   **PRECAST WALL PANELS:** Precast wall panels shall be connected to the foundation for Life Safety and the connections shall be able to develop the strength of the walls for Immediate Occupancy. (Tier 2: Sec. 4.6.3.7)

C   NC   N/A   **GIRDER/COLUMN CONNECTION:** There shall be a positive connection utilizing plates, connection hardware, or straps between the girder and the column support. (Tier 2: Sec. 4.6.4.1)

Screening Phase (Tier 1)

3.7.11AS Supplemental Structural Checklist for Building Type PC1A: Precast Tilt-Up Concrete Shear Walls with Stiff Diaphragms

This Supplemental Structural Checklist shall be completed where required by Table 3-2. The Basic Structural Checklist shall be completed prior to completing this Supplemental Structural Checklist.

Lateral-Force-Resisting System

C NC N/A   DEFLECTION COMPATIBILITY: Secondary components shall have the shear capacity to develop the flexural strength of the components for Life Safety and shall meet the requirements of Sections 4.4.1.4.9, 4.4.1.4.10, 4.4.1.4.11, 4.4.1.4.12 and 4.4.1.4.15 for Immediate Occupancy. (Tier 2: Sec. 4.4.1.6.2)

C NC N/A   COUPLING BEAMS: The stirrups in coupling beams over means of egress shall be spaced at or less than $d/2$ and shall be anchored into the confined core of the beam with hooks of 135° or more for Life Safety. All coupling beams shall comply with the requirements above and shall have the capacity in shear to develop the uplift capacity of the adjacent wall for Immediate Occupancy. (Tier 2: Sec. 4.4.2.2.3)

C NC N/A   WALL OPENINGS: The total width of openings along any perimeter wall line shall constitute less than 75 percent of the length of any perimeter wall for Life Safety and 50 percent for Immediate Occupancy with the wall piers having aspect ratios of less than 2-to-1 for Life Safety and Immediate Occupancy. (Tier 2: Sec. 4.4.2.3.3)

C NC N/A   CORNER OPENINGS: Walls with openings at a building corner larger than the width of a typical panel shall be connected to the remainder of the wall with collector reinforcing. (Tier 2: Sec. 4.4.2.3.4)

C NC N/A   PANEL-TO-PANEL CONNECTIONS: Adjacent wall panels shall be interconnected to transfer overturning forces between panels by methods other than welded steel inserts. This statement shall apply to the Immediate Occupancy Performance Level only. (Tier 2: Sec. 4.4.2.3.5)

C NC N/A   WALL THICKNESS: Thickness of bearing walls shall not be less than 1/25 the unsupported height or length, whichever is shorter, nor less than 4 inches. This statement shall apply to the Immediate Occupancy Performance Level only. (Tier 2: Sec. 4.4.2.3.6)

Diaphragms

C NC N/A   PLAN IRREGULARITIES: There shall be tensile capacity to develop the strength of the diaphragm at re-entrant corners or other locations of plan irregularities. This statement shall apply to the Immediate Occupancy Performance Level only. (Tier 2: Sec. 4.5.1.7)

C NC N/A   DIAPHRAGM REINFORCEMENT AT OPENINGS: There shall be reinforcing around all diaphragm openings larger than 50 percent of the building width in either major plan dimension. This statement shall apply to the Immediate Occupancy Performance Level only. (Tier 2: Sec. 4.5.1.8)

Connections

C NC N/A   PRECAST PANEL CONNECTIONS: There shall be at least two anchors from each precast wall panel into the diaphragm elements for Life Safety and the anchors shall be able to develop the strength of the panels for Immediate Occupancy. (Tier 2: Sec. 4.6.1.3)

C NC N/A   UPLIFT AT PILE CAPS: Pile caps shall have top reinforcement and piles shall be anchored to the pile caps for Life Safety, and the pile cap reinforcement and pile anchorage shall be able to develop the tensile capacity of the piles for Immediate Occupancy. (Tier 2: Sec. 4.6.3.10)

C NC N/A   GIRDERS: Girders supported by walls or pilasters shall have at least two ties securing the anchor bolts for Life Safety and Immediate Occupancy. (Tier 2: Sec. 4.6.4.2)

Screening Phase (Tier 1)

3.7.12 Basic Structural Checklist for Building Type PC2: Precast Concrete Frames with Shear Walls

This Basic Structural Checklist shall be completed where required by Table 3-2.

Each of the evaluation statements on this checklist shall be marked Compliant (C), Non-compliant (NC), or Not Applicable (N/A) for a Tier 1 Evaluation. Compliant statements identify issues that are acceptable according to the criteria of this standard, while non-compliant statements identify issues that require further investigation. Certain statements may not apply to the buildings being evaluated. For non-compliant evaluation statements, the design professional may choose to conduct further investigation using the corresponding Tier 2 Evaluation procedure; corresponding section numbers are in parentheses following each evaluation statement.

> **C3.7.12 Basic Structural Checklist for Building Type PC2**
>
> These buildings consist of a frame assembly of precast concrete girders and columns with the presence of shear walls. Floor and roof framing consists of precast concrete planks, tees, or double-tees supported on precast concrete girders and columns. Lateral forces are resisted by precast or cast-in-place concrete shear walls. Diaphragms consist of precast elements interconnected with welded inserts, cast-in-place closure strips, or reinforced concrete topping slabs.

Building System

C NC N/A   **LOAD PATH:** The structure shall contain a minimum of one complete load path for Life Safety and Immediate Occupancy for seismic force effects from any horizontal direction that serves to transfer the inertial forces from the mass to the foundation. (Tier 2: Sec. 4.3.1.1)

C NC N/A   **MEZZANINES:** Interior mezzanine levels shall be braced independently from the main structure, or shall be anchored to the lateral-force-resisting elements of the main structure. (Tier 2: Sec. 4.3.1.3)

C NC N/A   **WEAK STORY:** The strength of the lateral-force-resisting system in any story shall not be less than 80 percent of the strength in an adjacent story, above or below, for Life Safety and Immediate Occupancy. (Tier 2: Sec. 4.3.2.1)

C NC N/A   **SOFT STORY:** The stiffness of the lateral-force-resisting system in any story shall not be less than 70 percent of the lateral-force-resisting system stiffness in an adjacent story above or below, or less than 80 percent of the average lateral-force-resisting system stiffness of the three stories above or below for Life Safety and Immediate Occupancy. (Tier 2: Sec. 4.3.2.2)

C NC N/A   **GEOMETRY:** There shall be no changes in horizontal dimension of the lateral-force-resisting system of more than 30 percent in a story relative to adjacent stories for Life Safety and Immediate Occupancy, excluding one-story penthouses and mezzanines. (Tier 2: Sec. 4.3.2.3)

C NC N/A   **VERTICAL DISCONTINUITIES:** All vertical elements in the lateral-force-resisting system shall be continuous to the foundation. (Tier 2: Sec. 4.3.2.4)

C NC N/A   **MASS:** There shall be no change in effective mass more than 50 percent from one story to the next for Life Safety and Immediate Occupancy. Light roofs, penthouses, and mezzanines need not be considered. (Tier 2: Sec. 4.3.2.5)

C NC N/A   **TORSION:** The estimated distance between the story center of mass and the story center of rigidity shall be less than 20 percent of the building width in either plan dimension for Life Safety and Immediate Occupancy. (Tier 2: Sec. 4.3.2.6)

Screening Phase (Tier 1)

C NC N/A **DETERIORATION OF CONCRETE:** There shall be no visible deterioration of concrete or reinforcing steel in any of the vertical- or lateral-force-resisting elements. (Tier 2: Sec. 4.3.3.4)

C NC N/A **POST-TENSIONING ANCHORS:** There shall be no evidence of corrosion or spalling in the vicinity of post-tensioning or end fittings. Coil anchors shall not have been used. (Tier 2: Sec. 4.3.3.5)

C NC N/A **CONCRETE WALL CRACKS:** All existing diagonal cracks in wall elements shall be less than 1/8 inch for Life Safety and 1/16 inch for Immediate Occupancy, shall not be concentrated in one location, and shall not form an X pattern. (Tier 2: Sec. 4.3.3.9)

Lateral-Force-Resisting System

C NC N/A **COMPLETE FRAMES:** Steel or concrete frames classified as secondary components shall form a complete vertical-load-carrying system. (Tier 2: Sec. 4.4.1.6.1)

C NC N/A **REDUNDANCY:** The number of lines of shear walls in each principal direction shall be greater than or equal to 2 for Life Safety and Immediate Occupancy. (Tier 2: Sec. 4.4.2.1.1)

C NC N/A **SHEAR STRESS CHECK:** The shear stress in the concrete shear walls, calculated using the Quick Check procedure of Section 3.5.3.3, shall be less than the greater of 100 psi or $2\sqrt{f'c}$ for Life Safety and Immediate Occupancy. (Tier 2: Sec. 4.4.2.2.1)

C NC N/A **REINFORCING STEEL:** The ratio of reinforcing steel area to gross concrete area shall be not less than 0.0015 in the vertical direction and 0.0025 in the horizontal direction for Life Safety and Immediate Occupancy. The spacing of reinforcing steel shall be equal to or less than 18 inches for Life Safety and Immediate Occupancy. (Tier 2: Sec. 4.4.2.2.2)

Diaphragms

C NC N/A **TOPPING SLAB:** Precast concrete diaphragm elements shall be interconnected by a continuous reinforced concrete topping slab. (Tier 2: Sec. 4.5.5.1)

Connections

C NC N/A **WALL ANCHORAGE:** Exterior concrete or masonry walls that are dependent on the diaphragm for lateral support shall be anchored for out-of-plane forces at each diaphragm level with steel anchors, reinforcing dowels, or straps that are developed into the diaphragm. Connections shall have adequate strength to resist the connection force calculated in the Quick Check procedure of Section 3.5.3.7. (Tier 2: Sec. 4.6.1.1)

C NC N/A **TRANSFER TO SHEAR WALLS:** Diaphragms shall be connected for transfer of loads to the shear walls for Life Safety and the connections shall be able to develop the lesser of the shear strength of the walls or diaphragms for Immediate Occupancy. (Tier 2: Sec. 4.6.2.1)

C NC N/A **TOPPING SLAB TO WALLS OR FRAMES:** Reinforced concrete topping slabs that interconnect the precast concrete diaphragm elements shall be doweled for transfer of forces into the shear wall or frame elements for Life Safety, and the dowels shall be able to develop the lesser of the shear strength of the walls, frames, or slabs for Immediate Occupancy. (Tier 2: Sec. 4.6.2.3)

C NC N/A **FOUNDATION DOWELS:** Wall reinforcement shall be doweled into the foundation for Life Safety, and the dowels shall be able to develop the lesser of the strength of the walls or the uplift capacity of the foundation for Immediate Occupancy. (Tier 2: Sec. 4.6.3.5)

C NC N/A **GIRDER/COLUMN CONNECTION:** There shall be a positive connection utilizing plates, connection hardware, or straps between the girder and the column support. (Tier 2: Sec. 4.6.4.1)

3.7.12S Supplemental Structural Checklist for Building Type PC2: Precast Concrete Frames with Shear Walls

This Supplemental Structural Checklist shall be completed where required by Table 3-2. The Basic Structural Checklist shall be completed prior to completing this Supplemental Structural Checklist.

Lateral-Force-Resisting System

C NC N/A **PRECAST FRAMES:** For buildings with concrete shear walls, precast concrete frame elements shall not be considered as primary components for resisting lateral forces. (Tier 2: Sec. 4.4.1.5.2)

C NC N/A **PRECAST CONNECTIONS:** For buildings with concrete shear walls, the connection between precast frame elements such as chords, ties, and collectors in the lateral-force-resisting system shall develop the capacity of the connected members. (Tier 2: Sec. 4.4.1.5.3)

C NC N/A **DEFLECTION COMPATIBILITY:** Secondary components shall have the shear capacity to develop the flexural strength of the components for Life Safety and shall meet the requirements of Sections 4.4.1.4.9, 4.4.1.4.10, 4.4.1.4.11, 4.4.1.4.12 and 4.4.1.4.15 for Immediate Occupancy. (Tier 2: Sec. 4.4.1.6.2)

C NC N/A **COUPLING BEAMS:** The stirrups in coupling beams over means of egress shall be spaced at or less than $d/2$ and shall be anchored into the confined core of the beam with hooks of 135° or more for Life Safety. All coupling beams shall comply with the requirements above and shall have the capacity in shear to develop the uplift capacity of the adjacent wall for Immediate Occupancy. (Tier 2: Sec. 4.4.2.2.3)

C NC N/A **OVERTURNING:** All shear walls shall have aspect ratios less than 4-to-1. Wall piers need not be considered. This statement shall apply to the Immediate Occupancy Performance Level only. (Tier 2: Sec. 4.4.2.2.4)

C NC N/A **CONFINEMENT REINFORCING:** For shear walls with aspect ratios greater than 2-to-1, the boundary elements shall be confined with spirals or ties with spacing less than $8d_b$. This statement shall apply to the Immediate Occupancy Performance Level only. (Tier 2: Sec. 4.4.2.2.5)

C NC N/A **REINFORCING AT OPENINGS:** There shall be added trim reinforcement around all wall openings with a dimension greater than three times the thickness of the wall. This statement shall apply to the Immediate Occupancy Performance Level only. (Tier 2: Sec. 4.4.2.2.6)

C NC N/A **WALL THICKNESS:** Thickness of bearing walls shall not be less than 1/25 the unsupported height or length, whichever is shorter, nor less than 4 inches. This statement shall apply to the Immediate Occupancy Performance Level only. (Tier 2: Sec. 4.4.2.2.7)

Diaphragms

C NC N/A **OPENINGS AT SHEAR WALLS:** Diaphragm openings immediately adjacent to the shear walls shall be less than 25 percent of the wall length for Life Safety and 15 percent of the wall length for Immediate Occupancy. (Tier 2: Sec. 4.5.1.4)

C NC N/A **PLAN IRREGULARITIES:** There shall be tensile capacity to develop the strength of the diaphragm at re-entrant corners or other locations of plan irregularities. This statement shall apply to the Immediate Occupancy Performance Level only. (Tier 2: Sec. 4.5.1.7)

C NC N/A **DIAPHRAGM REINFORCEMENT AT OPENINGS:** There shall be reinforcing around all diaphragm openings larger than 50 percent of the building width in either major plan dimension. This statement shall apply to the Immediate Occupancy Performance Level only. (Tier 2: Sec. 4.5.1.8)

Screening Phase (Tier 1)

Connections

C　NC　N/A　　UPLIFT AT PILE CAPS: Pile caps shall have top reinforcement and piles shall be anchored to the pile caps for Life Safety, and the pile cap reinforcement and pile anchorage shall be able to develop the tensile capacity of the piles for Immediate Occupancy. (Tier 2: Sec. 4.6.3.10)

C　NC　N/A　　CORBEL BEARING: If the frame girders bear on column corbels, the length of bearing shall be greater than 3 inches for Life Safety and Immediate Occupancy. (Tier 2: Sec. 4.6.4.3)

C　NC　N/A　　CORBEL CONNECTIONS: The frame girders shall not be connected to corbels with welded elements. (Tier 2: Sec. 4.6.4.4)

3.7.12A Basic Structural Checklist for Building Type PC2A: Precast Concrete Frames without Shear Walls

This Basic Structural Checklist shall be completed where required by Table 3-2.

Each of the evaluation statements on this checklist shall be marked Compliant (C), Non-compliant (NC), or Not Applicable (N/A) for a Tier 1 Evaluation. Compliant statements identify issues that are acceptable according to the criteria of this standard, while non-compliant statements identify issues that require further investigation. Certain statements may not apply to the buildings being evaluated. For non-compliant evaluation statements, the design professional may choose to conduct further investigation using the corresponding Tier 2 Evaluation procedure; corresponding section numbers are in parentheses following each evaluation statement.

C3.7.12A Basic Structural Checklist for Building Type PC2A

These buildings are similar to PC2 buildings, except that concrete shear walls are not present. Lateral forces are resisted by precast concrete moment frames that develop their stiffness through beam-column joints rigidly connected by welded inserts or cast-in-place concrete closures. Diaphragms consist of precast elements interconnected with welded inserts, cast-in-place closure strips, or reinforced concrete topping slabs.

Building System

C   NC   N/A   LOAD PATH: The structure shall contain a minimum of one complete load path for Life Safety and Immediate Occupancy for seismic force effects from any horizontal direction that serves to transfer the inertial forces from the mass to the foundation. (Tier 2: Sec. 4.3.1.1)

C   NC   N/A   ADJACENT BUILDINGS: The clear distance between the building being evaluated and any adjacent building shall be greater than 4 percent of the height of the shorter building for Life Safety and Immediate Occupancy. (Tier 2: Sec. 4.3.1.2)

C   NC   N/A   MEZZANINES: Interior mezzanine levels shall be braced independently from the main structure, or shall be anchored to the lateral-force-resisting elements of the main structure. (Tier 2: Sec. 4.3.1.3)

C   NC   N/A   WEAK STORY: The strength of the lateral-force-resisting system in any story shall not be less than 80 percent of the strength in an adjacent story, above or below, for Life Safety and Immediate Occupancy. (Tier 2: Sec. 4.3.2.1)

C   NC   N/A   SOFT STORY: The stiffness of the lateral-force-resisting system in any story shall not be less than 70 percent of the lateral-force-resisting system stiffness in an adjacent story above or below, or less than 80 percent of the average lateral-force-resisting system stiffness of the three stories above or below for Life Safety and Immediate Occupancy. (Tier 2: Sec. 4.3.2.2)

C   NC   N/A   GEOMETRY: There shall be no changes in horizontal dimension of the lateral-force-resisting system of more than 30 percent in a story relative to adjacent stories for Life Safety and Immediate Occupancy, excluding one-story penthouses and mezzanines. (Tier 2: Sec. 4.3.2.3)

C   NC   N/A   VERTICAL DISCONTINUITIES: All vertical elements in the lateral-force-resisting system shall be continuous to the foundation. (Tier 2: Sec. 4.3.2.4)

C   NC   N/A   MASS: There shall be no change in effective mass more than 50 percent from one story to the next for Life Safety and Immediate Occupancy. Light roofs, penthouses, and mezzanines need not be considered. (Tier 2: Sec. 4.3.2.5)

Screening Phase (Tier 1)

C NC N/A   TORSION: The estimated distance between the story center of mass and the story center of rigidity shall be less than 20 percent of the building width in either plan dimension for Life Safety and Immediate Occupancy. (Tier 2: Sec. 4.3.2.6)

C NC N/A   DETERIORATION OF CONCRETE: There shall be no visible deterioration of concrete or reinforcing steel in any of the vertical- or lateral-force-resisting elements. (Tier 2: Sec. 4.3.3.4)

C NC N/A   POST-TENSIONING ANCHORS: There shall be no evidence of corrosion or spalling in the vicinity of post-tensioning or end fittings. Coil anchors shall not have been used. (Tier 2: Sec. 4.3.3.5)

Lateral-Force-Resisting System

C NC N/A   REDUNDANCY: The number of lines of moment frames in each principal direction shall be greater than or equal to 2 for Life Safety and Immediate Occupancy. The number of bays of moment frames in each line shall be greater than or equal to 2 for Life Safety and 3 for Immediate Occupancy. (Tier 2: Sec. 4.4.1.1.1)

C NC N/A   SHEAR STRESS CHECK: The shear stress in the concrete columns, calculated using the Quick Check procedure of Section 3.5.3.2, shall be less than the greater of 100 psi or $2\sqrt{f'c}$ for Life Safety and Immediate Occupancy. (Tier 2: Sec. 4.4.1.4.1)

C NC N/A   AXIAL STRESS CHECK: The axial stress due to gravity loads in columns subjected to overturning forces shall be less than $0.10f'_c$ for Life Safety and Immediate Occupancy. Alternatively, the axial stresses due to overturning forces alone, calculated using the Quick Check procedure of Section 3.5.3.6, shall be less than $0.30f'_c$ for Life Safety and Immediate Occupancy. (Tier 2: Sec. 4.4.1.4.2)

C NC N/A   PRECAST CONNECTION CHECK: The precast connections at frame joints shall have the capacity to resist the shear and moment demands calculated using the Quick Check procedure of Section 3.5.3.5. (Tier 2: Sec. 4.4.1.5.1)

Diaphragms

C NC N/A   TOPPING SLAB: Precast concrete diaphragm elements shall be interconnected by a continuous reinforced concrete topping slab. (Tier 2: Sec. 4.5.5.1)

Connections

C NC N/A   TOPPING SLAB TO WALLS OR FRAMES: Reinforced concrete topping slabs that interconnect the precast concrete diaphragm elements shall be doweled for transfer of forces into the shear wall or frame elements for Life Safety, and the dowels shall be able to develop the lesser of the shear strength of the walls, frames, or slabs for Immediate Occupancy. (Tier 2: Sec. 4.6.2.3)

C NC N/A   GIRDER/COLUMN CONNECTION: There shall be a positive connection between the girder and the column support. (Tier 2: Sec. 4.6.4.1)

3.7.12AS Supplemental Structural Checklist for Building Type PC2A: Precast Concrete Frames without Shear Walls

This Supplemental Structural Checklist shall be completed where required by Table 3-2. The Basic Structural Checklist shall be completed prior to completing this Supplemental Structural Checklist.

Lateral-Force-Resisting System

C   NC   N/A   PRESTRESSED FRAME ELEMENTS: The lateral-force-resisting frames shall not include any prestressed or post-tensioned elements where the average prestress exceeds the lesser of 700 psi or $f'_c/6$ at potential hinge locations. The average prestress shall be calculated in accordance with the Quick Check procedure of Section 3.5.3.8. (Tier 2: Sec. 4.4.1.4.4)

C   NC   N/A   CAPTIVE COLUMNS: There shall be no columns at a level with height/depth ratios less than 50 percent of the nominal height/depth ratio of the typical columns at that level for Life Safety and 75 percent for Immediate Occupancy. (Tier 2: Sec. 4.4.1.4.5)

C   NC   N/A   JOINT REINFORCING: Beam-column joints shall have ties spaced at or less than $8d_b$ for Life Safety and Immediate Occupancy. (Tier 2: Sec. 4.4.1.4.13)

C   NC   N/A   DEFLECTION COMPATIBILITY: Secondary components shall have the shear capacity to develop the flexural strength of the components for Life Safety and shall meet the requirements of Sections 4.4.1.4.9, 4.4.1.4.10, 4.4.1.4.11, 4.4.1.4.12 and 4.4.1.4.15 for Immediate Occupancy. (Tier 2: Sec. 4.4.1.6.2)

Diaphragms

C   NC   N/A   PLAN IRREGULARITIES: There shall be tensile capacity to develop the strength of the diaphragm at re-entrant corners or other locations of plan irregularities. This statement shall apply to the Immediate Occupancy Performance Level only. (Tier 2: Sec. 4.5.1.7)

C   NC   N/A   DIAPHRAGM REINFORCEMENT AT OPENINGS: There shall be reinforcing around all diaphragm openings larger than 50 percent of the building width in either major plan dimension. This statement shall apply to the Immediate Occupancy Performance Level only. (Tier 2: Sec. 4.5.1.8)

Connections

C   NC   N/A   UPLIFT AT PILE CAPS: Pile caps shall have top reinforcement and piles shall be anchored to the pile caps for Life Safety, and the pile cap reinforcement and pile anchorage shall be able to develop the tensile capacity of the piles for Immediate Occupancy. (Tier 2: Sec. 4.6.3.10)

C   NC   N/A   GIRDERS: Girders supported by walls or pilasters shall have at least two ties securing the anchor bolts for Life Safety and Immediate Occupancy. (Tier 2: Sec. 4.6.4.2)

C   NC   N/A   CORBEL BEARING: If the frame girders bear on column corbels, the length of bearing shall be greater than 3 inches for Life Safety and Immediate Occupancy. (Tier 2: Sec. 4.6.4.3)

C   NC   N/A   CORBEL CONNECTIONS: The frame girders shall not be connected to corbels with welded elements. (Tier 2: Sec. 4.6.4.4)

3.7.13 Basic Structural Checklist for Building Type RM1: Reinforced Masonry Bearing Walls with Flexible Diaphragms

This Basic Structural Checklist shall be completed where required by Table 3-2.

Each of the evaluation statements on this checklist shall be marked Compliant (C), Non-compliant (NC), or Not Applicable (N/A) for a Tier 1 Evaluation. Compliant statements identify issues that are acceptable according to the criteria of this standard, while non-compliant statements identify issues that require further investigation. Certain statements may not apply to the buildings being evaluated. For non-compliant evaluation statements, the design professional may choose to conduct further investigation using the corresponding Tier 2 Evaluation procedure; corresponding section numbers are in parentheses following each evaluation statement.

C3.7.13 Basic Structural Checklist for Building Type RM1

These buildings have bearing walls that consist of reinforced brick or concrete block masonry. Wood floor and roof framing consists of wood joists, glulam beams, and wood posts or small steel columns. Steel floor and roof framing consists of steel beams or open web joists, steel girders, and steel columns. Lateral forces are resisted by the reinforced brick or concrete block masonry shear walls. Diaphragms consist of straight or diagonal wood sheathing, plywood, or untopped metal deck, and are flexible relative to the walls. Foundations consist of brick or concrete spread footings or deep foundations.

Building System

C NC N/A LOAD PATH: The structure shall contain a minimum of one complete load path for Life Safety and Immediate Occupancy for seismic force effects from any horizontal direction that serves to transfer the inertial forces from the mass to the foundation. (Tier 2: Sec. 4.3.1.1)

C NC N/A ADJACENT BUILDINGS: The clear distance between the building being evaluated and any adjacent building shall be greater than 4 percent of the height of the shorter building for Life Safety and Immediate Occupancy. (Tier 2: Sec. 4.3.1.2)

C NC N/A MEZZANINES: Interior mezzanine levels shall be braced independently from the main structure, or shall be anchored to the lateral-force-resisting elements of the main structure. (Tier 2: Sec. 4.3.1.3)

C NC N/A WEAK STORY: The strength of the lateral-force-resisting system in any story shall not be less than 80 percent of the strength in an adjacent story, above or below, for Life Safety and Immediate Occupancy. (Tier 2: Sec. 4.3.2.1)

C NC N/A SOFT STORY: The stiffness of the lateral-force-resisting system in any story shall not be less than 70 percent of the lateral-force-resisting system stiffness in an adjacent story above or below, or less than 80 percent of the average lateral-force-resisting system stiffness of the three stories above or below for Life Safety and Immediate Occupancy. (Tier 2: Sec. 4.3.2.2)

C NC N/A GEOMETRY: There shall be no changes in horizontal dimension of the lateral-force-resisting system of more than 30 percent in a story relative to adjacent stories for Life Safety and Immediate Occupancy, excluding one-story penthouses and mezzanines. (Tier 2: Sec. 4.3.2.3)

C NC N/A VERTICAL DISCONTINUITIES: All vertical elements in the lateral-force-resisting system shall be continuous to the foundation. (Tier 2: Sec. 4.3.2.4)

Screening Phase (Tier 1)

C NC N/A **MASS:** There shall be no change in effective mass more than 50 percent from one story to the next for Life Safety and Immediate Occupancy. Light roofs, penthouses, and mezzanines need not be considered. (Tier 2: Sec. 4.3.2.5)

C NC N/A **DETERIORATION OF WOOD:** There shall be no signs of decay, shrinkage, splitting, fire damage, or sagging in any of the wood members, and none of the metal connection hardware shall be deteriorated, broken, or loose. (Tier 2: Sec. 4.3.3.1)

C NC N/A **MASONRY UNITS:** There shall be no visible deterioration of masonry units. (Tier 2: Sec. 4.3.3.7)

C NC N/A **MASONRY JOINTS:** The mortar shall not be easily scraped away from the joints by hand with a metal tool, and there shall be no areas of eroded mortar. (Tier 2: Sec. 4.3.3.8)

C NC N/A **REINFORCED MASONRY WALL CRACKS:** All existing diagonal cracks in wall elements shall be less than 1/8 inch for Life Safety and 1/16 inch for Immediate Occupancy, shall not be concentrated in one location, and shall not form an X pattern. (Tier 2: Sec. 4.3.3.10)

Lateral-Force-Resisting System

C NC N/A **REDUNDANCY:** The number of lines of shear walls in each principal direction shall be greater than or equal to 2 for Life Safety and Immediate Occupancy. (Tier 2: Sec. 4.4.2.1.1)

C NC N/A **SHEAR STRESS CHECK:** The shear stress in the reinforced masonry shear walls, calculated using the Quick Check procedure of Section 3.5.3.3, shall be less than 70 psi for Life Safety and Immediate Occupancy. (Tier 2: Sec. 4.4.2.4.1)

C NC N/A **REINFORCING STEEL:** The total vertical and horizontal reinforcing steel ratio in reinforced masonry walls shall be greater than 0.002 for Life Safety and Immediate Occupancy of the wall with the minimum of 0.0007 for Life Safety and Immediate Occupancy in either of the two directions; the spacing of reinforcing steel shall be less than 48 inches for Life Safety and Immediate Occupancy; and all vertical bars shall extend to the top of the walls. (Tier 2: Sec. 4.4.2.4.2)

Connections

C NC N/A **WALL ANCHORAGE:** Exterior concrete or masonry walls that are dependent on the diaphragm for lateral support shall be anchored for out-of-plane forces at each diaphragm level with steel anchors, reinforcing dowels, or straps that are developed into the diaphragm. Connections shall have adequate strength to resist the connection force calculated in the Quick Check procedure of Section 3.5.3.7. (Tier 2: Sec. 4.6.1.1)

C NC N/A **WOOD LEDGERS:** The connection between the wall panels and the diaphragm shall not induce cross-grain bending or tension in the wood ledgers. (Tier 2: Sec. 4.6.1.2)

C NC N/A **TRANSFER TO SHEAR WALLS:** Diaphragms shall be connected for transfer of loads to the shear walls for Life Safety and the connections shall be able to develop the lesser of the shear strength of the walls or diaphragms for Immediate Occupancy. (Tier 2: Sec. 4.6.2.1)

C NC N/A **FOUNDATION DOWELS:** Wall reinforcement shall be doweled into the foundation for Life Safety, and the dowels shall be able to develop the lesser of the strength of the walls or the uplift capacity of the foundation for Immediate Occupancy. (Tier 2: Sec. 4.6.3.5)

C NC N/A **GIRDER/COLUMN CONNECTION:** There shall be a positive connection utilizing plates, connection hardware, or straps between the girder and the column support. (Tier 2: Sec. 4.6.4.1)

Screening Phase (Tier 1)

3.7.13S Supplemental Structural Checklist for Building Type RM1: Reinforced Masonry Bearing Walls with Flexible Diaphragms

This Supplemental Structural Checklist shall be completed where required by Table 3-2. The Basic Structural Checklist shall be completed prior to completing this Supplemental Structural Checklist.

Lateral-Force-Resisting System

C   NC   N/A   REINFORCING AT OPENINGS: All wall openings that interrupt rebar shall have trim reinforcing on all sides. This statement shall apply to the Immediate Occupancy Performance Level only. (Tier 2: Sec. 4.4.2.4.3)

C   NC   N/A   PROPORTIONS: The height-to-thickness ratio of the shear walls at each story shall be less than 30. This statement shall apply to the Immediate Occupancy Performance Level only. (Tier 2: Sec. 4.4.2.4.4)

Diaphragms

C   NC   N/A   CROSS TIES: There shall be continuous cross ties between diaphragm chords. (Tier 2: Sec. 4.5.1.2)

C   NC   N/A   OPENINGS AT SHEAR WALLS: Diaphragm openings immediately adjacent to the shear walls shall be less than 25 percent of the wall length for Life Safety and 15 percent of the wall length for Immediate Occupancy. (Tier 2: Sec. 4.5.1.4)

C   NC   N/A   OPENINGS AT EXTERIOR MASONRY SHEAR WALLS: Diaphragm openings immediately adjacent to exterior masonry shear walls shall not be greater than 8 feet long for Life Safety and 4 feet long for Immediate Occupancy. (Tier 2: Sec. 4.5.1.6)

C   NC   N/A   PLAN IRREGULARITIES: There shall be tensile capacity to develop the strength of the diaphragm at re-entrant corners or other locations of plan irregularities. This statement shall apply to the Immediate Occupancy Performance Level only. (Tier 2: Sec. 4.5.1.7)

C   NC   N/A   DIAPHRAGM REINFORCEMENT AT OPENINGS: There shall be reinforcing around all diaphragm openings larger than 50 percent of the building width in either major plan dimension. This statement shall apply to the Immediate Occupancy Performance Level only. (Tier 2: Sec. 4.5.1.8)

C   NC   N/A   STRAIGHT SHEATHING: All straight sheathed diaphragms shall have aspect ratios less than 2-to-1 for Life Safety and 1-to-1 for Immediate Occupancy in the direction being considered. (Tier 2: Sec. 4.5.2.1)

C   NC   N/A   SPANS: All wood diaphragms with spans greater than 24 feet for Life Safety and 12 feet for Immediate Occupancy shall consist of wood structural panels or diagonal sheathing. (Tier 2: Sec. 4.5.2.2)

C   NC   N/A   UNBLOCKED DIAPHRAGMS: All diagonally sheathed or unblocked wood structural panel diaphragms shall have horizontal spans less than 40 feet for Life Safety and 30 feet for Immediate Occupancy and shall have aspect ratios less than or equal to 4-to-1 for Life Safety and 3-to-1 for Immediate Occupancy. (Tier 2: Sec. 4.5.2.3)

C   NC   N/A   NON-CONCRETE FILLED DIAPHRAGMS: Untopped metal deck diaphragms or metal deck diaphragms with fill other than concrete shall consist of horizontal spans of less than 40 feet and shall have span/depth ratios less than 4-to-1. This statement shall apply to the Immediate Occupancy Performance Level only. (Tier 2: Sec. 4.5.3.1)

Screening Phase (Tier 1)

C NC N/A OTHER DIAPHRAGMS: The diaphragm shall not consist of a system other than wood, metal deck, concrete, or horizontal bracing. (Tier 2: Sec. 4.5.7.1)

Connections

C NC N/A STIFFNESS OF WALL ANCHORS: Anchors of concrete or masonry walls to wood structural elements shall be installed taut and shall be stiff enough to limit the relative movement between the wall and the diaphragm to no greater than 1/8 inch prior to engagement of the anchors. (Tier 2: Sec. 4.6.1.4)

Screening Phase (Tier 1)

3.7.14 Basic Structural Checklist for Building Type RM2: Reinforced Masonry Bearing Walls with Stiff Diaphragms

This Basic Structural Checklist shall be completed where required by Table 3-2.

Each of the evaluation statements on this checklist shall be marked Compliant (C), Non-compliant (NC), or Not Applicable (N/A) for a Tier 1 Evaluation. Compliant statements identify issues that are acceptable according to the criteria of this standard, while non-compliant statements identify issues that require further investigation. Certain statements may not apply to the buildings being evaluated. For non-compliant evaluation statements, the design professional may choose to conduct further investigation using the corresponding Tier 2 Evaluation procedure; corresponding section numbers are in parentheses following each evaluation statement.

C3.7.14 Basic Structural Checklist for Building Type RM2

These buildings have bearing walls that consist of reinforced brick or concrete block masonry. Diaphragms consist of metal deck with concrete fill, precast concrete planks, tees, or double-tees, with or without a cast-in-place concrete topping slab, and are stiff relative to the walls. The floor and roof framing is supported on interior steel or concrete frames or interior reinforced masonry walls.

Building System

C NC N/A   **LOAD PATH:** The structure shall contain a minimum of one complete load path for Life Safety and Immediate Occupancy for seismic force effects from any horizontal direction that serves to transfer the inertial forces from the mass to the foundation. (Tier 2: Sec. 4.3.1.1)

C NC N/A   **MEZZANINES:** Interior mezzanine levels shall be braced independently from the main structure, or shall be anchored to the lateral-force-resisting elements of the main structure. (Tier 2: Sec. 4.3.1.3)

C NC N/A   **WEAK STORY:** The strength of the lateral-force-resisting system in any story shall not be less than 80 percent of the strength in an adjacent story, above or below, for Life Safety and Immediate Occupancy. (Tier 2: Sec. 4.3.2.1)

C NC N/A   **SOFT STORY:** The stiffness of the lateral-force-resisting system in any story shall not be less than 70 percent of the lateral-force-resisting system stiffness in an adjacent story above or below, or less than 80 percent of the average lateral-force-resisting system stiffness of the three stories above or below for Life Safety and Immediate Occupancy. (Tier 2: Sec. 4.3.2.2)

C NC N/A   **GEOMETRY:** There shall be no changes in horizontal dimension of the lateral-force-resisting system of more than 30 percent in a story relative to adjacent stories for Life Safety and Immediate Occupancy, excluding one-story penthouses and mezzanines. (Tier 2: Sec. 4.3.2.3)

C NC N/A   **VERTICAL DISCONTINUITIES:** All vertical elements in the lateral-force-resisting system shall be continuous to the foundation. (Tier 2: Sec. 4.3.2.4)

C NC N/A   **MASS:** There shall be no change in effective mass more than 50 percent from one story to the next for Life Safety and Immediate Occupancy. Light roofs, penthouses, and mezzanines need not be considered. (Tier 2: Sec. 4.3.2.5)

C NC N/A   **TORSION:** The estimated distance between the story center of mass and the story center of rigidity shall be less than 20 percent of the building width in either plan dimension for Life Safety and Immediate Occupancy. (Tier 2: Sec. 4.3.2.6)

Screening Phase (Tier 1)

C NC N/A DETERIORATION OF CONCRETE: There shall be no visible deterioration of concrete or reinforcing steel in any of the vertical- or lateral-force-resisting elements. (Tier 2: Sec. 4.3.3.4)

C NC N/A MASONRY UNITS: There shall be no visible deterioration of masonry units. (Tier 2: Sec. 4.3.3.7)

C NC N/A MASONRY JOINTS: The mortar shall not be easily scraped away from the joints by hand with a metal tool, and there shall be no areas of eroded mortar. (Tier 2: Sec. 4.3.3.8)

C NC N/A REINFORCED MASONRY WALL CRACKS: All existing diagonal cracks in wall elements shall be less than 1/8 inch for Life Safety and 1/16 inch for Immediate Occupancy, shall not be concentrated in one location, and shall not form an X pattern. (Tier 2: Sec. 4.3.3.10)

Lateral-Force-Resisting System

C NC N/A REDUNDANCY: The number of lines of shear walls in each principal direction shall be greater than or equal to 2 for Life Safety and Immediate Occupancy. (Tier 2: Sec. 4.4.2.1.1)

C NC N/A SHEAR STRESS CHECK: The shear stress in the reinforced masonry shear walls, calculated using the Quick Check procedure of Section 3.5.3.3, shall be less than 70 psi for Life Safety and Immediate Occupancy. (Tier 2: Sec. 4.4.2.4.1)

C NC N/A REINFORCING STEEL: The total vertical and horizontal reinforcing steel ratio in reinforced masonry walls shall be greater than 0.002 for Life Safety and Immediate Occupancy of the wall with the minimum of 0.0007 for Life Safety and Immediate Occupancy in either of the two directions; the spacing of reinforcing steel shall be less than 48 inches for Life Safety and Immediate Occupancy; and all vertical bars shall extend to the top of the walls. (Tier 2: Sec. 4.4.2.4.2)

Diaphragms

C NC N/A TOPPING SLAB: Precast concrete diaphragm elements shall be interconnected by a continuous reinforced concrete topping slab. (Tier 2: Sec. 4.5.5.1)

Connections

C NC N/A WALL ANCHORAGE: Exterior concrete or masonry walls that are dependent on the diaphragm for lateral support shall be anchored for out-of-plane forces at each diaphragm level with steel anchors, reinforcing dowels, or straps that are developed into the diaphragm. Connections shall have adequate strength to resist the connection force calculated in the Quick Check procedure of Section 3.5.3.7. (Tier 2: Sec. 4.6.1.1)

C NC N/A TRANSFER TO SHEAR WALLS: Diaphragms shall be connected for transfer of loads to the shear walls for Life Safety and the connections shall be able to develop the lesser of the shear strength of the walls or diaphragms for Immediate Occupancy. (Tier 2: Sec. 4.6.2.1)

C NC N/A TOPPING SLAB TO WALLS OR FRAMES: Reinforced concrete topping slabs that interconnect the precast concrete diaphragm elements shall be doweled for transfer of forces into the shear wall or frame elements for Life Safety, and the dowels shall be able to develop the lesser of the shear strength of the walls, frames, or slabs for Immediate Occupancy. (Tier 2: Sec. 4.6.2.3)

C NC N/A FOUNDATION DOWELS: Wall reinforcement shall be doweled into the foundation for Life Safety, and the dowels shall be able to develop the lesser of the strength of the walls or the uplift capacity of the foundation for Immediate Occupancy. (Tier 2: Sec. 4.6.3.5)

C NC N/A GIRDER/COLUMN CONNECTION: There shall be a positive connection utilizing plates, connection hardware, or straps between the girder and the column support. (Tier 2: Sec. 4.6.4.1)

3.7.14S Supplemental Structural Checklist for Building Type RM2: Reinforced Masonry Bearing Walls with Stiff Diaphragms

This Supplemental Structural Checklist shall be completed where required by Table 3-2. The Basic Structural Checklist shall be completed prior to completing this Supplemental Structural Checklist.

Lateral-Force-Resisting System

C   NC   N/A   REINFORCING AT OPENINGS: There shall be added trim reinforcement around all wall openings with a dimension greater than three times the thickness of the wall. This statement shall apply to the Immediate Occupancy Performance Level only. (Tier 2: Sec. 4.4.2.2.6)

C   NC   N/A   PROPORTIONS: The height-to-thickness ratio of the shear walls at each story shall be less than 30. This statement shall apply to the Immediate Occupancy Performance Level only. (Tier 2: Sec. 4.4.2.4.4)

Diaphragms

C   NC   N/A   OPENINGS AT SHEAR WALLS: Diaphragm openings immediately adjacent to the shear walls shall be less than 25 percent of the wall length for Life Safety and 15 percent of the wall length for Immediate Occupancy. (Tier 2: Sec. 4.5.1.4)

C   NC   N/A   OPENINGS AT EXTERIOR MASONRY SHEAR WALLS: Diaphragm openings immediately adjacent to exterior masonry shear walls shall not be greater than 8 feet long for Life Safety and 4 feet long for Immediate Occupancy. (Tier 2: Sec. 4.5.1.6)

C   NC   N/A   PLAN IRREGULARITIES: There shall be tensile capacity to develop the strength of the diaphragm at re-entrant corners or other locations of plan irregularities. This statement shall apply to the Immediate Occupancy Performance Level only. (Tier 2: Sec. 4.5.1.7)

C   NC   N/A   DIAPHRAGM REINFORCEMENT AT OPENINGS: There shall be reinforcing around all diaphragm openings larger than 50 percent of the building width in either major plan dimension. This statement shall apply to the Immediate Occupancy Performance Level only. (Tier 2: Sec. 4.5.1.8)

3.7.15 Basic Structural Checklist for Building Type URM: Unreinforced Masonry Bearing Walls with Flexible Diaphragms

This Basic Structural Checklist shall be completed where required by Table 3-2.

Each of the evaluation statements on this checklist shall be marked Compliant (C), Non-compliant (NC), or Not Applicable (N/A) for a Tier 1 Evaluation. Compliant statements identify issues that are acceptable according to the criteria of this standard, while non-compliant statements identify issues that require further investigation. Certain statements may not apply to the buildings being evaluated. For non-compliant evaluation statements, the design professional may choose to conduct further investigation using the Tier 2 Special Procedure for Unreinforced Masonry or the Tier 3 Evaluation Procedure.

> **C3.7.15 Basic Structural Checklist for Building Type URM**
>
> These buildings have bearing walls that consist of unreinforced (or lightly reinforced) brick, stone, or concrete block masonry. Wood floor and roof framing consists of wood joists, glulam beams, and wood posts or small steel columns. Steel floor and roof framing consists of steel beams or open web joists, steel girders, and steel columns. Lateral forces are resisted by the brick or concrete block masonry shear walls. Diaphragms consist of straight or diagonal lumber sheathing, structural wood panels, or untopped metal deck, and are flexible relative to the walls. Foundations consist of brick or concrete spread footings or deep foundations.

Building System

C   NC   N/A   **LOAD PATH:** The structure shall contain a minimum of one complete load path for Life Safety and Immediate Occupancy for seismic force effects from any horizontal direction that serves to transfer the inertial forces from the mass to the foundation. (Tier 2: Sec. 4.3.1.1)

C   NC   N/A   **ADJACENT BUILDINGS:** The clear distance between the building being evaluated and any adjacent building shall be greater than 4 percent of the height of the shorter building for Life Safety and Immediate Occupancy. (Tier 2: Sec. 4.3.1.2)

C   NC   N/A   **MEZZANINES:** Interior mezzanine levels shall be braced independently from the main structure, or shall be anchored to the lateral-force-resisting elements of the main structure. (Tier 2: Sec. 4.3.1.3)

C   NC   N/A   **WEAK STORY:** The strength of the lateral-force-resisting system in any story shall not be less than 80 percent of the strength in an adjacent story, above or below, for Life Safety and Immediate Occupancy. (Tier 2: Sec. 4.3.2.1)

C   NC   N/A   **SOFT STORY:** The stiffness of the lateral-force-resisting system in any story shall not be less than 70 percent of the lateral-force-resisting system stiffness in an adjacent story above or below, or less than 80 percent of the average lateral-force-resisting system stiffness of the three stories above or below for Life Safety and Immediate Occupancy. (Tier 2: Sec. 4.3.2.2)

C   NC   N/A   **GEOMETRY:** There shall be no changes in horizontal dimension of the lateral-force-resisting system of more than 30 percent in a story relative to adjacent stories for Life Safety and Immediate Occupancy, excluding one-story penthouses and mezzanines. (Tier 2: Sec. 4.3.2.3)

C   NC   N/A   **VERTICAL DISCONTINUITIES:** All vertical elements in the lateral-force-resisting system shall be continuous to the foundation. (Tier 2: Sec. 4.3.2.4)

Screening Phase (Tier 1)

C NC N/A   MASS: There shall be no change in effective mass more than 50 percent from one story to the next for Life Safety and Immediate Occupancy. Light roofs, penthouses, and mezzanines need not be considered. (Tier 2: Sec. 4.3.2.5)

C NC N/A   DETERIORATION OF WOOD: There shall be no signs of decay, shrinkage, splitting, fire damage, or sagging in any of the wood members, and none of the metal connection hardware shall be deteriorated, broken, or loose. (Tier 2: Sec. 4.3.3.1)

C NC N/A   MASONRY UNITS: There shall be no visible deterioration of masonry units. (Tier 2: Sec. 4.3.3.7)

C NC N/A   MASONRY JOINTS: The mortar shall not be easily scraped away from the joints by hand with a metal tool, and there shall be no areas of eroded mortar. (Tier 2: Sec. 4.3.3.8)

C NC N/A   UNREINFORCED MASONRY WALL CRACKS: There shall be no existing diagonal cracks in the wall elements greater than 1/8 inch for Life Safety and 1/16 inch for Immediate Occupancy, or out-of-plane offsets in the bed joint greater than 1/8 inch for Life Safety and 1/16 inch for Immediate Occupancy, and shall not form an X pattern. (Tier 2: Sec. 4.3.3.11)

Lateral-Force-Resisting System

C NC N/A   REDUNDANCY: The number of lines of shear walls in each principal direction shall be greater than or equal to 2 for Life Safety and Immediate Occupancy. (Tier 2: Sec. 4.4.2.1.1)

C NC N/A   SHEAR STRESS CHECK: The shear stress in the unreinforced masonry shear walls, calculated using the Quick Check procedure of Section 3.5.3.3, shall be less than 30 psi for clay units and 70 psi for concrete units for Life Safety and Immediate Occupancy. (Tier 2: Sec. 4.4.2.5.1)

Connections

C NC N/A   WALL ANCHORAGE: Exterior concrete or masonry walls that are dependent on the diaphragm for lateral support shall be anchored for out-of-plane forces at each diaphragm level with steel anchors, reinforcing dowels, or straps that are developed into the diaphragm. Connections shall have adequate strength to resist the connection force calculated in the Quick Check procedure of Section 3.5.3.7. (Tier 2: Sec. 4.6.1.1)

C NC N/A   WOOD LEDGERS: The connection between the wall panels and the diaphragm shall not induce cross-grain bending or tension in the wood ledgers. (Tier 2: Sec. 4.6.1.2)

C NC N/A   TRANSFER TO SHEAR WALLS: Diaphragms shall be connected for transfer of loads to the shear walls for Life Safety and the connections shall be able to develop the lesser of the shear strength of the walls or diaphragms for Immediate Occupancy. (Tier 2 Sec. 4.6.2.1)

C NC N/A   GIRDER/COLUMN CONNECTION: There shall be a positive connection utilizing plates, connection hardware, or straps between the girder and the column support. (Tier 2: Sec. 4.6.4.1)

3.7.15S Supplemental Structural Checklist for Building Type URM: Unreinforced Masonry Bearing Walls with Flexible Diaphragms

This Supplemental Structural Checklist shall be completed where required by Table 3-2. The Basic Structural Checklist shall be completed prior to completing this Supplemental Structural Checklist.

Lateral-Force-Resisting System

C   NC   N/A   PROPORTIONS: The height-to-thickness ratio of the shear walls at each story shall be less than the following for Life Safety and Immediate Occupancy (Tier 2: Sec. 4.4.2.5.2):

Top story of multi-story building	9
First story of multi-story building	15
All other conditions	13

C   NC   N/A   MASONRY LAY-UP: Filled collar joints of multi-wythe masonry walls shall have negligible voids. (Tier 2: Sec. 4.4.2.5.3)

Diaphragms

C   NC   N/A   CROSS TIES: There shall be continuous cross ties between diaphragm chords. (Tier 2: Sec. 4.5.1.2)

C   NC   N/A   OPENINGS AT SHEAR WALLS: Diaphragm openings immediately adjacent to the shear walls shall be less than 25 percent of the wall length for Life Safety and 15 percent of the wall length for Immediate Occupancy. (Tier 2: Sec. 4.5.1.4)

C   NC   N/A   OPENINGS AT EXTERIOR MASONRY SHEAR WALLS: Diaphragm openings immediately adjacent to exterior masonry shear walls shall not be greater than 8 feet long for Life Safety and 4 feet long for Immediate Occupancy. (Tier 2: Sec. 4.5.1.6)

C   NC   N/A   PLAN IRREGULARITIES: There shall be tensile capacity to develop the strength of the diaphragm at re-entrant corners or other locations of plan irregularities. This statement shall apply to the Immediate Occupancy Performance Level only. (Tier 2: Sec. 4.5.1.7)

C   NC   N/A   DIAPHRAGM REINFORCEMENT AT OPENINGS: There shall be reinforcing around all diaphragm openings larger than 50 percent of the building width in either major plan dimension. This statement shall apply to the Immediate Occupancy Performance Level only. (Tier 2: Sec. 4.5.1.8)

C   NC   N/A   STRAIGHT SHEATHING: All straight sheathed diaphragms shall have aspect ratios less than 2-to-1 for Life Safety and 1-to-1 for Immediate Occupancy in the direction being considered. (Tier 2: Sec. 4.5.2.1)

C   NC   N/A   SPANS: All wood diaphragms with spans greater than 24 feet for Life Safety and 12 feet for Immediate Occupancy shall consist of wood structural panels or diagonal sheathing (Tier 2: Sec. 4.5.2.2)

C   NC   N/A   UNBLOCKED DIAPHRAGMS: All diagonally sheathed or unblocked wood structural panel diaphragms shall have horizontal spans less than 40 feet for Life Safety and 30 feet for Immediate Occupancy and shall have aspect ratios less than or equal to 4-to-1 for Life Safety and 3-to-1 for Immediate Occupancy. (Tier 2: Sec. 4.5.2.3)

C   NC   N/A   NON-CONCRETE FILLED DIAPHRAGMS: Untopped metal deck diaphragms or metal deck diaphragms with fill other than concrete shall consist of horizontal spans of less than 40 feet and shall have span/depth ratios less than 4-to-1. This statement shall apply to the Immediate Occupancy Performance Level only. (Tier 2: Sec. 4.5.3.1)

Screening Phase (Tier 1)

C NC N/A OTHER DIAPHRAGMS: The diaphragm shall not consist of a system other than wood, metal deck, concrete, or horizontal bracing. (Tier 2: Sec. 4.5.7.1)

Connections

C NC N/A STIFFNESS OF WALL ANCHORS: Anchors of concrete or masonry walls to wood structural elements shall be installed taut and shall be stiff enough to limit the relative movement between the wall and the diaphragm to no greater than 1/8 inch prior to engagement of the anchors. (Tier 2: Sec. 4.6.1.4)

C NC N/A BEAM, GIRDER, AND TRUSS SUPPORTS: Beams, girders, and trusses supported by unreinforced masonry walls or pilasters shall have independent secondary columns for support of vertical loads. (Tier 2: Sec. 4.6.4.5)

Screening Phase (Tier 1)

3.7.15A Basic Structural Checklist for Building Type URMA: Unreinforced Masonry Bearing Walls with Stiff Diaphragms

This Basic Structural Checklist shall be completed where required by Table 3-2.

Each of the evaluation statements on this checklist shall be marked Compliant (C), Non-compliant (NC), or Not Applicable (N/A) for a Tier 1 Evaluation. Compliant statements identify issues that are acceptable according to the criteria of this standard, while non-compliant statements identify issues that require further investigation. Certain statements may not apply to the buildings being evaluated. For non-compliant evaluation statements, the design professional may choose to conduct further investigation using the corresponding Tier 2 Evaluation procedure; corresponding section numbers are in parentheses following each evaluation statement.

> **C3.7.15A Basic Structural Checklist for Building Type URMA**
>
> These buildings have perimeter bearing walls that consist of unreinforced clay brick, stone, or concrete masonry. Interior bearing walls, where present, also consist of unreinforced clay brick, stone, or concrete masonry. Diaphragms are stiff relative to the unreinforced masonry walls and interior framing. In older construction or large, multi-story buildings, diaphragms consist of cast-in-place concrete. In levels of low seismicity, more recent construction consists of metal deck and concrete fill supported on steel framing.

Building System

C   NC   N/A   **LOAD PATH:** The structure shall contain a minimum of one complete load path for Life Safety and Immediate Occupancy for seismic force effects from any horizontal direction that serves to transfer the inertial forces from the mass to the foundation. (Tier 2: Sec. 4.3.1.1)

C   NC   N/A   **MEZZANINES:** Interior mezzanine levels shall be braced independently from the main structure, or shall be anchored to the lateral-force-resisting elements of the main structure. (Tier 2: Sec. 4.3.1.3)

C   NC   N/A   **WEAK STORY:** The strength of the lateral-force-resisting system in any story shall not be less than 80 percent of the strength in an adjacent story, above or below, for Life Safety and Immediate Occupancy. (Tier 2: Sec. 4.3.2.1)

C   NC   N/A   **SOFT STORY:** The stiffness of the lateral-force-resisting system in any story shall not be less than 70 percent of the lateral-force-resisting system stiffness in an adjacent story above or below, or less than 80 percent of the average lateral-force-resisting system stiffness of the three stories above or below for Life Safety and Immediate Occupancy. (Tier 2: Sec. 4.3.2.2)

C   NC   N/A   **GEOMETRY:** There shall be no changes in horizontal dimension of the lateral-force-resisting system of more than 30 percent in a story relative to adjacent stories for Life Safety and Immediate Occupancy, excluding one-story penthouses and mezzanines. (Tier 2: Sec. 4.3.2.3)

C   NC   N/A   **VERTICAL DISCONTINUITIES:** All vertical elements in the lateral-force-resisting system shall be continuous to the foundation. (Tier 2: Sec. 4.3.2.4)

C   NC   N/A   **MASS:** There shall be no change in effective mass more than 50 percent from one story to the next for Life Safety and Immediate Occupancy. Light roofs, penthouses, and mezzanines need not be considered. (Tier 2: Sec. 4.3.2.5)

Screening Phase (Tier 1)

C NC N/A   **TORSION:** The estimated distance between the story center of mass and the story center of rigidity shall be less than 20 percent of the building width in either plan dimension for Life Safety and Immediate Occupancy. (Tier 2: Sec. 4.3.2.6)

C NC N/A   **DETERIORATION OF CONCRETE:** There shall be no visible deterioration of concrete or reinforcing steel in any of the vertical- or lateral-force-resisting elements. (Tier 2: Sec. 4.3.3.4)

C NC N/A   **MASONRY UNITS:** There shall be no visible deterioration of masonry units. (Tier 2: Sec. 4.3.3.7)

C NC N/A   **MASONRY JOINTS:** The mortar shall not be easily scraped away from the joints by hand with a metal tool, and there shall be no areas of eroded mortar. (Tier 2: Sec. 4.3.3.8)

C NC N/A   **UNREINFORCED MASONRY WALL CRACKS:** There shall be no existing diagonal cracks in wall elements greater than 1/8 inch for Life Safety and 1/16 inch for Immediate Occupancy or out-of-plane offsets in the bed joint greater than 1/8 inch for Life Safety and 1/16 inch for Immediate Occupancy, and shall not form an X pattern. (Tier 2: Sec. 4.3.3.11)

Lateral-Force-Resisting System

C NC N/A   **REDUNDANCY:** The number of lines of shear walls in each principal direction shall be greater than or equal to 2 for Life Safety and Immediate Occupancy. (Tier 2: Sec. 4.4.2.1.1)

C NC N/A   **SHEAR STRESS CHECK:** The shear stress in the unreinforced masonry shear walls, calculated using the Quick Check procedure of Section 3.5.3.3, shall be less than 30 psi for clay units and 70 psi for concrete units for Life Safety and Immediate Occupancy. (Tier 2: Sec. 4.4.2.5.1)

Connections

C NC N/A   **WALL ANCHORAGE:** Exterior concrete or masonry walls that are dependent on the diaphragm for lateral support shall be anchored for out-of-plane forces at each diaphragm level with steel anchors, reinforcing dowels, or straps that are developed into the diaphragm. Connections shall have adequate strength to resist the connection force calculated in the Quick Check procedure of Section 3.5.3.7. (Tier 2: Sec. 4.6.1.1)

C NC N/A   **TRANSFER TO SHEAR WALLS:** Diaphragms shall be connected for transfer of loads to the shear walls for Life Safety and the connections shall be able to develop the lesser of the shear strength of the walls or diaphragms for Immediate Occupancy. (Tier 2: Sec. 4.6.2.1)

C NC N/A   **GIRDER/COLUMN CONNECTION:** There shall be a positive connection utilizing plates, connection hardware, or straps between the girder and the column support. (Tier 2: Sec. 4.6.4.1)

Screening Phase (Tier 1)

3.7.15AS Supplemental Structural Checklist for Building Type URMA: Unreinforced Masonry Bearing Walls with Stiff Diaphragms

This Supplemental Structural Checklist shall be completed where required by Table 3-2. The Basic Structural Checklist shall be completed prior to completing this Supplemental Structural Checklist.

Lateral-Force-Resisting System

C NC N/A   PROPORTIONS: The height-to-thickness ratio of the shear walls at each story shall be less than the following for Life Safety and Immediate Occupancy (Tier 2: Sec. 4.4.2.5.2):

Top story of multi-story building	9
First story of multi-story building	15
All other conditions	13

C NC N/A   MASONRY LAY-UP: Filled collar joints of multi-wythe masonry walls shall have negligible voids. (Tier 2: Sec. 4.4.2.5.3)

Diaphragms

General

C NC N/A   OPENINGS AT SHEAR WALLS: Diaphragm openings immediately adjacent to the shear walls shall be less than 25 percent of the wall length for Life Safety and 15 percent of the wall length for Immediate Occupancy. (Tier 2: Sec. 4.5.1.4)

C NC N/A   OPENINGS AT EXTERIOR MASONRY SHEAR WALLS: Diaphragm openings immediately adjacent to exterior masonry shear walls shall not be greater than 8 feet long for Life Safety and 4 feet long for Immediate Occupancy. (Tier 2: Sec. 4.5.1.6)

C NC N/A   PLAN IRREGULARITIES: There shall be tensile capacity to develop the strength of the diaphragm at re-entrant corners or other locations of plan irregularities. This statement shall apply to the Immediate Occupancy Performance Level only. (Tier 2: Sec. 4.5.1.7)

C NC N/A   DIAPHRAGM REINFORCEMENT AT OPENINGS: There shall be reinforcing around all diaphragm openings larger than 50 percent of the building width in either major plan dimension. This statement shall apply to the Immediate Occupancy Performance Level only. (Tier 2: Sec. 4.5.1.8)

Connections

C NC N/A   BEAM, GIRDER, AND TRUSS SUPPORTS: Beams, girders, and trusses supported by unreinforced masonry walls or pilasters shall have independent secondary columns for support of vertical loads. (Tier 2: Sec. 4.6.4.5)

Screening Phase (Tier 1)

3.7.16 General Basic Structural Checklist

This General Basic Structural Checklist shall be completed where required by Table 3-2.

Each of the evaluation statements on this checklist shall be marked Compliant (C), Non-compliant (NC), or Not Applicable (N/A) for a Tier 1 Evaluation. Compliant statements identify issues that are acceptable according to the criteria of this standard, while non-compliant statements identify issues that require further investigation. Certain statements may not apply to the buildings being evaluated. For non-compliant evaluation statements, the design professional may choose to conduct further investigation using the corresponding Tier 2 Evaluation procedure; corresponding section numbers are in parentheses following each evaluation statement.

BUILDING SYSTEM

General

C NC N/A LOAD PATH: The structure shall contain a minimum of one complete load path for Life Safety and Immediate Occupancy for seismic force effects from any horizontal direction that serves to transfer the inertial forces from the mass to the foundation. (Tier 2: Sec. 4.3.1.1)

C NC N/A ADJACENT BUILDINGS: The clear distance between the building being evaluated and any adjacent building shall be greater than 4 percent of the height of the shorter building for Life Safety and Immediate Occupancy. (Tier 2: Sec. 4.3.1.2)

C NC N/A MEZZANINES: Interior mezzanine levels shall be braced independently from the main structure, or shall be anchored to the lateral-force-resisting elements of the main structure. (Tier 2: Sec. 4.3.1.3)

Configuration

C NC N/A WEAK STORY: The strength of the lateral-force-resisting system in any story shall not be less than 80 percent of the strength in an adjacent story, above or below, for Life Safety and Immediate Occupancy. (Tier 2: Sec. 4.3.2.1)

C NC N/A SOFT STORY: The stiffness of the lateral-force-resisting system in any story shall not be less than 70 percent of the lateral-force-resisting system stiffness in an adjacent story above or below, or less than 80 percent of the average lateral-force-resisting system stiffness of the three stories above or below for Life Safety and Immediate Occupancy. (Tier 2: Sec. 4.3.2.2)

C NC N/A GEOMETRY: There shall be no changes in horizontal dimension of the lateral-force-resisting system of more than 30 percent in a story relative to adjacent stories for Life Safety and Immediate Occupancy, excluding one-story penthouses and mezzanines. (Tier 2: Sec. 4.3.2.3)

C NC N/A VERTICAL DISCONTINUITIES: All vertical elements in the lateral-force-resisting system shall be continuous to the foundation. (Tier 2: Sec. 4.3.2.4)

C NC N/A MASS: There shall be no change in effective mass more than 50 percent from one story to the next for Life Safety and Immediate Occupancy. Light roofs, penthouses, and mezzanines need not be considered. (Tier 2: Sec. 4.3.2.5)

C NC N/A TORSION: The estimated distance between the story center of mass and the story center of rigidity shall be less than 20 percent of the building width in either plan dimension for Life Safety and Immediate Occupancy. (Tier 2: Sec. 4.3.2.6)

Screening Phase (Tier 1)

Condition of Materials

C NC N/A   **DETERIORATION OF WOOD:** There shall be no signs of decay, shrinkage, splitting, fire damage, or sagging in any of the wood members, and none of the metal connection hardware shall be deteriorated, broken, or loose. (Tier 2: Sec. 4.3.3.1)

C NC N/A   **WOOD STRUCTURAL PANEL SHEAR WALL FASTENERS:** There shall be no more than 15 percent of inadequate fastening such as overdriven fasteners, omitted blocking, excessive fastening spacing, or inadequate edge distance. This statement shall apply to the Immediate Occupancy Performance Level only. (Tier 2: Sec. 4.3.3.2)

C NC N/A   **DETERIORATION OF STEEL:** There shall be no visible rusting, corrosion, cracking, or other deterioration in any of the steel elements or connections in the vertical- or lateral-force-resisting systems. (Tier 2: Sec. 4.3.3.3)

C NC N/A   **DETERIORATION OF CONCRETE:** There shall be no visible deterioration of concrete or reinforcing steel in any of the vertical- or lateral-force-resisting elements. (Tier 2: Sec. 4.3.3.4)

C NC N/A   **POST-TENSIONING ANCHORS:** There shall be no evidence of corrosion or spalling in the vicinity of post-tensioning or end fittings. Coil anchors shall not have been used. (Tier 2: Sec. 4.3.3.5)

C NC N/A   **PRECAST CONCRETE WALLS:** There shall be no visible deterioration of concrete or reinforcing steel or evidence of distress, especially at the connections. (Tier 2: Sec. 4.3.3.6)

C NC N/A   **MASONRY UNITS:** There shall be no visible deterioration of masonry units. (Tier 2: Sec. 4.3.3.7)

C NC N/A   **MASONRY JOINTS:** The mortar shall not be easily scraped away from the joints by hand with a metal tool, and there shall be no areas of eroded mortar. (Tier 2: Sec. 4.3.3.8)

C NC N/A   **CONCRETE WALL CRACKS:** All existing diagonal cracks in wall elements shall be less than 1/8 inch for Life Safety and 1/16 inch for Immediate Occupancy, shall not be concentrated in one location, and shall not form an X pattern. (Tier 2: Sec. 4.3.3.9)

C NC N/A   **REINFORCED MASONRY WALL CRACKS:** All existing diagonal cracks in wall elements shall be less than 1/8 inch for Life Safety and 1/16 inch for Immediate Occupancy, shall not be concentrated in one location, and shall not form an X pattern. (Tier 2: Sec. 4.3.3.10)

C NC N/A   **UNREINFORCED MASONRY WALL CRACKS:** There shall be no existing diagonal cracks in wall elements greater than 1/8 inch for Life Safety and 1/16 inch for Immediate Occupancy or out-of-plane offsets in the bed joint greater than 1/8 inch for Life Safety and 1/16 inch for Immediate Occupancy, and shall not form an X pattern. (Tier 2: Sec. 4.3.3.11)

C NC N/A   **CRACKS IN INFILL WALLS:** There shall be no existing diagonal cracks in the infilled walls that extend throughout a panel greater than 1/8 inch for Life Safety and 1/16 inch for Immediate Occupancy, or out-of-plane offsets in the bed joint greater than 1/8 inch for Life Safety and 1/16 inch for Immediate Occupancy. (Tier 2: Sec. 4.3.3.12)

C NC N/A   **CRACKS IN BOUNDARY COLUMNS:** There shall be no existing diagonal cracks wider than 1/8 inch for Life Safety and 1/16 inch for Immediate Occupancy in concrete columns that encase masonry infills. (Tier 2: Sec. 4.3.3.13)

Screening Phase (Tier 1)

LATERAL-FORCE-RESISTING SYSTEM

Moment Frames

General

C   NC   N/A   REDUNDANCY: The number of lines of moment frames in each principal direction shall be greater than or equal to 2 for Life Safety and Immediate Occupancy. The number of bays of moment frames in each line shall be greater than or equal to 2 for Life Safety and 3 for Immediate Occupancy. (Tier 2: Sec. 4.4.1.1.1)

Moment Frames with Infill Walls

C   NC   N/A   INTERFERING WALLS: All concrete and masonry infill walls placed in moment frames shall be isolated from structural elements. (Tier 2: Sec. 4.4.1.2.1)

Steel Moment Frames

C   NC   N/A   DRIFT CHECK: The drift ratio of the steel moment frames, calculated using the Quick Check procedure of Section 3.5.3.1, shall be less than 0.025 for Life Safety and 0.015 for Immediate Occupancy. (Tier 2: Sec. 4.4.1.3.1)

C   NC   N/A   AXIAL STRESS CHECK: The axial stress due to gravity loads in columns subjected to overturning forces shall be less than $0.10F_y$ for Life Safety and Immediate Occupancy. Alternatively, the axial stress due to overturning forces alone, calculated using the Quick Check procedure of Section 3.5.3.6, shall be less than $0.30F_y$ for Life Safety and Immediate Occupancy. (Tier 2: Sec. 4.4.1.3.2)

Concrete Moment Frames

C   NC   N/A   SHEAR STRESS CHECK: The shear stress in the concrete columns, calculated using the Quick Check procedure of Section 3.5.3.2, shall be less than the greater of 100 psi or $2\sqrt{f'c}$ for Life Safety and Immediate Occupancy. (Tier 2: Sec. 4.4.1.4.1)

C   NC   N/A   AXIAL STRESS CHECK: The axial stress due to gravity loads in columns subjected to overturning forces shall be less than $0.10f'_c$ for Life Safety and Immediate Occupancy. Alternatively, the axial stresses due to overturning forces alone, calculated using the Quick Check procedure of Section 3.5.3.6, shall be less than $0.30f'_c$ for Life Safety and Immediate Occupancy. (Tier 2: Sec. 4.4.1.4.2)

Precast Concrete Moment Frames

C   NC   N/A   PRECAST CONNECTION CHECK: The precast connections at frame joints shall have the capacity to resist the shear and moment demands calculated using the Quick Check procedure of Section 3.5.3.5. (Tier 2: Sec. 4.4.1.5.1)

Frames Not Part of the Lateral-Force-Resisting System

C   NC   N/A   COMPLETE FRAMES: Steel or concrete frames classified as secondary components shall form a complete vertical-load-carrying system. (Tier 2: Sec. 4.4.1.6.1)

Shear Walls

General

C   NC   N/A   REDUNDANCY: The number of lines of shear walls in each principal direction shall be greater than or equal to 2 for Life Safety and Immediate Occupancy. (Tier 2: Sec. 4.4.2.1.1)

Screening Phase (Tier 1)

Concrete Shear Walls

C NC N/A SHEAR STRESS CHECK: The shear stress in the concrete shear walls, calculated using the Quick Check procedure of Section 3.5.3.3, shall be less than the greater of 100 psi or $2\sqrt{f'c}$ for Life Safety and Immediate Occupancy. (Tier 2: Sec. 4.4.2.2.1)

C NC N/A REINFORCING STEEL: The ratio of reinforcing steel area to gross concrete area shall be not less than 0.0015 in the vertical direction and 0.0025 in the horizontal direction for Life Safety and Immediate Occupancy. The spacing of reinforcing steel shall be equal to or less than 18 inches for Life Safety and Immediate Occupancy. (Tier 2: Sec. 4.4.2.2.2)

C NC N/A COLUMN SPLICES: Steel columns encased in shear-wall-boundary elements shall have splices that develop the tensile strength of the column. This statement shall apply to the Immediate Occupancy Performance Level only. (Tier 2: Sec. 4.4.2.2.9)

Precast Concrete Shear Walls

C NC N/A SHEAR STRESS CHECK: The shear stress in the precast panels, calculated using the Quick Check procedure of Section 3.5.3.3, shall be less than the greater of 100 psi or $2\sqrt{f'c}$ for Life Safety and Immediate Occupancy. (Tier 2: Sec. 4.4.2.3.1)

C NC N/A REINFORCING STEEL: The ratio of reinforcing steel area to gross concrete area shall be not less than 0.0015 in the vertical direction and 0.0025 in the horizontal direction for Life Safety and Immediate Occupancy. The spacing of reinforcing steel shall be equal to or less than 18 inches for Life Safety and Immediate Occupancy. (Tier 2: Sec. 4.4.2.3.2)

Reinforced Masonry Shear Walls

C NC N/A SHEAR STRESS CHECK: The shear stress in the reinforced masonry shear walls, calculated using the Quick Check procedure of Section 3.5.3.3, shall be less than 70 psi for Life Safety and Immediate Occupancy. (Tier 2: Sec. 4.4.2.4.1)

C NC N/A REINFORCING STEEL: The total vertical and horizontal reinforcing steel ratio in reinforced masonry walls shall be greater than 0.002 for Life Safety and Immediate Occupancy of the wall with the minimum of 0.0007 for Life Safety and Immediate Occupancy in either of the two directions; the spacing of reinforcing steel shall be less than 48 inches for Life Safety and Immediate Occupancy; and all vertical bars shall extend to the top of the walls. (Tier 2: Sec. 4.4.2.4.2)

Unreinforced Masonry Shear Walls

C NC N/A SHEAR STRESS CHECK: The shear stress in the unreinforced masonry shear walls, calculated using the Quick Check procedure of Section 3.5.3.3, shall be less than 30 psi for clay units and 70 psi for concrete units for Life Safety and Immediate Occupancy. (Tier 2: Sec. 4.4.2.5.1)

Infill Walls in Frames

C NC N/A WALL CONNECTIONS: Masonry shall be in full contact with frame for Life Safety and Immediate Occupancy. (Tier 2: Sec. 4.4.2.6.1)

Screening Phase (Tier 1)

Walls in Wood-Frame Buildings

C NC N/A SHEAR STRESS CHECK: The shear stress in the shear walls, calculated using the Quick Check procedure of Section 3.5.3.3, shall be less than the following values for Life Safety and Immediate Occupancy (Tier 2: Sec. 4.4.2.7.1):

Structural panel sheathing:	1,000 plf
Diagonal sheathing:	700 plf
Straight sheathing:	100 plf
All other conditions:	100 plf

C NC N/A STUCCO (EXTERIOR PLASTER) SHEAR WALLS: Multi-story buildings shall not rely on exterior stucco walls as the primary lateral-force-resisting system. (Tier 2: Sec. 4.4.2.7.2)

C NC N/A GYPSUM WALLBOARD OR PLASTER SHEAR WALLS: Interior plaster or gypsum wallboard shall not be used as shear walls on buildings over one story in height with the exception of the uppermost level of a multi-story building. (Tier 2: Sec. 4.4.2.7.3)

C NC N/A NARROW WOOD SHEAR WALLS: Narrow wood shear walls with an aspect ratio greater than 2-to-1 for Life Safety and 1.5-to-1 for Immediate Occupancy shall not be used to resist lateral forces developed in the building in levels of moderate and high seismicity. Narrow wood shear walls with an aspect ratio greater than 2-to-1 for Immediate Occupancy shall not be used to resist lateral forces developed in the building in levels of low seismicity. (Tier 2: Sec. 4.4.2.7.4)

C NC N/A WALLS CONNECTED THROUGH FLOORS: Shear walls shall have interconnection between stories to transfer overturning and shear forces through the floor. (Tier 2: Sec. 4.4.2.7.5)

C NC N/A HILLSIDE SITE: For structures that are taller on at least one side by more than one-half story due to a sloping site, all shear walls on the downhill slope shall have an aspect ratio less than 1-to-1 for Life Safety and 1-to-2 for Immediate Occupancy. (Tier 2: Sec. 4.4.2.7.6)

C NC N/A CRIPPLE WALLS: Cripple walls below first-floor-level shear walls shall be braced to the foundation with wood structural panels. (Tier 2: Sec. 4.4.2.7.7)

C NC N/A OPENINGS: Walls with openings greater than 80 percent of the length shall be braced with wood structural panel shear walls with aspect ratios of not more than 1.5-to-1 or shall be supported by adjacent construction through positive ties capable of transferring the lateral forces. (Tier 2: Sec. 4.4.2.7.8)

Braced Frames

General

C NC N/A REDUNDANCY: The number of lines of braced frames in each principal direction shall be greater than or equal to 2 for Life Safety and Immediate Occupancy. The number of braced bays in each line shall be greater than 2 for Life Safety and 3 for Immediate Occupancy. (Tier 2: Sec. 4.4.3.1.1)

C NC N/A AXIAL STRESS CHECK: The axial stress in the diagonals, calculated using the Quick Check procedure of Section 3.5.3.4, shall be less than $0.50F_y$ for Life Safety and for Immediate Occupancy. (Tier 2: Sec. 4.4.3.1.2)

C NC N/A COLUMN SPLICES: All column splice details located in braced frames shall develop the tensile strength of the column. This statement shall apply to the Immediate Occupancy Performance Level only. (Tier 2: Sec. 4.4.3.1.3)

DIAPHRAGMS

Precast Concrete Diaphragms

C   NC   N/A   **TOPPING SLAB:** Precast concrete diaphragm elements shall be interconnected by a continuous reinforced concrete topping slab. (Tier 2: Sec. 4.5.5.1)

CONNECTIONS

Anchorage for Normal Forces

C   NC   N/A   **WALL ANCHORAGE:** Exterior concrete or masonry walls that are dependent on the diaphragm for lateral support shall be anchored for out-of-plane forces at each diaphragm level with steel anchors, reinforcing dowels, or straps that are developed into the diaphragm. Connections shall have adequate strength to resist the connection force calculated in the Quick Check procedure of Section 3.5.3.7. (Tier 2: Sec. 4.6.1.1)

C   NC   N/A   **WOOD LEDGERS:** The connection between the wall panels and the diaphragm shall not induce cross-grain bending or tension in the wood ledgers. (Tier 2: Sec. 4.6.1.2)

Shear Transfer

C   NC   N/A   **TRANSFER TO SHEAR WALLS:** Diaphragms shall be connected for transfer of loads to the shear walls for Life Safety and the connections shall be able to develop the lesser of the shear strength of the walls or diaphragms for Immediate Occupancy. (Tier 2 Sec. 4.6.2.1)

C   NC   N/A   **TRANSFER TO STEEL FRAMES:** Diaphragms shall be connected for transfer of loads to the steel frames for Life Safety, and the connections shall be able to develop the lesser of the strength of the frames or the diaphragms for Immediate Occupancy. (Tier 2: Sec. 4.6.2.2)

C   NC   N/A   **TOPPING SLAB TO WALLS OR FRAMES:** Reinforced concrete topping slabs that interconnect the precast concrete diaphragm elements shall be doweled for transfer of forces into the shear wall or frame elements for Life Safety, and the dowels shall be able to develop the lesser of the shear strength of the walls, frames, or slabs for Immediate Occupancy. (Tier 2: Sec. 4.6.2.3)

Vertical Components

C   NC   N/A   **STEEL COLUMNS:** The columns in lateral-force-resisting frames shall be anchored to the building foundation for Life Safety, and the anchorage shall be able to develop the lesser of the tensile capacity of the column, the tensile capacity of the lowest level column splice (if any), or the uplift capacity of the foundation, for Immediate Occupancy. (Tier 2: Sec. 4.6.3.1)

C   NC   N/A   **CONCRETE COLUMNS:** All concrete columns shall be doweled into the foundation for Life Safety, and the dowels shall be able to develop the tensile capacity of reinforcement in columns of lateral-force-resisting system for Immediate Occupancy. (Tier 2: Sec. 4.6.3.2)

C   NC   N/A   **WOOD POSTS:** There shall be a positive connection of wood posts to the foundation. (Tier 2: Sec. 4.6.3.3)

C   NC   N/A   **WOOD SILLS:** All wood sills shall be bolted to the foundation. (Tier 2: Sec. 4.6.3.4)

C   NC   N/A   **FOUNDATION DOWELS:** Wall reinforcement shall be doweled into the foundation for Life Safety, and the dowels shall be able to develop the lesser of the strength of the walls or the uplift capacity of the foundation for Immediate Occupancy. (Tier 2: Sec. 4.6.3.5)

C   NC   N/A   **SHEAR-WALL-BOUNDARY COLUMNS:** The shear-wall-boundary columns shall be anchored to the building foundation for Life Safety, and the anchorage shall be able to develop the tensile capacity of the column for Immediate Occupancy. (Tier 2: Sec. 4.6.3.6)

Screening Phase (Tier 1)

C   NC   N/A   PRECAST WALL PANELS: Precast wall panels shall be connected to the foundation for Life Safety and the connections shall be able to develop the strength of the walls for Immediate Occupancy. (Tier 2: Sec. 4.6.3.7)

C   NC   N/A   WALL PANELS: Metal, fiberglass, or cementitious wall panels shall be positively attached to the foundation for Life Safety and Immediate Occupancy. (Tier 2: Sec. 4.6.3.8)

Interconnection of Elements

C   NC   N/A   GIRDER/COLUMN CONNECTION: There shall be a positive connection utilizing plates, connection hardware, or straps between the girder and the column support. (Tier 2: Sec. 4.6.4.1)

Panel Connections

C   NC   N/A   ROOF PANELS: Metal, plastic, or cementitious roof panels shall be positively attached to the roof framing to resist seismic forces for Life Safety and Immediate Occupancy. (Tier 2: Sec. 4.6.5.1)

C   NC   N/A   WALL PANELS: Metal, fiberglass, or cementitious wall panels shall be positively attached to the framing to resist seismic forces for Life Safety and Immediate Occupancy. (Tier 2: Sec. 4.6.5.2)

Screening Phase (Tier 1)

3.7.16S General Supplemental Structural Checklist

This General Supplemental Structural Checklist shall be completed where required by Table 3-2. The General Basic Structural Checklist shall be completed prior to completing this General Supplemental Structural Checklist.

LATERAL-FORCE-RESISTING SYSTEM

Moment Frames

Steel Moment Frames

C NC N/A   **MOMENT-RESISTING CONNECTIONS:** All moment connections shall be able to develop the strength of the adjoining members or panel zones. (Tier 2: Sec. 4.4.1.3.3)

C NC N/A   **PANEL ZONES:** All panel zones shall have the shear capacity to resist the shear demand required to develop 0.8 times the sum of the flexural strengths of the girders framing in at the face of the column. (Tier 2: Sec. 4.4.1.3.4)

C NC N/A   **COLUMN SPLICES:** All column splice details located in moment-resisting frames shall include connection of both flanges and the web for Life Safety, and the splice shall develop the strength of the column for Immediate Occupancy. (Tier 2: Sec. 4.4.1.3.5)

C NC N/A   **STRONG COLUMN/WEAK BEAM:** The percentage of strong column/weak beam joints in each story of each line of moment-resisting frames shall be greater than 50 percent for Life Safety and Immediate Occupancy. (Tier 2: Sec. 4.4.1.3.6)

C NC N/A   **COMPACT MEMBERS:** All frame elements shall meet section requirements set forth by *Seismic Provisions for Structural Steel Buildings* Table I-9-1 (AISC, 1997). (Tier 2: Sec. 4.4.1.3.7)

C NC N/A   **BEAM PENETRATIONS:** All openings in frame-beam webs shall be less than ¼ of the beam depth and shall be located in the center half of the beams. This statement shall apply to the Immediate Occupancy Performance Level only. (Tier 2: Sec. 4.4.1.3.8)

C NC N/A   **GIRDER FLANGE CONTINUITY PLATES:** There shall be girder flange continuity plates at all moment-resisting frame joints. This statement shall apply to the Immediate Occupancy Performance Level only. (Tier 2: Sec. 4.4.1.3.9)

C NC N/A   **OUT-OF-PLANE BRACING:** Beam-column joints shall be braced out-of-plane. This statement shall apply to the Immediate Occupancy Performance Level only. (Tier 2: Sec. 4.4.1.3.10

C NC N/A   **BOTTOM FLANGE BRACING:** The bottom flanges of beams shall be braced out-of-plane. This statement shall apply to the Immediate Occupancy Performance Level only. (Tier 2: Sec. 4.4.1.3.11)

Concrete Moment Frames

C NC N/A   **FLAT SLAB FRAMES:** The lateral-force-resisting system shall not be a frame consisting of columns and a flat slab/plate without beams. (Tier 2: Sec. 4.4.1.4.3)

C NC N/A   **PRESTRESSED FRAME ELEMENTS:** The lateral-force-resisting frames shall not include any prestressed or post-tensioned elements where the average prestress exceeds the lesser of 700 psi or $f'_c/6$ at potential hinge locations. The average prestress shall be calculated in accordance with the Quick Check procedure of Section 3.5.3.8. (Tier 2: Sec. 4.4.1.4.4)

C NC N/A   **CAPTIVE COLUMNS:** There shall be no columns at a level with height/depth ratios less than 50 percent of the nominal height/depth ratio of the typical columns at that level for Life Safety and 75 percent for Immediate Occupancy. (Tier 2: Sec. 4.4.1.4.5)

Screening Phase (Tier 1)

C NC N/A   **NO SHEAR FAILURES:** The shear capacity of frame members shall be able to develop the moment capacity at the ends of the members. (Tier 2: Sec. 4.4.1.4.6)

C NC N/A   **STRONG COLUMN/WEAK BEAM:** The sum of the moment capacity of the columns shall be 20 percent greater than that of the beams at frame joints. (Tier 2: Sec. 4.4.1.4.7)

C NC N/A   **BEAM BARS:** At least two longitudinal top and two longitudinal bottom bars shall extend continuously throughout the length of each frame beam. At least 25 percent of the longitudinal bars provided at the joints for either positive or negative moment shall be continuous throughout the length of the members for Life Safety and Immediate Occupancy. (Tier 2: Sec. 4.4.1.4.8)

C NC N/A   **COLUMN-BAR SPLICES:** All column bar lap splice lengths shall be greater than $35d_b$ for Life Safety and $50d_b$ for Immediate Occupancy, and shall be enclosed by ties spaced at or less than $8d_b$ for Life Safety and Immediate Occupancy. Alternatively, column bars shall be spliced with mechanical couplers with a capacity of at least 1.25 times the nominal yield strength of the spliced bar. (Tier 2: Sec. 4.4.1.4.9)

C NC N/A   **BEAM-BAR SPLICES:** The lap splices or mechanical couplers for longitudinal beam reinforcing shall not be located within $l_b/4$ of the joints and shall not be located in the vicinity of potential plastic hinge locations. (Tier 2: Sec. 4.4.1.4.10)

C NC N/A   **COLUMN-TIE SPACING:** Frame columns shall have ties spaced at or less than $d/4$ for Life Safety and Immediate Occupancy throughout their length and at or less than $8d_b$ for Life Safety and Immediate Occupancy at all potential plastic hinge locations. (Tier 2: Sec. 4.4.1.4.11)

C NC N/A   **STIRRUP SPACING:** All beams shall have stirrups spaced at or less than $d/2$ for Life Safety and Immediate Occupancy throughout their length. At potential plastic hinge locations, stirrups shall be spaced at or less than the minimum of $8d_b$ or $d/4$ for Life Safety and Immediate Occupancy. (Tier 2: Sec. 4.4.1.4.12)

C NC N/A   **JOINT REINFORCING:** Beam-column joints shall have ties spaced at or less than $8d_b$ for Life Safety and Immediate Occupancy. (Tier 2: Sec. 4.4.1.4.13)

C NC N/A   **JOINT ECCENTRICITY:** There shall be no eccentricities larger than 20 percent of the smallest column plan dimension between girder and column centerlines. This statement shall apply to the Immediate Occupancy Performance Level only. (Tier 2: Sec. 4.4.1.4.14)

C NC N/A   **STIRRUP AND TIE HOOKS:** The beam stirrups and column ties shall be anchored into the member cores with hooks of 135° or more. This statement shall apply to the Immediate Occupancy Performance Level only. (Tier 2: Sec. 4.4.1.4.15)

Precast Concrete Moment Frames

C NC N/A   **PRECAST FRAMES:** For buildings with concrete shear walls, precast concrete frame elements shall not be considered as primary components for resisting lateral forces. (Tier 2: Sec. 4.4.1.5.2)

C NC N/A   **PRECAST CONNECTIONS:** For buildings with concrete shear walls, the connection between precast frame elements such as chords, ties, and collectors in the lateral-force-resisting system shall develop the capacity of the connected members. (Tier 2: Sec. 4.4.1.5.3)

Frames Not Part of the Lateral-Force-Resisting System

C NC N/A   **DEFLECTION COMPATIBILITY:** Secondary components shall have the shear capacity to develop the flexural strength of the components for Life Safety and shall meet the requirements of Sections 4.4.1.4.9, 4.4.1.4.10, 4.4.1.4.11, 4.4.1.4.12 and 4.4.1.4.15 for Immediate Occupancy. (Tier 2: Sec. 4.4.1.6.2)

Screening Phase (Tier 1)

C   NC   N/A   **FLAT SLABS:** Flat slabs/plates not part of lateral-force-resisting system shall have continuous bottom steel through the column joints for Life Safety and Immediate Occupancy. (Tier 2: Sec. 4.4.1.6.3)

Shear Walls

Concrete Shear Walls

C   NC   N/A   **COUPLING BEAMS:** The stirrups in coupling beams over means of egress shall be spaced at or less than $d/2$ and shall be anchored into the confined core of the beam with hooks of 135° or more for Life Safety. All coupling beams shall comply with the requirements above and shall have the capacity in shear to develop the uplift capacity of the adjacent wall for Immediate Occupancy. (Tier 2: Sec. 4.4.2.2.3)

C   NC   N/A   **OVERTURNING:** All shear walls shall have aspect ratios less than 4-to-1. Wall piers need not be considered. This statement shall apply to the Immediate Occupancy Performance Level only. (Tier 2: Sec. 4.4.2.2.4)

C   NC   N/A   **CONFINEMENT REINFORCING:** For shear walls with aspect ratios greater than 2-to-1, the boundary elements shall be confined with spirals or ties with spacing less than $8d_b$. This statement shall apply to the Immediate Occupancy Performance Level only. (Tier 2: Sec. 4.4.2.2.5)

C   NC   N/A   **REINFORCING AT OPENINGS:** There shall be added trim reinforcement around all wall openings with a dimension greater than three times the thickness of the wall. This statement shall apply to the Immediate Occupancy Performance Level only. (Tier 2: Sec. 4.4.2.2.6)

C   NC   N/A   **WALL THICKNESS:** Thickness of bearing walls shall not be less than 1/25 the unsupported height or length, whichever is shorter, nor less than 4 inches. This statement shall apply to the Immediate Occupancy Performance Level only. (Tier 2: Sec. 4.4.2.2.7)

C   NC   N/A   **WALL CONNECTIONS:** There shall be a positive connection between the shear walls and the steel beams and columns for Life Safety and the connection shall be able to develop the strength of the walls for Immediate Occupancy. (Tier 2: Sec. 4.4.2.2.8)

Precast Concrete Shear Walls

C   NC   N/A   **WALL OPENINGS:** The total width of openings along any perimeter wall line shall constitute less than 75 percent of the length of any perimeter wall for Life Safety and 50 percent for Immediate Occupancy with the wall piers having aspect ratios of less than 2-to-1 for Life Safety and Immediate Occupancy. (Tier 2: Sec. 4.4.2.3.3)

C   NC   N/A   **CORNER OPENINGS:** Walls with openings at a building corner larger than the width of a typical panel shall be connected to the remainder of the wall with collector reinforcing. (Tier 2: Sec. 4.4.2.3.4)

C   NC   N/A   **PANEL-TO-PANEL CONNECTIONS:** Adjacent wall panels shall be interconnected to transfer overturning forces between panels by methods other than welded steel inserts. This statement shall apply to the Immediate Occupancy Performance Level only. (Tier 2: Sec. 4.4.2.3.5)

C   NC   N/A   **WALL THICKNESS:** Thickness of bearing walls shall not be less than 1/25 the unsupported height or length, whichever is shorter, nor less than 4 inches. This statement shall apply to the Immediate Occupancy Performance Level only. (Tier 2: Sec. 4.4.2.3.6)

Reinforced Masonry Shear Walls

C   NC   N/A   **REINFORCING AT OPENINGS:** All wall openings that interrupt rebar shall have trim reinforcing on all sides. This statement shall apply to the Immediate Occupancy Performance Level only. (Tier 2: Sec. 4.4.2.4.3)

Screening Phase (Tier 1)

C NC N/A PROPORTIONS: The height-to-thickness ratio of the shear walls at each story shall be less than 30. This statement shall apply to the Immediate Occupancy Performance Level only. (Tier 2: Sec. 4.4.2.4.4)

Unreinforced Masonry Shear Walls

C NC N/A PROPORTIONS: The height-to-thickness ratio of the shear walls at each story shall be less than the following for Life Safety and Immediate Occupancy (Tier 2: Sec. 4.4.2.5.2):

Top story of multi-story building:	9
First story of multi-story building:	15
All other conditions:	13

C NC N/A MASONRY LAY-UP: Filled collar joints of multi-wythe masonry walls shall have negligible voids. (Tier 2: Sec. 4.4.2.5.3)

Infill Walls in Frames

C NC N/A PROPORTIONS: The height-to-thickness ratio of the infill walls at each story shall be less than 9 for Life Safety in levels of high seismicity, 13 for Immediate Occupancy in levels of moderate seismicity, and 8 for Immediate Occupancy in levels of high seismicity. (Tier 2: Sec. 4.4.2.6.2)

C NC N/A SOLID WALLS: The infill walls shall not be of cavity construction. (Tier 2: Sec. 4.4.2.6.3)

C NC N/A INFILL WALLS: The infill walls shall be continuous to the soffits of the frame beams and to the columns to either side. (Tier 2: Sec. 4.4.2.6.4)

Walls in Wood-Frame Buildings

C NC N/A HOLD-DOWN ANCHORS: All shear walls shall have hold-down anchors constructed per acceptable construction practices, attached to the end studs. This statement shall apply to the Immediate Occupancy Performance Level only. (Tier 2: Sec. 4.4.2.7.9)

Braced Frames

General

C NC N/A SLENDERNESS OF DIAGONALS: All diagonal elements required to carry compression shall have Kl/r ratios less than 120. (Tier 2: Sec. 4.4.3.1.4)

C NC N/A CONNECTION STRENGTH: All the brace connections shall develop the yield capacity of the diagonals. (Tier 2: Sec. 4.4.3.1.5)

C NC N/A OUT-OF-PLANE BRACING: Braced frame connections attached to beam bottom flanges located away from beam-column joints shall be braced out-of-plane at the bottom flange of the beams. This statement shall apply to the Immediate Occupancy Performance Level only. (Tier 2: Sec. 4.4.3.1.6)

Screening Phase (Tier 1)

Concentrically Braced Frames

C   NC   N/A   **K-BRACING:** The bracing system shall not include K-braced bays. (Tier 2: Sec. 4.4.3.2.1)

C   NC   N/A   **TENSION-ONLY BRACES:** Tension-only braces shall not comprise more than 70 percent of the total lateral-force-resisting capacity in structures over two stories in height. This statement shall apply to the Immediate Occupancy Performance Level only. (Tier 2: Sec. 4.4.3.2.2)

C   NC   N/A   **CHEVRON BRACING:** The bracing system shall not include chevron, or V-braced, bays. This statement shall apply to the Immediate Occupancy Performance Level only. (Tier 2: Sec. 4.4.3.2.3)

C   NC   N/A   **CONCENTRICALLY BRACED FRAME JOINTS:** All the diagonal braces shall frame into the beam-column joints concentrically. This statement shall apply to the Immediate Occupancy Performance Level only. (Tier 2: Sec. 4.4.3.2.4)

DIAPHRAGMS

General

C   NC   N/A   **DIAPHRAGM CONTINUITY:** The diaphragms shall not be composed of split-level floors and shall not have expansion joints. (Tier 2: Sec. 4.5.1.1)

C   NC   N/A   **CROSS TIES:** There shall be continuous cross ties between diaphragm chords. (Tier 2: Sec. 4.5.1.2)

C   NC   N/A   **ROOF CHORD CONTINUITY:** All chord elements shall be continuous, regardless of changes in roof elevation. (Tier 2: Sec. 4.5.1.3)

C   NC   N/A   **OPENINGS AT SHEAR WALLS:** Diaphragm openings immediately adjacent to the shear walls shall be less than 25 percent of the wall length for Life Safety and 15 percent of the wall length for Immediate Occupancy. (Tier 2: Sec. 4.5.1.4)

C   NC   N/A   **OPENINGS AT BRACED FRAMES:** Diaphragm openings immediately adjacent to the braced frames shall extend less than 25 percent of the frame length for Life Safety and 15 percent of the frame length for Immediate Occupancy. (Tier 2: Sec. 4.5.1.5)

C   NC   N/A   **OPENINGS AT EXTERIOR MASONRY SHEAR WALLS:** Diaphragm openings immediately adjacent to exterior masonry shear walls shall not be greater than 8 feet long for Life Safety and 4 feet long for Immediate Occupancy. (Tier 2: Sec. 4.5.1.6)

C   NC   N/A   **PLAN IRREGULARITIES:** There shall be tensile capacity to develop the strength of the diaphragm at re-entrant corners or other locations of plan irregularities. This statement shall apply to the Immediate Occupancy Performance Level only. (Tier 2: Sec. 4.5.1.7)

C   NC   N/A   **DIAPHRAGM REINFORCEMENT AT OPENINGS:** There shall be reinforcing around all diaphragm openings larger than 50 percent of the building width in either major plan dimension. This statement shall apply to the Immediate Occupancy Performance Level only. (Tier 2: Sec. 4.5.1.8)

Screening Phase (Tier 1)

Wood Diaphragms

C   NC   N/A   **STRAIGHT SHEATHING:** All straight sheathed diaphragms shall have aspect ratios less than 2-to-1 for Life Safety and 1-to-1 for Immediate Occupancy in the direction being considered. (Tier 2: Sec. 4.5.2.1)

C   NC   N/A   **SPANS:** All wood diaphragms with spans greater than 24 feet for Life Safety and 12 feet for Immediate Occupancy shall consist of wood structural panels or diagonal sheathing. Wood commercial and industrial buildings may have rod-braced systems. (Tier 2: Sec. 4.5.2.2)

C   NC   N/A   **UNBLOCKED DIAPHRAGMS:** All diagonally sheathed or unblocked wood structural panel diaphragms shall have horizontal spans less than 40 feet for Life Safety and 30 feet for Immediate Occupancy and shall have aspect ratios less than or equal to 4-to-1 for Life Safety and 3-to-1 for Immediate Occupancy. (Tier 2: Sec. 4.5.2.3)

Metal Deck Diaphragms

C   NC   N/A   **NON-CONCRETE FILLED DIAPHRAGMS:** Untopped metal deck diaphragms or metal deck diaphragms with fill other than concrete shall consist of horizontal spans of less than 40 feet and shall have span/depth ratios less than 4-to-1. This statement shall apply to the Immediate Occupancy Performance Level only. (Tier 2: Sec. 4.5.3.1)

Other Diaphragms

C   NC   N/A   **OTHER DIAPHRAGMS:** The diaphragm shall not consist of a system other than wood, metal deck, concrete, or horizontal bracing. (Tier 2: Sec. 4.5.7.1)

CONNECTIONS

Anchorage For Normal Forces

C   NC   N/A   **PRECAST PANEL CONNECTIONS:** There shall be at least two anchors from each precast wall panel into the diaphragm elements for Life Safety and the anchors shall be able to develop the strength of the panels for Immediate Occupancy. (Tier 2: Sec. 4.6.1.3)

C   NC   N/A   **STIFFNESS OF WALL ANCHORS:** Anchors of concrete or masonry walls to wood structural elements shall be installed taut and shall be stiff enough to limit the relative movement between the wall and the diaphragm to no greater than 1/8 inch prior to engagement of the anchors. (Tier 2: Sec. 4.6.1.4)

Vertical Components

C   NC   N/A   **WOOD SILL BOLTS:** Sill bolts shall be spaced at 6 feet or less for Life Safety and 4 feet or less for Immediate Occupancy, with proper edge and end distance provided for wood and concrete. (Tier 2: Sec. 4.6.3.9)

C   NC   N/A   **UPLIFT AT PILE CAPS:** Pile caps shall have top reinforcement and piles shall be anchored to the pile caps for Life Safety, and the pile cap reinforcement and pile anchorage shall be able to develop the tensile capacity of the piles for Immediate Occupancy. (Tier 2: Sec. 4.6.3.10)

Screening Phase (Tier 1)

Interconnection Of Elements

C NC N/A **GIRDERS:** Girders supported by walls or pilasters shall have at least two ties securing the anchor bolts for Life Safety and Immediate Occupancy. (Tier 2: Sec. 4.6.4.2)

C NC N/A **CORBEL BEARING:** If the frame girders bear on column corbels, the length of bearing shall be greater than 3 inches for Life Safety and Immediate Occupancy. (Tier 2: Sec. 4.6.4.3)

C NC N/A **CORBEL CONNECTIONS:** The frame girders shall not be connected to corbels with welded elements. (Tier 2: Sec. 4.6.4.4)

C NC N/A **BEAM, GIRDER, AND TRUSS SUPPORTS:** Beams, girders, and trusses supported by unreinforced masonry walls or pilasters shall have independent secondary columns for support of vertical loads. (Tier 2: Sec. 4.6.4.5)

Panel Connections

C NC N/A **ROOF PANEL CONNECTIONS:** Roof panel connections shall be spaced at or less than 12 inches for Life Safety and 8 inches for Immediate Occupancy. (Tier 2: Sec. 4.6.5.3)

Screening Phase (Tier 1)

3.8 Geologic Site Hazards and Foundations Checklist

This Geologic Site Hazards and Foundations Checklist shall be completed where required by Table 3-2.

Each of the evaluation statements on this checklist shall be marked Compliant (C), Non-compliant (NC), or Not Applicable (N/A) for a Tier 1 Evaluation. Compliant statements identify issues that are acceptable according to the criteria of this standard, while non-compliant statements identify issues that require further investigation. Certain statements may not apply to the buildings being evaluated. For non-compliant evaluation statements, the design professional may choose to conduct further investigation using the corresponding Tier 2 Evaluation procedure; corresponding section numbers are in parentheses following each evaluation statement.

Geologic Site Hazards

The following statements shall be completed for buildings in levels of high or moderate seismicity.

C   NC   N/A   LIQUEFACTION: Liquefaction-susceptible, saturated, loose granular soils that could jeopardize the building's seismic performance shall not exist in the foundation soils at depths within 50 feet under the building for Life Safety and Immediate Occupancy. (Tier 2: Sec. 4.7.1.1)

C   NC   N/A   SLOPE FAILURE: The building site shall be sufficiently remote from potential earthquake-induced slope failures or rockfalls to be unaffected by such failures or shall be capable of accommodating any predicted movements without failure. (Tier 2: Sec. 4.7.1.2)

C   NC   N/A   SURFACE FAULT RUPTURE: Surface fault rupture and surface displacement at the building site is not anticipated. (Tier 2: Sec. 4.7.1.3)

Condition of Foundations

The following statement shall be completed for all Tier 1 building evaluations.

C   NC   N/A   FOUNDATION PERFORMANCE: There shall be no evidence of excessive foundation movement such as settlement or heave that would affect the integrity or strength of the structure. (Tier 2: Sec. 4.7.2.1)

The following statement shall be completed for buildings in levels of high or moderate seismicity being evaluated to the Immediate Occupancy Performance Level.

C   NC   N/A   DETERIORATION: There shall not be evidence that foundation elements have deteriorated due to corrosion, sulfate attack, material breakdown, or other reasons in a manner that would affect the integrity or strength of the structure. (Tier 2: Sec. 4.7.2.2)

Capacity of Foundations

The following statement shall be completed for all Tier 1 building evaluations.

C   NC   N/A   POLE FOUNDATIONS: Pole foundations shall have a minimum embedment depth of 4 feet for Life Safety and Immediate Occupancy. (Tier 2: Sec. 4.7.3.1)

The following statements shall be completed for buildings in levels of moderate seismicity being evaluated to the Immediate Occupancy Performance Level and for buildings in levels of high seismicity.

C   NC   N/A   OVERTURNING: The ratio of the horizontal dimension of the lateral-force-resisting system at the foundation level to the building height (base/height) shall be greater than $0.6S_a$. (Tier 2: Sec. 4.7.3.2)

Screening Phase (Tier 1)

C NC N/A   **TIES BETWEEN FOUNDATION ELEMENTS:** The foundation shall have ties adequate to resist seismic forces where footings, piles, and piers are not restrained by beams, slabs, or soils classified as Class A, B, or C. (Section 3.5.2.3.1, Tier 2: Sec. 4.7.3.3)

C NC N/A   **DEEP FOUNDATIONS:** Piles and piers shall be capable of transferring the lateral forces between the structure and the soil. This statement shall apply to the Immediate Occupancy Performance Level only. (Tier 2: Sec. 4.7.3.4)

C NC N/A   **SLOPING SITES:** The difference in foundation embedment depth from one side of the building to another shall not exceed one story in height. This statement shall apply to the Immediate Occupancy Performance Level only. (Tier 2: Sec. 4.7.3.5)

3.9 Nonstructural Checklists

The following checklists are included in this section:

- Basic Nonstructural Component Checklist (Section 3.9.1)
- Intermediate Nonstructural Component Checklist (Section 3.9.2)
- Supplemental Nonstructural Component Checklist (Section 3.9.3)

These checklists shall be completed where required by Table 3-2. The Basic Nonstructural Component Checklist shall be completed prior to completing the Intermediate Nonstructural Component Checklist. The Intermediate Nonstructural Component Checklist shall be completed prior to completing the Supplemental Nonstructural Component Checklist.

Screening Phase (Tier 1)

3.9.1 Basic Nonstructural Component Checklist

This Basic Nonstructural Component Checklist shall be completed where required by Table 3-2.

Each of the evaluation statements on this checklist shall be marked Compliant (C), Non-compliant (NC), or Not Applicable (N/A) for a Tier 1 Evaluation. Compliant statements identify issues that are acceptable according to the criteria of this standard, while non-compliant statements identify issues that require further investigation. Certain statements may not apply to the buildings being evaluated. For non-compliant evaluation statements, the design professional may choose to conduct further investigation using the corresponding Tier 2 Evaluation procedure; corresponding section numbers are in parentheses following each evaluation statement.

Partitions

C NC N/A UNREINFORCED MASONRY: Unreinforced masonry or hollow clay tile partitions shall be braced at a spacing equal to or less than 10 feet in levels of low or moderate seismicity and 6 feet in levels of high seismicity. (Tier 2: Sec. 4.8.1.1)

Ceiling Systems

C NC N/A SUPPORT: The integrated suspended ceiling system shall not be used to laterally support the tops of gypsum board, masonry, or hollow clay tile partitions. Gypsum board partitions need not be evaluated where only the Basic Nonstructural Component Checklist is required by Table 3-2. (Tier 2: Sec. 4.8.2.1)

Light Fixtures

C NC N/A EMERGENCY LIGHTING: Emergency lighting shall be anchored or braced to prevent falling during an earthquake. (Tier 2: Sec. 4.8.3.1)

Cladding and Glazing

C NC N/A CLADDING ANCHORS: Cladding components weighing more than 10 psf shall be mechanically anchored to the exterior wall framing at a spacing equal to or less than 4 feet. A spacing of up to 6 feet is permitted where only the Basic Nonstructural Component Checklist is required by Table 3-2. (Tier 2: Sec. 4.8.4.1)

C NC N/A DETERIORATION: There shall be no evidence of deterioration, damage or corrosion in any of the connection elements. (Tier 2: Sec. 4.8.4.2)

C NC N/A CLADDING ISOLATION: For moment frame buildings of steel or concrete, panel connections shall be detailed to accommodate a story drift ratio of 0.02. Panel connection detailing for a story drift ratio of 0.01 is permitted where only the Basic Nonstructural Component Checklist is required by Table 3-2. (Tier 2: Sec. 4.8.4.3)

C NC N/A MULTI-STORY PANELS: For multi-story panels attached at each floor level, panel connections shall be detailed to accommodate a story drift ratio of 0.02. Panel connection detailing for a story drift ratio of 0.01 is permitted where only the Basic Nonstructural Component Checklist is required by Table 3-2. (Tier 2: Sec. 4.8.4.4)

C NC N/A BEARING CONNECTIONS: Where bearing connections are required, there shall be a minimum of two bearing connections for each wall panel. (Tier 2: Sec. 4.8.4.5)

Screening Phase (Tier 1)

C   NC   N/A   INSERTS: Where inserts are used in concrete connections, the inserts shall be anchored to reinforcing steel or other positive anchorage. (Tier 2: Sec. 4.8.4.6)

C   NC   N/A   PANEL CONNECTIONS: Exterior cladding panels shall be anchored out-of-plane with a minimum of 4 connections for each wall panel. Two connections per wall panel are permitted where only the Basic Nonstructural Component Checklist is required by Table 3-2. (Tier 2: Sec. 4.8.4.7)

Masonry Veneer

C   NC   N/A   SHELF ANGLES: Masonry veneer shall be supported by shelf angles or other elements at each floor 30 feet or more above ground for Life Safety and at each floor above the first floor for Immediate Occupancy. (Tier 2: Sec. 4.8.5.1)

C   NC   N/A   TIES: Masonry veneer shall be connected to the back-up with corrosion-resistant ties. The ties shall have a spacing equal to or less than 24 inches with a minimum of one tie for every 2-2/3 square feet. A spacing of up to 36 inches is permitted where only the Basic Nonstructural Component Checklist is required by Table 3-2. (Tier 2: Sec. 4.8.5.2)

C   NC   N/A   WEAKENED PLANES: Masonry veneer shall be anchored to the back-up adjacent to weakened planes, such as at the locations of flashing. (Tier 2: Sec. 4.8.5.3)

C   NC   N/A   DETERIORATION: There shall be no evidence of deterioration, damage, or corrosion in any of the connection elements. (Tier 2: Sec. 4.8.5.4)

Parapets, Cornices, Ornamentation, and Appendages

C   NC   N/A   URM PARAPETS: There shall be no laterally unsupported unreinforced masonry parapets or cornices with height-to-thickness ratios greater than 1.5. A height-to-thickness ratio of up to 2.5 is permitted where only the Basic Nonstructural Component Checklist is required by Table 3-2. (Tier 2: Sec. 4.8.8.1)

C   NC   N/A   CANOPIES: Canopies located at building exits shall be anchored to the structural framing at a spacing of 6 feet or less. An anchorage spacing of up to 10 feet is permitted where only the Basic Nonstructural Component Checklist is required by Table 3-2. (Tier 2: Sec. 4.8.8.2)

Masonry Chimneys

C   NC   N/A   URM CHIMNEYS: No unreinforced masonry chimney shall extend above the roof surface more than twice the least dimension of the chimney. A height above the roof surface of up to three times the least dimension of the chimney is permitted where only the Basic Nonstructural Component Checklist is required by Table 3-2. (Tier 2: Sec. 4.8.9.1)

Stairs

C   NC   N/A   URM WALLS: Walls around stair enclosures shall not consist of unbraced hollow clay tile or unreinforced masonry with a height-to-thickness ratio greater than 12-to-1. A height-to-thickness ratio of up to 15-to-1 is permitted where only the Basic Nonstructural Component Checklist is required by Table 3-2. (Tier 2: Sec. 4.8.10.1)

C   NC   N/A   STAIR DETAILS: In moment frame structures, the connection between the stairs and the structure shall not rely on shallow anchors in concrete. Alternatively, the stair details shall be capable of accommodating the drift calculated using the Quick Check procedure of Section 3.5.3.1 without including tension in the anchors. (Tier 2: Sec. 4.8.10.2)

Screening Phase (Tier 1)

Building Contents and Furnishing

C   NC   N/A   **TALL NARROW CONTENTS:** Contents over 4 feet in height with a height-to-depth or height-to-width ratio greater than 3-to-1 shall be anchored to the floor slab or adjacent structural walls. A height-to-depth or height-to-width ratio of up to 4-to-1 is permitted where only the Basic Nonstructural Component Checklist is required by Table 3-2. (Tier 2: Sec. 4.8.11.1)

Mechanical and Electrical Equipment

C   NC   N/A   **EMERGENCY POWER:** Equipment used as part of an emergency power system shall be mounted to maintain continued operation after an earthquake. (Tier 2: Sec. 4.8.12.1)

C   NC   N/A   **HAZARDOUS MATERIAL EQUIPMENT:** HVAC or other equipment containing hazardous material shall not have damaged supply lines or unbraced isolation supports. (Tier 2: Sec. 4.8.12.2)

C   NC   N/A   **DETERIORATION:** There shall be no evidence of deterioration, damage, or corrosion in any of the anchorage or supports of mechanical or electrical equipment. (Tier 2: Sec. 4.8.12.3)

C   NC   N/A   **ATTACHED EQUIPMENT:** Equipment weighing over 20 lb that is attached to ceilings, walls, or other supports 4 feet above the floor level shall be braced. (Tier 2: Sec. 4.8.12.4)

Piping

C   NC   N/A   **FIRE SUPPRESSION PIPING:** Fire suppression piping shall be anchored and braced in accordance with NFPA-13 (NFPA, 1996). (Tier 2: Sec. 4.8.13.1)

C   NC   N/A   **FLEXIBLE COUPLINGS:** Fluid, gas, and fire suppression piping shall have flexible couplings. (Tier 2: Sec. 4.8.13.2)

Hazardous Materials Storage and Distribution

C   NC   N/A   **TOXIC SUBSTANCES:** Toxic and hazardous substances stored in breakable containers shall be restrained from falling by latched doors, shelf lips, wires, or other methods. (Tier 2: Sec. 4.8.15.1)

3.9.2 Intermediate Nonstructural Component Checklist

This Intermediate Nonstructural Component Checklist shall be completed where required by Table 3-2. The Basic Nonstructural Component Checklist shall be completed prior to completing this Intermediate Nonstructural Component Checklist.

Ceiling Systems

C   NC   N/A   **LAY-IN TILES:** Lay-in tiles used in ceiling panels located at exits and corridors shall be secured with clips. (Tier 2: Sec. 4.8.2.2)

C   NC   N/A   **INTEGRATED CEILINGS:** Integrated suspended ceilings at exits and corridors or weighing more than 2 pounds per square foot shall be laterally restrained with a minimum of four diagonal wires or rigid members attached to the structure above at a spacing equal to or less than 12 feet. (Tier 2: Sec. 4.8.2.3)

C   NC   N/A   **SUSPENDED LATH AND PLASTER:** Ceilings consisting of suspended lath and plaster or gypsum board shall be attached to resist seismic forces for every 12 square feet of area. (Tier 2: Sec. 4.8.2.4)

Light Fixtures

C   NC   N/A   **INDEPENDENT SUPPORT:** Light fixtures in suspended grid ceilings shall be supported independently of the ceiling suspension system by a minimum of two wires at diagonally opposite corners of the fixtures. (Tier 2: Sec. 4.8.3.2)

Cladding and Glazing

C   NC   N/A   **GLAZING:** Glazing in curtain walls and individual panes over 16 square feet in area, located up to a height of 10 feet above an exterior walking surface, shall have safety glazing. Such glazing located over 10 feet above an exterior walking surface shall be laminated annealed or laminated heat-strengthened safety glass or other glazing system that will remain in the frame when glass is cracked. (Tier 2: Sec. 4.8.4.8)

Parapets, Cornices, Ornamentation, and Appendages

C   NC   N/A   **CONCRETE PARAPETS:** Concrete parapets with height-to-thickness ratios greater than 2.5 shall have vertical reinforcement. (Tier 2: Sec. 4.8.8.3)

C   NC   N/A   **APPENDAGES:** Cornices, parapets, signs, and other appendages that extend above the highest point of anchorage to the structure or cantilever from exterior wall faces and other exterior wall ornamentation shall be reinforced and anchored to the structural system at a spacing equal to or less than 10 feet for Life Safety and 6 feet for Immediate Occupancy. This requirement need not apply to parapets or cornices compliant with Section 4.8.8.1 or 4.8.8.3. (Tier 2: Sec. 4.8.8.4)

Masonry Chimneys

C   NC   N/A   **ANCHORAGE:** Masonry chimneys shall be anchored at each floor level and the roof. (Tier 2: Sec. 4.8.9.2)

Mechanical and Electrical Equipment

C NC N/A   VIBRATION ISOLATORS: Equipment mounted on vibration isolators shall be equipped with restraints or snubbers. (Tier 2: Sec. 4.8.12.5)

Ducts

C NC N/A   STAIR AND SMOKE DUCTS: Stair pressurization and smoke control ducts shall be braced and shall have flexible connections at seismic joints. (Tier 2: Sec. 4.8.14.1)

Screening Phase (Tier 1)

3.9.3 Supplemental Nonstructural Component Checklist

This Supplemental Nonstructural Component Checklist shall be completed where required by Table 3-2. The Basic and Intermediate Nonstructural Component Checklists shall be completed prior to completing this Supplemental Nonstructural Component Checklist.

> **C3.9.3 Supplemental Nonstructural Component Checklist**
>
> The statements in this checklist are intended to evaluate elements that may prevent or limit use of a building following an earthquake. While this checklist is required only for buildings in levels of high seismicity being evaluated to the Immediate Occupancy Performance Level, it may be used as a guide to evaluate potential disruption to building use following an earthquake.

Partitions

C   NC   N/A   **DRIFT:** Rigid cementititous partitions shall be detailed to accommodate a drift ratio of 0.02 in steel moment frame, concrete moment frame, and wood frame buildings. Rigid cementititous partitions shall be detailed to accommodate a drift ratio of 0.005 in other buildings. (Tier 2: Sec. 4.8.1.2)

C   NC   N/A   **STRUCTURAL SEPARATIONS:** Partitions at structural separations shall have seismic or control joints. (Tier 2: Sec. 4.8.1.3)

C   NC   N/A   **TOPS:** The tops of framed or panelized partitions that only extend to the ceiling line shall have lateral bracing to the building structure at a spacing equal to or less than 6 feet. (Tier 2: Sec. 4.8.1.4)

Ceiling Systems

C   NC   N/A   **EDGES:** The edges of integrated suspended ceilings shall be separated from enclosing walls by a minimum of 1/2 inch. (Tier 2: Sec. 4.8.2.5)

C   NC   N/A   **SEISMIC JOINT:** The ceiling system shall not extend continuously across any seismic joint. (Tier 2: Sec. 4.8.2.6)

Light Fixtures

C   NC   N/A   **PENDANT SUPPORTS:** Light fixtures on pendant supports shall be attached at a spacing equal to or less than 6 feet and, if rigidly supported, shall be free to move with the structure to which they are attached without damaging adjoining materials. (Tier 2: Sec. 4.8.3.3)

C   NC   N/A   **LENS COVERS:** Lens covers on light fixtures shall be attached or supplied with safety devices. (Tier 2: Sec. 4.8.3.4)

Cladding and Glazing

C   NC   N/A   **GLAZING:** All exterior glazing shall be laminated, annealed or laminated heat-strengthened safety glass or other glazing system that will remain in the frame when glass is cracked. (Tier 2: Sec. 4.8.4.9)

Screening Phase (Tier 1)

Masonry Veneer

C NC N/A **MORTAR:** The mortar in masonry veneer shall not be easily scraped away from the joints by hand with a metal tool, and there shall not be significant areas of eroded mortar. (Tier 2: Sec. 4.8.5.5)

C NC N/A **WEEP HOLES:** In veneer braced by stud walls, functioning weep holes and base flashing shall be present. (Tier 2: Sec. 4.8.5.6)

C NC N/A **STONE CRACKS:** There shall no be visible cracks or signs of visible distortion in the stone. (Tier 2: Sec. 4.8.5.7)

Metal Stud Back-Up Systems

C NC N/A **STUD TRACKS:** Stud tracks shall be fastened to structural framing at a spacing equal to or less than 24 inches on center. (Tier 2: Sec. 4.8.6.1)

C NC N/A **OPENINGS:** Steel studs shall frame window and door openings. (Tier 2: Sec. 4.8.6.2)

Concrete Block and Masonry Back-Up Systems

C NC N/A **ANCHORAGE:** Back-up shall have a positive anchorage to the structural framing at a spacing equal to or less than 4 feet along the floors and roof. (Tier 2: Sec. 4.8.7.1)

C NC N/A **URM BACK-UP:** There shall be no unreinforced masonry back-up. (Tier 2: Sec. 4.8.7.2)

Building Contents and Furnishing

C NC N/A **FILE CABINETS:** File cabinets arranged in groups shall be attached to one another. (Tier 2: Sec. 4.8.11.2)

C NC N/A **CABINET DOORS AND DRAWERS:** Cabinet doors and drawers shall have latches to keep them closed during an earthquake. (Tier 2: Sec. 4.8.11.3)

C NC N/A **ACCESS FLOORS:** Access floors over 9 inches in height shall be braced. (Tier 2: Sec. 4.8.11.4)

C NC N/A **EQUIPMENT ON ACCESS FLOORS:** Equipment and computers supported on access floor systems shall be either attached to the structure or fastened to a laterally braced floor system. (Tier 2: Sec. 4.8.11.5)

Mechanical and Electrical Equipment

C NC N/A **HEAVY EQUIPMENT:** Equipment weighing over 100 pounds shall be anchored to the structure or foundation. (Tier 2: Sec. 4.8.12.6)

C NC N/A **ELECTRICAL EQUIPMENT:** Electrical equipment and associated wiring shall be laterally braced to the structural system. (Tier 2: Sec. 4.8.12.7)

C NC N/A **DOORS:** Mechanically operated doors shall be detailed to operate at a story drift ratio of 0.01. (Tier 2: Sec. 4.8.12.8)

Piping

C NC N/A **FLUID AND GAS PIPING:** Fluid and gas piping shall be anchored and braced to the structure to prevent breakage in piping. (Tier 2: Sec. 4.8.13.3)

Screening Phase (Tier 1)

C NC N/A SHUT-OFF VALVES: Shut-off devices shall be present at building utility interfaces to shut off the flow of gas and high-temperature energy in the event of earthquake-induced failure. (Tier 2: Sec. 4.8.13.4)

C NC N/A C-CLAMPS: One-sided C-clamps that support piping greater than 2.5 inches in diameter shall be restrained. (Tier 2: Sec. 4.8.13.5)

Ducts

C NC N/A DUCT BRACING: Rectangular ductwork exceeding 6 square feet in cross-sectional area, and round ducts exceeding 28 inches in diameter, shall be braced. Maximum spacing of transverse bracing shall not exceed 30 feet. Maximum spacing of longitudinal bracing shall not exceed 60 feet. Intermediate supports shall not be considered part of the lateral-force-resisting system. (Tier 2: Sec. 4.8.14.2)

C NC N/A DUCT SUPPORT: Ducts shall not be supported by piping or electrical conduit. (Tier 2: Sec. 4.8.14.3)

Hazardous Materials Storage and Distribution

C NC N/A GAS CYLINDERS: Compressed gas cylinders shall be restrained. (Tier 2: Sec. 4.8.15.2)

C NC N/A HAZARDOUS MATERIALS: Piping containing hazardous materials shall have shut-off valves or other devices to prevent major spills or leaks. (Tier 2: Sec. 4.8.15.3)

Elevators

C NC N/A SUPPORT SYSTEM: All elements of the elevator system shall be anchored. (Tier 2: Sec. 4.8.16.1)

C NC N/A SEISMIC SWITCH: All elevators shall be equipped with seismic switches that will terminate operations when the ground motion exceeds 0.10g. (Tier 2: Sec. 4.8.16.2)

C NC N/A SHAFT WALLS: All elevator shaft walls shall be anchored and reinforced to prevent toppling into the shaft during strong shaking. (Tier 2: Sec. 4.8.16.3)

C NC N/A RETAINER GUARDS: Cable retainer guards on sheaves and drums shall be present to inhibit the displacement of cables. (Tier 2: Sec. 4.8.16.4)

C NC N/A RETAINER PLATE: A retainer plate shall be present at the top and bottom of both car and counterweight. (Tier 2: Sec. 4.8.16.5)

C NC N/A COUNTERWEIGHT RAILS: All counterweight rails and divider beams shall be sized in accordance with ASME A17.1. (Tier 2: Sec. 4.8.16.6)

C NC N/A BRACKETS: The brackets that tie the car rails and the counterweight rail to the building structure shall be sized in accordance with ASME A17.1. (Tier 2: Sec. 4.8.16.7)

C NC N/A SPREADER BRACKET: Spreader brackets shall not be used to resist seismic forces. (Tier 2: Sec. 4.8.16.8)

C NC N/A GO-SLOW ELEVATORS: The building shall have a go-slow elevator system. (Tier 2: Sec. 4.8.16.9)

4.0 Evaluation Phase (Tier 2)

4.1 General

A Tier 1 Evaluation shall be completed for all buildings prior to performing a Tier 2 Evaluation. A Full-Building Tier 2 Analysis and Evaluation of the adequacy of the lateral-force-resisting system shall be performed for all buildings designated as "T2" in Table 3-3. For all other buildings, the design professional may choose to perform a Deficiency-Only Tier 2 Evaluation that addresses only the deficiencies identified in Tier 1. Each Tier 1 Checklist evaluation statement in Chapter 3 provides a reference to the corresponding Tier 2 Procedures for further evaluation of Tier 1 deficiencies.

A Tier 2 Evaluation shall include an analysis using one of the following linear methods: Linear Static Procedure, Linear Dynamic Procedure, or Special Procedure. Analysis procedures and component acceptance criteria are specified in Section 4.2. Unless otherwise designated in Table 3-3, the analysis shall address, at a minimum, all of the potential deficiencies identified in Tier 1, using the procedures specified in Sections 4.3 to 4.8.

If deficiencies are identified in a Tier 2 Evaluation, the design professional may perform a Tier 3 Evaluation in accordance with the requirements of Chapter 5. Alternatively, the design professional may choose to end the investigation and report the deficiencies in accordance with Chapter 1.

> **C4.1 General**
>
> The procedures for evaluating potential deficiencies have been completely revised from FEMA 178. The new procedures represent the most current available techniques and are consistent with procedures used in FEMA 356.
>
> The original evaluation process defined in ATC-14 was based on the *Uniform Building Code* (ICBO, 1997) equivalent lateral force procedure, a working-stress-based process using R_w factors, allowable stress design, and capacity over demand ratios that accounted for the lack of modern detailing.
>
> FEMA 178 used an analysis procedure based on the 1988 NEHRP Provisions equivalent lateral force procedure using R factors and ultimate strength design. Nonconforming structural systems that did not have proper detailing were assigned lower R factors to account for their lack of ductility.
>
> This standard uses a displacement-based lateral force procedure and m-factors on an element-by-element basis. It represents the most direct method for considering nonconforming systems. The lateral forces related to each of these approaches are radically different and cannot be directly compared.

4.2 Tier 2 Analysis

4.2.1 General

Four analysis procedures are provided in this section:
- Linear Static Procedure (LSP)
- Linear Dynamic Procedure (LDP)
- Special Procedure
- Procedures for Nonstructural Components

Evaluation Phase (Tier 2)

All building structures, except unreinforced masonry (URM) bearing wall buildings with flexible diaphragms, shall be evaluated by either the Linear Static Procedure of Section 4.2.2.1 or the Linear Dynamic Procedure of Section 4.2.2.2. The acceptance criteria for both the LSP and LDP are provided in Section 4.2.4. Out-of-plane forces on walls shall be calculated in accordance with Section 4.2.5.

If original design calculations are available, the results may be used; however, an appropriate scaling factor shall be applied to relate the original design base shear to the pseudo lateral force of this standard.

Unreinforced masonry bearing wall buildings with flexible diaphragms shall be evaluated in accordance with the requirements of the Special Procedure defined in Section 4.2.6 directly.

The demands on nonstructural components shall be calculated in accordance with Section 4.2.7 where triggered by the Procedures for Nonstructural Components in Section 4.8.

4.2.2 Analysis Procedures for LSP and LDP

The Linear Static or Linear Dynamic Procedure shall be performed as required by the procedures of Sections 4.3 through 4.6.

The Linear Static Procedure is applicable for all buildings unless a Linear Dynamic Procedure or Special Procedure is required.

The Linear Dynamic Procedure shall be used for:
- Buildings taller than 100 feet, or
- Buildings with mass, stiffness, or geometric irregularities as specified in Sections 4.3.2.2, 4.3.2.3, and 4.3.2.5.

4.2.2.1 Linear Static Procedure (LSP)

The Linear Static Procedure shall be performed as follows:

- Develop a mathematical building model in accordance with Section 4.2.3.
- Calculate the pseudo lateral force in accordance with Section 4.2.2.1.1.
- Calculate the lateral forces to be distributed vertically in accordance with Section 4.2.2.1.2.
- Calculate the building or component forces and displacements using linear, elastic analysis methods.
- Calculate diaphragm forces in accordance with Section 4.2.2.1.4, if required.
- Compare the component actions with the acceptance criteria of Section 4.2.4.5.

C4.2.2.1 Linear Static Procedure (LSP)

In the Linear Static Procedure, the building is modeled with linearly elastic stiffness and equivalent viscous damping that approximate values expected for loading to near the yield point. Design earthquake demands for the Linear Static Procedure are represented by static lateral forces whose sum is equal to the pseudo lateral force defined by Equation (3-1). The magnitude of the pseudo lateral force has been selected with the intention that where it is applied to the linearly elastic model of the building it will result in design displacement amplitudes approximating maximum displacements that are expected during the design earthquake. If the building responds essentially elastically to the design earthquake, the calculated internal forces will be reasonable approximations to those expected during the design earthquake. If the building responds inelastically to the design earthquake, as will commonly be the case, the calculated internal forces will exceed those that would develop in the yielding building.

Evaluation Phase (Tier 2)

> The component forces in yielding structures calculated from linear analysis represent the total (linear and nonlinear) deformation of the component. The acceptance criteria reconcile the calculated forces with component capacities using component ductility related factors, m. The linear procedures represent a rough approximation of the nonlinear behavior of the actual structure and ignore redistribution of forces and other nonlinear effects. In certain cases, alternative acceptable approaches are presented that may provide wide variation in the results. This is expected, considering the limitations of the linear analysis procedures.

4.2.2.1.1 Pseudo Lateral Force

The pseudo lateral force applied in a Linear Static Procedure shall be calculated in accordance with Section 3.5.2.1.

4.2.2.1.2 Period

The fundamental period of vibration of the building for use in Equation (3-1) shall be determined using one of the following calculations:

- For a one-story building with a single-span flexible diaphragm, in accordance with Equation (4-1):

$$T = (0.1\Delta_w + 0.078\Delta_d)^{0.5} \quad (4\text{-}1)$$

  where:
  Δ_w and Δ_d are in-plane wall and diaphragm displacements in inches due to a lateral force equal to the weight tributary to diaphragm in the direction under consideration.

- For multiple-span flexible diaphragms, a lateral force equal to the weight tributary to the diaphragm span under consideration shall be applied to each span of the diaphragm to calculate a separate period for each diaphragm span. The period that maximizes the pseudo lateral force shall be used for design of all walls and diaphragm spans in the building.

- Based on an eigenvalue (dynamic) analysis of the mathematical model of the building.

- In accordance with Section 3.5.2.4.

> **C4.2.2.1.2 Period**
>
> Equation (4-1) is derived from an assumed first-mode shape for the building for single-span flexible diaphragms. For multiple-span diaphragms with widely varying aspect ratios, this approach may be unduly conservative. It is recommended a dynamic analysis be performed for such cases to determine the period.

4.2.2.1.3 Vertical Distribution of Seismic Forces

The pseudo lateral force calculated in accordance with Section 4.2.2.1.1 shall be distributed vertically in accordance with Equations (4-2) and (4-3).

$$F_x = C_{vx} V \quad (4\text{-}2)$$

$$C_{vx} = \frac{w_x h_x^k}{\sum_{i=1}^{n} w_i h_i^k} \quad (4\text{-}3)$$

where:
- k = 1.0 for $T < 0.5$ second
-     = 2.0 for $T > 2.5$ seconds (linear interpolation shall be used for intermediate values of k)
- C_{vx} = Vertical distribution factor at floor level x
- V = Pseudo lateral force (Section 4.2.2.1.1)
- w_i = Portion of the total building weight W located on or assigned to floor level i
- w_x = Portion of the total building weight W located on or assigned to floor level x
- h_i = Height (ft) from the base to floor level i
- h_x = Height (ft) from the base to floor level x

4.2.2.1.4 Diaphragms

Diaphragms shall be designed to resist the combined effects of the inertial force, F_{px}, calculated in accordance with Equation (4-4), and horizontal forces resulting from offsets in or changes in the stiffness of the vertical lateral-force-resisting elements above and below the diaphragm. Forces resulting from offsets in or changes in the stiffness of the vertical lateral-force-resisting elements shall be taken as the forces due to the psuedo lateral force of Equation (3-1) without reduction, unless smaller forces are justified by a limit-state or other rational analysis, and shall be added directly to the diaphragm inertial forces.

$$F_{px} = \frac{1}{C} \frac{\sum_{i=x}^{n} F_i}{\sum_{i=x}^{n} w_i} w_x \qquad (4\text{-}4)$$

where:
- F_{px} = Total diaphragm force at level x
- F_i = Lateral force applied at floor level i defined by Equation (4-2)
- w_i = Portion of the effective seismic weight W located on or assigned to floor level i
- w_x = Portion of the effective seismic weight W located on or assigned to floor level x
- C = Modification factor defined in Table 3-4

The seismic force on each flexible diaphragm shall be distributed along the span of that diaphragm, proportional to its expected displaced shape.

Diaphragms receiving horizontal forces from discontinuous vertical elements shall be taken as force controlled as defined in Section 4.2.4.3. Actions on other diaphragms shall be considered force controlled or deformation controlled as specified for diaphragm components in Section 4.5.

C4.2.2.1.4 Diaphragms

Traditionally, diaphragm forces have been distributed along the span in a uniform manner or in accordance with the distributed mass of the diaphragm. However, for flexible diaphragms, the center of the diaphragm will be more flexible than the two edges, resulting in a higher acceleration at the center. Therefore, the distribution of the seismic forces for flexible diaphragms should be based on the expected displaced shape rather than the distributed mass. Preliminary loading patterns can include a triangular or parabolic force distribution with verification against the actual displaced shape. A uniform distribution can be assumed if the diaphragm is expected to yield.

Evaluation Phase (Tier 2)

4.2.2.1.5 Determination of Deformations

Structural deformations and story drifts shall be calculated using lateral forces in accordance with Equations (3-1), (4-2), and (4-4).

4.2.2.2 Linear Dynamic Procedure (LDP)

The Linear Dynamic Procedure shall be performed as follows:

- Develop a mathematical building model in accordance with Section 4.2.3.
- Develop a response spectrum for the site in accordance with Section 4.2.2.2.2.
- Perform a response spectrum analysis of the building.
- Modify the actions and deformations in accordance with Section 4.2.2.2.3.
- Compute diaphragm forces in accordance with Section 4.2.2.2.4, if required.
- Compute the component actions in accordance with Section 4.2.4.3.
- Compare the component actions with the acceptance criteria of Section 4.2.4.5.

> **C4.2.2.2 Linear Dynamic Procedure (LDP)**
>
> Note that, in contrast to the NEHRP Provisions and the *Uniform Building Code*, the results of the response spectrum analysis are not scaled to the pseudo lateral force of the LSP. Such scaling is unnecessary since the LSP is based on the use of actual spectral acceleration values from proper response spectra and is not reduced by R values used in traditional code design.

4.2.2.2.1 Modal Responses

Modal responses shall be combined using the square root sum of the squares (SRSS) or complete quadratic combination (CQC) method to estimate the response quantities. The CQC shall be used where modal periods associated with motion in a given direction are within 25 percent. The number of modes considered in the response spectrum analysis shall be sufficient to capture at least 90 percent of the participating mass of the building in each of the building's principal horizontal axes.

Multidirectional excitation effects shall be considered in accordance with Section 4.2.3.5. Alternatively, the SRSS method may be used to combine multidirectional effects. The CQC method shall not be used for combination of multidirectional effects.

4.2.2.2.2 Ground Motion Characterization

The seismic ground motions shall be characterized for use in the LDP by developing:

- A mapped response spectrum in accordance with Section 3.5.2.3.1, or
- A site-specific response spectrum in accordance with Section 3.5.2.3.2.

4.2.2.2.3 Modification of Demands

With the exception of diaphragm actions and deformations, all actions and deformations calculated using the LDP shall be multiplied by the modification factor, C, defined in Table 3-4. For dynamic analyses using a site-specific spectrum, all actions and deformations shall be permitted to be multiplied by a factor of 2/3.

4.2.2.2.4 Diaphragms

Diaphragms shall be analyzed for the sum of (1) the seismic forces calculated by dynamic analysis, but not less than 85 percent of the forces calculated using Equation (4-4); and (2) the horizontal forces resulting from offsets in, or changes in stiffness of, the vertical seismic framing elements above and below the diaphragm. Forces resulting from offsets in, or changes in stiffness of, the vertical lateral-force-resisting elements shall be taken to be equal to the elastic forces without reduction, unless smaller forces can be justified by rational analysis.

4.2.3 Mathematical Model for LSP and LDP

4.2.3.1 Basic Assumptions

A two-dimensional model for buildings with stiff diaphragms shall be developed if torsional effects are insignificant or indirectly captured; otherwise, a three-dimensional model shall be developed.

Lateral-force-resisting frames in buildings with flexible diaphragms shall be modeled and analyzed as two-dimensional assemblies of components; alternatively, a three-dimensional model shall be used with the diaphragms modeled as flexible elements.

4.2.3.2 Horizontal Torsion

The effects of horizontal torsion shall be considered in a Tier 2 Analysis. The total torsional moment at a given floor level shall be equal to the sum of the following two torsional moments:

1. Actual torsion resulting from the eccentricity between the centers of mass and the centers of rigidity of all floors above and including the given floor, and
2. Accidental torsion produced by horizontal offset in the centers of mass, at all floors above and including the given floor, equal to a minimum of 5 percent of the horizontal dimension at the given floor level measured perpendicular to the direction of the applied load.

The effects of accidental torsion shall not be used to reduce force and deformation demands on building components.

A building is considered torsionally irregular if the building has stiff diaphragms and the ratio $\eta = \delta_{max}/\delta_{avg}$ due to total torsional moment exceeds 1.2. In torsionally irregular buildings, the effect of accidental torsion shall be amplified by the factor, A_x, given in Equation (4-5).

$$A_x = \left(\frac{\delta_{max}}{1.2\,\delta_{avg}}\right)^2 \qquad (4-5)$$

where:
δ_{max} = The maximum displacement at any point of diaphragm at level x
δ_{avg} = The algebraic average of displacements at the extreme points of the diaphragm at level x
A_x = Shall be greater than or equal to 1.0 and need not exceed 3.0

If the ratio, η, including torsional amplification, exceeds 1.50, a three-dimensional model shall be developed for a Tier 2 Analysis. Where $\eta < 1.5$, the forces and displacements calculated using two-dimensional models shall be increased by the maximum value of η calculated for the building.

> **C4.2.3.2 Horizontal Torsion**
>
> Although the torsional response of buildings with flexible diaphragms is not subject to the amplification factor, A_x, this section does require the consideration of actual and accidental torsion for such buildings.

4.2.3.3 Primary and Secondary Components

Components shall be classified as either primary or secondary. A primary component is an element that is considered to resist the seismic forces in order for the structure to achieve the selected performance level. A secondary component is an element that may attract seismic forces but is not required to resist seismic forces in order for the structure to achieve the selected performance level. Any component assumed in the design to be a secondary component, but whose failure under lateral forces causes failure of a primary component, shall be reclassified as a primary component. Only components classified as primary components need to be included in the mathematical model for the lateral force analysis unless the interaction of secondary components may result in less desirable seismic performance, in which case the interacting secondary components shall be included. All secondary components shall be capable of maintaining support for gravity loads for the displacements calculated using the procedures of Section 4.2. If the combined lateral stiffness of secondary components exceeds 25 percent of the total lateral stiffness of the primary components at a level of the building, some secondary components shall be reclassified as primary components such that the combined lateral stiffness of secondary components does not exceed the limit of 25 percent.

> **C4.2.3.3 Primary and Secondary Components**
>
> The classification of components and elements should not result in a change in the regularity of a building. That is, components and elements should not be selectively assigned as either primary or secondary to change the configuration of a building from irregular to regular.
>
> This standard requires that no more than 25 percent of the lateral resistance be provided by secondary components. The main reason for this limitation is that sudden loss of lateral-force-resisting components or elements can result in irregular response of a building that is difficult to detect. An example is a masonry infill wall that, if it collapses from one story of an infilled frame, may result in a severe strength and stiffness irregularity in the building. A secondary reason is to prevent the engineer from manipulating the analysis model to minimize design actions on critical components and elements. In the linear models, this 25 percent criterion can be checked by including the secondary components in the analysis model and examining their stiffness contribution.

4.2.3.4 Diaphragms

Mathematical models of buildings with stiff diaphragms shall explicitly include diaphragm flexibility. Mathematical models of buildings with rigid diaphragms shall explicitly account for the rigidity of the diaphragms. For buildings with flexible diaphragms at each floor level, the vertical lines of seismic framing shall be permitted to be considered independently, with seismic masses assigned on the basis of tributary area.

The in-plane deflection of the diaphragm shall be calculated for an in-plane distribution of lateral force consistent with the distribution of mass, as well as all in-plane lateral forces associated with offsets in the vertical seismic framing.

4.2.3.5 Multidirectional Excitation Effects

Buildings shall be analyzed for seismic forces in any horizontal direction except as allowed in Section 4.2.3.5. Seismic displacements and forces shall be assumed to act non-concurrently in the direction of each principal axis of a building, unless the building is torsionally irregular as defined in Section 4.2.3.2 or one or more components form part of two or more intersecting elements, in which case multidirectional excitation effects shall be considered.

Multidirectional (orthogonal) excitation shall be evaluated by applying 100 percent of the seismic forces in one horizontal direction plus 30 percent of the seismic forces in the perpendicular horizontal direction. Alternatively, the effects of the two orthogonal directions are permitted to be combined on an SRSS basis.

4.2.3.6 Vertical Acceleration

The effects of vertical excitation on horizontal cantilevers and prestressed elements shall be considered using static or dynamic analysis methods. Vertical earthquake motions shall be characterized by a spectrum with ordinates equal to 67 percent of those of the horizontal spectrum in Section 3.5.2.3.1. Alternatively, vertical response spectra developed using site-specific analysis may be used.

4.2.4 Acceptance Criteria for LSP and LDP

4.2.4.1 General Requirements

Component actions shall be computed according to Section 4.2.4.3; gravity loads calculated in accordance with Section 4.2.4.2 as well as seismic forces shall be considered. Component strengths shall be computed in accordance with Section 4.2.4.4. Component actions and strengths then shall be compared with the acceptance criteria in Section 4.2.4.5.

4.2.4.2 Component Gravity Loads

Component gravity loads shall be calculated in accordance with Equations (4-6) and (4-7).

$$Q_G = 1.1(Q_D + Q_L + Q_S) \qquad (4\text{-}6)$$
$$Q_G = 0.9\,Q_D \qquad (4\text{-}7)$$

where:
Q_D = Dead load.
Q_L = Effective live load, equal to 25 percent of the unreduced design live load but not less than the measured live load.
Q_S = Effective snow load, equal to 20 percent of the design flat roof snow load calculated in accordance with ASCE 7-02. Where the design flat roof snow load is less than 30 psf, the effective snow load shall be permitted to be zero.

C4.2.4.2 Component Gravity Loads

The minimum live load specification equal to 0.25 of the unreduced design live load is a traditionally applied value used in design to represent the likely live load acting in a structure. Where the load is likely to be larger, use this larger load value.

Evaluation Phase (Tier 2)

4.2.4.3 Component Actions

Actions shall be classified as either deformation controlled or force controlled. A deformation-controlled action shall be defined as an action that has an associated deformation that is allowed to exceed the yield value; the maximum associated deformation is limited by the ductility capacity of the component. A force-controlled action shall be defined as an action that has an associated force or moment that is not allowed to exceed the yield value.

C4.2.4.3 Component Actions

Global deformation of a structure is primarily due to the elastic and inelastic deformations associated with the deformation-controlled actions. The maximum forces in force-controlled components are governed by the capacity of deformation-controlled components.

Consider actions in beams and columns of a reinforced concrete moment frame. Flexural moments are typically a deformation-controlled action. Shear forces in beams and axial forces in columns are force-controlled actions. The yielding of deformation-controlled actions (in this example, beam moment) controls the forces that can be delivered to the force-controlled actions (in this example, beam shear and column axial force).

Consider a braced frame structure. The axial forces in the diagonal braces are deformation-controlled actions. The force in brace connections and axial force in columns are force-controlled actions. Yielding and buckling of braces control the maximum force that can be delivered to the connections and columns. The action is not directly related to the pseudo seismic forces used in the evaluation. Instead, it is based on the maximum action that can be delivered to the element by the yielding structural system.

Table C4-1 shows examples of possible deformation- and force-controlled actions.

Table C4-1. Examples of Possible Deformation-Controlled and Force-Controlled Actions

Component	Deformation-Controlled Actions	Force-Controlled Actions
Moment Frames:		
Beams	M	V
Columns	M	P, V
Joints	--	V^1
Shear Walls	M, V	P
Braced Frames:		
Braces	P	--
Beams	--	P
Columns	--	P
Shear Link	V	P, M
Connections	P, V, M^3	P, V, M
Diaphragms	M, V^2	P, V, M

M = moment, V = shear force, P = axial load

[1] Shear may be a deformation-controlled action in steel moment frame construction.

[2] If the diaphragm carries lateral loads from vertical seismic resisting elements above the diaphragm level, then M and V shall be considered force-controlled actions.

[3] Axial, shear, and moment may be deformation-controlled actions for certain steel and wood connections.

Evaluation Phase (Tier 2)

4.2.4.3.1 Deformation-Controlled Actions

Deformation-controlled design actions, Q_{UD}, shall be calculated according to Equation (4-8).

$$Q_{UD} = Q_G \pm Q_E \tag{4-8}$$

where:
Q_{UD} = Action due to gravity loads and earthquake forces
Q_G = Action due to gravity forces as defined in Section 4.2.4.2
Q_E = Action due to earthquake forces calculated using forces and analysis models described in either Section 4.2.2.1 or Section 4.2.2.2

4.2.4.3.2 Force-Controlled Actions

There are three methods for determining force-controlled actions:

- **Method 1:** Force-controlled actions, Q_{UF}, shall be calculated as the sum of forces due to gravity and the maximum force that can be delivered by deformation-controlled actions.

- **Method 2:** Alternatively, Equation (4-9) shall be used where the forces contributing to Q_{UF} are delivered by yielding components of the seismic framing system, and Equation (4-10) shall be used for all other evaluations.

$$Q_{UF} = Q_G \pm \frac{Q_E}{CJ} \tag{4-9}$$

$$Q_{UF} = Q_G \pm \frac{Q_E}{C} \tag{4-10}$$

where:
Q_{UF} = Actions due to gravity loads and earthquake forces.
C = Modification factor defined in Table 3-4, except for diaphragms where horizontal forces are calculated by Equation (4-4) in which case C shall be taken equal to 1.0.
J = Force-delivery reduction factor, greater than or equal to 1.0, taken as the smallest demand-capacity ratio (DCR) of the components in the load path delivering force to the component in question. Alternatively, values of J equal to 2.5 in levels of high seismicity, 2.0 in levels of moderate seismicity, and 1.5 in levels of low seismicity shall be permitted when not based on calculated $DCRs$. J shall be taken as 1.5 for the Immediate Occupancy Structural Performance Level. In any case where the forces contributing to Q_{UF} are delivered by components of the lateral-force-resisting system that remain elastic, J shall be taken as 1.0.

- **Method 3:** For the evaluation of buildings analyzed using pseudo lateral force of Equation (3-2), Equation (4-10) with $C = 1.0$ shall be used.

C4.2.4.3.2 Force-Controlled Actions

Force-controlled actions provide little deformation to the entire building through inelastic behavior. Because of the limited ductility associated with force-controlled actions, inelastic action in these elements may cause a sudden partial or total collapse of the structure.

> There are three methods for determining force-controlled actions. In the first method, Q_{UF} for a brace connection would be equal to the axial force capacity for the brace member. Q_{UF} for shear in a beam would be equal to gravity shear plus the shear force associated with development of flexural moment capacity at the ends of the beam. Q_{UF} for axial force in a moment frame column would be equal to the sum of maximum shear forces that can be developed in the beams supported by the columns. If it can be shown that the deformation-controlled action can be developed before the failure of the associated force-controlled action, then the failure will not occur. This is due to the fact that the yielding of the deformation-controlled components will limit the demand on the force-controlled component. This is the method recommended for evaluating force-controlled components.
>
> The second and third methods provide conservative estimates of force-controlled actions due to a design earthquake. Equation (4-9) may be used if other yielding elements in the building will limit the amount of force that can be delivered to the force-controlled component. Equation (4-10) is used if the force-controlled component is the "weak link" and, thus, it must be evaluated for full earthquake force. Equation (4-10) also must be used if foundation sliding controls the behavior of the building as assumed by Equation (3-2).

4.2.4.3.3 Connections

Connections shall be evaluated as force-controlled actions. Alternatively, hold-down anchors used to resist overturning forces in wood shear wall buildings shall be permitted to be evaluated as deformation-controlled actions using the appropriate m-factors specified in Table 4-8.

4.2.4.3.4 Foundation/Soil Interface

Actions at the foundation/soil interface shall be considered force-controlled as defined in Section 4.2.4.3.2. Alternatively, actions at the foundation/soil interface may be reduced by R_{OT}, where R_{OT} shall be taken as 8.0 for Life Safety and 4.0 for Immediate Occupancy. R_{OT} and J factors shall not be permitted to be used concurrently.

> #### C4.2.4.3.4 Foundation/Soil Interface
>
> Evaluation of overturning effects using the pseudo lateral force procedure typically results in larger forces than current code-based analytical procedures that reduce earthquake forces by an R-factor. In spite of this force reduction, however, code-based design procedures have yielded satisfactory performance with regard to overturning. Therefore, it seems unnecessary to require buildings to be evaluated for full pseudo lateral force levels. Thus, for evaluating the actions at the foundation/soil interface, the alternative R_{OT} procedure is intended to provide a method that is consistent with prevailing practice specified in current codes for new buildings.

4.2.4.4 Component Strength

Component strength shall be taken as the expected strength, Q_{CE}, for deformation-controlled actions, and as the nominal strength, Q_{CN}, for force-controlled actions. Unless calculated otherwise, the expected strength shall be assumed equal to the nominal strength multiplied by 1.25. Alternatively, if allowable stresses are used, nominal strengths shall be taken as the allowable values multiplied by the following values:

Steel	1.7
Masonry	2.5
Wood	2.0

Evaluation Phase (Tier 2)

Except for wood diaphragms and wood and masonry shear walls, the allowable values shall not include a one-third increase for short-term loading.

Default values defined in Section 2.2 are to be assumed unless otherwise indicated by the available documents.

Where calculating capacities of deteriorated elements, the evaluating design professional shall account for the deterioration by making reductions in the material strength, section properties, and other parameters as approved by the authority having jurisdiction.

C4.2.4.4   Component Strength

The *NEHRP Recommended Provisions for Seismic Regulations of New Buildings and Other Structures* (BSSC, 2000) and the *Guidelines for Seismic Retrofit of Existing Buildings* provide component capacities for use in strength design or load and resistance factor design. These include nominal strength for wood, concrete, masonry, and steel. Note that the resistance factors, ϕ, which are used in ultimate strength code design, are not used in calculating capacities of members where the LSP or LDP is used.

The 1.25 factor for deformation-controlled actions can be applied because the force levels are intended for design. The design earthquake motion is deemed too conservative for an existing building. The 1.25 factor is intended to remove this conservatism since the actual strength of the components is typically greater than the design strength. Note that the 1.25 factor applies only to deformation-controlled actions since the nominal strength is a lower bound limit for components with little or no ductility.

4.2.4.5   Acceptance Criteria for the LSP and LDP

4.2.4.5.1   Deformation-Controlled Actions

The acceptability of deformation-controlled primary and secondary components shall be determined in accordance with Equation (4-11).

$$Q_{CE} \geq \frac{Q_{UD}}{m} \qquad (4-11)$$

where:
Q_{UD} =   Action due to gravity and earthquake loading per Section 4.2.4.3.1.
m =   Component demand modifier to account for the expected ductility of the component. The appropriate m-factor shall be chosen from Tables 4-5 to 4-8, based on the level of performance and component characteristics. Interpolation shall be permitted in Tables 4-5 to 4-8. $m = 1.0$ for all components in buildings analyzed using Equation (3-2).
Q_{CE} =   Expected strength of the component at the deformation level under consideration. Q_{CE} shall be calculated in accordance with Section 4.2.4.4 considering all co-existing actions due to gravity and earthquake loads.

C4.2.4.5.1 Deformation-Controlled Actions

The m-factors in Tables 4-5 to 4-8 were developed using the values in FEMA 356 as a starting point and then modified so that this standard provides comparable results to FEMA 178 for the Life Safety Performance Level. Considering the effect of factor C (for short period structures) and different capacities used in the two documents, it can be shown that, for equivalent results with FEMA 178, the value of m for Life Safety Performance Level should be in the range of 0.7 to 0.9 times the value of R.

Evaluation Phase (Tier 2)

> Note that the acceptability criteria and use of *m*-factors is applicable to the LSP and LDP only. *m*-factors are not used in conjunction with evaluating walls for out-of-plane forces or nonstructural elements, or where using the Special Procedures for unreinforced masonry bearing walls with flexible diaphragms.

4.2.4.5.1.1 Steel Beams

Values for the *m*-factor for steel beams shall be as specified in Table 4-5. If $L_p < L_b = L_r$, then m shall be replaced by m_e, calculated in accordance with Equation (4-12).

$$m_e = C_b \left[m - (m-1)\frac{L_b - L_p}{L_r - L_p} \right] \qquad (4-12)$$

where:

L_b = Distance between points braced against lateral displacement of the compression flange, or between points braced to prevent twist of the cross-section (*LRFD Specifications* [AISC, 1999])

L_p = Limiting unbraced length between points of lateral restraint for the full plastic moment capacity to be effective (AISC, 1997)

L_r = Limiting unbraced length between points of lateral support beyond which elastic lateral torsional buckling of the beam is the failure mode (AISC, 1997)

m = Value of *m* given in Table 4-5

m_e = Effective *m* computed in accordance with Equation (4-12)

C_b = Coefficient to account for effect of non-uniform moment (AISC, 1997)

4.2.4.5.2 Force-Controlled Actions

The acceptability of force-controlled primary and secondary components shall be determined in accordance with Equation (4-13).

$$Q_{CN} \geq Q_{UF} \qquad (4-13)$$

where:

Q_{UF} = Action due to gravity and earthquake loading; Q_{UF} shall be calculated in accordance with Section 4.2.4.3.2

Q_{CN} = Nominal strength of the component at the deformation level under consideration; Q_{CN} shall be calculated in accordance with Section 4.2.4.4 considering all co-existing actions due to gravity and earthquake loads

4.2.5 Out-of-Plane Wall Forces

Walls shall be evaluated for out-of-plane inertial forces as required by this section when triggered by the procedures of Section 4.3 through 4.6.

4.2.5.1 Out-of-Plane Anchorage to Diaphragms

Walls shall be positively anchored to all diaphragms that provide lateral support for the wall or are vertically supported by the wall. Anchorage of walls to diaphragms shall be evaluated for forces calculated using Equation (4-14), which shall be developed in the diaphragm. If sub-diaphragms are used, each sub-diaphragm shall be capable of transmitting the shear forces due to wall anchorage to a continuous diaphragm tie. Sub-diaphragms shall have length-to-depth ratios not exceeding 3-to-1. Where wall panels are stiffened for out-of-plane behavior by pilasters or similar elements, anchors shall be provided at each such element, and the distribution of out-of-plane forces to wall anchors and diaphragm ties shall consider the stiffening effect and accumulation of forces at these elements. Wall anchor connections shall be considered force controlled.

$$F_p = ?S_{DS} W \qquad (4\text{-}14)$$

where:

F_p = Design force for anchorage of walls to diaphragms
? = 0.3 for Life Safety and 0.45 for Immediate Occupancy
S_{DS} = Design short-period spectral response acceleration parameter
W = Weight of the wall tributary to the anchor

Exceptions:

1. F_p shall not be less than the minimum of 400 lb/ft or 400 S_{DS} (lb/ft).
2. Wall anchorage to flexible diaphragms shall have a minimum strength of three times the anchorage forces specified by Equation (4-14).

4.2.5.2 Out-of-Plane Strength

Wall components shall have adequate strength to span between locations of out-of-plane support when subjected to out-of-plane forces calculated using Equation (4-15).

$$F_p = ?S_{DS} W \qquad (4\text{-}15)$$

where:

F_p = Out-of-plane force per unit area for design of a wall spanning between two out-of-plane supports
? = 0.3 for Life Safety and 0.45 for Immediate Occupancy
S_{DS} = Design short-period spectral response acceleration parameter
W = Weight of the wall per unit area

C4.2.5 Out-of-Plane Wall Forces

Inadequate anchorage of heavy masonry and concrete walls to diaphragms for out-of-plane inertial loads has been a frequent cause of building collapse in past earthquakes. Following the 1971 San Fernando earthquake, the *Uniform Building Code* adopted requirements for positive direct connection of wall panels to diaphragms, with anchorage designed for a minimum force equal to ZIC_pW_p. In this equation, the quantity, ZIC_p, represents the equivalent out-of-plane inertial loading on the wall panel and typically had a value that was 75 percent of the effective peak ground acceleration for the site. This standard uses design provisions based on observation made following the 1994 Northridge earthquake. Failures occurred in a number of buildings meeting the requirements of the building code in effect at that time. Actual strong motion recordings in buildings with flexible diaphragms indicated that these diaphragms amplify the effective peak ground accelerations by as much as three times.

Evaluation Phase (Tier 2)

4.2.6 Special Procedure for Unreinforced Masonry

4.2.6.1 General

Unreinforced masonry bearing wall buildings with flexible diaphragms being evaluated to the Life Safety Performance Level shall be evaluated in accordance with the requirements of this section.

The evaluation requirements of Chapter 2 shall be met prior to conducting this special procedure.

This special procedure shall apply to unreinforced masonry bearing wall buildings with the following characteristics:

- Flexible diaphragms at all levels above the base of the structure
- A minimum of two lines of walls in each principal direction, except for single-story buildings with an open front on one side
- A maximum of six stories above the base of the structure

A Tier 3 Evaluation shall be conducted for buildings not meeting the requirements of this section.

4.2.6.2 Evaluation Requirements

4.2.6.2.1 Condition of Materials

All existing materials used as part of the required vertical-load-carrying and/or lateral-force-resisting system shall meet the requirements of Section 4.3.3.

In addition to meeting the requirements of Section 4.4.2.5.3, facing and backing of multi-wythe walls shall be bonded so that not less than 10 percent of the exposed face area is composed of solid headers extending not less than 4 inches into the backing. Where backing consists of two or more wythes, the headers shall extend not less than 4 inches into the most distant wythe, or the backing wythes shall be bonded together with separate headers for which the area and spacing conform to the foregoing. Wythes of walls not meeting these requirements shall be considered veneer and shall not be included in the effective thickness used in calculation of the height-to-thickness ratio and shear strength of the wall.

> **Exception:** Where S_{DS} is 0.50 or less, veneer wythes anchored in accordance with the jurisdiction having authority and made composite with backup masonry is permitted for the calculation of the effective thickness.

4.2.6.2.2 Testing

All unreinforced masonry (URM) walls used to carry vertical loads or resist lateral forces parallel and perpendicular to the wall plane shall be tested. The shear tests shall be taken at locations representative of the mortar conditions throughout the building. Test locations shall be determined by the design professional in charge. Results of all tests and their location shall be recorded.

The minimum number of tests per masonry class shall be determined as follows:

- At each of both the first and top stories, not less than two tests per wall or line of wall elements providing a common line of resistance to lateral forces.
- At each of all other stories, not less than one test per wall or line of wall elements providing a common line of resistance to lateral forces.
- Not less than one test per 1,500 square feet of wall surface or less than a total of eight tests.

Evaluation Phase (Tier 2)

For masonry walls that use high shear strength mortar, masonry testing shall be performed in accordance with Section 4.2.6.2.2.2. The quality of mortar in all other coursed masonry walls shall be determined by performing tests in accordance with Section 4.2.6.2.2.1.

C4.2.6.2.2 Testing

Location of Tests. Test locations should consider factors such as workmanship at different building height levels, weathering of exterior surfaces, condition of interior surfaces, and deterioration due to water or other substances contained within the building.

Pointing. Nothing should prevent pointing with mortar of all the masonry wall joints before the tests are made. All deteriorated mortar joints in URM walls should be pointed. Pointing should be performed under a permit and with special inspection. Any raking of mortar joints or drilling in URM structures should be done using nonimpact tools.

Collar Joints of Multi-Wythe Masonry. The collar joints should be inspected at the test locations during each in-place shear test, and estimates of the percentage of the surfaces of adjacent wythes that are covered with mortar should be reported with the results of the in-place shear tests.

Unreinforced Masonry Classes. All existing unreinforced masonry should be categorized into one or more classes based on strength, quality of construction, state of repair, deterioration, and weathering. A class should be characterized by the masonry strength determined in accordance with Section 4.2.6.2.3. Classes should be defined for whole walls, not for small areas of masonry within a wall. Discretion in the definition of classes of masonry is permitted to avoid unnecessary testing.

4.2.6.2.2.1 In-Place Mortar Test

Mortar shear test values, v_{to}, shall be calculated for each in-place shear test in accordance with Equation (4-16). Individual unreinforced masonry walls with 50 percent of mortar test values less than 30 psi shall be pointed and retested.

$$v_{to} = \frac{V_{test}}{A_b} - p_{D+L} \tag{4-16}$$

where:
V_{test} = Load at first observed movement
A_b = Total area of the bed joints above and below the test specimen
p_{D+L} = Stress resulting from actual dead plus live loads in place at the time of testing

The mortar shear strength, v_{te}, is defined as the value exceeded by 80 percent of all mortar shear test values, v_{to}. Unreinforced masonry with mortar shear strength, v_{te}, less than 30 psi shall be pointed and retested.

C4.2.6.2.2.1 In-Place Mortar Test

The available standard for masonry shear strength test is *Uniform Building Code Standard* 21-6 (ICBO, 1997). Multi-wythe masonry laid with headers should use the in-place shear push test. The bed joints of the outer wythe of the masonry should be tested in shear by laterally displacing a single brick relative to the adjacent bricks in the same wythe. The head joint opposite the loaded end of the test brick should be excavated and cleared. The brick adjacent to the loaded end of the test brick should be removed and excavated to provide space for a hydraulic ram and steel loading blocks. Steel blocks, the size of the end of the brick, should be used on each end of the ram to distribute the load to the brick. The blocks should not contact the mortar joints. The load should be applied horizontally, in the plane of the wythe. Load should be recorded at first sign of movement of the test brick as indicated by spalling of the face of the mortar bed joints. The strength of the mortar should be calculated by dividing the load at the first movement of the test brick by the nominal gross area of the sum of the two bed joints.

Evaluation Phase (Tier 2)

4.2.6.2.2.2 Masonry

The tensile-splitting strength, f_{sp}, of existing masonry using high-strength mortar shall be determined in accordance with the one of the following:

- The tensile-splitting strength of a core sample shall be determined in accordance with ASTM C496-96 The tensile-splitting strength shall be calculated in accordance with Equation (4-17).

$$f_{sp} = \frac{2P_{test}}{\pi a_n} \qquad (4\text{-}17)$$

- The tensile-splitting strength of a sawn rectangle shall be determined in accordance with ASTM E519-74. The tensile-splitting strength shall be calculated in accordance with Equation (4-18).

$$f_{sp} = \frac{0.494 P_{test}}{a_n} \qquad (4\text{-}18)$$

where:
P_{test} = Splitting test load
a_n = Diameter of core multiplied by its length or area of the side of a square prism

The minimum average value of tensile splitting strength, f_{sp}, as calculated by Equation (4-17) or (4-18), shall be 50 psi. Individual unreinforced masonry walls with an average tensile splitting strength of less than 50 psi shall be pointed and retested.

Alternatively, default material properties as defined in Section 2.2 may be assumed for the prism strength of masonry, f'_m, or material property data may be obtained from building codes from the building's year of construction or as-built plans for determination of f'_m.

C4.2.6.2.2.2 Masonry

Different types of masonry require different tests to determine the shear strength. As a general guide for selecting the correct test method for modern masonry, the design professional should consider using a core tested as prescribed in ASTM C496-96 to determine the tensile-splitting stress. The tensile-splitting stress is the same as the horizontal shear stress. Wythes of solid masonry units should be tested by sampling the masonry by drilled cores of not less than 8 inches in diameter. A bed joint intersection with a head joint should be in the center of the core. The core shall be placed in the test apparatus with the bed joint 45 degrees from the horizontal.

Another method is to use a square prism extracted from the wall that is tested as prescribed in ASTM E519-74 to determine the tensile-splitting stress. Hollow unit masonry constructed of through-the-wall units should be tested by sampling the masonry by a sawn prism of not less than 18 inches square. The diagonal of the prism should be placed in a vertical position. The effect of axial loading on the tensile-splitting stress must be added for the expected horizontal shear stress.

Estimation of f'_m should be limited to recently constructed masonry. The determination of f'_m requires the unit corresponding to a specification of the unit by an ASTM Standard and classification of the mortar by type. The source of the masonry units should be traced for the unit compressive strength. Then the unit compressive strength with the mortar class on the available construction documents should be used to determine f'_m.

Evaluation Phase (Tier 2)

4.2.6.2.2.3 Wall Anchors

Wall anchors used as part of the required tension anchors shall be tested in pull-out. Results of all tests shall be reported. A minimum of four anchors per floor shall be tested but not less than 10 percent of the total number of tension anchors at each level. Two tests per floor shall occur at walls with joists framing into the wall, and two tests per floor shall occur at walls with joists parallel to the wall. The strength of the wall anchors shall be calculated as the average of the tension tests values of anchors having the same wall thickness and framing orientation.

C4.2.6.2.2.3 Wall Anchors

The test apparatus for testing wall anchors should be supported by the masonry wall. The distance between the anchor and the test apparatus support should not be less than the wall thickness. Existing wall anchors should be given a preload of 300 pounds prior to establishing a datum for recording elongation. The tension test load reported should be recorded at 1/8-inch relative movement of the anchor and the adjacent masonry surface. The test report should include the test results as related to the wall thickness and framing orientation.

4.2.6.2.3 Masonry Strength

4.2.6.2.3.1 Shear Strength

The expected unreinforced masonry strength, v_{me}, shall be determined for each masonry class in accordance with the following:

- When testing in accordance with Section 4.2.6.2.2.1 is performed, v_{me} shall be calculated in accordance with Equation (4-19).

$$v_{me} = 0.56 v_{te} + \frac{0.75 P_D}{A_n} \qquad (4\text{-}19)$$

- When testing in accordance with Section 4.2.6.2.2.2 is performed, v_{me} shall be calculated in accordance with Equation (4-20).

$$v_{me} = 0.8 f_{sp} + \frac{0.5 P_D}{A_n} \qquad (4\text{-}20)$$

where:
- v_{te} = Mortar shear strength calculated in Section 4.2.6.2.2.1
- f_{sp} = Tensile splitting strength calculate in Section 4.2.6.2.2.2
- P_D = Superimposed dead load at the top of the pier under consideration (lb)
- A_n = Area of net mortared/grouted section (in.²)

- When the value of f'_m is assumed in accordance with Section 4.2.6.2.2.2, v_{me} shall be taken as the minimum of:

$$2.5 \sqrt{f'_m}$$

200 psi

$$v + \frac{0.75 P_D}{A_n}$$

where:
- v = 62.5 psi for running bond masonry not grouted solid
- v = 100 psi for running bond masonry grouted solid
- v = 25 psi for stack bond masonry grouted solid
- $f'm$ = compressive strength of masonry

4.2.6.2.3.2 Axial Strength

The allowable compressive stress in unreinforced masonry due to dead plus live loads shall be taken as 300 psi. Tensile stress is not permitted in unreinforced masonry.

C4.2.6.2.3.2 Axial Strength

There is no specific check for axial loads in this procedure. However, axial loads are used in determining the shear strength values (Equations 4-16, 4-19, and 4-20). Also, loss of masonry capacity due to seismic forces also may result in a loss of gravity carrying support (Section 4.8.4.5). Therefore, the design professional should be aware of any heavily loaded walls during the evaluation.

4.2.6.3 Analysis

4.2.6.3.1 Cross Walls

4.2.6.3.1.1 General

Only wood-framed walls sheathed with materials listed in Table 4-1 may be considered as cross walls. Cross walls shall not be spaced more than 40 feet on center measured perpendicular to the direction under consideration and should be present in each story of the building. Cross walls shall extend the full story height between diaphragms. Cross walls shall have a length-to-height ratio between openings equal to or greater than 1.5.

Exceptions:
- Cross walls need not be present at all levels in accordance with Section 4.2.6.3.2.2, Equation (4-25)
- Cross walls that meet the following requirements need not be continuous:
  - Shear connections and anchorage at all edges of the diaphragm shall meet the requirements of Section 4.2.6.3.2.6
  - Cross walls shall have a shear strength of $0.5 S_{DI} S W_d$ and shall interconnect the diaphragm to the foundation
  - Diaphragms spanning between cross walls that are continuous shall comply with the following equation:

$$\frac{2.1 S_{DI} W_d + V_{ca}}{2 v_u D} \leq 2.5 \qquad (4\text{-}21)$$

where:
- S_{DI} = Design spectral response acceleration parameter at a 1-second period
- W_d = Total dead load tributary to the diaphragm (lb)
- V_{ca} = Total shear strength of cross walls in the direction of analysis immediately above the diaphragm level being evaluated (lb)
- v_u = Unit shear strength of diaphragm (plf)
- D = Depth of diaphragm (ft)

Evaluation Phase (Tier 2)

Table 4-1. Cross Wall Shear Strengths[1,2,3]

Material and Configuration	Seismic Shear Strength (plf)
Plaster on wood or metal lath	600
Plaster on gypsum lath	550
Gypsum wall board, unblocked edges	200
Gypsum wall board, blocked edges	400
Plywood sheathing applied directly over wood studs	600
Plywood sheathing applied over wood sheathing	600
Plywood sheathing applied over existing plaster	0
Drywall or plaster applied directly over wood studs	230
Drywall or plaster applied to sheathing over existing wood studs	0

[1] Materials shall meet the requirements of Section 4.3.3.
[2] Shear values are permitted to be combined. However, total combined value shall not exceed 900 plf.
[3] No increase in stress allowed.

4.2.6.3.1.2 Shear Strength

Within any 40 feet measured along the span of the diaphragm, the sum of the cross wall shear strengths shall be greater than or equal to 30 percent of the diaphragm shear strength of the strongest diaphragm at or above the level under consideration. The values in Table 4-1 may be assumed for cross wall strengths. Strengths only apply to the provisions of Section 4.2.6.

4.2.6.3.2 Diaphragms

4.2.6.3.2.1 Shear Strength

The values in Table 4-2 may be assumed for diaphragm strengths. Strengths apply only to the provisions of Section 4.2.6.

Table 4-2. Diaphragm Shear Strengths[1,4]

Material and Configuration	Seismic Shear Strength (plf)
Roofs with straight sheathing and roofing applied directly to sheathing	300
Roofs with diagonal sheathing and roofing applied directly to sheathing	750
Floors with straight tongue-and-groove sheathing	300
Floors w/ straight sheathing and finished wood flooring w/ board edges offset or perpendicular	1,500
Floors with diagonal sheathing and finished wood flooring	1,800
Metal deck[2]	1,800
Metal deck welded for seismic resistance[3]	3,000
Plywood sheathing applied directly over existing straight sheathing with ends of plywood sheets bearing on joists or rafters and edges of plywood located on center of individual sheathing boards	675

[1] Materials shall meet the requirements of Section 4.3.3.
[2] Minimum 22-gage steel deck with welds to supports at a maximum average spacing of 12 inches.
[3] Minimum 22-gage steel deck with ¾-inch F plug welds at a maximum average spacing of 8 inches and with sidelap welds, screws, or button punches at a spacing of 24 inches or less.
[4] Values taken from ABK Methodology (ABK, 1981).

Evaluation Phase (Tier 2)

4.2.6.3.2.2 Demand-Capacity Ratios

Demand-capacity ratios, *DCR*, shall be evaluated when S_{D1} exceeds 0.20. Demand-capacity ratios shall be calculated for a diaphragm at any level in accordance with the following equations:

- Diaphragms without cross walls at levels immediately above or below:

$$DCR = \frac{2.1 S_{D1} W_d}{\sum v_u D} \quad (4\text{-}22)$$

- Diaphragms in a one-story building with cross walls:

$$DCR = \frac{2.1 S_{D1} W_d}{\sum v_u D + V_{cb}} \quad (4\text{-}23)$$

- Diaphragms in a multi-story building with cross walls at all levels:

$$DCR = \frac{2.1 S_{D1} \sum W_d}{\sum (\sum v_u D + V_{cb})} \quad (4\text{-}24)$$

- Roof diaphragms and the diaphragms directly below if coupled by cross walls:

$$DCR = \frac{2.1 S_{D1} \sum W_d}{\sum (\sum v_u D)} \quad (4\text{-}25)$$

where:
S_{D1} = Design spectral response acceleration parameter at a 1-second period
W_d = Total dead load tributary to the diaphragm (lb)
V_{cb} = Total shear strength of cross walls in the direction of analysis immediately below the diaphragm level being evaluated (lb)
v_u = Unit shear strength of diaphragm (plf)
D = Depth of diaphragm (ft)

4.2.6.3.2.3 Acceptability Criteria

The intersection of diaphragm span between walls, *L*, and the demand-capacity ratio, *DCR*, shall be located within Region 1, 2, or 3 on Figure 4-1.

4.2.6.3.2.4 Chords and Collectors

An analysis for diaphragm flexure need not be made and chords need not be present.

Where walls do not extend the length of the diaphragm, collectors shall be present. The collectors shall be able to transfer diaphragm shears calculated in accordance with Section 4.2.6.3.2.6 into the shear walls.

4.2.6.3.2.5 Diaphragm Openings

Diaphragm forces at corners of openings shall be investigated.

There shall be sufficient capacity to develop the strength of the diaphragm at opening corners.

The demand-capacity ratio shall be calculated and evaluated in accordance with Sections 4.2.6.3.2.2 and 4.2.6.3.2.3 for the portion of the diaphragm adjacent to an opening using the opening dimension as the diaphragm span.

The demand-capacity ratio shall be calculated and evaluated in accordance with Sections 4.2.6.3.2.2 and 4.2.6.3.2.3 for openings occurring in the end quarter of the diaphragm span. The diaphragm strength, $v_u D$, shall be based on the net depth of the diaphragm.

Evaluation Phase (Tier 2)

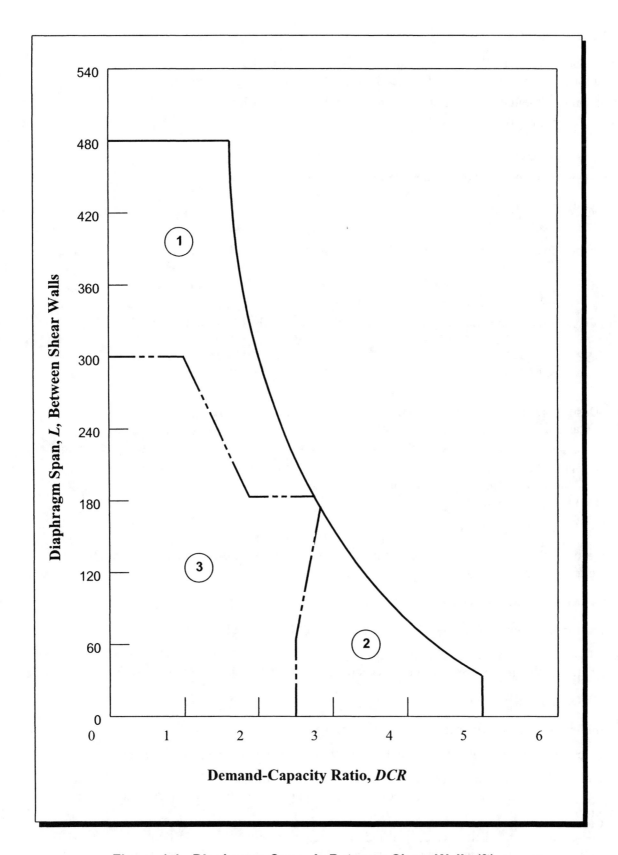

Figure 4-1. Diaphragm Span, *L*, Between Shear Walls (ft)

Evaluation Phase (Tier 2)

4.2.6.3.2.6 Diaphragm Shear Transfer

Diaphragm shear transfer shall be evaluated when S_{DI} exceeds 0.133. Diaphragms shall be connected to shear walls at each end and shall be able to develop the minimum of the forces calculated in accordance with Equations (4-26) and (4-27).

$$V_d = 1.25\, S_{DI}\, C_p W_d \qquad (4\text{-}26)$$

$$V_d = v_u D \qquad (4\text{-}27)$$

where:
S_{DI} = Design spectral response acceleration parameter at a 1-second period
W_d = Total dead load tributary to the diaphragm (lb)
v_u = Unit shear strength of diaphragm (plf)
D = Depth of diaphragm (ft)
C_p = Horizontal force factor (see Table 4-3)

Table 4-3. Horizontal Force Factor, C_p

Configuration of Materials	C_p
Roofs with straight or diagonal sheathing and roofing applied directly to the sheathing, or floors with straight tongue-and-groove sheathing	0.50
Diaphragm with double or multiple layers of boards with edges offset, and blocked structural panel systems	0.75

4.2.6.3.3 Shear Walls

4.2.6.3.3.1 Shear Wall Actions

In-plane shear shall be evaluated when S_{DI} exceeds 0.133. The story force distributed to a shear wall at any diaphragm level shall be determined in accordance with the following equations:

- For buildings without cross walls:

$$F_{wx} = 0.8\, S_{DI}\, (W_{wx} + 0.5 W_d) \qquad (4\text{-}28)$$

but not exceeding,

$$F_{wx} = 0.8\, S_{DI}\, W_{wx} + v_u D \qquad (4\text{-}29)$$

- For buildings with cross walls in all levels:

$$F_{wx} = 0.75\, S_{DI}\, (W_{wx} + 0.5 W_d) \qquad (4\text{-}30)$$

but need not exceed,

$$F_{wx} = 0.75\, S_{DI}\, (W_{wx} + \Sigma W_d (\frac{v_u D}{\Sigma(\Sigma v_u D)})) \qquad (4\text{-}31)$$

and need not exceed,

$$F_{wx} = 0.75\, S_{DI}\, W_{wx} + v_u D \qquad (4\text{-}32)$$

The wall story shear shall be calculated in accordance with Equation (4-33).

$$V_{wx} = \Sigma F_{wx} \qquad (4\text{-}33)$$

where:
S_{D1} = Design spectral response acceleration parameter at a 1-second period
W_{wx} = Dead load of an unreinforced masonry wall assigned to level x, taken from mid-story below level x to mid-story above level x (lb)
W_d = Total dead load tributary to the diaphragm (lb)
v_u = Unit shear strength of diaphragm (plf)
D = Depth of diaphragm (ft)

4.2.6.3.3.2 Shear Wall Strengths

The shear wall strength shall be calculated in accordance with Equation (4-34).

$$V_a = 0.67 v_{me} D t \tag{4-34}$$

where:
v_{me} = Expected masonry shear strength (psi) calculated in accordance with Section 4.2.6.2.3.1
D = In-plane width dimension of masonry (in.)
t = Thickness of wall (in.)

The rocking shear strength shall be calculated in accordance with Equations (4-35) and (4-36):

- For walls without openings:

$$V_r = 0.9(P_D + 0.5 P_w) \frac{D}{H} \tag{4-35}$$

- For walls with openings:

$$V_r = 0.9 P_D \frac{D}{H} \tag{4-36}$$

where:
P_D = Superimposed dead load at the top of the pier under consideration (lb)
P_W = Weight of wall (lb)
D = In-plane width dimension of masonry (in.)
H = Least clear height of opening on either side of pier (in.)

4.2.6.3.3.3 Shear Wall Acceptance Criteria

The acceptability of unreinforced masonry shear walls shall be determined in accordance with Equations (4-37), (4-38), and (4-39).

- When $V_r < V_a$,

$$0.7 V_{wx} < \Sigma V_r \tag{4-37}$$

- When $V_a < V_r$, V_{wx} shall be distributed to the individual wall piers, V_p, in proportion to D/H, and Equations (4-38) and (4-39) shall be met.

$$V_p < V_a \tag{4-38}$$
$$V_p < V_r \tag{4-39}$$

If $V_p < V_a$ and $V_p > V_r$ for any pier, the pier shall be omitted from the analysis and the procedure repeated.

4.2.6.3.4 Out-of-Plane Demands

Out-of-plane demands shall be evaluated when S_{D1} exceeds 0.133. The unreinforced masonry wall height-to-thickness ratios shall be less than those set forth in Table 4-4.

Evaluation Phase (Tier 2)

The following limitations shall apply to Table 4-4 when S_{D1} exceeds 0.4:

- For buildings within Region 1 of Figure 4-1 as defined in Section 4.2.6.3.2.3, height-to-thickness ratios in column A of Table 4-4 may be used if cross walls comply with the requirements of Section 4.2.6.3.1 and are present in all stories.
- For buildings within Region 2 of Figure 4-1, as defined in Section 4.2.6.3.2.3, height-to-thickness ratios in column A may be used.
- For buildings within Region 3 of Figure 4-1, as defined in Section 4.2.6.3.2.3, height-to-thickness ratios in column B may be used.

Table 4-4. Allowable Height-to Thickness Ratios of Unreinforced Masonry Walls

Wall Type	$0.13 = S_{D1} < 0.25$	$0.25 = S_{D1} < 0.4$	$S_{D1} = 0.4$ A	$S_{D1} = 0.4$ B
Walls of one-story buildings	20	16	16[1,2]	13
Top story of multi-story building	14	14	14[1,2]	9
First story of multi-story building	20	18	16	15
All other conditions	20	16	16	13

[1] Value may be used when in-plane shear tests in accordance with Section 4.2.6.2.2.1 have a minimum v_{te} of 100 psi or a minimum v_{te} of 60 psi and a minimum of 50 percent mortar coverage of the collar joint.

[2] Values may be interpolated between column A and B where in-plane shear tests in accordance with Section 4.2.6.2.2.1 have a v_{te} of between 30 and 60 psi and a minimum of 50 percent mortar coverage of the collar joint.

4.2.6.3.5 Wall Anchorage

Wall anchorage shall be evaluated when S_{D1} exceeds 0.067. Anchors shall be capable of developing the maximum of:

- 2.1 S_{D1} times the weight of the wall, or
- 200 pounds per lineal foot, acting normal to the wall at the level of the floor or roof.

Walls shall be anchored at the roof and all floor levels at a spacing equal to or less than 6 feet on center.

At the roof and all floor levels, anchors shall be provided within 2 feet horizontally from the inside corners of the wall.

The connection between the walls and the diaphragm shall not induce cross-grain bending or tension in the wood ledgers.

4.2.6.3.6 Buildings with Open Fronts

Single-story buildings with an open front on one side shall have cross walls parallel to the open front. The effective diaphragm span, L_i, for use in Figure 4-1, shall be calculated in accordance with Equation (4-40).

$$L_i = 2L\left(\frac{W_w}{W_d} + 1\right) \tag{4-40}$$

where:
L = Span of diaphragm between shear wall and open front (ft)
W_w = Total weight of wall above open front
W_d = Total dead load tributary to the diaphragm (lb)

Evaluation Phase (Tier 2)

The diaphragm demand-capacity ratio, DCR, shall be calculated in accordance with Equation (4-41).

$$DCR = \frac{2.1 S_{D1} (W_d + W_w)}{(v_u D + V_{cb})} \quad (4\text{-}41)$$

where:
- S_{D1} = Design spectral response acceleration parameter at a 1-second period
- v_u = Unit shear strength of diaphragm (plf)
- D = Depth of diaphragm (ft)
- V_{cb} = Total shear strength of cross walls in the direction of analysis immediately below the diaphragm level being evaluated (lb)
- W_w = Total weight of wall above open front
- W_d = Total dead load tributary to the diaphragm (lb)

4.2.7 Nonstructural Components

4.2.7.1 Component Demands

The seismic forces on nonstructural components shall be calculated in accordance with Equations (4-42), (4-43), and (4-44) where triggered by the procedures in Section 4.8.

$$F_p = 0.4 \, a_p S_{DS} I_p W_p (1 + 2x/h)/R_p \quad (4\text{-}42)$$

F_p shall not be greater than:

$$F_p = 1.6 \, S_{DS} I_p W_p \quad (4\text{-}43)$$

and F_p shall not be taken as less than:

$$F_p = 0.3 \, S_{DS} I_p W_p \quad (4\text{-}44)$$

where:
- F_p = Seismic design force centered at the component's center of gravity and distributed relative to the component's mass distribution.
- S_{DS} = Design short-period spectral acceleration, as determined from Section 3.5.2.3.1.
- a_p = Component amplification factor from Table 4-9.
- W_p = Component operating weight.
- R_p = Component response modification factor, which varies from 1.0 to 6.0 (select appropriate value from Table 4-9).
- x = Height in structure at point of attachment of component. For items at or below the base, x shall be taken as 0. For items at or above the roof, x is not required to be taken as greater than the roof height h. For items attached at multiple locations, x shall be taken as the average height of attachment.
- h = Average roof height of structure relative to grade.
- I_p = Component performance factor taken as 1.0 for the Life Safety Performance Level and 1.5 for the Immediate Occupancy Performance Level.

The force, F_p, shall be applied independently, longitudinally, and laterally in combination with service loads associated with the component. Where positive and negative wind loads exceed F_p for nonstructural exterior walls, these wind loads shall govern the analysis. Similarly, where the building code horizontal loads exceed F_p for interior partitions, these building code loads shall govern the analysis.

The nonstructural component meets the criteria of this standard if the nominal strength of the nonstructural component exceeds F_p.

Drift ratios, D, shall be determined in accordance with Equations (4-45) or (4-46) where triggered by the procedures in Section 4.8.

For two connecting points on the same building or structural system:

$$D_r = (\delta_{xA} - \delta_{yA})/(X - Y) \tag{4-45}$$

For two connection points on separate buildings or structural systems:

$$D_p = \delta_{xA} + \delta_{xB} \tag{4-46}$$

where:
- D_p = Relative displacement
- D_r = Drift ratio
- X = Height of upper support attachment at level x as measured from grade
- Y = Height of lower support attachment at level y as measured from grade
- δ_{xA} = Deflection at building level x of building A, determined by elastic analysis
- δ_{yA} = Deflection at building level y of building A, determined by elastic analysis
- δ_{xB} = Deflection at building level x of building B, determined by elastic analysis

The effects of seismic displacements shall be calculated in accordance with displacements caused by other loads, as appropriate. Seismic displacements shall be calculated from the application of pseudo lateral force (Section 3.5.2.1) on the elastic structure.

4.2.7.2 Component Strengths

Component strength shall be taken as the nominal strength, Q_{CN}, as defined in Section 4.2.4.4.

4.2.7.3 Acceptance Criteria

The acceptability of nonstructural components shall be determined in accordance with Equation (4-47).

$$Q_{CN} = F_p \tag{4-47}$$

where:
- F_p = Demand on nonstructural components determined in accordance with Section 4.2.7.1
- Q_{CN} = Nominal strength of the component calculated in accordance with Section 4.2.4.4

Evaluation Phase (Tier 2)

Table 4-5. *m*-factors for Steel Components

Component/Conditions	Primary		Secondary	
	LS	IO	LS	IO
Fully Restrained Moment Frames:				
Beams[6]				
$\quad \dfrac{b}{2t_f} < \dfrac{52}{\sqrt{F_{ye}}}$	8.0	3.0	13.0	3.0
$\quad \dfrac{b}{2t_f} > \dfrac{95}{\sqrt{F_{ye}}}$	3.0	2.0	4.0	2.0
Columns ($P^4 < 0.2P_y$)				
$\quad \dfrac{b}{2t_f} < \dfrac{52}{\sqrt{F_{ye}}}$	8.0	3.0	13.0	3.0
$\quad \dfrac{b}{2t_f} > \dfrac{95}{\sqrt{F_{ye}}}$	2.0	2.0	3.0	2.0
Columns ($0.2P_y < P < 0.5P_y$)				
$\quad \dfrac{b}{2t_f} < \dfrac{52}{\sqrt{F_{ye}}}$	(1)	2.0	(2)	2.0
$\quad \dfrac{b}{2t_f} > \dfrac{95}{\sqrt{F_{ye}}}$	2.0	2.0	3.0	2.0
Columns ($P > 0.5P_y$)	(3)	(3)	(3)	(3)
Panel Zones	10.0	3.0	14.0	3.0
Welded Moment Connections[5]	2.0	1.0	2.0	1.0
Partially Restrained Moment Connections:				
Bolts or Welds in Tension	2.5	1.5	3.5	1.5
Other	4.0	2.0	6.0	2.0
Braced Frames				
Columns[3]				
Eccentric Braced Frames:	Same as beams in fully restrained frames			
Link Beam				
Brace and Column[3]				
Braces in Compression:				
$\quad$ Tubes: $\dfrac{d}{t} \le \dfrac{90}{\sqrt{F_{ye}}}$; Pipes: $\dfrac{d}{t} \le \dfrac{1{,}500}{F_{ye}}$	6.0	2.5	9.0	2.5
$\quad$ Tubes: $\dfrac{d}{t} \ge \dfrac{190}{\sqrt{F_{ye}}}$; Pipes: $\dfrac{d}{t} \ge \dfrac{6{,}000}{F_{ye}}$	3.0	1.5	3.0	1.5
Other Shapes	6.0	2.5	9.0	2.5
Braces in Tension:				
Tension-Compression Brace	6.0	2.5	11.0	2.5
Tension-Only Brace	3.0	1.5	11.0	1.5
Metal Deck	4.0	2.0	--	--

F_{ye} = 1.25F_y, expected yield stress.
[1]m = 12(1-1.7P/P_y).
[2]m = 20(1-1.7P/P_y).
[3]Force controlled.
[4]Axial load due to gravity and earthquake calculated as force-controlled action per Section 4.2.4.3.2.
[5]Alternatively, these connections may be considered force controlled if connections and joint web shear can be shown to develop the capacity of the beam.
[6]$L_b < L_p$; for $L_p < L_b < L_r$ see Section 4.2.4.5.1.1; $L_b > L_r$ beam shall be treated as force-controlled.

Evaluation Phase (Tier 2)

Table 4-6. *m*-factors for Concrete Components

Component/Conditions	Primary		Secondary	
	LS	IO	LS	IO
Beams, Flexure:				
Ductile[1]				
$? \leq 3\sqrt{f'_c}$	8.0	3.0	8.0	3.0
$v \geq 6\sqrt{f'_c}$	4.0	2.5	4.0	2.5
Nonductile	2.5	1.5	3.0	1.5
Columns, Flexure:				
Ductile[1]				
$\dfrac{P^{(4)}}{A_g f'_c} \leq 0.1$	5.0	3.0	5.0	3.0
$\dfrac{P}{A_g f'_c} \geq 0.4$	2.0	1.5	2.0	1.5
Nonductile				
$\dfrac{P}{A_g f'_c} \leq 0.1$	2.5	1.5	3.0	1.5
$\dfrac{P}{A_g f'_c} \geq 0.4$	1.5	1.5	1.5	1.5
Beams Controlled by Shear	2.0	1.5	3.5	1.5
Beam-Column Joints	(2)	(2)	(2)	(2)
Slab-Column Systems:[5]				
$\dfrac{V_g}{V_o} \leq 0.1$	3.0	3.0	3.0	3.0
$\dfrac{V_g}{V_o} \geq 0.4$	1.5	1.5	1.5	1.5
Infilled Frame Columns Modeled as Chords:				
Confined Along Entire Length	4.0	1.5	5.0	1.5
Not Confined	1.5	1.5	1.5	1.5
Shear Walls Controlled by Flexure:				
With Confined Boundary				
$a \leq 0.1$[3]	5.0	3.0	6.0	3.0
$a \geq 0.25$	3.0	1.5	4.0	1.5
Without Confined Boundary				
$a \leq 0.1$	3.0	2.0	4.0	2.0
$a \geq 0.25$	2.0	1.5	2.5	1.5
Coupling Beams	2.5	1.5	4.0	1.5
Shear Walls Controlled by Shear	2.5	2.0	3.0	2.0
Diaphragms	2.5	1.5	--	--

[1] Ductile beams and columns shall conform to the following requirements: (a) beams and columns shall meet the requirements of Sections 4.4.1.4.9, 4.4.1.4.10, 4.4.1.4.11, 4.4.1.4.12, and 4.4.1.4.15; (b) within the plastic region, closed stirrups shall be spaced at $<d/3$; (c) strength provided by stirrups shall be at least 3/4 of the design shear; (d) longitudinal reinforcement shall not be lapped within the plastic hinge region; (e) $(\rho-\rho')/\rho_{bal}<0.5$; and (f) column flexural capacity exceeds beam flexural capacity.

[2] These joints shall be considered force controlled.

[3] $a=[(A_s-A_s')f_y+P]/A_w f_c'$

[4] P=Axial load due to gravity and earthquake calculated as a force-controlled action per Section 4.2.4.3.2.

[5] V_g=gravity shear; V_o=punching shear capacity.

Evaluation Phase (Tier 2)

Table 4-7. m-factors for Masonry Components

Component/Conditions	Primary		Secondary	
	LS	IO	LS	IO
Unreinforced Masonry[1]	1.5	1.0	3.0	1.0
Reinforced Masonry in Flexure:[2]				
$f_a \leq 0.04 f'_m$				
$\quad \rho f_y / f'_m = 0.01$[3]	6.0	3.0	8.0	3.0
$\quad \rho f_y / f'_m = 0.05$	4.5	2.5	7.0	2.5
$\quad \rho f_y / f'_m = 0.20$	2.5	1.5	4.0	1.5
$0.04 f'_m < f_a \leq 0.075 f'_m$				
$\quad \rho f_y / f'_m = 0.01$	4.0	2.5	7.0	2.5
$\quad \rho f_y / f'_m = 0.05$	3.0	2.0	6.0	2.0
$\quad \rho f_y / f'_m = 0.20$	2.5	1.5	4.0	1.5
Reinforced Masonry in Shear	2.5	2.0	4.0	2.0
Masonry Infill[4]	3.0	1.0	--	--

[1] Applicable to building with rigid diaphragms; for flexible diaphragms see Special Procedure.
[2] f_a = Axial stress due to gravity loads per Equation (4-6).
[3] ρ = Percentage of total vertical reinforcement including boundary elements, if any.
[4] Capacity based on bed joint shear strength for zero vertical compressive stress.

Table 4-8. m-factors for Wood Components

Component/Conditions	Primary		Secondary	
	LS	IO	LS	IO
Straight Sheathing, Diagonal Sheathing, and Double Diagonal Sheathing[1]	3.0	1.5	4.0	1.5
Gypsum Sheathing/Wallboard[1]	4.0	2.0	5.0	2.0
Structural Panel Sheathing				
$\quad$ Shear Walls				
$\quad\quad h/L \leq 1.0$	4.5	2.0	5.5	2.0
$\quad\quad 3.5 \geq h/L \geq 2.0$[2]	3.5	1.7	4.5	1.7
$\quad$ Diaphragms	3.5	2.0	4.0	2.0
Hold-Down Anchors	3.5	2.0	4.0	2.0

[1] For $h/L \geq 2.0$, the component shall not be considered effective as a primary component.
[2] For $h/L \geq 3.5$, the component shall not be considered effective as a primary component.

Table 4-9. Nonstructural Component Amplification and Response Modification Factors

Component		a_p[1]	R_p
A. ARCHITECTURAL			
1.	**Interior nonstructural walls and partitions**		
	Plain (unreinforced) masonry walls	1.0	1.25
	Other walls and partitions	1.0	2.5
2.	**Cantilever elements (unbraced or braced to structural frame below its center of mass)**		
	Parapets and cantilever interior nonstructural walls	2.5	2.5
	Chimneys and stacks when laterally braced or supported by the structural frame	2.5	2.5
3.	**Cantilever elements (braced to structural frame above its center of mass)**		
	Parapets	1.0	2.5
	Chimneys and stacks	1.0	2.5
	Exterior nonstructural walls	1.0	2.5
4.	**Exterior nonstructural wall elements and connections**		
	Wall element	1.0	2.5
	Body of wall panel connections	1.0	2.5
	Fasteners of the connection system	1.25	1.0
5.	**Veneer**		
	Limited deformability elements and attachments	1.0	2.5
	Low deformability elements of attachments	1.0	1.25
6.	**Penthouses (except when framed by an extension of structural frame)**	2.5	3.5
7.	**Ceilings**	1.0	2.5
8.	**Cabinets**		
	Storage cabinets and laboratory equipment	1.0	2.5
9.	**Access floors**		
	Access floors designed to resist seismic forces	1.0	2.5
	All other	1.0	1.25
10.	**Appendages and ornamentations**	2.5	2.5
11.	**Signs and billboards**	2.5	2.5
12.	**Other rigid components**		
	High deformability elements and attachments	1.0	3.5
	Limited deformability elements and attachments	1.0	2.5
	Low deformability elements and attachments	1.0	1.25
13.	**Other flexible components**		
	High deformability elements and attachments	1.0	3.5
	Limited deformability elements and attachments	2.5	2.5
	Low deformability elements and attachments	2.5	1.25
B. MECHANICAL AND ELECTRICAL EQUIPMENT			
1.	**General mechanical**		
	Boilers and furnaces	1.0	2.5
	Pressure vessels on skirts or free standing	2.5	2.5
	Stacks	2.5	2.5
	Cantilevered chimneys	2.5	2.5
	Other	1.0	2.5
2.	**Manufacturing and process machinery**		
	General	1.0	2.5
	Conveyors (non-personnel)	2.5	2.5

Table 4-9. Nonstructural Component Amplification and Response Modification Factors

Component		a_p^1	R_p
3.	**Piping systems**		
	High deformability elements and attachments	1.0	3.5
	Limited deformability elements and attachments	1.0	2.5
	Low deformability elements and attachments	1.0	1.25
4.	**HVAC system equipment**		
	Vibration isolated	2.5	2.5
	Non-vibration isolated	1.0	2.5
	Mounted in line with ductwork	1.0	2.5
	Other	1.0	2.5
5.	**Elevator components**	1.0	2.5
6.	**Escalator components**	1.0	2.5
7.	**Trussed towers (free standing or guyed)**	2.5	2.5
8.	**General electrical**		
	Distributed systems (bus ducts, conduit, cable tray)	1.0	3.5
	Equipment	1.0	2.5
9.	**Light fixtures**	1.0	1.25

[1]Where justified by detailed dynamic analysis, a lower value for a_p is permitted, but it shall not be less than 1. The reduced value of a_p shall be between 2.5, assigned to flexible or flexibly attached equipment, and 1, assigned to rigid or rigidly attached equipment. Refer to the definitions (Section 1.3) for explanations of "Component, Flexible" and "Component, Rigid."

4.3 Procedures for Building Systems

This section provides Tier 2 Evaluation procedures that apply to all building systems: general, configuration and condition of the materials.

4.3.1 General

4.3.1.1 LOAD PATH: The structure shall contain a minimum of one complete load path for Life Safety and Immediate Occupancy for seismic force effects from any horizontal direction that serves to transfer the inertial forces from the mass to the foundation.

Tier 2 Evaluation Procedure: No Tier 2 Evaluation procedure is available for load paths in non-compliance.

C4.3.1.1   Load Path

There must be a complete lateral-force-resisting system that forms a continuous load path between the foundation, all diaphragm levels, and all portions of the building for proper seismic performance. The general load path is as follows: seismic forces originating throughout the building are delivered through structural connections to horizontal diaphragms; the diaphragms distribute these forces to vertical lateral-force-resisting elements such as shear walls and frames; the vertical elements transfer the forces into the foundation; and the foundation transfers the forces into the supporting soil. Compliance of this statement indicates only the existence of a complete load path. The adequacy of the load path is checked in subsequent statements.

If there is a discontinuity in the load path, the building is unable to resist seismic forces regardless of the strength of the existing elements. Mitigation with elements or connections needed to complete the load path is necessary to achieve the selected performance level. The design professional should be watchful for gaps in the load path. Examples would include a shear wall that does not extend to the foundation, a missing shear transfer connection between a diaphragm and vertical element, a discontinuous chord at a diaphragm notch, or a missing collector.

In cases where there is a structural discontinuity, a load path may exist but it may be a very undesirable one. At discontinuous shear walls, for example, the diaphragm may transfer the forces to frames not intended to be part of the lateral-force-resisting system. While not ideal, the load path is compliant and it may be possible to show that the load path is acceptable. Another compliant load path that may be undesirable is where seismic forces are transferred between lateral-force-resisting elements through friction.

A complete load path is a basic requirement for all buildings. The remaining evaluation statements in this standard target specific components of the load path and are intended to assist the design professional in locating potential gaps in the load path. While non-compliant statements later in the procedure might indicate a potential discontinuity or inadequacy in the load path, the identification of a complete load path is a necessary first step before continuing with the evaluation.

Evaluation Phase (Tier 2)

4.3.1.2 ADJACENT BUILDINGS: The clear distance between the building being evaluated and any adjacent building shall be greater than 4 percent of the height of the shorter building for Life Safety and Immediate Occupancy.

Tier 2 Evaluation Procedure: The drifts in the structure being evaluated shall be calculated using the Linear Static Procedure in Section 4.2. The drifts in the adjacent building shall be estimated using available information and the procedures of this standard. The SRSS combination of this assumed drift and the calculated drift of the structure being evaluated shall be less than the total separation at each level. The design professional shall note any obvious potential hazards posed by adjacent buildings.

C4.3.1.2 Adjacent Buildings

Buildings are often built right up to property lines in order to make maximum use of space, and historically buildings have been designed as if the adjacent buildings do not exist. As a result, the buildings may impact each other, or pound, during an earthquake. Building pounding can alter the dynamic response of both buildings and impart additional inertial loads on both structures.

Where one or both buildings have setbacks, the minimum separation should be evaluated based on the common height between the two buildings. Above the level of the setback, the separation should be evaluated based on the total height of the shorter building.

Buildings that are the same height and have matching floors will exhibit similar dynamic behavior. If the buildings pound, floors will impact other floors, so damage due to pounding usually will be limited to nonstructural components. Where the floors of adjacent buildings are at different elevations, floors will impact the columns of the adjacent building and can cause structural damage (see Figure C4-1). Where the buildings are of different heights, the shorter building can act as a buttress for the taller building. The shorter building receives an unexpected load while the taller building suffers from a major stiffness discontinuity that alters its dynamic response (see Figure C4-2). Since neither building is designed for these conditions, there is a potential for extensive damage and possible collapse.

Many buildings that are built tight to each other appear to survive earthquakes by acting as a solid block. However, the end buildings of the block may have pronounced pounding. An example of this condition was the downtown area of San Francisco during the Loma Prieta earthquake. End block buildings with unmatching floors have the greatest life safety concern.

Buildings that are the same height and have matching floor levels need not comply with this statement. Non-compliant separations between buildings that do not have matching floors must be checked using calculated drifts for both buildings. The SSRS combination is used because of the low probability that maximum drifts in both buildings will occur simultaneously and out of phase. Where information on the adjacent building is not available, conservative estimates for drift should be made in the evaluation.

The potential hazard of the adjacent building also must be evaluated. If a neighbor is a potential collapse hazard, this must be reported.

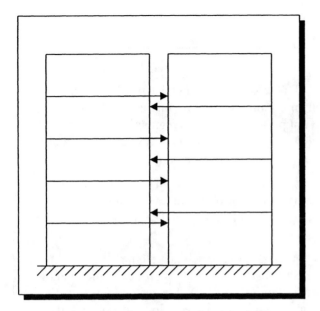
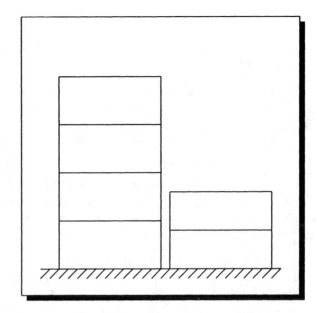

Figure C4-1. Unmatching Floors   Figure C4-2. Buildings of Different Heights

4.3.1.3 MEZZANINES: Interior mezzanine levels shall be braced independently from the main structure, or shall be anchored to the lateral-force-resisting elements of the main structure.

Tier 2 Evaluation Procedure: The load path from the mezzanine to the main structure shall be identified. The adequacy of the load path shall be evaluated for the forces in Section 4.2 considering the effect of the magnitude and location of any forces imparted by the mezzanine on the main structure.

C4.3.1.3 Mezzanines

It is very common for mezzanines to lack a lateral-force-resisting system. Often, mezzanines are added on by the building owner. Unbraced mezzanines can be a potential collapse hazard and should be checked for stability.

Lateral-force-resisting elements must be present in both directions to provide bracing. Where the mezzanine is attached to the main structure, the supporting elements of the main structure should be evaluated, considering both the magnitude and location of the additional forces imparted by the mezzanine.

If the load path is incomplete or nonexistent, mitigation with elements or connections needed to complete the load path is necessary to achieve the selected performance level.

Evaluation Phase (Tier 2)

4.3.2 Configuration

> **C4.3.2 Configuration**
>
> Good details and construction quality are of secondary value if a building has an odd shape that was not properly considered in the design. Although a building with an irregular configuration may be designed to meet all code requirements, irregular buildings generally do not perform as well as regular buildings in an earthquake. Typical building configuration deficiencies include an irregular geometry, a weakness in a given story, a concentration of mass, or a discontinuity in the lateral-force-resisting system.
>
> Vertical irregularities are defined in terms of strength, stiffness, geometry, and mass. These quantities are evaluated separately, but they are related and may occur simultaneously. For example, the frame in Figure C4-3 has a tall first story. It can be a weak story, a soft story, or both, depending on the relative strength and stiffness of this story and the stories above.
>
> One of the basic goals in the design of a building is efficient use of materials such that all members are stressed about equally. In seismic design, this goal is modified so that stresses within groups of members will be about the same. For example, in moment frames (as discussed in Section 4.4), it is desirable to have the beams weaker than the columns but to have all of the beams at the same stress level. In such a design, the members will yield at about the same level of earthquake forces; there will be no single weak link. Code provisions regarding vertical irregularities are intended to achieve this result. Significant irregularities that would cause damage to be concentrated in certain areas require special treatment.
>
> Horizontal irregularities involve the horizontal distribution of lateral forces to the resisting frames or shear walls. Irregularities in the shape of the diaphragm itself (i.e., diaphragms that are L-shaped or have notches) are discussed in Section 4.5.

4.3.2.1 WEAK STORY: The strength of the lateral-force-resisting system in any story shall not be less than 80 percent of the strength in an adjacent story, above or below, for Life Safety and Immediate Occupancy.

Tier 2 Evaluation Procedure: An analysis in accordance with the procedures in Section 4.2 shall be performed. The story strength shall be calculated, and the adequacy of the lateral-force-resisting elements in the non-compliant story shall be checked for the capacity to resist one-half of the total pseudo lateral force.

> **C4.3.2.1 Weak Story**
>
> The story strength is the total strength of all the lateral-force-resisting elements in a given story for the direction under consideration. It is the shear capacity of columns or shear walls, or the horizontal component of the capacity of diagonal braces. If the columns are flexure controlled, the shear strength is the shear corresponding to the flexural strength. Weak stories are usually found where vertical discontinuities exist or where member size or reinforcement has been reduced. It is necessary to calculate the story strengths and compare them. The result of a weak story is a concentration of inelastic activity that may result in the partial or total collapse of the story. By showing that the story strength is greater than $V/2$, the side-sway mechanism will most likely not be the story mechanism.
>
> Generally an examination of the building elevations can determine if a weak story exists without the need for calculation. A reduction in the number or length of lateral-force-resisting elements or a change in the type of lateral-force-resisting system are obvious indications that a weak story might exist. A gradual reduction of

Evaluation Phase (Tier 2)

> lateral-force-resisting elements as the building increases in height is typical and is not considered a weak story condition.
>
> An examination of recent earthquake damage revealed a number of buildings that suffered mid-height collapses. It appears that this situation occurred most often in the near field area of major earthquakes and only affected mid-rise buildings between five and fifteen stories tall. These types of buildings are typically designed for primary mode effects but have a significant strength and stiffness reduction at one level up the height of the structure. This reduction in strength and stiffness coupled with unexpected higher mode effects may have been the potential cause of the mid-height collapses.
>
> A dynamic analysis should be performed to determine if there are unexpectedly high seismic demands at locations of strength discontinuities. Compliance can be achieved if the elements of the weak story can be shown to have adequate capacity near elastic levels.

4.3.2.2 SOFT STORY: The stiffness of the lateral-force-resisting system in any story shall not be less than 70 percent of the lateral-force-resisting system stiffness in an adjacent story above or below, or less than 80 percent of the average lateral-force-resisting system stiffness of the three stories above or below for Life Safety and Immediate Occupancy.

Tier 2 Evaluation Procedure: An analysis in accordance with the Linear Dynamic Procedure of Section 4.2 shall be performed. The adequacy of the elements in the lateral-force-resisting system shall be evaluated.

> **C4.3.2.2 Soft Story**
>
> This condition commonly occurs in commercial buildings with open fronts at ground-floor storefronts, and hotels or office buildings with particularly tall first stories. Figure C4-3 shows an example of a tall story. Such cases are not necessarily soft stories because the tall columns may have been designed with appropriate stiffness, but they are likely to be soft stories if they have been designed without consideration for story drift. Soft stories usually are revealed by an abrupt change in story drift. Generally an examination of the building elevations can determine if a soft story exists without the need for calculation. A tall story or a change in the type of lateral-force-resisting system are obvious indications that a soft story might exist. A gradual reduction of lateral-force-resisting elements as the building increases in height is typical and is not considered a soft story condition. Another simple first step might be to plot and compare the story drifts, as indicated in Figure C4-4, if analysis results happen to be available.
>
> The difference between "soft" and "weak" stories is the difference between stiffness and strength. A column may be limber but strong, or stiff but weak. A change in column size can affect strength and stiffness, and both need to be considered.
>
> An examination of recent earthquake damage revealed a number of buildings that suffered mid-height collapses. It appears that this situation occurs most often in the near field area of major earthquakes and only affects mid-rise buildings between five and fifteen stories tall. These types of buildings are typically designed for primary mode effects but have a significant strength and stiffness reduction at one level up the height of the structure. This reduction in strength and/or stiffness coupled with unexpected higher mode effects may have the potential to cause mid-height collapses. A dynamic analysis should be performed to determine if there are unexpectedly high seismic demands at locations of stiffness discontinuities.

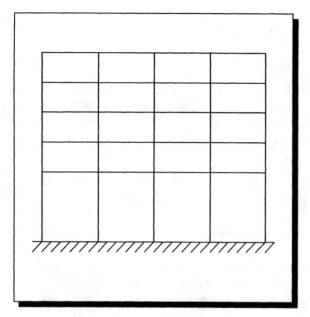

Figure C4-3. Tall Story

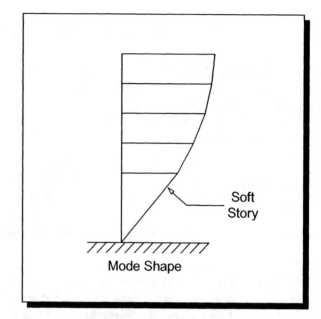

Figure C4-4. Soft Story

4.3.2.3 GEOMETRY: There shall be no change in horizontal dimension of the lateral-force-resisting system of more than 30 percent in a story relative to adjacent stories for Life Safety and Immediate Occupancy, excluding one-story penthouses and mezzanines.

Tier 2 Evaluation Procedure: An analysis in accordance with the Linear Dynamic Procedure of Section 4.2 shall be performed. The adequacy of the lateral-force-resisting elements shall be evaluated.

C4.3.2.3 Geometry

Geometric irregularities are usually detected in an examination of the story-to-story variation in the dimensions of the lateral-force-resisting system (see Figure C4-5). A building with upper stories set back from a broader base structure is a common example. Another example is a story in a high-rise that is set back for architectural reasons. It should be noted that the irregularity of concern is in the dimensions of the lateral-force-resisting system, not in the dimensions of the envelope of the building, and, as such, it may not be obvious.

Geometric irregularities affect the dynamic response of the structure and may lead to unexpected higher mode effects and concentrations of demand. A dynamic analysis is required to more accurately calculate the distribution of seismic forces. One-story penthouses need not be considered except for the added mass.

4.3.2.4 VERTICAL DISCONTINUITIES: All vertical elements in the lateral-force-resisting system shall be continuous to the foundation.

Tier 2 Evaluation Procedure: The adequacy of elements below vertical discontinuities shall be evaluated to support gravity forces and overturning forces generated by the capacity of the discontinuous elements above. The adequacy of struts and diaphragms to transfer load from discontinuous elements to adjacent elements shall be evaluated.

Evaluation Phase (Tier 2)

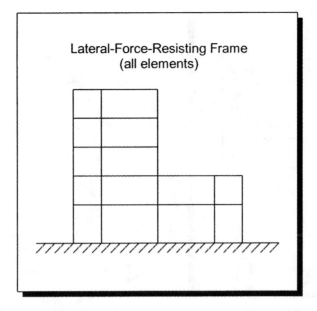

Figure C4-5. Geometric Irregularities

C4.3.2.4 Vertical Discontinuities

Vertical discontinuities are usually detected by visual observation. The most common example is a discontinuous shear wall or braced frame. The element is not continuous to the foundation; rather, it stops at an upper level. The shear at this level is transferred through the diaphragm to other resisting elements below. This force transfer can be accomplished through a strut if the elements are on the same plane (see Figure C4-6) or through a connecting diaphragm if the elements are not in the same plane (see Figure C4-7). In either case, the overturning forces that develop in the element continue down through the supporting columns.

This issue is a local strength and ductility problem below the discontinuous elements, not a global story strength or stiffness irregularity. The concern is that the wall or braced frame may have more shear capacity than considered in the design. These capacities impose overturning forces that could overwhelm the columns. While the strut or connecting diaphragm may be adequate to transfer the shear forces to adjacent elements, the columns that support vertical loads are the most critical. It should be noted that moment frames can have the same kind of discontinuity.

Compliance can be achieved if an adequate load path exists to transfer seismic force, and if the supporting columns can be demonstrated to have adequate capacity to resist the overturning forces generated by the shear capacity of the discontinuous elements.

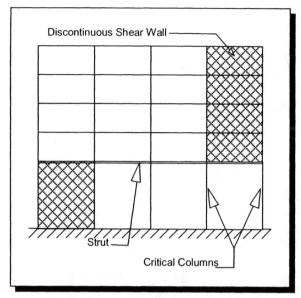
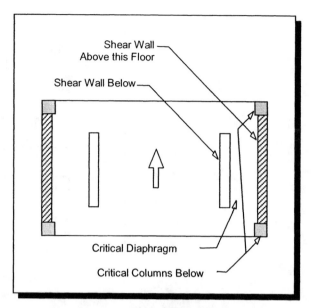

Figure C4-6. Vertical Discontinuity In-Plane

Figure C4-7. Vertical Discontinuity Out-of-Plane

4.3.2.5 MASS: There shall be no change in effective mass more than 50 percent from one story to the next for Life Safety and for Immediate Occupancy. Light roofs, penthouses, and mezzanines need not be considered.

Tier 2 Evaluation Procedure: An analysis in accordance with the Linear Dynamic Procedure of Section 4.2 shall be performed. The adequacy of the lateral-force-resisting elements shall be evaluated.

C4.3.2.5 Mass

Mass irregularities can be detected by comparison of the story weights (see Figure C4-8). The effective mass consists of the dead load of the structure tributary to each level, plus the actual weights of partitions and permanent equipment at each floor. Buildings are typically designed for primary mode effects. The validity of this approximation is dependent on the vertical distribution of mass and stiffness in the building. Mass irregularities affect the dynamic response of the structure and may lead to unexpected higher mode effects and concentrations of demand.

A dynamic analysis is required to more accurately calculate the distribution of seismic forces. Light roofs and penthouses need not be considered.

Evaluation Phase (Tier 2)

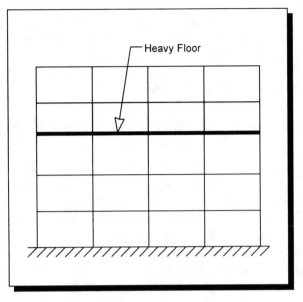

Figure C4-8. Heavy Floor

4.3.2.6 TORSION: The estimated distance between the story center of mass and the story center of rigidity is less than 20 percent of the building width in either plan dimension for Life Safety and Immediate Occupancy.

Tier 2 Evaluation Procedure: An analysis in accordance with the procedures in Section 4.2 shall be performed, including the effects of horizontal torsion. The adequacy of the lateral-force-resisting system including torsional demands shall be evaluated. The maximum story drift including the additional displacement due to torsion shall be calculated. The adequacy of the vertical-load-carrying elements under the calculated drift, including P-Δ effects, shall be evaluated.

C4.3.2.6 Torsion

Wherever there is significant torsion in a building, the concern is for additional seismic demands and lateral drifts imposed on the vertical elements by rotation of the diaphragm. Buildings can be designed to meet code forces including torsion, but buildings with severe torsion are less likely to perform well in earthquakes. It is best to provide a balanced system at the start, rather than design torsion into the system.

One concern is for columns that support the diaphragm, especially if the columns are not intended to be part of the lateral-force-resisting system. The columns are forced to drift laterally with the diaphragm, inducing lateral forces and P-Δ effects. Such columns often have not been designed to resist these movements.

Another concern is the strength of the vertical elements of the lateral-force-resisting system that will experience additional seismic demands due to torsion.

In the Case A building shown in Figure C4-9, the center of gravity is near the center of the diaphragm, while the center of rigidity is also near the centerline but close to wall A. Under longitudinal loading, the eccentricity, $e1$, between the center of gravity (center of earthquake load) and the center of rigidity (center of resistance) causes a torsional moment. The entire earthquake force is resisted directly by wall A, and the torsional moment is resisted by a couple consisting of equal and opposite forces in walls B and C. These two walls have displacements in opposite directions, and the diaphragm rotates.

Evaluation Phase (Tier 2)

These are very simple cases for analysis and design, and if the systems are designed and detailed properly, they should perform well. With the ample portions suggested by the length of the walls in Figure C4-9, stresses will be low and there will be little rotation of the diaphragm. The hazard appears where the diaphragm, and, consequently, the diaphragm stresses, become large; where the stiffness of the walls is reduced; or where the walls have substantial differences in stiffnesses.

The Case C building, shown in Figure C4-10, has a more serious torsional condition than the ones in Figure C4-9. Wall A has much greater rigidity than wall D, as indicated by their relative lengths.

For transverse loading, the center of rigidity is close to wall A, and there is a significant torsional movement. Walls B, C, and D, although strong enough for design forces, have little rigidity, and that allows substantial rotation of the diaphragm. There are two concerns here. First, because of the rotation of the diaphragm, there is a displacement at E and F that induces side-sway moments in the columns that may not have been recognized in the design. Their failure could lead to a collapse. Second, the stability of the building under transverse loading depends on wall D. The Case D building shown in Figure C4-10 is shown with wall D failed. The remaining walls, A, B, and C, are in the configuration of Figure C4-9, and now there is a very large eccentricity that may cause walls B and C to fail. Note that this is an example of a building that lacks redundancy.

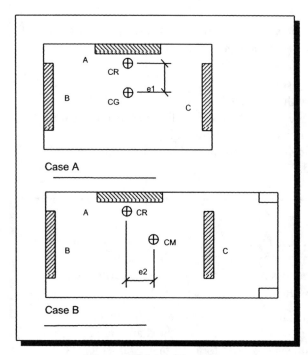

Figure C4-9. Torsion: Cases A and B

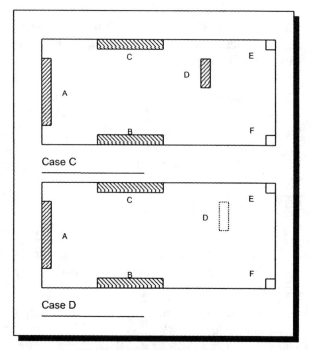

Figure C4-10. Torsion: Cases C and D

4.3.3 Condition of Materials

> **C4.3.3 Condition of Materials**
>
> Deteriorated structural materials may jeopardize the capacity of the vertical- and lateral-force-resisting systems. The most common type of deterioration is caused by the intrusion of water. Stains may be a clue to water-caused deterioration where the structure is visible on the exterior, but the deterioration may be hidden where the structure is concealed by finishes. In the latter case, the design professional may have to find a way into attics, plenums, and crawl spaces in order to assess the structural systems and their condition.
>
> The design professional should be careful where dealing with a building that appears to be in good condition and is known to have been subjected to earthquakes in the past. One is tempted to say that the building has "withstood the test of time"; however, the earthquakes the building was subjected to may not have been significant, or the good appearance may only be a good cosmetic repair that hides damage that was not repaired. Examples of problems include cracked concrete walls and frames, torn steel connections, bent fasteners or torn plywood in diaphragms and walls, and loose anchors in masonry. Evaluations should include consideration for long-term impacts, especially if deterioration is currently minor and repair to the source of deterioration is not completed in a timely manner.

4.3.3.1 DETERIORATION OF WOOD: There shall be no signs of decay, shrinkage, splitting, fire damage, or sagging in any of the wood members, and none of the metal connection hardware shall be deteriorated, broken, or loose.

Tier 2 Evaluation Procedure: The cause and extent of damage shall be identified by visual inspection. The consequences of this damage to the lateral-force-resisting system shall be determined. The adequacy of damaged lateral-force-resisting elements shall be evaluated considering the extent of the damage and impact on the capacity of each damaged element.

> **C4.3.3.1 Deterioration of Wood**
>
> The condition of the wood in a structure has a direct relationship to its performance in a seismic event. Wood that is split, rotten, or has insect damage may have a very low capacity to resist loads imposed by earthquakes. Structures with wood elements depend to a large extent on the connections between members. If the wood at a bolted connection is split, the connection will possess only a fraction of the capacity of a similar connection in undamaged wood. Limited intrusive investigation may be required to determine the cause and relative magnitude of the damage.

4.3.3.2 WOOD STRUCTURAL PANEL SHEAR WALL FASTENERS: There shall be no more than 15 percent of inadequate fastening such as overdriven fasteners, omitted blocking, excessive fastening spacing, or inadequate edge distance. This statement shall apply to the Immediate Occupancy Performance Level only.

Tier 2 Evaluation Procedure: The extent of inadequate shear wall fasteners shall be identified. The consequences of inadequate shear wall fasteners to the lateral-force-resisting system shall be determined. The adequacy of these shear walls shall be evaluated considering the extent of inadequate shear wall fasteners and impact on the capacity.

> **C4.3.3.2 Wood Structural Panel Shear Wall Fasteners**
>
> Fasteners connecting structural panels to the framing are supposed to be driven flush with, but should not penetrate the surface of the sheathing. This effectively reduces the shear capacity of the fastener and increases the potential for the fastener to fail by pulling through the sheathing.
>
> For structures built prior to the wide use of nailing guns (pre-1970), the problem is generally not present. More recent projects are often constructed with alternate fasteners, such as staples, T-nails, clipped head nails, or cooler nails, which, where installed with pneumatic nail guns, are often overdriven, completely penetrating one or more panel plies.
>
> Other issues regarding fasteners that could reduce the capacity of shear wall include omitted blocking, excessive fastening spacing, and inadequate edge distance.

4.3.3.3 DETERIORATION OF STEEL: There shall be no visible rusting, corrosion, cracking, or other deterioration in any of the steel elements or connections in the vertical- or lateral-force-resisting systems.

Tier 2 Evaluation Procedure: The cause and extent of damage shall be identified. The consequences of this damage to the lateral-force-resisting system shall be determined. The adequacy of damaged lateral-force-resisting elements shall be evaluated considering the extent of the damage and impact on the capacity of each damaged element.

> **C4.3.3.3 Deterioration of Steel**
>
> Environmental effects over prolonged periods of time may lead to deterioration of steel elements. Significant rusting or corrosion can substantially reduce the member cross sections, with a corresponding reduction in capacity.
>
> Often steel elements have surface corrosion that looks worse than it is, and is likely not a concern. Where corrosion is present, care should be taken to determine the actual loss in cross section. Such deterioration must be considered in the evaluation where it occurs at critical locations in the lateral-force-resisting system.

4.3.3.4 DETERIORATION OF CONCRETE: There shall be no visible deterioration of concrete or reinforcing steel in any of the vertical- or lateral-force-resisting elements.

Tier 2 Evaluation Procedure: The cause and extent of damage shall be identified. The consequences of this damage to the lateral-force-resisting system shall be determined. The adequacy of damaged lateral-force-resisting elements shall be evaluated considering the extent of the damage and impact on the capacity of each damaged element.

C4.3.3.4 Deterioration of Concrete

Deteriorated concrete and reinforcing steel can significantly reduce the strength of concrete elements. This statement is concerned with deterioration such as spalled concrete associated with rebar corrosion and water intrusion. Cracks in concrete are covered elsewhere in this standard. Spalled concrete over reinforcing bars reduces the available surface for bond between the concrete and steel. Bar corrosion may significantly reduce the cross section of the bar.

Deterioration is a concern where the concrete cover has begun to spall and there is evidence of rusting at critical locations.

4.3.3.5 POST-TENSIONING ANCHORS: There shall be no evidence of corrosion or spalling in the vicinity of post-tensioning or end fittings. Coil anchors shall not have been used.

Tier 2 Evaluation Procedure: The cause and extent of damage shall be identified. The consequences of this damage to the lateral-force-resisting system shall be determined. The adequacy of damaged lateral-force-resisting elements shall be evaluated considering the extent of the damage and impact on the capacity of each damaged element.

C4.3.3.5 Post-Tensioning Anchors

Corrosion in post-tensioning anchors can lead to failure of the gravity load system if ground motion causes a release or slip of prestressing strands. Coil anchors (see Figure C4-11), with or without corrosion, have performed poorly under cyclic loads and are no longer allowed by current standards. The deficiency is the ability of the coil anchor to maintain its grip under cyclic loading. There is no Tier 2 procedure for coil anchors.

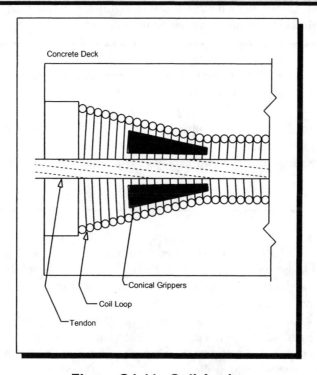

Figure C4-11. Coil Anchor

4.3.3.6   PRECAST CONCRETE WALLS: There shall be no visible deterioration of concrete or reinforcing steel or evidence of distress, especially at the connections.

Tier 2 Evaluation Procedure: The cause and extent of damage shall be identified. The consequences of this damage to the lateral-force-resisting system shall be determined. The adequacy of damaged walls shall be evaluated considering the extent of the damage and the impact on the capacity of each damaged wall.

C4.3.3.6   Precast Concrete Walls

Precast concrete elements are sometimes only nominally interconnected and may be subject to shrinkage, creep, or temperature stresses that were not adequately considered in design. Distress caused by these factors could directly affect the lateral strength of the building. The most common damage is cracking and spalling at embedded connections between panels. This includes both the nominal connections along the vertical edges and the chord connections at the level of the diaphragm. The performance of precast concrete wall systems is completely dependent on the condition of the connections.

4.3.3.7   MASONRY UNITS: There shall be no visible deterioration of masonry units.

Tier 2 Evaluation Procedure: The cause and extent of damage shall be identified. The consequences of this damage to the lateral-force-resisting system shall be determined. The adequacy of damaged lateral-force-resisting elements shall be evaluated considering the extent of the damage and impact on the capacity of each damaged element.

C4.3.3.7   Masonry Units

Deteriorated or poor-quality masonry elements can result in significant reductions in the strength of structural elements. Damaged or deteriorated masonry may not be readily observable.

4.3.3.8   MASONRY JOINTS: The mortar shall not be easily scraped away from the joints by hand with a metal tool, and there shall be no areas of eroded mortar.

Tier 2 Evaluation Procedure: The extent of loose or eroded mortar shall be identified. Walls with loose mortar shall be omitted from the analysis, and the adequacy of the lateral-force-resisting system shall be evaluated. Alternatively, the adequacy of the walls may be evaluated with shear strength determined by testing.

C4.3.3.8   Masonry Joints

Older buildings constructed with lime mortar may have surface repointing but still have deteriorated mortar in the main part of the joint. One test is to tap a small hole with a nail in the repointing and, if it breaks through, powdery lime mortar shows on the nail. If it does not break through after moderate-to-hard blows, the wall probably is repointed full depth. This also can be seen by looking behind exterior trim or wall fixtures where the new repointing never reached. Mortar that is severely eroded or can be easily scraped away has been found to have low shear strength, which results in low wall strength. Destructive or in-plane shear tests, such as those in Section 4.2.6, are required to measure the strength of the bond between the brick and mortar in order to determine the shear capacity of the walls.

Evaluation Phase (Tier 2)

4.3.3.9 CONCRETE WALL CRACKS: All existing diagonal cracks in the wall elements shall be less than 1/8 inch for Life Safety and 1/16 inch for Immediate Occupancy, shall not be concentrated in one location, and shall not form an X pattern.

Tier 2 Evaluation Procedure: The cause and extent of damage shall be identified. The consequences of the damage to the lateral-force-resisting system shall be determined. The adequacy of damaged walls shall be evaluated considering the extent of the damage and impact on the capacity of each damaged wall.

> **C4.3.3.9 Concrete Wall Cracks**
>
> Cracks in concrete elements have little effect on strength of well-reinforced wall elements. A significant reduction in strength is usually the result of large displacements or crushing of concrete. Only where the cracks are large enough to prevent aggregate interlock or to allow for the potential for buckling of the reinforcing steel does the adequacy of the concrete capacity become a concern.
>
> Cracks in unusual patterns, such as concentrated on one floor or at one end of the wall, usually indicate a specific cause. The cause of observed cracking needs to be identified to determine whether future cracking will affect the capacity of the wall.
>
> Crack width is commonly used as a convenient indicator of damage to a wall. However, it should be noted that recent studies (FEMA 306 and 307, *Evaluation of Earthquake Damaged Concrete and Masonry Wall Buildings* [FEMA, 1999a and 1999b]) list other factors, such as location, orientation, number, distribution, and pattern of the cracks, to be equally important in measuring the extent of damage present in the shear walls. All these factors should be considered where evaluating the reduced capacity of a cracked element.

4.3.3.10 REINFORCED MASONRY WALL CRACKS: All existing diagonal cracks in the wall elements shall be less than 1/8 inch for Life Safety and 1/16 inch for Immediate Occupancy, shall not be concentrated in one location, and shall not form an X pattern.

Tier 2 Evaluation Procedure: The cause and extent of damage shall be identified. The consequences of the damage to the lateral-force-resisting system shall be determined. The adequacy of damaged lateral-force-resisting elements shall be evaluated considering the extent of the damage and impact on the capacity of each damaged element.

> **C4.3.3.10 Reinforced Masonry Wall Cracks**
>
> Diagonal wall cracks, especially along the masonry joints, may affect the interaction of the masonry units, leading to a reduction of strength and stiffness. The cracks may indicate distress in the wall from past seismic events, foundation settlement, or other causes.
>
> Cracks in unusual patterns, such as concentrated on one floor or at one end of the wall, usually indicate a specific cause. The cause of observed cracking needs to be identified to determine whether future cracking will affect the capacity of the wall.
>
> Crack width is commonly used as a convenient indicator of damage to a wall. However, it should be noted that recent studies (FEMA 306 and 307) list other factors, such as location, orientation, number, distribution, and pattern of the cracks, to be equally important in measuring the extent of damage present in the shear walls. All these factors should be considered where evaluating the reduced capacity of a cracked element.

Evaluation Phase (Tier 2)

4.3.3.11   UNREINFORCED MASONRY WALL CRACKS: There shall be no existing diagonal cracks in the wall elements greater than 1/8 inch for Life Safety and 1/16 inch for Immediate Occupancy, or out-of-plane offsets in the bed joint greater than 1/8 inch for Life Safety and 1/16 inch for Immediate Occupancy, and shall not form an X pattern.

Tier 2 Evaluation Procedure: The cause and extent of damage shall be identified. Damaged walls or portions of walls shall be omitted from the analysis, and the adequacy of the lateral-force-resisting system shall be evaluated.

> **C4.3.3.11   Unreinforced Masonry Wall Cracks**
>
> Diagonal wall cracks, especially along the masonry joints, may affect the interaction of the masonry units, leading to a reduction of strength and stiffness. The cracks may indicate distress in the wall from past seismic events, foundation settlement, or other causes.
>
> Crack width is commonly used as a convenient indicator of damage to a wall, but it should be noted that recent studies (FEMA 306 and 307) list other factors, such as location, orientation, number, distribution, and pattern of the cracks, to be equally important in measuring the extent of damage present in the shear walls. All these factors should be considered where evaluating the reduced capacity of a cracked element.

4.3.3.12   CRACKS IN INFILL WALLS: There shall be no existing diagonal cracks in the infilled walls that extend throughout a panel greater than 1/8 inch for Life Safety and 1/16 inch for Immediate Occupancy, or out-of-plane offsets in the bed joint greater than 1/8 inch for Life Safety and 1/16 inch for Immediate Occupancy.

Tier 2 Evaluation Procedure: The cause and extent of damage shall be identified. The consequences of the damage to the lateral-force-resisting system shall be determined. The adequacy of damaged lateral-force-resisting elements shall be evaluated considering the extent of the damage and impact on the capacity of each damaged element.

> **C4.3.3.12   Cracks in Infill Walls**
>
> Diagonal wall cracks, especially along the masonry joints, may affect the interaction of the masonry units, leading to a reduction of strength and stiffness. The cracks may indicate distress in the wall from past seismic events, foundation settlement, or other causes.
>
> Offsets in the bed joint along the masonry joints may affect the interaction of the masonry units in resisting out-of-plane forces. The offsets may indicate distress in the wall from past seismic events, or just poor construction.
>
> Crack width is commonly used as a convenient indicator of damage to a wall, but it should be noted that recent studies (FEMA 306 and 307) list other factors, such as location, orientation, number, distribution, and pattern of the cracks, to be equally important in measuring the extent of damage present in the shear walls. All these factors should be considered where evaluating the reduced capacity of a cracked element.

Evaluation Phase (Tier 2)

4.3.3.13 CRACKS IN BOUNDARY COLUMNS: There shall be no existing diagonal cracks wider than 1/8 inch for Life Safety and 1/16 inch for Immediate Occupancy in concrete columns that encase masonry infills.

Tier 2 Evaluation Procedure: The cause and extent of damage shall be identified. The consequences of the damage to the lateral-force-resisting system shall be determined. The adequacy of damaged lateral-force-resisting elements shall be evaluated considering the extent of the damage and impact on the capacity of each damaged element.

C4.3.3.13 Cracks in Boundary Columns

Small cracks in concrete elements have little effect on strength. A significant reduction in strength is usually the result of large displacements or crushing of concrete. Only where the cracks are large enough to prevent aggregate interlock or to allow for the potential for buckling of the reinforcing steel does the adequacy of the concrete element capacity become a concern.

Columns are required to resist diagonal compression strut forces that develop in infill wall panels. Vertical components induce axial forces in the columns. The eccentricity between horizontal components and the beams is resisted by the columns. Extensive cracking in the columns may indicate locations of possible weakness. Such columns may not be able to function in conjunction with the infill panel as expected.

Evaluation Phase (Tier 2)

4.4 Procedures for Lateral-Force-Resisting Systems

This section provides Tier 2 evaluation Tier 2 Evaluation procedures that apply to lateral-force-resisting systems: moment frames, shear walls, and braced frames.

4.4.1 Moment Frames

> **C4.4.1 Moment Frames**
>
> Moment frames develop their resistance to forces primarily through the flexural strength of the beam and column elements.
>
> In an earthquake, a frame with suitable proportions and details can develop plastic hinges that will absorb energy and allow the frame to survive actual displacements that are larger than calculated in an elastic-based design.
>
> In "special" moment frames, the ends of beams and columns, being the locations of maximum seismic moment, are designed to sustain inelastic behavior associated with plastic hinging over many cycles and load reversals.
>
> Frames without special seismic detailing depend on the reserve strength inherent in the design of the members. The basis of this reserve strength is the load factors in strength design or the factors of safety in working-stress design. Such frames are called "ordinary" moment frames. For ordinary moment frames, failure usually occurs due to a sudden brittle mechanism, such as shear failure in concrete members.
>
> For evaluations using this standard, it is not necessary to determine the type of frame in the building. The performance issue is addressed by appropriate acceptance criteria in the specified procedures. The fundamental requirements for all ductile moment frames are that:
>
> 1. They should have sufficient strength to resist seismic demands,
> 2. They should have sufficient stiffness to limit inter-story drift,
> 3. Beam-column joints should have the shear capacity to resist the shear demand and to develop the strength of the connected members,
> 4. Elements can form plastic hinges which have the ductility to sustain the rotations to which they are subjected, and
> 5. Beams will develop hinges before the columns at locations distributed throughout the structure (the strong column/weak beam concept).
>
> These items are covered in more detail in the evaluation statements that follow.
>
> It is expected that the combined action of gravity loads and seismic forces will cause the formation of plastic hinges in the structure. However, a concentration of plastic hinge formation at undesirable locations can severely undermine the stability of the structure. For example, the lower sketch in Figure C4-12 shows a story mechanism in which hinges form at the tops and bottoms of all the columns in a particular story. This condition results in a concentration of ductility demand and displacement in a single story that can lead to collapse.

> In a strong column situation (see upper sketch, Figure C4-12), the beams hinge first, yielding is distributed throughout the structure, and the ductility demand is more dispersed.

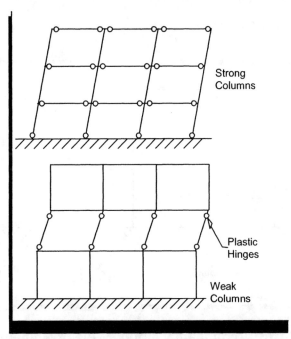

Figure C4-12. Plastic Hinge Formation

4.4.1.1 General

4.4.1.1.1 REDUNDANCY: The number of lines of moment frames in each direction shall be greater than or equal to 2.0 for Life Safety and for Immediate Occupancy. The number of bays of moment frames in each line shall be greater than or equal to 2.0 for Life Safety and 3.0 for Immediate Occupancy.

Tier 2 Evaluation Procedure: An analysis in accordance with the procedures in Section 4.2 shall be performed. The adequacy of all elements and connections in the frames shall be evaluated.

> **C4.4.1.1.1 Redundancy**
>
> Redundancy is a fundamental characteristic of lateral-force-resisting systems with superior seismic performance. Redundancy in the structure will ensure that if an element in the lateral-force-resisting system fails for any reason, there is another element present that can provide lateral force resistance. Redundancy also provides multiple locations for potential yielding, distributing inelastic activity throughout the structure and improving ductility and energy absorption. Typical characteristics of redundancy include multiple lines of resistance to distribute the lateral forces uniformly throughout the structure, and multiple bays in each line of resistance to reduce the shear and axial demands on any one element (see Figure C4-13).

> A distinction should be made between redundancy and adequacy. For the purpose of this standard, redundancy is intended to mean simply "more than one." That is not to say that for large buildings two elements is adequate, or for small buildings one is not enough. Separate evaluation statements are present in the standard to determine the adequacy of the elements provided.
>
> Where redundancy is not present in the structure, an analysis that demonstrates the adequacy of the lateral force elements is required.

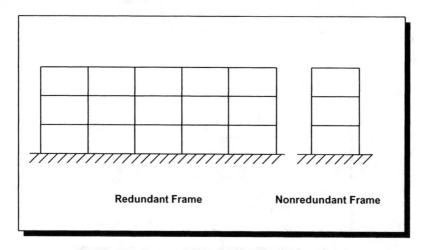

Figure C4-13. Redundancy along a Line of Moment Frame

4.4.1.2 Moment Frames with Infill Walls

> **C4.4.1.2 Moment Frames with Infill Walls**
>
> Infill walls used for partitions, cladding or shaft walls that enclose stairs and elevators should be isolated from the frames. If not isolated, they will alter the response of the frames and change the behavior of the entire structural system. Lateral drifts of the frame will induce forces on walls that interfere with this movement. Cladding connections must allow for this relative movement. Stiff infill walls confined by the frame will develop compression struts that will impart loads to the frame and cause damage to the walls. This is particularly important around stairs or other means of egress from the building.

4.4.1.2.1 INTERFERING WALLS: All concrete and masonry infill walls placed in moment frames shall be isolated from structural elements.

Tier 2 Evaluation Procedure: An analysis in accordance with Section 4.2 shall be performed. The demands imparted by the structure to the interfering walls, and the demands induced on the frame, shall be calculated. The adequacy of the interfering walls and the frame to resist the induced forces shall be evaluated.

Evaluation Phase (Tier 2)

C4.4.1.2.1 Interfering Walls

Where an infill wall interferes with the moment frame, the wall becomes an unintended part of the lateral-force-resisting system. Typically these walls are not designed and detailed to participate in the lateral-force-resisting system and may be subject to significant damage. The amount of isolation must be able to accommodate the inter-story drift of the moment frame.

Interfering walls should be checked for forces induced by the frame, particularly where damage to these walls can lead to falling hazards near means of egress. The frames should be checked for forces induced by contact with the walls, particularly if the walls are not full height or do not completely fill the bay.

It should be noted that it is impossible to simultaneously satisfy this section and Section 4.4.2.6.1.

4.4.1.3 Steel Moment Frames

C4.4.1.3 Steel Moment Frames

The following are characteristics of steel moment frames that have demonstrated acceptable seismic performance:

1. The beam end connections develop the plastic moment capacity of the beam or panel zone.
2. There is a high level of redundancy in the number of moment connections.
3. The column web has sufficient strength to sustain the stresses in the beam-column joint.
4. The lower flanges have lateral bracing sufficient to maintain stability of the frame.
5. There is flange continuity through the column.

Prior to the 1994 Northridge earthquake, steel moment-resisting frame connections generally consisted of complete penetration flange welds and a bolted or welded shear tab connection at the web. This type of connection, which was an industry standard from 1970 to 1995, was thought to be ductile and capable of developing the full capacity of the beam sections. However, a large number of buildings experienced extensive brittle damage to this type of connection during the Northridge earthquake. As a result, an emergency code change was made to the 1994 *Uniform Building Code* (ICBO, 1994a) to remove the prequalification of this type of connection. For a full discussion of these connections, please refer to FEMA 351, *Recommended Seismic Evaluation and Upgrade Criteria for Existing Steel Moment-Frame Buildings* (FEMA, 2000b).

4.4.1.3.1 DRIFT CHECK: The drift ratio of the steel moment frames, calculated using the Quick Check procedure of Section 3.5.3.1, shall be less than 0.025 for Life Safety and 0.015 for Immediate Occupancy.

Tier 2 Evaluation Procedure: An analysis in accordance with Section 4.2 shall be performed. The adequacy of the beams and columns, including $P\text{-}\Delta$ effects, shall be evaluated using the m-factors in Table 4-5.

C4.4.1.3.1 Drift Check

Moment-resisting frames are more flexible than shear wall or braced frame structures. This flexibility can lead to large inter-story drifts that may potentially cause extensive structural and nonstructural damage to welded beam-column connections, partitions, and cladding. Drifts also may induce large P-Δ demands and pounding where adjacent buildings are present.

An analysis of non-compliant frames is required to demonstrate the adequacy of frame elements subjected to excessive lateral drifts.

4.4.1.3.2 AXIAL STRESS CHECK: The axial stress due to gravity loads in columns subjected to overturning forces shall be less than $0.10F_y$ for Life Safety and Immediate Occupancy. Alternatively, the axial stress due to overturning forces alone, calculated using the Quick Check procedure of Section 3.5.3.6, shall be less than $0.30F_y$ for Life Safety and Immediate Occupancy.

Tier 2 Evaluation Procedure: An analysis in accordance with Section 4.2 shall be performed. The gravity and overturning demands for non-compliant columns shall be calculated, and the adequacy of the columns to resist overturning forces shall be evaluated using the m-factors in Table 4-5.

C4.4.1.3.2 Axial Stress Check

Columns that carry a substantial amount of gravity load may have limited additional capacity to resist seismic forces. Where axial forces due to seismic overturning moments are added, the columns may buckle in a nonductile manner due to excessive axial compression.

The alternative calculation of overturning stresses due to seismic forces alone is intended to provide a means of identifying frames that are likely to be adequate: frames with high gravity loads, but small seismic overturning forces.

Where both demands are large, the combined effect of gravity and seismic forces must be calculated to demonstrate compliance.

Evaluation Phase (Tier 2)

4.4.1.3.3 MOMENT-RESISTING CONNECTIONS: All moment connections shall be able to develop the strength of the adjoining members or panel zones.

Tier 2 Evaluation Procedure: An analysis in accordance with Section 4.2 shall be performed. The adequacy of the members and connections shall be evaluated using the *m*-factors in Table 4-5.

> **C4.4.1.3.3 Moment-Resisting Connections**
>
> Prior to the 1994 Northridge earthquake, steel moment-resisting frame connections generally consisted of complete penetration flange welds and a bolted or welded shear tab connection at the web (see Figure C4-14). This type of connection, which was an industry standard from 1970 to 1995, was thought to be ductile and capable of developing the full capacity of the beam sections. However, a large number of buildings experienced extensive brittle damage to this type of connection during the Northridge earthquake. As a result, an emergency code change was made to the 1994 *Uniform Building Code* removing the prequalification of this type of connection. For a full discussion of these connections, please refer to FEMA 351.
>
> For this standard, the Tier 1 Evaluation statement is considered non-complaint for full penetration flange welds and a more detailed analysis is required to determine the adequacy of these moment-resisting connections.

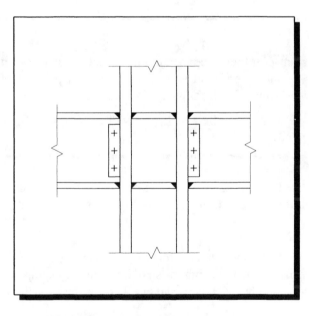

Figure C4-14. Pre–Northridge-Type Connection

Evaluation Phase (Tier 2)

4.4.1.3.4 PANEL ZONES: All panel zones shall have the shear capacity to resist the shear demand required to develop 0.8 times the sum of the flexural strengths of the girders framing in at the face of the column.

Tier 2 Evaluation Procedure: An analysis in accordance with Section 4.2 shall be performed. The demands in non-compliant joints shall be calculated, and the adequacy of the panel zones for web shear shall be evaluated using the m-factors in Table 4-5.

> **C4.4.1.3.4 Panel Zones**
>
> Panel zones with thin webs may yield or buckle before developing the capacity of the adjoining members, reducing the inelastic performance and ductility of the moment frames.
>
> Where panel zones cannot develop the strength of the beams, compliance can be demonstrated by checking the panel zones for actual shear demands.

4.4.1.3.5 COLUMN SPLICES: All column splice details located in moment-resisting frames shall include connection of both flanges and the web for Life Safety, and the splice shall develop the strength of the column for Immediate Occupancy.

Tier 2 Evaluation Procedure: An analysis in accordance with Section 4.2 shall be performed. The gravity and seismic demands shall be calculated, and the adequacy of the splice connection shall be evaluated.

> **C4.4.1.3.5 Column Splices**
>
> The lack of a substantial connection at the splice location may lead to separation of the spliced sections and misalignment of the columns, resulting in loss of vertical support and partial or total collapse of the building. Tests on partial-penetration weld splices have shown limited ductility.
>
> An inadequate connection also reduces the effective capacity of the column. Splices are checked against calculated demands to demonstrate compliance.

4.4.1.3.6 STRONG COLUMN/WEAK BEAM: The percentage of strong column/weak beam joints in each story of each line of moment-resisting frames shall be greater than 50 percent for Life Safety and Immediate Occupancy.

Tier 2 Evaluation Procedure: An analysis in accordance with Section 4.2 shall be performed. The adequacy of the columns to resist calculated demands shall be evaluated using an m-factor equal to 2.5. Alternatively, the story strength shall be calculated and checked for the capacity to resist one-half of the total pseudo lateral force.

> **C4.4.1.3.6 Strong Column/Weak Beam**
>
> Where columns are not strong enough to force hinging in the beams, column hinging can lead to story mechanisms and a concentration of inelastic activity at a single level. Excessive story drifts may result in an instability of the frame due to P-Δ effects. Good post-elastic behavior consists of yielding distributed throughout the frame. A story mechanism will limit forces in the levels above, preventing the upper levels from yielding. Joints at the roof level need not be considered.

> If it can be demonstrated that non-compliant columns are strong enough to resist calculated demands with sufficient overstrength, acceptable behavior can be expected.
>
> The alternative procedure checks for the formation of a story mechanism. The story strength is the sum of the shear capacities of all the columns as limited by the controlling action. If the columns are shear critical, a shear mechanism forms at the shear capacity of the columns. If the columns are controlled by flexure, a flexural mechanism forms at a shear corresponding to the flexural capacity.
>
> Should additional study be required, a Tier 3 evaluation Tier 3 Evaluation would include a non-linear pushover analysis. The formation of a story mechanism would be acceptable, provided the target displacement is met.

4.4.1.3.7 COMPACT MEMBERS: All frame elements shall meet section requirements set forth by *Seismic Provisions for Structural Steel Buildings* Table I-9-1 (AISC, 1997).

Tier 2 Evaluation Procedure: An analysis in accordance with Section 4.2 shall be performed. The adequacy of non-compliant beams and columns shall be evaluated using the *m*-factors in Table 4-5.

> **C4.4.1.3.7 Compact Members**
>
> Noncompact frame elements may experience premature local buckling prior to development of their full moment capacities. This can lead to poor inelastic behavior and ductility. The 1997 AISC Seismic Provisions explicitly address the section requirements that should be considered.
>
> The adequacy of the frame elements can be demonstrated using reduced *m*-factors in consideration of reduced capacities for noncompact sections.

4.4.1.3.8 BEAM PENETRATIONS: All openings in frame-beam webs shall be less than one-fourth of the beam depth and shall be located in the center half of the beams. This statement shall apply to the Immediate Occupancy Performance Level only.

Tier 2 Evaluation Procedure: An analysis in accordance with Section 4.2 shall be performed. The shear and flexural demands on non-compliant beams shall be calculated. The adequacy of the beams considering the strength around the penetrations shall be evaluated.

> **C4.4.1.3.8 Beam Penetrations**
>
> Members with large beam penetrations may fail in shear prior to the development of their full moment capacity, resulting in poor inelastic behavior and ductility.
>
> The critical section is at the penetration with the highest shear demand. Shear transfer across the web opening will induce secondary moments in the beam sections above and below the opening that must be considered in the analysis.

Evaluation Phase (Tier 2)

4.4.1.3.9 GIRDER FLANGE CONTINUITY PLATES: There shall be girder flange continuity plates at all moment-resisting frame joints. This statement shall apply to the Immediate Occupancy Performance Level only.

Tier 2 Evaluation Procedure: The adequacy of the column flange to transfer girder flange forces to the panel zone without continuity plates shall be evaluated.

C4.4.1.3.9 Girder Flange Continuity Plates

The lack of girder flange continuity plates may lead to a premature failure at the column web or flange at the joint. Beam flange forces are transferred to the column web through the column flange, resulting in a high stress concentration at the base of the column web. The presence of continuity plates, on the other hand, transfers the beam flange forces along the entire length of the column web.

Adequate force transfer without continuity plates will depend on the strength and stiffness of the column flange in weak-way bending.

4.4.1.3.10 OUT-OF-PLANE BRACING: Beam-column joints shall be braced out-of-plane. This statement shall apply to the Immediate Occupancy Performance Level only.

Tier 2 Evaluation Procedure: An analysis in accordance with Section 4.2 shall be performed. The axial demands on non-compliant columns shall be calculated, and the adequacy of the column to resist buckling between points of lateral support shall be evaluated considering a horizontal out-of-plane force equal to 6 percent of the critical column flange compression force acting concurrently at the non-compliant joint.

C4.4.1.3.10 Out-of-Plane Bracing

Columns without proper bracing may buckle prematurely out-of-plane before the strength of the joint can be developed. This will limit the ability of the frame to resist seismic forces.

The combination of axial load and moment on the columns will result in higher compression forces in one of the column flanges. The tendency for highly loaded joints to twist out-of-plane is due to compression buckling of the critical column compression flange.

Compliance can be demonstrated if the column section can provide adequate lateral restraint for the joint between points of lateral support.

4.4.1.3.11 BOTTOM FLANGE BRACING: The bottom flanges of beams shall be braced out-of-plane. This statement shall apply to the Immediate Occupancy Performance Level only.

Tier 2 Evaluation Procedure: An analysis in accordance with Section 4.2 shall be performed. The adequacy of the beams shall be evaluated considering the potential for lateral torsional buckling of the bottom flange between points of lateral support.

C4.4.1.3.11 Bottom Flange Bracing

Beam flanges in compression require out-of-plane bracing to prevent lateral torsional buckling. Buckling will occur before the full strength of the beam is developed, and the ability of the frame to resist lateral forces will be limited.

> Top flanges are typically braced by connection to the diaphragm. Bottom flange bracing occurs at discrete locations, such as at connection points for supported beams. The spacing of bottom flange bracing may not be close enough to prevent premature lateral torsional buckling where seismic loads induce large compression forces in the bottom flange.

4.4.1.4 Concrete Moment Frames

> **C4.4.1.4 Concrete Moment Frames**
>
> Concrete moment frame buildings typically are more flexible than shear wall buildings. This flexibility can result in large inter-story drifts that may lead to extensive nonstructural damage and P-Δ effects. If a concrete column has a capacity in shear that is less than the shear associated with the flexural capacity of the column, brittle column shear failure may occur and result in collapse. This condition is common in buildings in zones of moderate seismicity and in older buildings in zones of high seismicity. The columns in these buildings often have ties at standard spacing equal to the depth of the column, whereas current American Concrete Institute (ACI) code maximum spacing for shear reinforcing is much smaller. The following are the characteristics of concrete moment frames that have demonstrated acceptable seismic performance:
>
> 1. Brittle failure is prevented by providing a sufficient number of beam stirrups, column ties, and joint ties to ensure that the shear capacity of all elements exceeds the shear associated with flexural capacity.
> 2. Concrete confinement is provided by beam stirrups and column ties in the form of closed hoops with 135-degree hooks at locations where plastic hinges are expected to occur.
> 3. Overall performance is enhanced by long lap splices that are restricted to favorable locations and protected with additional transverse reinforcement.
> 4. The strong column/weak beam requirement is achieved by suitable proportioning of the members and their longitudinal reinforcing.
>
> Older frame systems that are lightly reinforced, precast concrete frames, and flat slab frames usually do not meet the detail requirements for ductile behavior.

4.4.1.4.1 SHEAR STRESS CHECK: The shear stress in the concrete columns, calculated using the Quick Check procedure of Section 3.5.3.2, shall be less than the greater of 100 psi or $2\sqrt{f'c}$ for Life Safety and Immediate Occupancy.

Tier 2 Evaluation Procedure: An analysis in accordance with Section 4.2 shall be performed. The adequacy of the concrete frame elements shall be evaluated using the *m*-factors in Table 4-6.

> **C4.4.1.4.1 Shear Stress Check**
>
> The shear stress check provides a quick assessment of the overall level of demand on the structure. The concern is the overall strength of the building.

Evaluation Phase (Tier 2)

4.4.1.4.2 AXIAL STRESS CHECK: The axial compressive stress due to gravity loads in columns subjected to overturning forces shall be less than $0.10f'_c$ for Life Safety and Immediate Occupancy. Alternatively, the axial compressive stress due to overturning forces alone, calculated using the Quick Check procedure of Section 3.5.3.6, shall be less than $0.30f'_c$ for Life Safety and Immediate Occupancy.

Tier 2 Evaluation Procedure: An analysis in accordance with Section 4.2 shall be performed. The gravity and overturning demands for non-compliant columns shall be calculated, and the adequacy of the columns to resist overturning forces shall be evaluated using the m-factors in Table 4-6.

C4.4.1.4.2 Axial Stress Check

Columns that carry a substantial amount of gravity load may have limited additional capacity to resist seismic forces. Where axial forces due to seismic overturning moments are added, the columns may crush in a nonductile manner due to excessive axial compression.

The alternative calculation of overturning stresses due to seismic forces alone is intended to provide a means of identifying frames that are likely to be adequate: frames with high gravity loads, but small seismic overturning forces.

Where both demands are large, the combined effect of gravity and seismic forces must be calculated to demonstrate compliance.

4.4.1.4.3 FLAT SLAB FRAMES: The lateral-force-resisting system shall not be a frame consisting of columns and a flat slab/plate without beams.

Tier 2 Evaluation Procedure: An analysis in accordance with Section 4.2 shall be performed. The adequacy of the concrete frame including prestressed elements shall be evaluated using the m-factors in Table 4.4.

C4.4.1.4.3 Flat Slab Frames

The concern is the transfer of the shear and bending forces between the slab and column, which could result in a punching shear failure and partial collapse. The flexibility of the lateral-force-resisting system will increase as the slab cracks.

Continuity of some bottom reinforcement through the column joint will assist in the transfer of forces and provide some resistance to collapse by catenary action in the event of a punching shear failure.

4.4.1.4.4 PRESTRESSED FRAME ELEMENTS: The lateral-force-resisting frames shall not include any prestressed or post-tensioned elements where the average prestress exceeds the lesser of 700 psi or $f'_c/6$ at potential hinge locations. The average prestress shall be calculated in accordance with the Quick Check procedure of Section 3.5.3.8.

Tier 2 Evaluation Procedure: An analysis in accordance with Section 4.2 shall be performed. The adequacy of the slab-column system for resisting seismic forces and punching shear shall be evaluated using the m-factors in Table 4-6.

C4.4.1.4.4 Prestressed Frame Elements

Frame elements that are prestressed or post-tensioned may not behave in a ductile manner. The concern is the inelastic behavior of prestressed elements.

4.4.1.4.5 CAPTIVE COLUMNS: There shall be no columns at a level with height/depth ratios less than 50 percent of the nominal height/depth ratio of the typical columns at that level for Life Safety and 75 percent for Immediate Occupancy.

Tier 2 Evaluation Procedure: The adequacy of the columns for the shear force required to develop the moment capacity at the top and bottom of the clear height of the columns shall be evaluated. Alternatively, the columns shall be evaluated as force-controlled elements in accordance with the alternative equations in Section 4.2.4.3.2.

C4.4.1.4.5 Captive Columns

Captive columns tend to attract seismic forces because of high stiffness relative to other columns in a story. Significant damage has been observed in parking structure columns adjacent to ramping slabs, even in structures with shear walls. Captive column behavior also may occur in buildings with clerestory windows or in buildings with partial height masonry infill panels.

If not adequately detailed, the columns may suffer a nonductile shear failure, which may result in partial collapse of the structure.

A captive column that can develop the shear capacity to develop the flexural strength over the clear height will have some ductility to prevent sudden nonductile failure of the vertical support system.

4.4.1.4.6 NO SHEAR FAILURES: The shear capacity of frame members shall be able to develop the moment capacity at the ends of the members.

Tier 2 Evaluation Procedure: An analysis in accordance with Section 4.2 shall be performed. The shear demands shall be calculated for non-compliant members, and the adequacy of the members for shear shall be evaluated.

C4.4.1.4.6 No Shear Failures

If the shear capacity of a member is reached before the moment capacity, there is a potential for a sudden nonductile failure of the member, leading to collapse.

Members that cannot develop the flexural capacity in shear should be checked for adequacy against calculated shear demands. Note that, for columns, the shear capacity is affected by the axial loads and should be based on the most critical combination of axial load and shear.

4.4.1.4.7 STRONG COLUMN/WEAK BEAM: The sum of the moment capacity of the columns shall be 20 percent greater than that of the beams at frame joints.

Tier 2 Evaluation Procedure: An analysis in accordance with Section 4.2 shall be performed. The adequacy of the columns to resist calculated demands shall be evaluated using an m-factor equal to 2.0. Alternatively, the story strength shall be calculated and checked for the capacity to resist one-half of the total pseudo lateral force.

> **C4.4.1.4.7 Strong Column/Weak Beam**
>
> Where columns are not strong enough to force hinging in the beams, column hinging can lead to story mechanisms and a concentration of inelastic activity at a single level. Excessive story drifts may result in an instability of the frame due to P-Δ effects. Good post-elastic behavior consists of yielding distributed throughout the frame. A story mechanism will limit forces in the levels above, preventing the upper levels from yielding. Joints at the roof level need not be considered.
>
> If it can be demonstrated that non-compliant columns are strong enough to resist calculated demands with sufficient overstrength, acceptable behavior can be expected. Reduced m-factors are used to check the columns at near elastic levels.
>
> The alternative procedure checks for the formation of a story mechanism. The story strength is the sum of the shear capacities of all the columns as limited by the controlling action. If the columns are shear critical, a shear mechanism forms at the shear capacity of the columns. If the columns are controlled by flexure, a flexural mechanism forms at a shear corresponding to the flexural capacity.

4.4.1.4.8 BEAM BARS: At least two longitudinal top and two longitudinal bottom bars shall extend continuously throughout the length of each frame beam. At least 25 percent of the longitudinal bars, provided at the joints for either positive or negative moment, shall be continuous throughout the length of the members for Life Safety and Immediate Occupancy.

Tier 2 Evaluation Procedure: An analysis in accordance with Section 4.2 shall be performed. The flexural demand at the ends and midspan of the non-compliant beams shall be calculated, and the adequacy of the beams shall be evaluated using an m-factor equal to 1.0.

> **C4.4.1.4.8 Beam Bars**
>
> The requirement for two continuous bars is a collapse prevention measure. In the event of complete beam failure, continuous bars will prevent total collapse of the supported floor, holding the beam in place by catenary action.
>
> Previous construction techniques used bent-up longitudinal bars as reinforcement. These bars transitioned from bottom to top reinforcement at the gravity load inflection point. Some amount of continuous top and bottom reinforcement is desired because moments due to seismic forces can shift the location of the inflection point.
>
> Because non-compliant beams are vulnerable to collapse, the beams are required to resist demands at an elastic level. Continuous slab reinforcement adjacent to the beam may be considered as continuous top reinforcement.

Evaluation Phase (Tier 2)

4.4.1.4.9 COLUMN-BAR SPLICES: All column bar lap splice lengths shall be greater than $35d_b$ for Life Safety and $50d_b$ for Immediate Occupancy, and shall be enclosed by ties spaced at or less than $8d_b$ for Life Safety and Immediate Occupancy. Alternatively, column bars shall be spliced with mechanical couplers with a capacity of at least 1.25 times the nominal yield strength of the spliced bar.

Tier 2 Evaluation Procedure: An analysis in accordance with Section 4.2 shall be performed. The flexural demands at non-compliant column splices shall be calculated, and the adequacy of the columns shall be evaluated using the *m*-factors in Table 4-6.

C4.4.1.4.9 Column-Bar Splices

Located just above the floor level, column bar splices are typically located in regions of potential plastic hinge formation. Short splices are subject to sudden loss of bond. Widely spaced ties can result in a spalling of the concrete cover and loss of bond. Splice failures are sudden and nonductile.

Columns with non-compliant lap splices are checked using reduced *m*-factors to account for this potential lack of ductility. If the members have sufficient capacity, the demands on the splices are less likely to exceed the capacity of the bond.

4.4.1.4.10 BEAM-BAR SPLICES: The lap splices or mechanical couplers for longitudinal beam reinforcing shall not be located within $l_b/4$ of the joints and shall not be located in the vicinity of potential plastic hinge locations.

Tier 2 Evaluation Procedure: An analysis in accordance with Section 4.2 shall be performed. The flexural demands in non-compliant beams shall be calculated, and the adequacy of the beams shall be evaluated using the *m*-factors for nonductile beams in Table 4-6.

C4.4.1.4.10 Beam-Bar Splices

Lap splices located at the end of beams and in the vicinity of potential plastic hinges may not be able to develop the full moment capacity of the beam as the concrete degrades during multiple cycles.

Beams with non-compliant lap splices are checked using reduced *m*-factors to account for this potential lack of ductility. If the members have sufficient capacity, the demands are less likely to cause degradation and loss of bond between concrete and the reinforcing steel.

4.4.1.4.11 COLUMN-TIE SPACING: Frame columns shall have ties spaced at or less than $d/4$ for Life Safety and Immediate Occupancy throughout their length and at or less than $8d_b$ for Life Safety and Immediate Occupancy at all potential plastic hinge locations.

Tier 2 Evaluation Procedure: An analysis in accordance with Section 4.2 shall be performed. The flexural demand in non-compliant columns shall be calculated, and the adequacy of the columns shall be evaluated using the *m*-factors in Table 4-6.

C4.4.1.4.11 Column-Tie Spacing

Widely spaced ties will reduce the ductility of the column, and the column may not be able to maintain full moment capacity through several cycles. Columns with widely spaced ties have limited shear capacity and nonductile shear failures may result.

Elements with non-compliant confinement are checked using reduced m-factors to account for this potential lack of ductility.

4.4.1.4.12 STIRRUP SPACING: All beams shall have stirrups spaced at or less than $d/2$ for Life Safety and Immediate Occupancy throughout their length. At potential plastic hinge locations, stirrups shall be spaced at or less than the minimum of $8d_b$ or $d/4$ for Life Safety and Immediate Occupancy.

Tier 2 Evaluation Procedure: An analysis in accordance with Section 4.2 shall be performed. The flexural demand in non-compliant beams shall be calculated, and the adequacy of the beams shall be evaluated using the m-factors in Table 4-6.

C4.4.1.4.12 Stirrup Spacing

Widely spaced stirrups will reduce the ductility of the beam, and the beam may not be able to maintain full moment capacity through several cycles. Beams with widely spaced stirrups have limited shear capacity and nonductile shear failures may result.

Elements with non-compliant confinement are checked using reduced m-factors to account for this potential lack of ductility.

4.4.1.4.13 JOINT REINFORCING: Beam-column joints shall have ties spaced at or less than $8d_b$ for Life Safety and Immediate Occupancy.

Tier 2 Evaluation Procedure: An analysis in accordance with Section 4.2 shall be performed. The joint shear demands shall be calculated, and the adequacy of the joint to develop the adjoining members' forces shall be evaluated.

C4.4.1.4.13 Joint Reinforcing

Beam-column joints without shear reinforcement may not be able to develop the strength of the connected members, leading to a nonductile failure of the joint. Perimeter columns are especially vulnerable because the confinement of joint is limited to three sides (along the exterior) or two sides (at a corner). Joints will have more capacity if transverse beams exist on both sides of the joint.

The shear capacity of the joint may be calculated as follows:

Evaluation Phase (Tier 2)

$Q_{cl} = \lambda \gamma A_j (f'_c)^{1/2}$ psi, where γ is:

	$\rho'' < 0.003$	$\rho'' \geq 0.003$
Interior joints with two transverse beams	12	20
Interior joints without two transverse beams	10	15
Exterior joints with two transverse beams	8	15
Exterior joints without two transverse beams	6	12
Corner joints	4	8

$\lambda = 0.75$ for lightweight concrete

A_j = joint cross sectional area

4.4.1.4.14 JOINT ECCENTRICITY: There shall be no eccentricities larger than 20 percent of the smallest column plan dimension between girder and column centerlines for Immediate Occupancy. This statement shall apply to the Immediate Occupancy Performance Level only.

Tier 2 Evaluation Procedure: An analysis in accordance with Section 4.2 shall be performed. The joint shear demands including additional shear stresses from joint torsion shall be calculated, and the adequacy of the beam-column joints shall be evaluated.

C4.4.1.4.14 Joint Eccentricity

Joint eccentricities can result in high torsional demands on the joint area, which will result in higher shear stresses. The smallest column plan dimension should be calculated for the column at each joint under consideration.

4.4.1.4.15 STIRRUP AND TIE HOOKS: The beam stirrups and column ties shall be anchored into the member cores with hooks of 135° or more. This statement shall apply to the Immediate Occupancy Performance Level only.

Tier 2 Evaluation Procedure: An analysis in accordance with Section 4.2 shall be performed. The shear and axial demands in non-compliant members shall be calculated, and the adequacy of the beams and columns shall be evaluated using the *m*-factors in Table 4-6.

C4.4.1.4.15 Stirrup and Tie Hooks

To be fully effective, stirrups and ties must be anchored into the confined core of the member. Ninety-degree hooks that are anchored within the concrete cover are unreliable if the cover spalls during plastic hinging. The amount of shear resistance and confinement will be reduced if the stirrups and ties are not well anchored.

Elements with non-compliant confinement are checked using reduced *m*-factors to account for this potential lack of ductility for Life Safety as well as Immediate Occupancy.

Evaluation Phase (Tier 2)

4.4.1.5 Precast Concrete Moment Frames

4.4.1.5.1 PRECAST CONNECTION CHECK: The precast connections at frame joints shall have the capacity to resist the shear and moment demands calculated using the Quick Check procedure of Section 3.5.3.5.

Tier 2 Evaluation Procedure: An analysis in accordance with Section 4.2 shall be performed. The adequacy of the precast connections shall be evaluated as force-controlled elements using the procedures in Section 4.2.4.3.2.

C4.4.1.5.1 Precast Connection Check

Precast frame elements may have sufficient strength to meet lateral force requirements, but connections often cannot develop the strength of the members and may be subject to premature nonductile failures. Failure mechanisms may include fractures in the welded connections between inserts, pull-out of embeds, and spalling of concrete.

Since full member capacities cannot be realized, the behavior of this system is entirely dependent on the performance of the connections.

4.4.1.5.2 PRECAST FRAMES: For buildings with concrete shear walls, precast concrete frame elements shall not be considered primary components for resisting lateral forces.

Tier 2 Evaluation Procedure: An analysis in accordance with Section 4.2 shall be performed. The adequacy of the precast frame elements shall be evaluated as force-controlled elements using the procedures in Section 4.2.4.3.2.

C4.4.1.5.2 Precast Frames

Precast frame elements may have sufficient strength to meet lateral force requirements, but connections often cannot develop the strength of the members and may be subject to premature nonductile failures. Failure mechanisms may include fractures in the welded connections between inserts, pull-out of embeds, and spalling of concrete.

Since full member capacities cannot be realized, the behavior of this system is entirely dependent on the performance of the connections.

4.4.1.5.3 PRECAST CONNECTIONS: For buildings with concrete shear walls, the connection between precast frame elements such as chords, ties, and collectors in the lateral-force-resisting system shall develop the capacity of the connected members.

Tier 2 Evaluation Procedure: An analysis in accordance with Section 4.2 shall be performed. The adequacy of the connections for seismic forces shall be evaluated as force-controlled elements using the procedures in Section 4.2.4.3.2.

C4.4.1.5.3 Precast Connections

Precast frame elements may have sufficient strength to meet lateral force requirements, but connections often cannot develop the strength of the members and may be subject to premature nonductile failures. Failure mechanisms may include fractures in the welded connections between inserts, pull-out of embeds, and spalling of concrete.

Since full member capacities cannot be realized, the behavior of this system is entirely dependent on the performance of the connections.

4.4.1.6 Frames Not Part of the Lateral-Force-Resisting System

C4.4.1.6 Frames Not Part of the Lateral-Force-Resisting System

This section deals with secondary components consisting of frames that were not designed to be part of the lateral-force-resisting system. These are basic structural frames of steel or concrete that are designed for gravity loads only. Shear walls or other vertical elements provide the resistance to lateral forces. In actuality, however, all frames act as part of the lateral-force-resisting system. Lateral drifts of the building will induce forces in the beams and columns of the secondary frames. Furthermore, in the event that the primary elements fail, the secondary frames become the primary lateral-force-resisting components of the building.

If the walls are concrete (infilled in steel frames or monolithic in concrete frames), the building should be treated as a concrete shear wall building (Types C2 or C2A) with the frame columns as boundary elements. If the walls are masonry infills, the frames should be treated as steel or concrete frames with infill walls of masonry (Types S5, S5A, C3, or C3A). Research is continuing on the behavior of infill frames. Lateral forces are resisted by compression struts that develop in the masonry infill and induce forces on the frame elements eccentric to the joints.

The concern for secondary frames is the potential loss of vertical-load-carrying capacity due to excessive deformations and P-Δ effects.

4.4.1.6.1 COMPLETE FRAMES: Steel or concrete frames classified as secondary components shall form a complete vertical-load-carrying system.

Tier 2 Evaluation Procedure: An analysis in accordance with Section 4.2 shall be performed. The gravity and seismic demands for the shear walls shall be calculated, and the adequacy of the shear walls shall be evaluated.

C4.4.1.6.1 Complete Frames

If the frame does not form a complete vertical-load-carrying system, the walls will be required to provide vertical support as bearing walls (see Figure C4-15). A frame is incomplete if there are no columns cast into the wall, there are no columns adjacent to the wall, and beams frame into the wall, supported solely by the wall.

During an earthquake, shear walls might become damaged by seismic forces, limiting their ability to support vertical loads. Loss of vertical support may lead to partial collapse.

Compliance can be demonstrated if the wall is judged adequate for combined vertical and seismic forces.

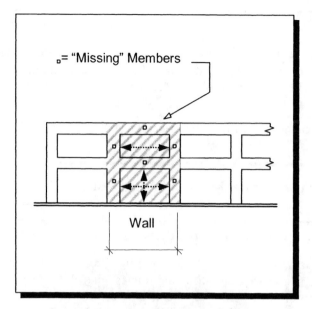

Figure C4-15. Incomplete Frame

4.4.1.6.2 DEFLECTION COMPATIBILITY: Secondary components shall have the shear capacity to develop the flexural strength of the components for Life Safety and shall meet the requirements of Sections 4.4.1.4.9, 4.4.1.4.10, 4.4.1.4.11, 4.4.1.4.12, and 4.4.1.4.15 for Immediate Occupancy.

Tier 2 Evaluation Procedure: An analysis in accordance with Section 4.2 shall be performed. The flexural and shear demands at maximum inter-story drifts for non-compliant elements shall be calculated, and the adequacy of the elements shall be evaluated.

C4.4.1.6.2 Deflection Compatibility

Frame components, especially columns, that are not specifically designed to participate in the lateral-force-resisting system will still undergo displacements associated with overall seismic inter-story drifts. If the columns are located some distance away from the lateral-force-resisting elements, the added deflections due to semi-rigid floor diaphragms will increase the drifts. Stiff columns, designed for potentially high gravity loads, may develop significant bending moments due to the imposed drifts. The moment/axial force interaction may lead to a nonductile failure of the columns and a collapse of the building.

4.4.1.6.3 FLAT SLABS: Flat slabs/plates not part of lateral-force-resisting system shall have continuous bottom steel through the column joints for Life Safety and Immediate Occupancy.

Tier 2 Evaluation Procedure: An analysis in accordance with Section 4.2 shall be performed. The adequacy of the joint for punching shear for all gravity and seismic demands, and shear transfer due to seismic moments, shall be evaluated.

C4.4.1.6.3 Flat Slabs

Flat slabs not designed to participate in the lateral-force-resisting system may still experience seismic forces due to displacements associated with overall building drift. The concern is the transfer of the shear and bending forces between the slab and column, which could result in a punching shear failure.

A problem with some slabs can occur when a small section of slab exists between two adjacent shear walls or braced frames. The section of slab can act as a coupling beam, even though it was not intended to do so. This can result in excessive damage to the slab and loss of vertical slab support if the slab is not properly detailed. Thin slabs and those with long spans have less tendency to act as a coupling beam and would attract less force.

Continuity of some bottom reinforcement through the column joint will assist in the transfer of forces and provide some resistance to collapse by catenary action in the event of a punching shear failure (see Figure C4-16). Bars can be considered continuous if they have proper lap splices, mechanical couplers, or are developed beyond the support.

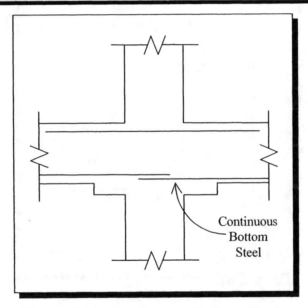

Figure 4-16. Continuous Bottom Steel

4.4.2 Shear Walls

4.4.2.1 General

C4.4.2.1 General

In the analysis of shear walls, it is customary to consider the shear taken by the length of the wall and the flexure taken by vertical reinforcement added at each end, much as flexure in diaphragms is designed to be taken by chords at the edges. Squat walls that are long compared to their height are dominated by shear behavior. Flexural forces require only a slight local reinforcement at each end. Slender walls that are tall compared to their length are usually dominated by flexural behavior and may require substantial boundary elements at each end.

Evaluation Phase (Tier 2)

> It is a good idea to sketch a complete free-body diagram of the wall (as indicated in Figure C4-17) so that no forces are inadvertently neglected. An error often made in the design of wood shear walls is to treat the walls one story at a time, considering only the shear force in the wall and overlooking the accumulation of overturning forces from the stories above.
>
> Where the earthquake direction being considered is parallel to a shear wall, the wall develops in-plane shear and flexural forces as described above. Where the earthquake direction is perpendicular to a shear wall, the wall contributes little to the lateral force resistance of the building and the wall is subjected to out-of-plane forces. This section addresses the in-plane behavior of shear walls. Out-of-plane strength and anchorage of shear walls to the structure is addressed in Section 4.5.
>
> Solid shear walls usually have sufficient strength, though they may be lightly reinforced. Problems with shear wall systems arise where walls are not continuous to the foundation, or where numerous openings break the walls up into small piers with limited shear and flexural capacity.

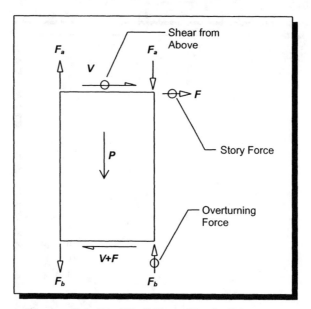

Figure C4-17. Wall Free-Body Diagram

4.4.2.1.1 REDUNDANCY: The number of lines of shear walls in each direction shall be greater than or equal to 2.0 for Life Safety and for Immediate Occupancy.

Tier 2 Evaluation Procedure: An analysis in accordance with the procedures in Section 4.2 shall be performed. The adequacy of all walls and connections shall be evaluated.

> **C4.4.2.1.1 Redundancy**
>
> Redundancy is a fundamental characteristic of lateral-force-resisting systems with superior seismic performance. Redundancy in the structure will ensure that, if an element in the lateral-force-resisting system fails for any reason, there is another element present that can provide lateral force resistance. Redundancy also provides multiple locations for potential yielding, distributing inelastic activity throughout the structure and improving ductility and energy absorption. Typical characteristics of redundancy include multiple lines of resistance to distribute the lateral forces uniformly throughout the structure (see Figure C4-18) and multiple bays in each line of resistance to reduce the shear and axial demands on any one element.

> A distinction should be made between redundancy and adequacy. For the purpose of this standard, redundancy is intended to mean simply "more than one." That is not to say that for large buildings two elements are adequate or for small buildings one is not enough. Separation evaluation statements are present in the standard to determine the adequacy of the elements provided.
>
> Where redundancy is not present in the structure, an analysis that demonstrates the adequacy of the lateral force elements is required.

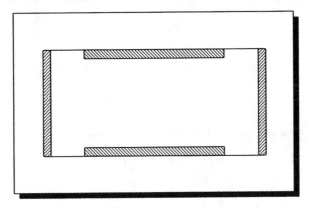

Figure C4-18. Redundancy in Shear Walls

4.4.2.2 Concrete Shear Walls

> **C4.4.2.2 Concrete Shear Walls**
>
> In highly redundant buildings with many long walls, stresses in concrete shear walls are usually low. In less redundant buildings with large openings and slender walls, the stresses can be high. In the ultimate state, where overturning forces are at their highest, a thin wall may fail in buckling along the compression edge, or it may fail in tension along the tension edge. Tension failures may consist of slippage in bar lap splices, or bar yield and fracture if adequate lap splices have been provided.
>
> In the past, designs have been based on liberal assumptions about compression capacity and have simply packed vertical rebar into the ends of the walls to resist the tensile forces. Recent codes, recognizing the importance of boundary members, have special requirements for proportions, bar splices, and transverse reinforcement. Examples of boundary members with varying amounts of reinforcing are shown in Figure C4-19. Existing buildings often do not have these elements, and the acceptance criteria are designed to allow for this.
>
> Another development in recent codes is the requirement to provide shear strength compatible with the flexural capacity of the wall to ensure ductile flexural yielding prior to brittle shear failure. Long continuous walls and walls with embedded steel or large boundary elements can have high flexural capacities with the potential to induce correspondingly high shear demands that are over and above the minimum design shear demands.

Evaluation Phase (Tier 2)

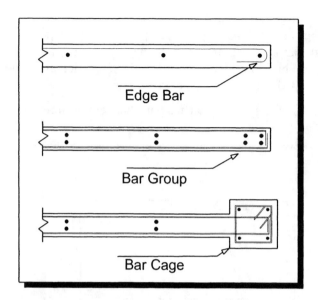

Figure C4-19. Boundary Elements

4.4.2.2.1 SHEAR STRESS CHECK: The shear stress in the concrete shear walls, calculated using the Quick Check procedure of Section 3.5.3.3, shall be less than the greater of 100 psi or $2\sqrt{f'c}$ for Life Safety and Immediate Occupancy.

Tier 2 Evaluation Procedure: An analysis in accordance with Section 4.2 shall be performed. The adequacy of the concrete shear wall elements shall be evaluated using the *m*-factors in Table 4-6.

> **C4.4.2.2.1 Shear Stress Check**
>
> The shear stress check provides a quick assessment of the overall level of demand on the structure. The concern is the overall strength of the building.

4.4.2.2.2 REINFORCING STEEL: The ratio of reinforcing steel area to gross concrete area shall be not less than 0.0015 in the vertical direction and 0.0025 in the horizontal direction for Life Safety and Immediate Occupancy. The spacing of reinforcing steel shall be equal to or less than 18" inches for Life Safety and for Immediate Occupancy.

Tier 2 Evaluation Procedure: An analysis in accordance with Section 4.2 shall be performed. The adequacy of the concrete shear wall elements shall be evaluated using the *m*-factors in Table 4-6.

> **C4.4.2.2.2 Reinforcing Steel**
>
> If the walls do not have sufficient reinforcing steel, they will have limited capacity in resisting seismic forces. The wall also will behave in a nonductile manner for inelastic forces.

Evaluation Phase (Tier 2)

4.4.2.2.3 COUPLING BEAMS: The stirrups in coupling beams over means of egress shall be spaced at or less than $d/2$ and shall be anchored into the confined core of the beam with hooks of 135° or more for Life Safety. All coupling beams shall comply with the requirements above and shall have the capacity in shear to develop the uplift capacity of the adjacent wall for Immediate Occupancy.

Tier 2 Evaluation Procedure: An analysis in accordance with Section 4.2 shall be performed. The shear and flexural demands on non-compliant coupling beams shall be calculated, and the adequacy of the coupling beams shall be evaluated. If the coupling beams are inadequate, the adequacy of the coupled walls shall be evaluated as if they were independent.

C4.4.2.2.3 Coupling Beams

Coupling beams with sufficient strength and stiffness can increase the lateral stiffness of the system significantly beyond the stiffnesses of the independent walls. When the walls deflect laterally, large moments and shears are induced in the coupling beams as they resist the imposed deformations. Coupling beams also link the coupled walls for overturning resistance (see Figure C4-20).

Coupling beam reinforcement is often inadequate for the demands that can be induced by the movement of the coupled walls. Seismic forces may damage and degrade the beams so severely that the system degenerates into a pair of independent walls. This changes the distribution of overturning forces which may result in potential stability problems for the independent walls. The boundary reinforcement also may be inadequate for flexural demands if the walls act independently.

If the beams are lightly reinforced, their degradation could result in falling debris that is a potential life safety hazard, especially at locations of egress.

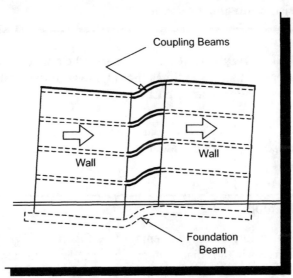

Figure C4-20. Coupled Walls

4.4.2.2.4 OVERTURNING: All shear walls shall have aspect ratios less than 4-to-1. Wall piers need not be considered. This statement shall apply to the Immediate Occupancy Performance Level only.

Evaluation Phase (Tier 2)

Tier 2 Evaluation Procedure: An analysis in accordance with Section 4.2 shall be performed. The overturning demands for non-compliant walls shall be calculated, and the adequacy of the shear walls shall be evaluated.

> **C4.4.2.2.4 Overturning**
>
> Tall, slender shear walls may have limited overturning resistance. Displacements at the top of the building will be greater than anticipated if overturning forces are not properly resisted.
>
> Often sufficient resistance can be found in immediately adjacent bays if a load path is present to activate the adjacent column dead loads.

4.4.2.2.5 CONFINEMENT REINFORCING: For shear walls with aspect ratios greater than 2-to-1, the boundary elements shall be confined with spirals or ties with spacing less than $8d_b$. This statement shall apply to the Immediate Occupancy Performance Level only.

Tier 2 Evaluation Procedure: An analysis in accordance with Section 4.2 shall be performed. The shear and flexural demands on the non-compliant walls shall be calculated, and the adequacy of the shear walls shall be evaluated.

> **C4.4.2.2.5 Confinement Reinforcing**
>
> Fully effective shear walls require boundary elements to be properly confined with closely spaced ties (see Figure C4-19). Degradation of the concrete in the vicinity of the boundary elements can result in buckling of rebar in compression and failure of lap splices in tension. Nonductile failure of the boundary elements will lead to reduced capacity to resist overturning forces.

4.4.2.2.6 REINFORCING AT OPENINGS: There shall be added trim reinforcement around all openings with a dimension greater than three times the thickness of the wall. This statement shall apply to the Immediate Occupancy Performance Level only.

Tier 2 Evaluation Procedure: An analysis in accordance with Section 4.2 shall be performed. The flexural and shear demands around the openings shall be calculated, and the adequacy of the piers and spandrels shall be evaluated.

> **C4.4.2.2.6 Reinforcing at Openings**
>
> Conventional trim steel is adequate only for small openings (see Figure C4-21). Large openings will cause significant shear and flexural stresses in the adjacent piers and spandrels. Inadequate reinforcing steel around these openings will lead to strength deficiencies, nonductile performance, and degradation of the wall.

Evaluation Phase (Tier 2)

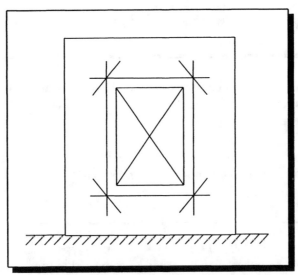

Figure C4-21. Conventional Trim Steel

4.4.2.2.7 WALL THICKNESS: Thickness of bearing walls shall not be less than 1/25 the unsupported height or length, whichever is shorter, nor less than 4 inches. This statement shall apply to the Immediate Occupancy Performance Level only.

Tier 2 Evaluation Procedure: The adequacy of the walls to resist out-of-plane forces in combination with vertical loads shall be evaluated.

C4.4.2.2.7 Wall Thickness

Slender bearing walls may have limited capacity for vertical loads and higher potential for damage due to out-of-plane forces and magnified moments. Note that this condition is not considered a life safety concern and need only be examined for the Immediate Occupancy Performance Level.

4.4.2.2.8 WALL CONNECTIONS: There shall be a positive connection between the shear walls and the steel beams and columns for Life Safety, and the connection shall be able to develop the strength of the walls for Immediate Occupancy.

Tier 2 Evaluation Procedure: An analysis in accordance with Section 4.2 shall be performed. The shear and flexural demands on the shear walls shall be calculated, and the adequacy of the connection to transfer shear between the walls and the steel frame shall be evaluated.

C4.4.2.2.8 Wall Connections

Insufficient shear transfer between the steel and concrete elements will limit the ability of the steel to contribute to the performance of the shear walls. The connections to the column are especially important because the columns will develop a portion of the shear wall overturning moment. The connections should include welded studs, welded reinforcing steel, or fully encased steel elements with longitudinal reinforcing and ties.

Shear friction between the concrete and steel should only be used where the steel is completely encased in the concrete.

4.4.2.2.9 COLUMN SPLICES: Steel columns encased in shear-wall-boundary elements shall have splices that develop the tensile strength of the column. This statement shall apply to the Immediate Occupancy Performance Level only.

Tier 2 Evaluation Procedure: An analysis in accordance with Section 4.2 shall be performed. The tension demands due to overturning forces on non-compliant columns shall be calculated, and the adequacy of the splice connections shall be evaluated.

C4.4.2.2.9 Column Splices

Columns encased in shear-wall-boundary elements may be subjected to high tensile forces due to shear wall overturning moments. If the splice cannot develop the strength of the column, the ability of the column to contribute to overturning resistance will be limited.

The presence of axial loads may reduce the net tensile demand on the boundary element columns to a level below the capacity of the splice.

4.4.2.3   Precast Concrete Shear Walls

C4.4.2.3   Precast Concrete Shear Walls

Precast concrete shear walls are constructed in segments that are usually interconnected by embedded steel elements. These connections usually possess little ductility but are important to the overall behavior of the wall assembly. Interconnection between panels increases the overturning capacity by transferring overturning demands to end panels. Panel connections at the diaphragm are often used to provide continuous diaphragm chords. Failure of these connections will reduce the capacity of the system.

4.4.2.3.1   SHEAR STRESS CHECK: The shear stress in the precast panels, calculated using the Quick Check procedure of Section 3.5.3.3, shall be less than the greater of 100 psi or $2\sqrt{f'c}$ for Life Safety and Immediate Occupancy.

Tier 2 Evaluation Procedure: An analysis in accordance with Section 4.2 shall be performed. The adequacy of the concrete shear wall elements shall be evaluated using the *m*-factors in Table 4-6.

C4.4.2.3.1 Shear Stress Check

The shear stress check provides a quick assessment of the overall level of demand on the structure. The concern is the overall strength of the building.

Evaluation Phase (Tier 2)

4.4.2.3.2 REINFORCING STEEL: The ratio of reinforcing steel area to gross concrete area shall be not less than 0.0015 in the vertical direction and 0.0025 in the horizontal direction for Life Safety and Immediate Occupancy. The spacing of reinforcing steel shall be equal to or less than 18 inches for Life Safety and Immediate Occupancy.

Tier 2 Evaluation Procedure: An analysis in accordance with Section 4.2 shall be performed. The adequacy of the concrete shear wall elements shall be evaluated using the m-factors in Table 4-6.

C4.4.2.3.2 Reinforcing Steel

If the walls do not have sufficient reinforcing steel, they will have limited capacity in resisting seismic forces. The wall also will behave in a nonductile manner for inelastic forces.

It should be noted that in tilt-up construction, the reinforcement ratios are typically reversed because the principal direction of bending is vertical rather than horizontal.

4.4.2.3.3 WALL OPENINGS: The total width of openings along any perimeter wall line shall constitute less than 75 percent of the length of any perimeter wall for Life Safety and 50 percent for Immediate Occupancy with the wall piers having aspect ratios of less than 2-to-1 for Life Safety and Immediate Occupancy.

Tier 2 Evaluation Procedure: An analysis in accordance with Section 4.2 shall be performed. The adequacy of the remaining wall shall be evaluated for shear and overturning resistance, and the adequacy of the shear transfer connection between the diaphragm and the wall shall be evaluated. The adequacy of the connection between any collector elements and the wall also shall be evaluated.

C4.4.2.3.3 Wall Openings

In tilt-up construction, typical wall panels are often of sufficient length that special detailing for collector elements, shear transfer, and overturning resistance is not provided. Perimeter walls that are substantially open, such as at loading docks, have limited wall length to resist seismic forces and may be subject to overturning or shear transfer problems that were not accounted for in the original design.

Walls will be compliant if an adequate load path for shear transfer, collector forces, and overturning resistance can be demonstrated.

4.4.2.3.4 CORNER OPENINGS: Walls with openings at a building corner larger than the width of a typical panel shall be connected to the remainder of the wall with collector reinforcing.

Tier 2 Evaluation Procedure: An analysis in accordance with Section 4.2 shall be performed. The adequacy of the diaphragm to transfer shear and spandrel panel forces to the remainder of the wall beyond the opening shall be evaluated.

C4.4.2.3.4 Corner Openings

Open corners often are designed as entrances with the typical wall panel replaced by a spandrel panel and a glass curtain wall. Seismic forces in these elements are resisted by adjacent panels and, therefore, must be delivered through collectors.

Evaluation Phase (Tier 2)

> If the spandrel and other wall elements are adequately tied to the diaphragm, panel forces can be transferred back to adjacent wall panels through collector elements in the diaphragm.

4.4.2.3.5 PANEL-TO-PANEL CONNECTIONS: Adjacent wall panels shall be interconnected to transfer overturning forces between panels by methods other than steel welded inserts. This statement shall apply to the Immediate Occupancy Performance Level only.

Tier 2 Evaluation Procedure: An analysis in accordance with Section 4.2 shall be performed. The overturning demands shall be calculated, and the adequacy of the welded inserts to transfer overturning forces shall be evaluated as force-controlled elements in accordance with Section 4.2.4.3.2. Alternatively, the panels shall be evaluated for seismic demands as independent elements without consideration of coupling between panels.

> **C4.4.2.3.5 Panel-to-Panel Connections**
>
> Welded steel inserts can be brittle and may not be able to transfer the overturning forces between panels. Latent stresses may be present due to shrinkage and temperature effects. Brittle failure may include weld fracture, pull-out of the embedded anchors, or spalling of the concrete.
>
> Failure of these connections will result in separation of the wall panels and a reduction in overturning resistance.

4.4.2.3.6 WALL THICKNESS: Thickness of bearing walls shall not be less than 1/25 the unsupported height or length, whichever is shorter, nor less than 4 inches. This statement shall apply to the Immediate Occupancy Performance Level only.

Tier 2 Evaluation Procedure: The adequacy of the walls to resist out-of-plane forces shall be evaluated.

> **C4.4.2.3.6 Wall Thickness**
>
> Slender bearing walls may have limited capacity for vertical loads and higher potential for damage due to out-of-plane forces and magnified moments. Note that this condition is not considered a life safety concern and only needs to be examined for the Immediate Occupancy Performance Level.

4.4.2.4 Reinforced Masonry Shear Walls

4.4.2.4.1 SHEAR STRESS CHECK: The shear stress in the reinforced masonry shear walls, calculated using the Quick Check procedure of Section 3.5.3.3, shall be less than 70 psi for Life Safety and Immediate Occupancy.

Tier 2 Evaluation Procedure: An analysis in accordance with Section 4.2 shall be performed. The adequacy of the reinforced masonry shear wall elements shall be evaluated using the m-factors in Table 4-6.

C4.4.2.4.1 Shear Stress Check

The shear stress check provides a quick assessment of the overall level of demand on the structure. The concern is the overall strength of the building. For partially grouted walls, the effective net section should be used in calculating the shear stress.

4.4.2.4.2 **REINFORCING STEEL: The total vertical and horizontal reinforcing steel ratio in reinforced masonry walls shall be greater than 0.002 for Life Safety and Immediate Occupancy, with the minimum of 0.0007 for Life Safety and Immediate Occupancy in either of the two directions; the spacing of reinforcing steel shall be less than 48 inches for Life Safety and Immediate Occupancy; and all vertical bars shall extend to the top of the walls.**

Tier 2 Evaluation Procedure: An analysis in accordance with Section 4.2 shall be performed. The adequacy of the reinforced masonry shear wall elements shall be evaluated using the m-factors in Table 4-6.

C4.4.2.4.2 Reinforcing Steel

If the walls do not have sufficient reinforcing steel, they will have limited capacity in resisting seismic forces. The wall also will behave in a nonductile manner for inelastic forces.

4.4.2.4.3 **REINFORCING AT OPENINGS: All wall openings that interrupt rebar shall have trim reinforcing on all sides. This statement shall apply to the Immediate Occupancy Performance Level only.**

Tier 2 Evaluation Procedure: An analysis in accordance with Section 4.2 shall be performed. The flexural and shear demands around the openings shall be calculated, and the adequacy of the walls shall be evaluated using only the length of the piers between reinforcing steel.

C4.4.2.4.3 Reinforcing at Openings

Conventional trim steel is adequate only for small openings. Large openings will cause significant shearing and flexural stresses in the adjacent piers and spandrels. Inadequate reinforcing steel around these openings will lead to strength deficiencies, nonductile performance, and degradation of the wall.

4.4.2.4.4 **PROPORTIONS: The height-to-thickness ratio of the shear walls at each story shall be less than 30. This statement shall apply to the Immediate Occupancy Performance Level only.**

Tier 2 Evaluation Procedure: The adequacy of the walls to resist out-of-plane forces in combination with vertical loads shall be evaluated.

C4.4.2.4.4 Proportions

Slender bearing walls may have limited capacity for vertical loads and higher potential for damage due to out-of-plane forces and magnified moments. Note that this condition is not considered a life safety concern and need only be examined for the Immediate Occupancy Performance Level.

Evaluation Phase (Tier 2)

4.4.2.5 Unreinforced Masonry Shear Walls

4.4.2.5.1 SHEAR STRESS CHECK: The shear stress in the unreinforced masonry shear walls, calculated using the Quick Check procedure of Section 3.5.3.3, shall be less than 30 psi for clay units and 70 psi for concrete units for Life Safety and Immediate Occupancy.

Tier 2 Evaluation Procedure: An analysis in accordance with Section 4.2 shall be performed. The adequacy of the unreinforced masonry shear wall elements shall be evaluated using the m-factors in Table 4-6.

C4.4.2.5.1 Shear Stress Check

The shear stress check provides a quick assessment of the overall level of demand on the structure. The concern is the overall strength of the building. For concrete units, the effective net shear area should be used in calculating the shear stress.

4.4.2.5.2 PROPORTIONS: The height-to-thickness ratio of the shear walls at each story shall be less than the following for Life Safety and for Immediate Occupancy:

Top story of multi-story building	9
First story of multi-story building	15
All other conditions	13

Tier 2 Evaluation Procedure: No Tier 2 Evaluation procedure is available for unreinforced masonry shear wall proportions in non-compliance. A Tier 3 Evaluation is necessary to achieve the selected performance level.

C4.4.2.5.2 Proportions

Slender unreinforced masonry bearing walls with large height-to-thickness ratios have a potential for damage due to out-of-plane forces which may result in falling hazards and potential collapse of the structure.

4.4.2.5.3 MASONRY LAY-UP: Filled collar joints of multi-wythe masonry walls shall have negligible voids.

Tier 2 Evaluation Procedure: An analysis in accordance with Section 4.2 shall be performed. For in-plane demands, only the inner wythe of the wall shall be considered to resist forces. For out-of-plane demands, each wythe shall be evaluated independently. Anchorage of the outer wythe shall be evaluated as veneer in accordance with Section 4.2.7.

C4.4.2.5.3 Masonry Lay-Up

Where walls have poor collar joints, the inner and outer wythes will act independently. The walls may be inadequate to resist out-of-plane forces due to a lack of composite action between the inner and outer wythes.

Mitigation to provide out-of-plane stability and anchorage of the wythes may be necessary to achieve the selected performance level.

Evaluation Phase (Tier 2)

4.4.2.6 Infill Walls in Frames

4.4.2.6.1 WALL CONNECTIONS: Masonry shall be in full contact with the frame for Life Safety and Immediate Occupancy.

Tier 2 Evaluation Procedure: An analysis in accordance with Section 4.2 shall be performed. The out-of-plane demands on the wall shall be calculated, and the adequacy of the connection to the frame shall be evaluated.

C4.4.2.6.1 Wall Connections

Performance of frame buildings with masonry infill walls is dependent on the interaction between the frame and infill panels. In-plane lateral force resistance is provided by a compression strut developing in the infill panel that extends diagonally between corners of the frame. If gaps exist between the frame and infill, this strut cannot be developed (see Figure C4-22). If the infill panels separate from the frame due to out-of-plane forces, the strength and stiffness of the system will be determined by the properties of the bare frame, which may not be detailed to resist seismic forces. Severe damage or partial collapse due to excessive drift and P-Δ effects may occur.

A positive connection is needed to anchor the infill panel for out-of-plane forces. In this case, a positive connection can consist of a fully grouted bed joint in full contact with the frame, or complete encasement of the frame by the brick masonry. The mechanism for out-of-plane resistance of infill panels is discussed in the commentary to Section 4.4.2.6.2.

If the connection is nonexistent, mitigation with adequate connection to the frame is necessary to achieve the selected performance level.

It should be noted that it is impossible to simultaneously satisfy this section and Section 4.4.1.2.1.

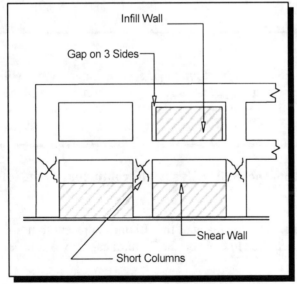

Figure C4-22. Infill Wall

4.4.2.6.2 PROPORTIONS: The height-to-thickness ratio of the infill walls at each story shall be less than 9.0 for Life Safety in levels of high seismicity, 13.0 for Immediate Occupancy in levels of moderate seismicity, and 8.0 for Immediate Occupancy in levels of high seismicity.

Tier 2 Evaluation Procedure: No Tier 2 Evaluation procedure is available for unreinforced masonry shear wall proportions in non-compliance. A Tier 3 Evaluation is necessary to demonstrate compliance with the selected performance level.

C4.4.2.6.2 Proportions

Slender masonry infill walls with large height-to-thickness ratios have a potential for damage due to out-of-plane forces. Failure of these walls out-of-plane will result in falling hazards and degradation of the strength and stiffness of the lateral-force-resisting system.

The out-of-plane stability of infill walls is dependent on many factors, including flexural strength of the wall and confinement provided by the surrounding frame. If the infill is unreinforced, the flexural strength is limited by the flexural tension capacity of the material. The surrounding frame will provide confinement, induce infill thrust forces, and develop arching action against out-of-plane forces. The height-to-thickness limits in the evaluation statement are based on arching action models that will exceed any plausible acceleration levels in various seismic zones.

Further investigation of non-compliant infill panels requires a Tier 3 Analysis.

4.4.2.6.3 SOLID WALLS: The infill walls shall not be of cavity construction.

Tier 2 Evaluation Procedure: An analysis in accordance with Section 4.2 shall be performed. For in-plane demands, only the inner wythe of the wall shall be considered to resist forces. For out-of-plane demands, each wythe shall be evaluated independently. Anchorage of the outer wythe shall be evaluated as veneer in accordance with Section 4.2.7.

C4.4.2.6.3 Solid Walls

Where the infill walls are of cavity construction, the inner and outer wythes will act independently due to a lack of composite action, increasing the potential for damage from out-of-plane forces. Failure of these walls out-of-plane will result in falling hazards and degradation of the strength and stiffness of the lateral-force-resisting system.

4.4.2.6.4 INFILL WALLS: The infill walls shall be continuous to the soffits of the frame beams and to the columns to either side.

Tier 2 Evaluation Procedure: The adequacy of the columns adjacent to non-conforming walls shall be evaluated for the shear force required to develop the flexural capacity of the column over the clear height above the infill.

> **C4.4.2.6.4 Infill Walls**
>
> Discontinuous infill walls occur where full bay windows or ventilation openings are provided between the top of the infill and bottom soffit of the frame beams. The portion of the column above the infill is a short captive column which may attract large shear forces due to increased stiffness relative to other columns (see Figure C4-22). Partial infill walls also will develop compression struts with horizontal components that are highly eccentric to the beam column joints. If not adequately detailed, concrete columns may suffer a nonductile shear failure which may result in partial collapse of the structure. Because steel columns are not subject to the same kind of brittle failure, this is not generally considered a concern in steel frame infill buildings.
>
> A column that can develop the shear capacity to develop the flexural strength over the clear height above the infill will have some ductility to prevent sudden catastrophic failure of the vertical support system.

4.4.2.7 Walls in Wood-Frame Buildings

4.4.2.7.1 SHEAR STRESS CHECK: The shear stress in the shear walls, calculated using the Quick Check procedure of 3.5.3.3, shall be less than the following values for Life Safety and Immediate Occupancy:

Structural panel sheathing	1,000 plf
Diagonal sheathing	700 plf
Straight sheathing	100 plf
All other conditions	100 plf

Tier 2 Evaluation Procedure: An analysis in accordance with Section 4.2 shall be performed. The adequacy of the wood shear wall elements shall be evaluated using the m-factors in Table 4-8.

> **C4.4.2.7.1 Shear Stress Check**
>
> The shear stress check provides a quick assessment of the overall level of demand on the structure. The concern is the overall strength of the building. The transfer of shear and overturning to the foundation also should be evaluated. The structural panel sheathing Quick Check capacity assumes the wall is compliant. Capacities should be reduced to account for deterioration or overdriven fasteners.

4.4.2.7.2 STUCCO (EXTERIOR PLASTER) SHEAR WALLS: Multi-story buildings shall not rely on exterior stucco walls as the primary lateral-force-resisting system.

Tier 2 Evaluation Procedure: An analysis in accordance with Section 4.2 shall be performed. The overturning and shear demands for non-compliant walls should be calculated, and the adequacy of the stucco shear walls shall be evaluated by using the m-factors in Table 4-8.

> **C4.4.2.7.2 Stucco (Exterior Plaster) Shear Walls**
>
> Exterior stucco walls are often used (intentionally and unintentionally) for resisting seismic forces. Stucco is relatively stiff but brittle, and the shear capacity is limited. Building movements due to differential settlement, temperature changes, and earthquake or wind forces can cause cracking in the stucco and loss of lateral strength. Lateral force resistance is unreliable because sometimes the stucco will delaminate from the framing and the system is lost. Multi-story buildings should not rely on stucco walls as the primary lateral-force-resisting system.
>
> Current research regarding the performance of wood-framed buildings is being funded by Consortium of Universities for Research in Earthquake Engineering (CUREE). Preliminary test results have indicated that older plaster wall construction can perform relatively well. For information regarding this research, go to www.curee.org/projects/woodframe_project/woodframe_intro.html.

4.4.2.7.3 GYPSUM WALLBOARD OR PLASTER SHEAR WALLS: Interior plaster or gypsum wallboard shall not be used as shear walls on buildings over one story in height with the exception of the uppermost level of a multi-story building.

Tier 2 Evaluation Procedure: An analysis in accordance with Section 4.2 shall be performed. The overturning and shear demands for non-compliant walls shall be calculated, and the adequacy of the gypsum wallboard or plaster shear walls shall be evaluated using the m-factors in Table 4-8.

> **C4.4.2.7.3 Gypsum Wallboard or Plaster Shear Walls**
>
> Gypsum wallboard or gypsum plaster sheathing tends to be easily damaged by differential foundation movement or earthquake shaking.
>
> Though the capacity of these walls is low, most residential buildings have numerous walls constructed with plaster or gypsum wallboard. As a result, plaster and gypsum wallboard walls may provide adequate resistance to moderate earthquake shaking.
>
> One problem that can occur is incompatibility with other lateral-forcing-resisting elements. For example, narrow plywood shear walls are more flexible than long stiff plaster walls; as a result, the plaster or gypsum walls will take all the seismic demand until they fail and then the plywood walls will start to resist the lateral forces. In multi-story buildings, plaster or gypsum wallboard walls should not be used for shear walls except in the top story.

4.4.2.7.4 NARROW WOOD SHEAR WALLS: Narrow wood shear walls with an aspect ratio greater than 2-to-1 for Life Safety and 1.5-to-1 for Immediate Occupancy shall not be used to resist lateral forces developed in the building in levels of moderate and high seismicity. Narrow wood shear walls with an aspect ratio greater than 2-to-1 for Immediate Occupancy shall not be used to resist lateral forces developed in the building in levels of low seismicity.

Tier 2 Evaluation Procedure: An analysis in accordance with Section 4.2 shall be performed. The overturning and shear demands for non-compliant walls shall be calculated, and the adequacy of the narrow shear walls shall be evaluated using the m-factors in Table 4-8.

Evaluation Phase (Tier 2)

> **C4.4.2.7.4 Narrow Wood Shear Walls**
>
> Narrow shear walls are highly stressed and subject to severe deformations that will reduce the capacity of the walls. Most of the damage occurs at the base and consists of sliding of the sill plate and deformation of hold-down anchors where present. As the deformation continues, the plywood pulls up on the sill plate, causing splitting. Splitting of the end studs at the bolted attachment of hold-down anchors is also common. Note that the aspect ratio for wood walls is the story height to wall length.

4.4.2.7.5 WALLS CONNECTED THROUGH FLOORS: Shear walls shall have interconnection between stories to transfer overturning and shear forces through the floor.

Tier 2 Evaluation Procedure: No Tier 2 Evaluation procedure is available for walls in non-compliance.

> **C4.4.2.7.5 Walls Connected Through Floors**
>
> In platform construction, wall framing is discontinuous at floor levels. The concern is that this discontinuity will prevent shear and overturning forces from being transferred between shear walls in adjacent stories.
>
> Mitigation with elements or connections needed to complete the load path is necessary to achieve the selected performance level.

4.4.2.7.6 HILLSIDE SITE: For structures that are taller on at least one side by more than half of a one-half story due to a sloping site, all shear walls on the downhill slope shall have an aspect ratio less than 1-to-1 for Life Safety and 1-to-2 for Immediate Occupancy.

Tier 2 Evaluation Procedure: An analysis in accordance with Section 4.2 shall be performed. The shear and overturning demands on the downhill slope walls shall be calculated, including the torsional effects of the hillside. The adequacy of the shear walls on the downhill slope shall be evaluated.

> **C4.4.2.7.6 Hillside Site**
>
> Buildings on a sloping site will experience significant torsion during an earthquake. Taller walls on the downhill slope are more flexible than the supports on the uphill slope. Therefore, significant displacement and racking of the shear walls on the downhill slope will occur. If the walls are narrow, significant damage or collapse may occur.

4.4.2.7.7 CRIPPLE WALLS: Cripple walls below first-floor-level shear walls shall be braced to the foundation with wood structural panels.

Tier 2 Evaluation Procedure: An analysis in accordance with Section 4.2 shall be performed. The shear demand for non-compliant walls shall be calculated, and the adequacy of the walls shall be evaluated using the m-factors in Table 4-8.

Evaluation Phase (Tier 2)

C4.4.2.7.7 Cripple Walls

Cripple walls are short stud walls that enclose a crawl space between the first floor and the ground. Often there are no other walls at this level, and these walls have no stiffening elements other than architectural finishes. If this sheathing fails, the building will experience significant damage and, in the extreme case, may fall off its foundation. To be effective, all exterior cripple walls below the first-floor level should have adequate shear strength, stiffness, and proper connection to the floor and foundation. Cripple walls that change height along their length, such as along sloping walls on hillside sites, will not have a uniform distribution of shear along the length of the wall, due to the varying stiffness. These walls may be subject to additional damage on the uphill side due to concentration of shear demand.

Mitigation with shear elements needed to complete the load path is necessary to achieve the selected performance level.

4.4.2.7.8 OPENINGS: Walls with openings greater than 80 percent of the length shall be braced with wood structural panel shear walls with aspect ratios of not more than 1.5 -to-1 or shall be supported by adjacent construction through positive ties capable of transferring the lateral forces.

Tier 2 Evaluation Procedure: An analysis in accordance with Section 4.2 shall be performed. The overturning and shear demands on non-compliant walls shall be calculated, and the adequacy of the shear walls shall be evaluated using the *m*-factors in Table 4-8.

C4.4.2.7.8 Openings

Walls with large openings, such as garage doors, may have little or no resistance to shear and overturning forces. They must be specially detailed to resist these forces or braced to other parts of the structure with collectors, such as metal straps developed into the adjacent construction. Special detailing and collectors are not part of conventional construction procedures. Lack of this bracing can lead to collapse of the wall.

4.4.2.7.9 HOLD-DOWN ANCHORS: All shear walls shall have hold-down anchors constructed per acceptable construction practices, attached to the end studs. This statement shall apply to the Immediate Occupancy Performance Level only.

Tier 2 Evaluation Procedure: An analysis in accordance with Section 4.2 shall be performed. The overturning and shear demands for non-compliant walls shall be calculated, and the adequacy of the shear walls shall be evaluated using the *m*-factors in Table 4-8.

C4.4.2.7.9 Hold-Down Anchors

Buildings without hold-down anchors may be subject to significant damage due to uplift and racking of the shear walls. Properly constructed hold-downs must connect the floors together and activate the weight of the foundation. They must be tightly connected to the boundary element in a manner such that the deformation of the shear wall does not destroy the integrity of the hold-downs. Building drawings and manufacturers' recommendations are helpful in determining the adequacy of the hold-downs.

Note that this condition is not considered a life safety concern and only needs to be examined for the Immediate Occupancy Performance Level.

Evaluation Phase (Tier 2)

4.4.3 Braced Frames

> **C4.4.3 Braced Frames**
>
> Braced frames develop their lateral force resistance through axial forces developed in the diagonal bracing members. The braces induce forces in the associated beams and columns, and all are subjected to stresses that are primarily axial. Where the braces are eccentric to beam-column joints, members are subjected to shear and flexure in addition to axial forces. A portal frame with knee braces near the frame joints is one example.
>
> Braced frames are classified as either concentrically braced frames or eccentrically braced frames (see Figure C4-23). Concentrically braced frames have braces that frame into beam-column joints or concentric connections with other braces. Minor connection eccentricities may be present and are accounted for in the design. Eccentrically braced frames have braces that are purposely located away from joints and connections that are intended to induce shear and flexure demands on the members. The eccentricity is intended to force a concentration of inelastic activity at a predetermined location that will control the behavior of the system. Modern eccentrically braced frames are designed with strict controls on member proportions and special out-of-plane bracing at the connections to ensure the frame behaves as intended.

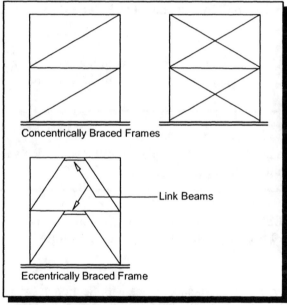

Figure C4-23. Braced Frames

4.4.3.1 General

4.4.3.1.1 REDUNDANCY: The number of lines of braced frames in each principal direction shall be greater than or equal to 2.0 for Life Safety and Immediate Occupancy. The number of braced bays in each line shall be greater than 2.0 for Life Safety and 3.0 for Immediate Occupancy.

Tier 2 Evaluation Procedure: An analysis in accordance with the procedures in Section 4.2 shall be performed. The adequacy of all elements and connections in the braced frames shall be evaluated.

C4.4.3.1.1 Redundancy

Redundancy is a fundamental characteristic of lateral-force-resisting systems with superior seismic performance. Redundancy in the structure will ensure that, if an element in the lateral-force-resisting system fails for any reason, there is another element present that can provide lateral force resistance. Redundancy also provides multiple locations for potential yielding, distributing inelastic activity throughout the structure and improving ductility and energy absorption. Typical characteristics of redundancy include multiple lines of resistance to distribute the lateral forces uniformly throughout the structure, and multiple bays in each line of resistance to reduce the shear and axial demands on any one element.

A distinction should be made between redundancy and adequacy. For the purpose of this standard, redundancy is intended to mean simply "more than one." That is not to say that for large buildings two elements is adequate or for small buildings one is not enough. Separate evaluation statements are present in the standard to determine the adequacy of the elements provided.

Where redundancy is not present in the structure, an analysis that demonstrates the adequacy of the lateral force elements is required.

4.4.3.1.2 AXIAL STRESS CHECK: The axial stress in the diagonals, calculated using the Quick Check procedure of Section 3.5.3.4, shall be less than $0.50F_y$ for Life Safety and for Immediate Occupancy.

Tier 2 Evaluation Procedure: An analysis in accordance with Section 4.2 shall be performed. The adequacy of the braced frame elements shall be evaluated using the *m*-factors in Table 4-5.

C4.4.3.1.2 Axial Stress Check

The axial stress check provides a quick assessment of the overall level of demand on the structure. The concern is the overall strength of the building.

4.4.3.1.3 COLUMN SPLICES: All column splice details located in braced frames shall develop the tensile strength of the column. This statement shall apply to the Immediate Occupancy Performance Level only.

Tier 2 Evaluation Procedure: An analysis in accordance with Section 4.2 shall be performed. The tension demands on non-compliant columns shall be calculated, and the adequacy of the splice connections shall be evaluated.

C4.4.3.1.3 Column Splices

Columns in braced frames may be subject to large tensile forces. A connection that is unable to resist this tension may limit the ability of the frame to resist lateral forces. Columns may uplift and slide off bearing supports, resulting in unexpected damage to the wall.

Evaluation Phase (Tier 2)

4.4.3.1.4 SLENDERNESS OF DIAGONALS: All diagonal elements required to carry compression shall have $Kl/rKl/r$ ratios less than 120.

Tier 2 Evaluation Procedure: An analysis in accordance with Section 4.2 shall be performed. The compression demands in non-compliant braces shall be calculated, and the adequacy of the braces shall be evaluated for buckling.

> **C4.4.3.1.4 Slenderness of Diagonals**
>
> Code design requirements have allowed compression diagonal braces to have Kl/r ratios of up to 200. Cyclic tests have demonstrated that elements with high Kl/r ratios are subjected to large buckling deformations, resulting in brace or connection fractures. They cannot be expected to provide adequate performance. Limited energy dissipation and premature buckling can significantly reduce strength, increase the building displacements, and jeopardize the performance of the framing system.

4.4.3.1.5 CONNECTION STRENGTH: All the brace connections shall develop the yield capacity of the diagonals.

Tier 2 Evaluation Procedure: An analysis in accordance with Section 4.2 shall be performed. The demands on the non-compliant connections shall be calculated, and the adequacy of the brace connections shall be evaluated.

> **C4.4.3.1.5 Connection Strength**
>
> Since connection failures are usually nonductile in nature, it is more desirable to have inelastic behavior in the members.

4.4.3.1.6 OUT-OF-PLANE BRACING: Braced frame connections attached to beam bottom flanges located away from beam-column joints shall be braced out-of-plane at the bottom flange of the beams. This statement shall apply to the Immediate Occupancy Performance Level only.

Tier 2 Evaluation Procedure: An analysis in accordance with Section 4.2 shall be performed. The demands shall be calculated, and the adequacy of the beam shall be evaluated considering a horizontal out-of-plane force equal to 2 percent of the brace compression force acting concurrently at the bottom flange of the beam.

> **C4.4.3.1.6 Out-of-Plane Bracing**
>
> Brace connections at beam bottom flanges that do not have proper bracing may have limited ability to resist seismic forces. Out-of-plane buckling may occur before the strength of the brace is developed. Connections to beam top flanges are braced by the diaphragm, so V-bracing need not be considered.
>
> This statement is intended to target chevron-type bracing, where braces intersect the beam from below at a location well away from a column. Here, only the beam can provide out-of-plane stability for the connection. At beam-column joints, the continuity of the column will provide stability for the connection.
>
> To demonstrate compliance, the beam is checked for the strength required to provide out-of-plane stability using the 2 percent rule.

4.4.3.2 Concentrically Braced Frames

> **C4.4.3.2 Concentrically Braced Frames**
>
> Common types of concentrically braced frames are shown in Figure C4-24.
>
> Braces can consist of light tension-only rod bracing, double angles, pipes, tubes, or heavy wide-flange sections.
>
> Concrete braced frames are rare and are not permitted in some jurisdictions because it is difficult to detail the joints with the kind of reinforcing that is required for ductile behavior.

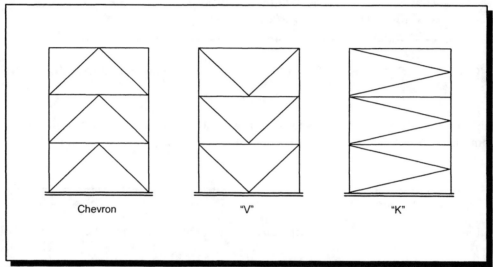

Figure C4-24. Bracing Types

4.4.3.2.1 K-BRACING: The bracing system shall not include K-braced bays.

Tier 2 Evaluation Procedure: An analysis in accordance with Section 4.2 shall be performed. The adequacy of the columns shall be evaluated for all demands, including concurrent application of the unbalanced force that can be applied to the column by the braces. The unbalanced force shall be taken as the horizontal component of the tensile capacity of one brace, assuming the other brace has buckled in compression. The m-factors in Table 4-5 shall be used.

> **C4.4.3.2.1 K-Bracing**
>
> In K-brace configurations, diagonal braces intersect the column between floor levels (see Figure C4-25). Where the compression brace buckles, the column will be loaded with the horizontal component of the adjacent tension brace. This will induce large mid-height demands that can jeopardize the stability of the column and vertical support of the building.
>
> In most cases, columns have not been designed to resist this force. The risk to the vertical support system makes this an undesirable bracing configuration.

Evaluation Phase (Tier 2)

4.4.3.2.2 TENSION-ONLY BRACES: Tension-only braces shall not comprise more than 70 percent of the total lateral-force-resisting capacity in structures over two stories in height. This statement shall apply to the Immediate Occupancy Performance Level only.

Tier 2 Evaluation Procedure: An analysis in accordance with Section 4.2 shall be performed. The adequacy of the tension-only braces shall be evaluated using the *m*-factors in Table 4-5.

C4.4.3.2.2 Tension-Only Braces

Tension-only brace systems may allow the brace to deform with large velocities during cyclic response after tension yielding cycles have occurred. Limited energy dissipation and premature fracture can significantly reduce the strength, increase the building displacements, and jeopardize the performance of the framing system.

4.4.3.2.3 CHEVRON BRACING: The bracing system shall not include chevron- or V-braced bays. This statement shall apply to the Immediate Occupancy Performance Level only.

Tier 2 Evaluation Procedure: An analysis in accordance with Section 4.2 shall be performed. The adequacy of the beams shall be evaluated for all demands, including concurrent application of the unbalanced force that can be applied to the beams by the braces. The unbalanced force shall be taken as the vertical component of the tensile capacity of one brace, assuming the other brace has buckled in compression. The *m*-factors in Table 4-5 shall be used.

C4.4.3.2.3 Chevron Bracing

In chevron- and V-brace configurations, diagonal braces intersect the beam between columns (see Figure C4-25). When the compression brace buckles, the beam will be loaded with the vertical component of the adjacent tension brace. This will induce large mid-span demands on the beam, resulting in structural damage to the beam. While the deformations of the frame will not result in a life safety hazard, the damage will probably prevent the structure from being used following the design earthquake.

4.4.3.2.4 CONCENTRICALLY BRACED FRAME JOINTS: All the diagonal braces shall frame into the beam-column joints concentrically. This statement shall apply to the Immediate Occupancy Performance Level only.

Tier 2 Evaluation Procedure: An analysis in accordance with Section 4.2 shall be performed. The axial, flexural, and shear demands, including the demands due to eccentricity of the braces, shall be calculated. The adequacy of the joints shall be evaluated.

C4.4.3.2.4 Concentric Joints

Frames that have been designed as concentrically braced frames may have local eccentricities within the joint. A local eccentricity is where the lines of action of the bracing members do not intersect the centerline of the connecting members. These eccentricities induce additional flexural and shear stresses in the members that may not have been accounted for in the design. Excessive eccentricity can cause premature yielding of the connecting members or failures in the connections, thereby reducing the strength of the frames.

Evaluation Phase (Tier 2)

4.4.3.3 Eccentrically Braced Frames

No evaluation statements or Tier 2 procedures have been provided specifically for eccentrically braced frames. Eccentrically braced frames shall be checked for the general braced frame evaluation statements and Tier 2 procedures in Section 4.4.3.1.

C4.4.3.3 Eccentrically Braced Frames

Eccentrically braced frames have braces that are purposely located away from joints, and connections that are intended to induce shear and flexure demands on the members. The eccentricity is intended to force a concentration of inelastic activity at a predetermined location that will control the behavior of the system. Modern eccentrically braced frames are designed with strict controls on member proportions and special out-of-plane bracing at the connections to ensure the frame behaves as intended.

The eccentrically braced frame is a relatively new type of frame that is recognizable by a diagonal with one end significantly offset from the joints (Figure C4-25). As with any braced frame, the function of the diagonal is to provide stiffness and transmit lateral forces from the upper to the lower level. The unique feature of eccentrically braced frames is an offset zone in the beam, called the "link." The link is specially detailed for controlled yielding. This detailing is subject to very specific requirements, so an ordinary braced frame that happens to have an offset zone that looks like a link may not necessarily behave like an eccentrically braced frame.

An eccentrically braced frame has the following essential features:

There is a link beam at one end of each brace.

The length of the link beam is limited to control shear deformations and rotations due to flexural yielding at the ends of the link.

The brace and the connections are designed to develop forces consistent with the strength of the link.

Where one end of a link beam is connected to a column, the connection is a full moment connection.

Lateral bracing is provided to prevent out-of-plane beam displacements that would compromise the intended action.

In most cases where eccentrically braced frames are used, the frames comprise the entire lateral-force-resisting system. In some tall buildings, eccentrically braced frames have been added as stiffening elements to help control drift in moment-resisting steel frames.

There are no evaluation statements for eccentrically braced frames because their history is so short, but the engineer is alerted to their possible presence in a building. For guidance in dealing with eccentrically braced frames, the evaluating engineer is referred to the *Recommended Lateral Force Requirements and Commentary* (SEAOC, 1996). It should be noted that some of the engineers familiar with current research designed eccentrically braced frames before the SEAOC provisions were developed. These frames may not satisfy all of the detailing requirements present in the current code. Any frame that was clearly designed to function as a proper eccentrically braced frame should be recognized and evaluated with due regard for any possible shortcomings that will affect the intended behavior. Acceptance criteria for eccentrically braced frames are provided in FEMA 356.

Evaluation Phase (Tier 2)

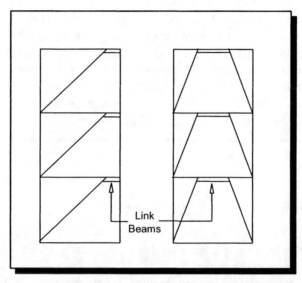

Figure C4-25. Eccentrically Braced Frames

Evaluation Phase (Tier 2)

4.5 Procedures for Diaphragms

This section provides Tier 2 Evaluation procedures that apply to diaphragms: general, wood, metal deck, concrete, precast concrete, horizontal bracing, and other diaphragms.

> **C4.5 Procedures for Diaphragms**
>
> Diaphragms are horizontal elements that distribute seismic forces to vertical lateral-force-resisting elements. They also provide lateral support for walls and parapets. Diaphragm forces are derived from the self-weight of the diaphragm and the weight of the elements and components that depend on the diaphragm for lateral support. Any roof, floor, or ceiling can participate in the distribution of lateral forces to vertical elements up to the limit of its strength. The degree to which it participates depends on relative stiffness and on connections. In order to function as a diaphragm, horizontal elements must be interconnected to transfer shear, with connections that have some degree of stiffness. An array of loose elements, such as ceiling tiles or metal-deck panels attached to beams with wind clips, does not qualify.

4.5.1 General

> **C4.5.1 General**
>
> It is customary to analyze diaphragms using a beam analogy. The floor, which is analogous to the web of a wide-flange beam, is assumed to carry the shear. The edge of the floor, which could be a spandrel or wall, is analogous to the flange, and is assumed to carry the flexural stress. A free-body diagram of these elements is shown in Figure C4-26. The diaphragm chord can consist of a line of edge beams that are connected to the floor, or reinforcing in the edge of a slab or in a spandrel. Examples of chords are shown in Figure C4-27.

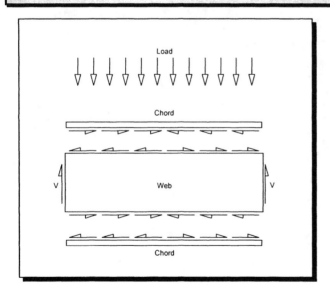

Figure C4-26. Diaphragm as a Beam

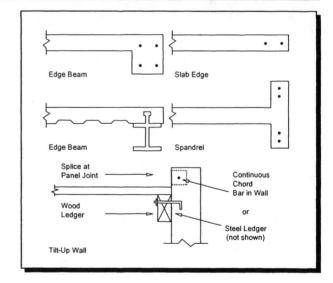

Figure C4-27. Chord Sections

Evaluation Phase (Tier 2)

Two essential requirements for the chord are continuity and connection with the slab. Almost any building with an edge beam has a potential diaphragm chord. Even if designed for vertical loads only, the beam end connections probably have some capacity to develop horizontal forces through the column.

The force in the chord is customarily determined by dividing the beam moment in the diaphragm by the depth of the diaphragm. This yields an upper bound on the chord force since it assumes elastic beam behavior in the diaphragm and neglects bending resistance provided by any other components of the diaphragm. A lack of diaphragm damage in post-earthquake observations provides some evidence that certain diaphragms may not require specific chords as determined by the beam analogy. For the purpose of this standard, the absence of chords is regarded as a deficiency that warrants further evaluation. Consideration may be given to the available evidence regarding the suitability of the beam analogy and the need for defined chords in the building being evaluated.

Consistent with the beam analogy, a stair or skylight opening may weaken the diaphragm just as a web opening for a pipe may weaken a beam. An opening at the edge of a floor may weaken the diaphragm just as a notch in a flange weakens a beam.

An important characteristic of diaphragms is flexibility, or its opposite, rigidity. In seismic design, rigidity means relative rigidity. Of importance is the in-plane rigidity of the diaphragm relative to the walls or frame elements that transmit the lateral forces to the ground (Figure C4-28). A concrete floor is relatively rigid compared to steel moment frames, whereas a metal deck roof is relatively flexible compared to concrete or masonry walls. Wood diaphragms are generally treated as flexible, but consideration must be given to rigidity of the vertical elements. Wood diaphragms may not be flexible compared to wood shear wall panels in a given building.

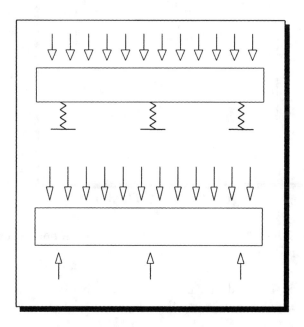

Figure C4-28. Rigid and Flexible Diaphragm

Evaluation Phase (Tier 2)

> Another consideration is continuity over intermediate supports. In a three-bay building, for example, the diaphragm has three spans and four supports. If the diaphragm is relatively rigid, the chords should be continuous over the supports like flanges of a continuous beam over intermediate supports. If the diaphragm is flexible, it may be designed as a simple beam spanning between walls without consideration of continuity of the chords. In the latter case, the design professional should remember that the diaphragm is really continuous, and that this continuity is simply being neglected.
>
> Figure C4-29 shows a diaphragm of two spans that may or may not be continuous over the intermediate support. If chord continuity is developed at the points marked X, these will be the locations of maximum chord force. If chord continuity is not provided at X, the spans will act as two simple beams. The maximum chord force will occur at the middle of each span, at the points marked Y. The end rotations of the two spans may cause local damage at points X.
>
> Finally, there must be an adequate mechanism for the transfer of diaphragm shear forces to the vertical elements. This topic is addressed in detail in Section 4.6. An important element related to diaphragm force transfer is the collector, or drag strut. In Figure C4-29, a member is added to collect the diaphragm shear and drag it into the short intermediate shear wall. The presence of a collector avoids a concentration of stress in the diaphragm at the short shear wall. Collectors must be continuous across any interrupting elements such as perpendicular beams, and must be adequately connected to the shear wall to deliver forces into the wall.
>
> In buildings of more than one story, the design professional must consider the effect of flexible diaphragms on walls perpendicular to the direction of seismic force under consideration.

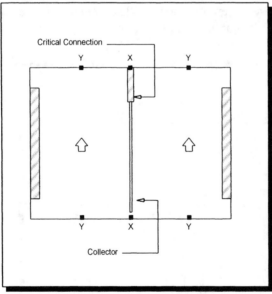

Figure C4-29. Collector

4.5.1.1 DIAPHRAGM CONTINUITY: The diaphragms shall not be composed of split-level floors and shall not have expansion joints.

Tier 2 Evaluation Procedure: The load path around the discontinuity shall be identified. The diaphragm shall be analyzed for the forces in Section 4.2 and the adequacy of the elements in the load path shall be evaluated.

Evaluation Phase (Tier 2)

C4.5.1.1 Diaphragm Continuity

Split level floors and roofs, or diaphragms interrupted by expansion joints, create discontinuities in the diaphragm. This condition is common in ramped parking structures. It is a problem unless special details are used, or lateral-force-resisting elements are provided at the vertical offset of the diaphragm or on both sides of the expansion joint. Such a discontinuity may cause the diaphragm to function as a cantilever element or three-sided diaphragm. If the diaphragm is not supported on at least three sides by lateral-force-resisting elements, torsional forces in the diaphragm may cause it to become unstable. In both the cantilever and three-sided cases, increased lateral deflection in the discontinuous diaphragm may cause increased damage to, or collapse of, the supporting elements.

If the load path is incomplete, mitigation with elements or connections required to complete the load path is necessary to achieve the selected performance level.

4.5.1.2 CROSS TIES: There shall be continuous cross ties between diaphragm chords.

Tier 2 Evaluation Procedure: Out-of-plane forces in accordance with Section 4.2 shall be calculated. The adequacy of the existing connections, including development of the forces into the diaphragm, shall be evaluated.

C4.5.1.2 Cross Ties

Continuous cross ties between diaphragm chords are needed to develop out-of-plane wall forces into the diaphragm (see Figure C4-30). The cross ties should have a positive and direct connection to the walls to keep the walls from separating from the building. The connection of the cross tie to the wall, and connections within the cross tie, must be detailed so that cross-grain bending or cross-grain tension does not occur in any wood member (see Section 4.6.1.2).

Sub-diaphragms may be used between continuous cross ties to reduce the number and length of additional cross ties.

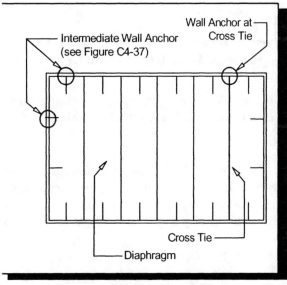

Figure C4-30. Cross Ties

Evaluation Phase (Tier 2)

4.5.1.3 ROOF CHORD CONTINUITY: All chord elements shall be continuous, regardless of changes in roof elevation.

Tier 2 Evaluation Procedure: The load path around the discontinuity shall be identified. The diaphragm shall be analyzed for the forces in Section 4.2, and the adequacy of the elements in the load path shall be evaluated.

C4.5.1.3   Roof Chord Continuity

Diaphragms with discontinuous chords will be more flexible and will experience more damage around the perimeter than properly detailed diaphragms. Vertical offsets or elevation changes in a diaphragm often cause a chord discontinuity (see Figure C4-31). To provide continuity, the following elements are required: a continuous chord element; lateral force resistance in plane X to connect the offset portions of the diaphragm; lateral force resistance in plane Y to develop the sloping diaphragm into the chord; and vertical supports (posts) to resist overturning forces generated by plane X.

If the load path is incomplete, mitigation with elements or connections required to complete the load path is necessary to achieve the selected performance level.

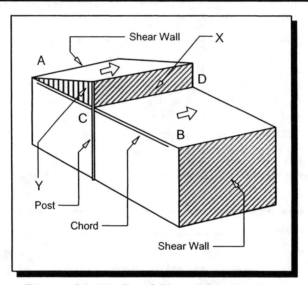

Figure C4-31. Roof Chord Continuity

4.5.1.4   OPENINGS AT SHEAR WALLS: Diaphragm openings immediately adjacent to the shear walls shall be less than 25 percent of the wall length for Life Safety and 15 percent of the wall length for Immediate Occupancy.

Tier 2 Evaluation Procedure: The in-plane shear transfer demand at the wall shall be calculated. The adequacy of the diaphragm to transfer loads to the wall shall be evaluated considering the available length and the presence of any drag struts. The adequacy of the walls to span out-of-plane between points of anchorage shall be evaluated, and the adequacy of the diaphragm connections to resist wall out-of-plane forces shall be evaluated.

C4.5.1.4 Openings at Shear Walls

Large openings at shear walls significantly limit the ability of the diaphragm to transfer lateral forces to the wall (see Figure C4-32). This can have a compounding effect if the opening is near one end of the wall and divides the diaphragm into small segments with limited stiffness that are ineffective in transferring shear to the wall. This might have the net effect of a much larger opening. Large openings also may limit the ability of the diaphragm to provide out-of-plane support for the wall.

The presence of drag struts developed into the diaphragm beyond the wall will help mitigate this effect.

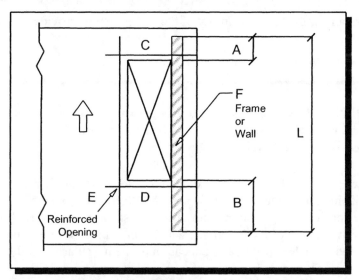

Figure C4-32. Opening at Exterior Wall

4.5.1.5 OPENINGS AT BRACED FRAMES: Diaphragm openings immediately adjacent to the braced frames shall extend less than 25 percent of the frame length for Life Safety and 15 percent of the frame length for Immediate Occupancy.

Tier 2 Evaluation Procedure: The in-plane shear transfer demand at the frame shall be calculated. The adequacy of the diaphragm to transfer loads to the frame shall be evaluated considering the available length and the presence of any drag struts.

C4.5.1.5 Openings at Braced Frames

Large openings at braced frames significantly limit the ability of the diaphragm to transfer lateral forces to the frame. This can have a compounding effect if the opening is near one end of the frame and divides the diaphragm into small segments with limited stiffness that are ineffective in transferring shear to the frame. This might have the net effect of a much larger opening.

The presence of drag struts developed into the diaphragm beyond the frame will help mitigate this effect.

Evaluation Phase (Tier 2)

4.5.1.6   OPENINGS AT EXTERIOR MASONRY SHEAR WALLS: Diaphragm openings immediately adjacent to exterior masonry walls shall not be greater than 8 feet long for Life Safety and 4 feet long for Immediate Occupancy

Tier 2 Evaluation Procedure: The adequacy of the walls to span out-of-plane between points of anchorage shall be evaluated, and the adequacy of the diaphragm connections to resist wall out-of-plane forces shall be evaluated.

C4.5.1.6   Openings at Exterior Masonry Shear Walls

Large openings at exterior masonry walls limit the ability of the diaphragm to provide out-of-plane support for the wall.

The presence of drag struts developed into the diaphragm beyond the wall will help mitigate this effect.

4.5.1.7   PLAN IRREGULARITIES: There shall be tensile capacity to develop the strength of the diaphragm at re-entrant corners or other locations of plan irregularities. This statement shall apply to the Immediate Occupancy Performance Level only.

Tier 2 Evaluation Procedure: The chord and collector demands at locations of plan irregularities shall be calculated by analyzing the diaphragm for the forces in Section 4.2. Relative movement of the projecting wings of the structure shall be considered by applying the static base shear, assuming each wing moves in the same direction or each wing moves in opposing directions, whichever is more severe. The adequacy of all elements that can contribute to the tensile capacity at the location of the irregularity shall be evaluated.

C4.5.1.7   Plan Irregularities

Diaphragms with plan irregularities such as extending wings, plan insets, or E-, T-, X-, L-, or C-shaped configurations have re-entrant corners where large tensile and compressive forces can develop (see Figure C4-33). Chords and collectors in the diaphragm may not have sufficient strength at these re-entrant corners to resist these tensile forces. Local damage may occur (see Figure C4-34). Chord reinforcing is typically required to be developed at the re-entrant corner. In some cases, the chord may be connected directly to a lateral-force-resisting element rather than developed into the diaphragm.

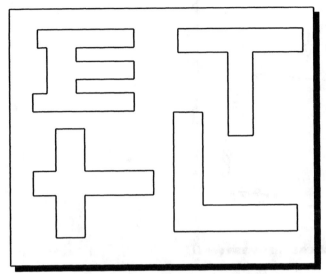

Figure C4-33. Plan Irregularities

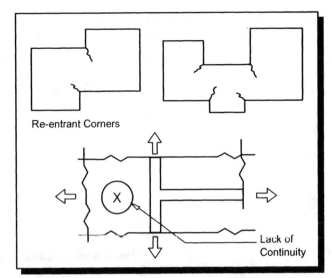

Figure C4-34. Re-entrant Corners

4.5.1.8 DIAPHRAGM REINFORCEMENT AT OPENINGS: There shall be reinforcing around all diaphragm openings larger than 50 percent of the building width in either major plan dimension. This statement shall apply to the Immediate Occupancy Performance Level only.

Tier 2 Evaluation Procedure: The diaphragm shall be analyzed for the forces in Section 4.2. The shear and flexural demands at major openings shall be calculated, and the resulting chord forces shall be determined. The adequacy of the diaphragm elements to transfer forces around the opening shall be evaluated.

C4.5.1.8 Diaphragm Reinforcement at Openings

Openings in diaphragms increase shear stresses and induce secondary moments in the diaphragm segments adjacent to the opening. Tension and compression forces are generated along the edges of these segments by the secondary moments and must be resisted by chord elements in the subdiaphragms around the openings.

Openings that are small relative to the diaphragm dimensions may have only a negligible impact. Openings that are large relative to the diaphragm dimensions can substantially reduce the stiffness of the diaphragm and induce large forces around the openings (see Figure C4-35).

Evaluation Phase (Tier 2)

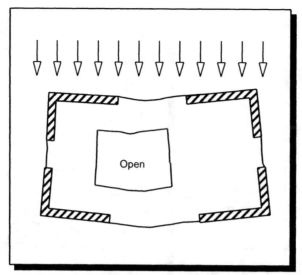

Figure C4-35. Diaphragm Opening

4.5.2 Wood Diaphragms

4.5.2.1 STRAIGHT SHEATHING: All straight sheathed diaphragms shall have aspect ratios less than 2-to-1 for Life Safety and 1-to-1 for Immediate Occupancy in the direction being considered.

Tier 2 Evaluation Procedure: An analysis in accordance with Section 4.2 shall be performed. The adequacy of the shear capacity of non-compliant diaphragms shall be evaluated.

C4.5.2.1 Straight Sheathing

Straight-sheathed diaphragms are flexible and weak relative to other types of wood diaphragms. Shear capacity is provided by a force couple between nails in the individual boards of the diaphragm and the supporting framing. Because of the limited strength and stiffness of these diaphragms, they are most suitable in applications with limited demand, such as in levels of low seismicity.

In levels of moderate and high seismicity, the span and aspect ratio of straight-sheathed diaphragms are limited to minimize shear demands. The aspect ratio (span/depth) must be calculated for the direction being considered.

Compliance can be achieved if the diaphragm has adequate capacity for the demands in the building being evaluated.

4.5.2.2 SPANS: All wood diaphragms with spans greater than 24 feet for Life Safety and 12 feet for Immediate Occupancy shall consist of wood structural panels or diagonal sheathing. Wood commercial and industrial buildings may have rod-braced systems.

Tier 2 Evaluation Procedure: An analysis in accordance with Section 4.2 shall be performed. The adequacy of the shear capacity of non-compliant diaphragms shall be evaluated. The diaphragm deflection shall be calculated, and the adequacy of the vertical-load-carrying elements shall be evaluated at maximum diaphragm deflection, including P-Δ effects.

C4.5.2.2 Spans

Long span diaphragms will often experience large lateral deflections and diaphragm shear demands. Large deflections in the diaphragm can result in increased damage or collapse of elements laterally supported by the diaphragm. Excessive diaphragm shear demands will cause damage and reduced stiffness in the diaphragm.

Compliance can be demonstrated if the diaphragm and vertical-load-carrying elements can be shown to have adequate capacity at maximum deflection.

Wood commercial and industrial buildings may have rod-braced systems in lieu of wood structural panels, and can be considered compliant.

4.5.2.3 UNBLOCKED DIAPHRAGMS: All diagonally sheathed or unblocked wood panel diaphragms shall have horizontal spans less than 40 feet for Life Safety and 30 feet for Immediate Occupancy and shall have aspect ratios less than or equal to 4-to-1 for Life Safety and 3-to-1 for Immediate Occupancy.

Tier 2 Evaluation Procedure: An analysis in accordance with Section 4.2 shall be performed. The adequacy of the shear capacity of non-compliant diaphragms shall be evaluated.

C4.5.2.3 Unblocked Diaphragms

Wood structural panel diaphragms may not have blocking below unsupported panel edges. Blocking may be necessary at diaphragm boundaries to prevent premature failure due to joist rolling. The shear capacity of diagonally sheathed or unblocked diaphragms is less than that of fully blocked wood structural panel diaphragms, due to the limited ability for direct shear transfer at unsupported panel edges. The span and aspect ratio of diaphragms is limited to minimize shear demands. The aspect ratio (span/depth) must be calculated for the direction being evaluated.

Compliance can be demonstrated if the diaphragm can be shown to have adequate capacity for the demands in the building being evaluated.

4.5.3 Metal Deck Diaphragms

C4.5.3 Metal Deck Diaphragms

Bare metal deck can be used as a roof diaphragm where the individual panels are adequately fastened to the supporting framing. The strength of the diaphragm depends on the profile and gage of the deck and the layout and size of the welds or fasteners. Allowable shear capacities for metal deck diaphragms are usually obtained from approved test data and analytical work developed by the industry.

Metal decks used in floors generally have concrete fill. In cases with structural concrete fill, the metal deck is considered to be a concrete form, and the diaphragm is treated as a reinforced concrete diaphragm. In some cases, however, the concrete fill is not structural. It may be a topping slab or an insulating layer that is used to encase conduits or provide a level wearing surface. This type of construction is considered to be an untopped metal deck diaphragm with a capacity determined by the metal deck alone. Nonstructural topping, however, is somewhat beneficial and has a stiffening effect on the metal deck.

> Metal deck diaphragm behavior is limited by buckling of the deck and by the attachment to the framing. Weld quality can be an issue because welding of light gage material requires special consideration. Care must be taken during construction to ensure the weld has proper fusion to the framing but did not burn through the deck material.
>
> Concrete-filled metal decks generally make excellent diaphragms and usually are not a problem as long as the basic requirements for chords, collectors, and reinforcement around openings are met. However, the evaluating engineer should look for conditions that can weaken the diaphragm such as troughs, gutters, and slab depressions that can have the effect of short-circuiting the system or of reducing the system to the bare deck.

4.5.3.1 NON-CONCRETE FILLED DIAPHRAGMS: Untopped metal deck diaphragms or metal deck diaphragms with fill other than concrete shall consist of horizontal spans of less than 40 feet and shall have span/depth ratios less than 4-to-1. This statement shall apply to the Immediate Occupancy Performance Level only.

Tier 2 Evaluation Procedure: Non-compliant diaphragms shall be evaluated for the forces in Section 4.2. The adequacy of the shear capacity of the metal deck diaphragm shall be evaluated.

> **C4.5.3.1 Non-Concrete Filled Diaphragms**
>
> Untopped metal deck diaphragms have limited strength and stiffness. Long span diaphragms with large aspect ratios will often experience large lateral deflections and high diaphragm shear demands. This is especially true for aspect ratios greater than 4-to-1.
>
> In levels of moderate and high seismicity, the span and aspect ratio of untopped metal deck diaphragms are limited to minimize shear demands. The aspect ratio (span/depth) must be calculated for the direction being considered.
>
> Compliance can be achieved if the diaphragm has adequate capacity for the demands in the building being evaluated.

4.5.4 Concrete Diaphragms

No evaluation statements or Tier 2 procedures specific to cast-in-place concrete diaphragms are included in this standard. Concrete diaphragms shall be evaluated for the general diaphragm evaluation statements and Tier 2 procedures in Section 4.5.1.

> **C4.5.4 Concrete Diaphragms**
>
> Concrete slab diaphragm systems have demonstrated good performance in past earthquakes. Building damage is rarely attributed to a failure of the concrete diaphragm itself, but rather to failure in related elements in the load path, such as collectors or connections between diaphragms and vertical elements. These issues are addressed elsewhere in this standard. The design professional should assess concrete diaphragms for general evaluation statements that will address configuration, irregularities, openings, and load path. The design professional also should carefully assess pan joist systems and other systems that have thin slabs.

Evaluation Phase (Tier 2)

4.5.5 Precast Concrete Diaphragms

> **C4.5.5 Precast Concrete Diaphragms**
>
> Precast concrete diaphragms consist of horizontal precast elements that may or may not have a cast-in-place topping slab. Precast elements may be precast planks laid on top of framing, or precast T-sections which consist of both the framing and the diaphragm surface cast in one piece.
>
> Because of the brittle nature of the connections between precast elements, special attention should be paid to eccentricities, adequacy of welds, and length of embedded bars. If a topping slab is provided, it should be capable of taking all the shear. Welded steel connections between precast elements, with low rigidity relative to the concrete topping, will not contribute significantly to the strength of the diaphragm where a topping slab is present.

4.5.5.1 TOPPING SLAB: Precast concrete diaphragm elements shall be interconnected by a continuous reinforced concrete topping slab.

Tier 2 Evaluation Procedure: Non-compliant diaphragms shall be evaluated for the forces in Section 4.2. The adequacy of the slab element interconnection shall be evaluated. The adequacy of the shear capacity of the diaphragm shall be evaluated.

> **C4.5.5.1 Topping Slab**
>
> Precast concrete diaphragm elements may be interconnected with welded steel inserts. These connections are susceptible to sudden failure such as weld fracture, pull-out of the embedment, or spalling of the concrete. Precast concrete diaphragms without topping slabs may be susceptible to damage unless they were specifically detailed with connections capable of yielding or of developing the strength of the connected elements.
>
> In precast construction, topping slabs may have been poured between elements without consideration for providing continuity. The topping slab may not be fully effective if it is interrupted at interior walls. The presence of dowels or continuous reinforcement is needed to provide continuity.
>
> Where the topping slab is not continuous, an evaluation considering the discontinuity is required to ensure a complete load path for shear transfer, collectors, and chords.

4.5.6 Horizontal Bracing

No evaluation statements or Tier 2 procedures have been provided for horizontal bracing. Horizontal bracing shall be evaluated for the general diaphragm evaluation statements and Tier 2 procedures in Section 4.5.1.

> **C4.5.6 Horizontal Bracing**
>
> Horizontal bracing usually is found in industrial buildings. These buildings often have very little mass, so wind considerations govern over seismic considerations. The wind design is probably adequate if the building shows no signs of distress. If bracing is present, the design professional should look for a complete load path with the ability to collect all tributary forces and deliver them to the walls or frames. Horizontal rod bracing should be investigated for eccentricities at the connections and sagging or looseness in the rods.

Evaluation Phase (Tier 2)

4.5.7 Other Diaphragms

4.5.7.1 OTHER DIAPHRAGMS: The diaphragm shall not consist of a system other than wood, metal deck, concrete, or horizontal bracing.

Tier 2 Evaluation Procedure: Non-compliant diaphragms shall be evaluated for the forces in Section 4.2. The adequacy of the non-compliant diaphragms shall be evaluated using available reference standards for the capacity of diaphragms not covered by this standard.

> **C4.5.7.1 Other Diaphragms**
>
> In some codes and standards, there are procedures and allowable diaphragm shear capacities for diaphragms not covered by this standard. Examples include thin planks and gypsum toppings, but these systems are brittle and have limited strength. As such, they may not be desirable elements in the lateral-force-resisting system. Another example is standing seam roofs or other metal roof systems that are designed to move to minimize thermal stresses. For seismic loading in certain directions, such roofs may not provide a diaphragm load path.
>
> The design professional should be watchful for systems that look like diaphragms but may not have the strength, stiffness, or interconnection between elements necessary to perform the intended function.

4.6 Procedures for Connections

This section provides Tier 2 Evaluation procedures that apply to structural connections: anchorage for normal forces, shear transfer, vertical components, interconnection of elements, and panel connections.

4.6.1 Anchorage for Normal Forces

> **C4.6.1 Anchorage for Normal Forces**
>
> Bearing walls that are not positively anchored to the diaphragms may separate from the structure. This may result in a loss of bearing support and partial collapse of the floors and roof. Non-bearing walls which separate from the structure may represent a significant falling hazard. The hazard increases with the height above the building base as the building response amplifies the ground motion. Amplification of the ground motion used to estimate the wall anchorage forces depends on the type and configuration of both the walls and the diaphragms, as well as the type of soil.

4.6.1.1 WALL ANCHORAGE: Exterior concrete or masonry walls, which are dependent on the diaphragm for lateral support, shall be anchored for out-of-plane forces at each diaphragm level with steel anchors, reinforcing dowels, or straps that are developed into the diaphragm. Connections shall have adequate strength to resist the connection force calculated in the Quick Check procedure of Section 3.5.3.7.

Tier 2 Evaluation Procedure: The adequacy of the walls to span between points of anchorage shall be evaluated. The adequacy of the existing connections for the wall forces in Section 4.2 shall be evaluated.

Evaluation Phase (Tier 2)

C4.6.1.1 Wall Anchorage

Bearing walls that are not positively anchored to the diaphragms may separate from the structure, causing partial collapse of the floors and roof. Non-bearing walls that separate from the structure may represent a significant falling hazard. The hazard amplifies with the height above the building base. Anchorage forces must be fully developed into the diaphragm to prevent pull-out failure of the anchor or local failure of the diaphragm (see Figure C4-36).

If the anchorage is nonexistent, mitigation with elements or connections needed to anchor the walls to the diaphragms is necessary to achieve the selected performance level.

4.6.1.2 WOOD LEDGERS: The connection between the wall panels and the diaphragm shall not induce cross-grain bending or tension in the wood ledgers.

Tier 2 Evaluation Procedure: No Tier 2 Evaluation procedure is available to demonstrate compliance of wood ledgers loaded in cross-grain bending.

C4.6.1.2 Wood Ledgers

Wood members in general have very little resistance to tension applied perpendicular to grain. Connections that rely on cross-grain bending in wood ledgers induce tension perpendicular to grain. Failure due to cross-grain bending results in the ledger breaking (Figure C4-37, top). Another significant failure mode caused by inadequate wall anchorage is the sheathing breaking at the line of nails (Figure C4-37, bottom). Failure of such connections is sudden and nonductile and can result in loss of bearing support and partial collapse of the floors and roof.

Mitigation with elements or connections needed to provide wall anchorage without inducing cross-grain bending is necessary to achieve the selected performance level.

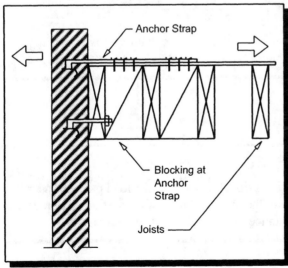

Figure C4-36. Wall Anchorage

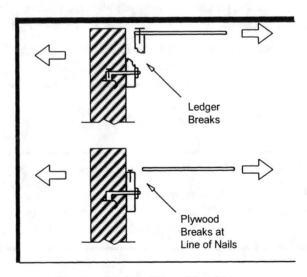

Figure C4-37. Wood Ledgers

Evaluation Phase (Tier 2)

4.6.1.3 PRECAST PANEL CONNECTIONS: There shall be at least two anchors from each precast wall panel into the diaphragm elements for Life Safety, and the anchors shall be able to develop the strength of the panels for Immediate Occupancy.

Tier 2 Evaluation Procedure: The stability of the wall panels for the out-of-plane forces in Section 4.2 shall be evaluated. The adequacy of the existing connections to deliver all forces into the diaphragm, including moments due to eccentricities between the panel center of mass and points of anchorage, shall be evaluated.

C4.6.1.3 Precast Panel Connections

At least two connections between each panel and the diaphragm are required for basic stability of the wall panel for out-of-plane forces. Many connection configurations are possible, including one anchor supporting two adjacent panels.

A single anchor, or line of anchors, near the panel center of mass should be evaluated for an accidental eccentricity of 5 percent of the critical panel dimension, as a minimum.

4.6.1.4 STIFFNESS OF WALL ANCHORS: Anchors of concrete or masonry walls to wood structural elements shall be installed taut and shall be stiff enough to limit the relative movement between the wall and the diaphragm to no greater than 1/8 inch prior to engagement of the anchors.

Tier 2 Evaluation Procedure: The amount of relative movement possible given the existing connection configuration shall be determined. The impact of this movement shall be evaluated by analyzing the elements of the connection for forces induced by the maximum potential movement.

C4.6.1.4 Stiffness of Wall Anchors

The concern is that flexibility or slip in wall anchorage connections requires relative movement between the wall and structure before the anchor is engaged. This relative movement can induce forces in elements not intended to be part of the load path for out-of-plane forces. It can be enough to cause a loss of bearing at vertical supports, or can induce cross-grain bending in wood ledger connections.

Compliance can be demonstrated if the movement has no detrimental affect on the connections. Forces generated by any additional eccentricity at bearing supports should be considered.

4.6.2 Shear Transfer

C4.6.2 Shear Transfer

The transfer of diaphragm shears into shear walls and frames is a critical element in the load path for lateral force resistance. If the connection is inadequate, or nonexistent, the ability of the walls and frames to receive lateral forces will be limited, and the overall lateral force resistance of the building will be reduced.

Evaluation Phase (Tier 2)

4.6.2.1 TRANSFER TO SHEAR WALLS: Diaphragms shall be connected for transfer of loads to the shear walls for Life Safety, and the connections shall be able to develop the lesser of the shear strength of the walls or diaphragms for Immediate Occupancy.

Tier 2 Evaluation Procedure: An analysis in accordance with Section 4.2 shall be performed. The diaphragm and wall demands shall be calculated, and the adequacy of the connection to transfer the demands to the shear walls shall be evaluated.

C4.6.2.1 Transfer to Shear Walls

The floor or roof diaphragms must be connected to the shear walls to provide a complete load path for the transfer of diaphragm shear forces to the walls. Where the wall does not extend the full depth of the diaphragm, this connection may include collectors or drag struts. Collectors and drag struts must be continuous across intersecting framing members, and must be adequately connected to the wall to deliver high tension and compression forces at a concentrated location.

In the case of frame buildings with infill walls (building types S5, S5A, C3, C3A), the seismic performance is dependent on the interaction between the frame and infill, and the behavior is more like that of a shear wall building. The load path between the diaphragms and the infill panels is most likely through the frame elements, which also may act as drag struts and collectors. In this case, the evaluation statement is addressing the connection between the diaphragm and the frame elements.

If the connection is nonexistent, mitigation with elements or connections needed to transfer diaphragm shear to the shear walls is necessary to achieve the selected performance level.

4.6.2.2 TRANSFER TO STEEL FRAMES: Diaphragms shall be connected for transfer of loads to the steel frames for Life Safety, and the connections shall be able to develop the lesser of the strength of the frames or the diaphragms for Immediate Occupancy.

Tier 2 Evaluation Procedure: An analysis in accordance with Section 4.2 shall be performed. The diaphragm and frame demands shall be calculated, and the adequacy of the connection to transfer the demands to the steel frames shall be evaluated.

C4.6.2.2 Transfer to Steel Frames

The floor and roof diaphragms must be adequately connected to the steel frames to provide a complete load path for shear transfer between the diaphragms and the frames. This connection may consist of shear studs or welds between the metal deck and steel framing. In older construction, steel framing may be encased in concrete. Direct force transfer between concrete and steel members by shear friction concepts should not be used unless the members are completely encased in concrete.

If the connection is nonexistent, mitigation with elements or connections needed to transfer diaphragm shear to the steel frames is necessary to achieve the selected performance level.

Evaluation Phase (Tier 2)

4.6.2.3 TOPPING SLAB TO WALLS OR FRAMES: Reinforced concrete topping slabs that interconnect the precast concrete diaphragm elements shall be doweled for transfer of forces into the shear wall or frame elements for Life Safety, and the dowels shall be able to develop the lesser of the shear strength of the walls, frames, or slabs for Immediate Occupancy.

Tier 2 Evaluation Procedure: An analysis in accordance with Section 4.2 shall be performed. The diaphragm and wall demands shall be calculated, and the adequacy of the connection to transfer the demands to the vertical elements shall be evaluated.

C4.6.2.3 Topping Slab to Walls or Frames

The topping slabs at each floor or roof must be connected to the shear walls or frame elements to provide a complete load path for the transfer of diaphragm shear forces to the vertical elements. Welded inserts between precast floor or roof elements are susceptible to weld fracture and spalling, and are likely not adequate to transfer these forces alone.

If a direct topping slab connection is nonexistent, mitigation with elements or connections needed to transfer diaphragm shear to the vertical elements is necessary to achieve the selected performance level.

4.6.3 Vertical Components

C4.6.3 Vertical Components

The following statements reflect a number of common concerns related to inadequate connections between elements. For example, members may be incapable of transferring forces into the foundation or may be displaced where uplifted, resulting in reduced support for vertical loads. A potential deficiency common to all of the following statements would be a nonexistent connection.

4.6.3.1 STEEL COLUMNS: The columns in lateral-force-resisting frames shall be anchored to the building foundation for Life Safety, and the anchorage shall be able to develop the lesser of the tensile capacity of the column, the tensile capacity of the lowest level column splice (if any), or the uplift capacity of the foundation, for Immediate Occupancy.

Tier 2 Evaluation Procedure: An analysis in accordance with Section 4.2 shall be performed. The column demands, including axial load due to overturning, shall be calculated, and the adequacy of the connections to transfer the demands to the foundation shall be evaluated.

C4.6.3.1 Steel Columns

Steel columns that are part of the lateral-force-resisting system must be connected for the transfer of uplift and shear forces at the foundation (see Figure C4-38). The absence of a substantial connection between the columns and the foundation may allow the column to uplift or slide off of bearing supports, which may limit the ability of the columns to support vertical loads or resist lateral forces.

As an upper bound limit for the Immediate Occupancy Performance Level, the connection is checked for the tensile capacity of the column, column splice, or the foundation, whichever is the weak link in the load path between the superstructure and the supporting soil. It could be the uplift capacity of the pile, the connection between the pile and the cap, or the foundation dead load that can be activated by the column, the column tensile capacity, or the splice capacity.

If the connection is nonexistent, mitigation with elements or connections needed to anchor the vertical elements to the foundation is necessary to achieve the selected performance level.

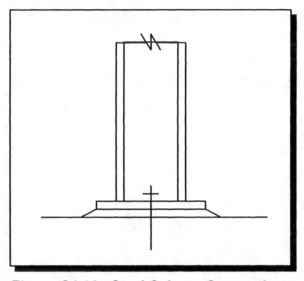

Figure C4-38. Steel Column Connection

4.6.3.2 CONCRETE COLUMNS: All concrete columns shall be doweled into the foundation for Life Safety, and the dowels shall be able to develop the tensile capacity of reinforcement in columns of lateral-force-resisting system for Immediate Occupancy.

Tier 2 Evaluation Procedure: An analysis in accordance with Section 4.2 shall be performed. The column demands shall be calculated, and the adequacy of the connection to transfer the demands to the foundation shall be evaluated.

C4.6.3.2 Concrete Columns

Concrete column that are part of the lateral-force-resisting system must be connected for the transfer of uplift and shear forces to the foundation (see Figure C4-39). The absence of a substantial connection between the columns and the foundation may allow the column to uplift or slide off of bearing supports, which will limit the ability of the columns to support vertical loads or resist lateral forces.

If the connection is nonexistent, mitigation with elements or connections needed to anchor the vertical elements to the foundation is necessary to achieve the selected performance level.

Evaluation Phase (Tier 2)

4.6.3.3 WOOD POSTS: There shall be a positive connection of wood posts to the foundation.

Tier 2 Evaluation Procedure: No Tier 2 Evaluation procedure is available for connections in non-compliance.

> **C4.6.3.3 Wood Posts**
>
> Typically, the base of wood posts are connected to a wood block embedded in a concrete footing. The use of two or more toenails connecting the post to the block is considered to be the minimum positive connection.
>
> The absence of a substantial connection between the wood posts and the foundation may allow the posts to slide off of bearing supports as the structure drifts in an earthquake.
>
> Mitigation with elements or connections needed to anchor the posts to the foundation is necessary to achieve the selected performance level.

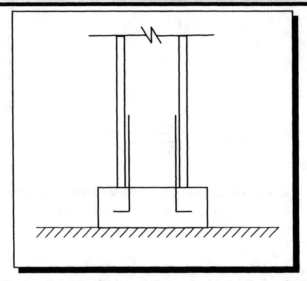

Figure C4-39. Column Doweled into Foundation

4.6.3.4 WOOD SILLS: All wood sills shall be bolted to the foundation.

Tier 2 Evaluation Procedure: The adequacy of the existing anchor bolts or other method of shear attachment for lateral forces in Section 4.2 shall be evaluated.

> **C4.6.3.4 Wood Sills**
>
> The absence of a connection between the wood sills and the foundation is a gap in the load path that will limit the ability of the shear walls to resist lateral forces. Structures may potentially slide off foundation supports.
>
> Where some, but not all, of the sill plates have been bolted or the sill is attached by shot pins or other types of shear connections, an evaluation can be performed to check the adequacy of existing elements. The evaluation should consider only those elements located below shear-resisting elements of the lateral-force-resisting system.
>
> Mitigation with elements or connections needed to anchor the sills to the foundation is necessary to achieve the selected performance level.

4.6.3.5 FOUNDATION DOWELS: Wall reinforcement shall be doweled into the foundation for Life Safety, and the dowels shall be able to develop the lesser of the strength of the walls or the uplift capacity of the foundation for Immediate Occupancy.

Tier 2 Evaluation Procedure: An analysis in accordance with Section 4.2 shall be performed. The wall demands shall be calculated, and the adequacy of the connection to transfer the demands to the foundation shall be evaluated.

C4.6.3.5 Foundation Dowels

The absence of an adequate connection between the shear walls and the foundation is a gap in the load path that will limit the ability of the shear walls to resist lateral forces.

If the connection is nonexistent, mitigation with elements or connections needed to anchor the walls to the foundation is necessary to achieve the selected performance level.

4.6.3.6 SHEAR-WALL-BOUNDARY COLUMNS: The shear-wall-boundary columns shall be anchored to the building foundation for Life Safety, and the anchorage shall be able to develop the tensile capacity of the column for Immediate Occupancy.

Tier 2 Evaluation Procedure: An analysis in accordance with Section 4.2 shall be performed. The overturning resistance of the shear wall considering the dead load above the foundation and the portion of the foundation dead load that can be activated by the boundary column anchorage connection shall be evaluated.

C4.6.3.6 Shear-Wall-Boundary Columns

Shear-wall-boundary column anchorage is necessary for overturning resistance of the shear walls. Boundary columns that are not substantially anchored to the foundation may not be able to activate foundation dead loads for overturning resistance.

4.6.3.7 PRECAST WALL PANELS: Precast wall panels shall be connected to the foundation for Life Safety, and the connections shall be able to develop the strength of the walls for Immediate Occupancy.

Tier 2 Evaluation Procedure: An analysis in accordance with Section 4.2 shall be performed. The wall panel demands shall be calculated, and the adequacy of the connection to transfer the demands to the foundation shall be evaluated.

C4.6.3.7 Precast Wall Panels

The absence of an adequate connection between the precast wall panels and the foundation is a gap in the load path that will limit the ability of the panels to resist lateral forces.

If the connection is nonexistent, mitigation with elements or connections needed to anchor the precast walls to the foundation is necessary to achieve the selected performance level.

4.6.3.8 WALL PANELS: Metal, fiberglass, or cementitious wall panels shall be positively attached to the foundation for Life Safety and Immediate Occupancy.

Tier 2 Evaluation Procedure: An analysis in accordance with Section 4.2 shall be performed. The wall panel demands shall be calculated, and the adequacy of the connection to transfer the demands to the foundation shall be evaluated.

C4.6.3.8 Wall Panels

The absence of a shear transfer connection between metal, fiberglass, or cementitious panel shear walls and the foundation is a gap in the load path that will limit the ability of the walls to resist lateral forces.

In some cases, these panels are not intended to be part of the lateral-force-resisting system. In this case, the evaluation should be limited to the anchorage forces and connections for the panels to prevent falling hazards. Consideration should be given to the ability of the connections to resist the deformations imposed by building movements.

If the connection is nonexistent, mitigation with elements or connections needed to anchor the vertical elements to the foundation is necessary to achieve the selected performance level.

4.6.3.9 WOOD SILL BOLTS: Sill bolts shall be spaced at 6 feet or less for Life Safety and 4 feet or less for Immediate Occupancy, with proper edge and end distance provided for wood and concrete.

Tier 2 Evaluation Procedure: The adequacy of the existing bolts for the lateral forces in Section 4.2 shall be evaluated. Reduced capacities shall be used where proper edge or end distance has not been provided.

C4.6.3.9 Wood Sill Bolts

The absence of an adequate connection between the wood sills and the foundation is a gap in the load path that will limit the ability of the shear walls to resist lateral forces. Structures may slide off foundation supports.

Sill bolt spacing has been limited in moderate and high seismic zones to limit the demand on individual bolts. Compliance can be demonstrated if the existing bolts are adequate to resist the demands in the building being evaluated.

4.6.3.10 UPLIFT AT PILE CAPS: Pile caps shall have top reinforcement and piles shall be anchored to the pile caps for Life Safety, and the pile cap reinforcement and pile anchorage shall be able to develop the tensile capacity of the piles for Immediate Occupancy.

Tier 2 Evaluation Procedure: An analysis in accordance with Section 4.2 shall be performed. The axial forces due to overturning and shear demands at the pile cap shall be calculated, and the adequacy of the pile cap reinforcement and pile connections to transfer uplift forces to the piles shall be evaluated.

C4.6.3.10 Uplift at Pile Caps

Pile foundations may have been designed considering downward gravity loads only. A potential problem is a lack of top reinforcement in the pile cap and a lack of a positive connection between the piles and the pile cap. The piles may be socketed into the cap without any connection to resist tension.

Seismic forces may induce uplift at the foundation which must be delivered into the piles for overturning stability. The absence of top reinforcement means the pile cap cannot distribute the uplift forces to the piles. The absence of pile tension connections means that the forces cannot be transferred to the piles. Piles also should be checked for confinement and spacing of ties and spirals.

4.6.4 Interconnection of Elements

4.6.4.1 GIRDER/COLUMN CONNECTION: There shall be a positive connection utilizing plates, connection hardware, or straps between the girder and the column support.

Tier 2 Evaluation Procedure: No Tier 2 Evaluation procedure is available for connections in non-compliance.

C4.6.4.1 Girder/Column Connection

The absence of a substantial connection between the girders and supporting columns may allow the girders to slide off bearing supports as the structure drifts in an earthquake.

Mitigation with elements or connections needed to connect the girders and columns is necessary to achieve the selected performance.

4.6.4.2 GIRDERS: Girders supported by walls or pilasters shall have at least two ties securing the anchor bolts for Life Safety and Immediate Occupancy.

Tier 2 Evaluation Procedure: A determination shall be made as to whether the girder connection at the pilaster will be required to resist wall out-of-plane forces. The adequacy of the connection to resist the forces in Section 4.2 without damage shall be evaluated.

C4.6.4.2 Girders

Girders supported on wall pilasters may be required to resist wall out-of-plane forces. Without adequate confinement, anchor bolts may pull out of the pilaster (see Figure C4-40). The potential for the pilaster to spall can lead to reduced bearing area or loss of bearing support for the girder.

Evaluation Phase (Tier 2)

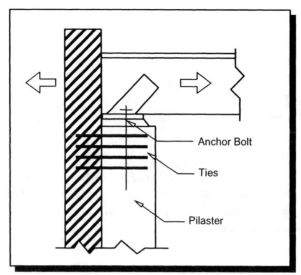

Figure C4-40. Girder Anchorage

4.6.4.3 CORBEL BEARING: If the frame girders bear on column corbels, the length of bearing shall be greater than 3 inches for Life Safety and for Immediate Occupancy.

Tier 2 Evaluation Procedure: The inter-story drift shall be calculated using the procedures in Section 4.2. The bearing length shall be sufficient to provide support for the girders at maximum drift. The adequacy of the bearing support for all loads, including any additional eccentricity at maximum drift, shall be evaluated.

C4.6.4.3 Corbel Bearing

If drifts are sufficiently large, girders can slide off bearing supports without adequate length. At maximum drift, the bearing support may experience additional eccentricity not considered in the design. The support should be evaluated for strength at this extreme condition.

4.6.4.4 CORBEL CONNECTIONS: The frame girders shall not be connected to corbels with welded elements.

Tier 2 Evaluation Procedure: The force in the welded connections induced by inter-story drift shall be calculated. The adequacy of the connections to resist these forces shall be evaluated. Calculated overstresses in these connections shall not jeopardize the vertical support of the girders or the lateral-force-resisting system.

C4.6.4.4 Corbel Connections

Precast elements that are interconnected at the supports may develop unintended frame action and attract seismic forces. The concern is that the welded connections are unable to develop the strength of the members and will be subject to sudden nonductile failure, possibly leading to partial collapse of the floor or roof.

Connections may be in compliance if failure of the connection will not jeopardize the vertical support of the girder.

4.6.4.5 BEAM, GIRDER, AND TRUSS SUPPORTS: Beams, girders, and trusses supported by unreinforced masonry walls or pilasters shall have independent secondary columns for support of vertical loads.

Tier 2 Evaluation Procedure: No Tier 2 Evaluation procedure is available for beam, girder, and truss supports in non-compliance.

> **C4.6.4.5 Beam, Girder, and Truss Supports**
>
> Loss of masonry capacity due to seismic forces also will result in loss of vertical support without a secondary gravity system.

4.6.5 Panel Connections

4.6.5.1 ROOF PANELS: Metal, plastic, or cementitious roof panels shall be positively attached to the roof framing to resist seismic forces for Life Safety and Immediate Occupancy.

Tier 2 Evaluation Procedure: An analysis in accordance with Section 4.2 shall be performed. The roof panel demands shall be calculated, and the adequacy of the wall panels to transfer the demands to the roof framing shall be evaluated.

> **C4.6.5.1 Roof Panels**
>
> The absence of a positive connection between metal, fiberglass, or cementitious panels and the roof framing is a gap in the load path that will limit the ability of the panels to act as a diaphragm.
>
> Panels not intended to be a part of the diaphragm represent a potential falling hazard if not positively attached to the framing. In this case, the evaluation should be limited to the anchorage forces and connections of the panels. Consideration should be given to the ability of the connections to resist the deformations imposed by building movements.
>
> If the connection is nonexistent, mitigation with elements or connections needed to attach the roof panels is necessary to achieve the selected performance level.

4.6.5.2 WALL PANELS: Metal, fiberglass, or cementitious wall panels shall be positively attached to the framing to resist seismic forces for Life Safety and Immediate Occupancy.

Tier 2 Evaluation Procedure: An analysis in accordance with Section 4.2 shall be performed. The wall panel demands shall be calculated, and the adequacy of the wall panels to transfer the demands to the framing shall be evaluated.

> **C4.6.5.2 Wall Panels**
>
> The absence of a positive connection between metal, fiberglass, or cementitious panels and the framing is a gap in the load path that will limit the ability of the panels to resist seismic forces.

> Panels not intended to be a part of the lateral-force-resisting system represent a potential falling hazard if not positively attached to the framing. In this case, the evaluation should be limited to the anchorage forces and connections of the panels. Consideration should be given to the ability of the connections to resist the deformations imposed by building movements.
>
> If the connection is nonexistent, mitigation with elements or connections needed to attach the panels is necessary to achieve the selected performance level.

4.6.5.3 ROOF PANEL CONNECTIONS: Roof panel connections shall be spaced at or less than 12 inches for Life Safety and 8 inches for Immediate Occupancy.

Tier 2 Evaluation Procedure: The adequacy of the existing connections for the lateral forces in Section 4.2 shall be evaluated.

> **C4.6.5.3 Roof Panel Connections**
>
> An insufficient number of connections between the panels and the framing will reduce the capacity of the panels to act as a diaphragm.

4.7 Procedures for Geologic Site Hazards and Foundations

This section provides Tier 2 Evaluation procedures that apply to foundations and supporting soils: geologic site hazards, condition of foundations, and capacity of foundations.

> **C4.7 Procedures for Geologic Site Hazards and Foundations**
>
> A thorough seismic evaluation of an existing building should include an examination of the foundation, an assessment of the capability of the soil beneath the foundation to withstand the forces applied during an earthquake, and consideration of nearby geologic hazards that may affect the stability of the building during an earthquake.
>
> To fully assess the potential hazard presented by local geologic site conditions, and to establish soil engineering parameters required for analysis of these hazards, it may be necessary to consult with a geotechnical design professional. The evaluating design professional is strongly urged to seek consultation with appropriate professionals wherever site conditions are beyond the experience or expertise of the design professional.

4.7.1 Geologic Site Hazards

> **C4.7.1 Geologic Site Hazards**
>
> Certain geologic and local site conditions can lead to structural damage in the event of an earthquake. Large foundation movements due to any number of causes can severely damage an otherwise seismic resistant building. Potential causes of significant foundation movement include settlement or lateral spreading due to liquefaction, slope failure, or surface ruptures. An evaluation of the building should include consideration for these effects and the impact they might have on the superstructure.

Evaluation Phase (Tier 2)

4.7.1.1 LIQUEFACTION: Liquefaction-susceptible, saturated, loose granular soils that could jeopardize the building's foundation support and seismic performance shall not exist in the foundation soils at depths within 50 feet under the building for Life Safety and Immediate Occupancy.

Tier 2 Evaluation Procedure: The potential for liquefaction and magnitude of differential settlement shall be evaluated. An analysis of the building in accordance with the procedures in Section 4.2 shall be performed. The adequacy of the structure shall be evaluated for all gravity and seismic forces in combination with the forces induced by the potential differential movement in the foundation.

> **C4.7.1.1 Liquefaction**
>
> Soils susceptible to liquefaction may lose all vertical-load-bearing capacity during an earthquake. Loss of vertical support for the foundation will cause large differential settlements and induce large forces in the building superstructure.
>
> These forces will be concurrent with all existing gravity loads and seismic forces during the earthquake.

4.7.1.2 SLOPE FAILURE: The building site shall be sufficiently remote from potential earthquake-induced slope failures or rockfalls to be unaffected by such failures or shall be capable of accommodating any predicted movements without failure.

Tier 2 Evaluation Procedure: The potential magnitude of differential movement in the foundation shall be evaluated. An analysis of the building in accordance with the procedures in Section 4.2 shall be performed. The adequacy of the structure shall be evaluated for all gravity and seismic forces in combination with the forces induced by the potential differential movement in the foundation.

> **C4.7.1.2 Slope Failure**
>
> Steep slopes are susceptible to slides during an earthquake. Slope failures are possible in rock or on other non-liquefiable soils on slopes that normally exceed 6 percent. Slopes that exhibit signs of prior landscapes require the most attention.
>
> The concern for buildings on the uphill side of slopes is lateral spreading of the downhill footings. The concern for buildings on the downhill side is impact by sliding soil and debris.

4.7.1.3 SURFACE FAULT RUPTURE: Surface fault rupture and surface displacement at the building site is not anticipated.

Tier 2 Evaluation Procedure: The proximity of the building to known active faults shall be determined. The potential for surface fault rupture and magnitude of rupture shall be determined. An analysis of the building in accordance with the procedures in Section 4.2 shall be performed. The adequacy of the structure shall be evaluated for all gravity and seismic forces in combination with the forces induced by the potential differential movement in the foundation.

> **C4.7.1.3   Surface Fault Rupture**
>
> In the near field of active faults there is a potential for large fissures and differential movement to occur in the surface soils. Foundations of buildings located above these ruptures will be subjected to large differential movements that will induce large forces in the building superstructure.
>
> These forces will be concurrent with all existing gravity loads and seismic forces during the earthquake.

4.7.2   Conditions of Foundations

> **C4.7.2   Conditions of Foundations**
>
> Foundation elements are usually below grade and concealed from view. Evaluations, however, should still include consideration of the foundation and the condition of the elements. Often signs of foundation performance are visible on the surface in the form of existing differential settlement, sloping floors, out-of-plumb walls, and cracking or distress in visible portions of the footings.

4.7.2.1   FOUNDATION PERFORMANCE: There shall be no evidence of excessive foundation movement such as settlement or heave that would affect the integrity or strength of the structure.

Tier 2 Evaluation Procedure: The magnitude of differential movement in the foundation shall be evaluated. An analysis of the building in accordance with the procedures in Section 4.2 shall be performed. The adequacy of the structure shall be evaluated for all gravity and seismic forces in combination with the forces induced by the potential differential movement in the foundation.

> **C4.7.2.1   Foundation Performance**
>
> The integrity and strength of foundation elements may be reduced by cracking, yielding, tipping, or buckling of the foundation. Such weakening may be critical in the event of an earthquake.
>
> Lower level walls, partitions, grade beams, visible footings, pile caps, and similar elements shall be visually examined for cracking, yielding, buckling, and out-of-level conditions. Any such signs should be identified and further evaluated.

4.7.2.2   DETERIORATION: There shall not be evidence that foundation elements have deteriorated due to corrosion, sulfate attack, material breakdown, or other reasons in a manner that would affect the integrity or strength of the structure.

Tier 2 Evaluation Procedure: The cause and extent of deterioration shall be identified. The consequences of this damage to the lateral-force-resisting system shall be determined. The adequacy of damaged lateral-force-resisting elements shall be evaluated considering the extent of the damage and impact on the capacity of each damaged element.

> **C4.7.2.2 Deterioration**
>
> Deterioration can cause weakening of the foundation elements, limiting their ability to support the building. Historical records of foundation performance in the local area may help assess the possibility of deterioration in the foundation of the building being evaluated.

4.7.3 Capacity of Foundations

> **C4.7.3 Capacity of Foundations**
>
> Building foundation elements normally have a capacity at least two times the gravity loads. If there are no signs of foundation distress due to settlement, erosion, corrosion, or other reasons, the foundations are likely to have adequate vertical capacity if the total gravity and seismic overturning loads do not exceed the allowable static capacity by more than a factor of 2.0.
>
> Foundations are considered to have adequate lateral capacity if the horizontal resistance of the foundation system exceeds the calculated lateral loads of Section 4.2.4.3.2. This means that horizontal resistance at the foundation is treated as a force-controlled action.
>
> Where the evaluation of foundation elements indicates significant problems, the evaluating design professional should consult with a qualified geotechnical design professional to establish rational criteria for foundation analysis and mitigation of unsatisfactory conditions.

4.7.3.1 POLE FOUNDATIONS: Pole foundations shall have minimum embedment of 4 feet for Life Safety and Immediate Occupancy.

Tier 2 Evaluation Procedure: The lateral force resistance of embedded poles shall be checked using conventional procedures; the lateral force resistance shall be compared with conventional allowable pressures times 1.5.

> **C4.7.3.1 Pole Foundations**
>
> Pole buildings are structures supported by poles or posts, usually found on rocky and hillside sites. Seismic resistance for a pole structure depends on the embedment depth of the poles and the resistance to active and passive soil pressures.

4.7.3.2 OVERTURNING: The ratio of the horizontal dimension of the lateral-force-resisting system at the foundation level to the building height (base/height) shall be greater than $0.6S_a$.

Tier 2 Evaluation Procedure: An analysis in accordance with the procedures in Section 4.2 shall be performed. The adequacy of the foundation, including all gravity and seismic overturning forces, shall be evaluated.

C4.7.3.2 Overturning

The concentration of seismic overturning forces in foundation elements may exceed the capacity of the soil, the foundation structure, or both. The effective horizontal dimension should be determined based on the ability of the lateral-force-resisting elements to act as a system. Therefore, the building dimension can be used if the elements are well connected. However, multiple checks may be required for elements isolated on opposite sides of the building.

For shallow foundations, the shear and moment capacity of the foundation elements should be evaluated for adequacy to resist calculated seismic forces. The vertical bearing pressure of the soil under seismic loading conditions due to the total gravity and overturning loads should be calculated and compared to two times the allowable static bearing pressure. For deep foundations, the ultimate vertical capacity of the pile or pier under seismic loads should be determined. The foundation capacity shall then be compared to the demands due to gravity loads plus overturning.

4.7.3.3 TIES BETWEEN FOUNDATION ELEMENTS: The foundation shall have ties adequate to resist seismic forces where footings, piles, and piers are not restrained by beams, slabs, or soils classified as Class A, B, or C.

Tier 2 Evaluation Procedure: The magnitude of differential movement in the foundation shall be determined. An analysis of the building in accordance with the procedures in Section 4.2 shall be performed. The adequacy of the structure shall be evaluated for all gravity and seismic forces in combination with the forces induced by the potential differential movement in the foundation.

C4.7.3.3 Ties Between Foundation Elements

Ties between discrete foundation elements, such as pile caps and pole footings, are required where the seismic ground motions are likely to cause significant lateral spreading of the foundations. Ties may consist of tie beams, grade beams, or slabs. If the foundations are restrained laterally by competent soils or rock, ties are not required.

4.7.3.4 DEEP FOUNDATIONS: Piles and piers shall be capable of transferring the lateral forces between the structure and the soil. This statement shall apply to the Immediate Occupancy Performance Level only.

Tier 2 Evaluation Procedure: The lateral capacity of the piles, as governed by the soil or pile construction, shall be determined. An analysis of the building in accordance with the procedures in Section 4.2 shall be performed. The adequacy of the piles shall be evaluated for all gravity and seismic forces.

C4.7.3.4 Deep Foundations

Common problems include flexural strength and ductility of the upper portions of piles or piers, or at the connection to the cap. Distinct changes in soil stiffness can create high bending stresses along the length of the pile.

> For concrete piles, the design professional should check for a minimal amount of longitudinal reinforcement in the upper portion of piles or piers, and for hoops or ties immediately beneath the caps. The design professional also should check for confining transverse reinforcement wherever bending moments might be high along the length of the pile, including changes in soil stiffness.

4.7.3.5 SLOPING SITES: The difference in foundation embedment depth from one side of the building to another shall not exceed one story in height. This statement shall apply to the Immediate Occupancy Performance Level only.

Tier 2 Evaluation Procedure: An analysis of the building in accordance with the procedures in Section 4.2 shall be performed. The adequacy of the foundation to resist sliding shall be evaluated including the horizontal force due to the grade difference.

C4.7.3.5 Sloping Sites

The transfer of seismic force is more difficult where a permanent horizontal force is present.

Evaluation Phase (Tier 2)

4.8 Procedures for Nonstructural Components

This section provides Tier 2 Evaluation procedures that apply to nonstructural components.

C4.8 Procedures for Nonstructural Components

Nonstructural Components

Nonstructural components refer to architectural, mechanical, and electrical components. Additional guidance may be requested from another design professional with expertise in structural evaluation and design.

Investigation of nonstructural components can be very time-consuming because they usually are not well detailed on plans and because they often are concealed. It is essential, however, to investigate these items because their seismic support may have been given little attention in the past and they are potentially dangerous. Of particular importance in nonstructural component evaluation efforts are site visits to identify the present status of nonstructural items.

For nonstructural component evaluation in general, the key issue is generally whether the component or piece of equipment is braced or anchored. This is generally immediately visible and is part of the Tier 1 Evaluation. If the component is braced or anchored, a Tier 2 Evaluation may be necessary (based on the design professional's judgment) to establish the capacity of the components. Evaluation of cladding, exterior veneers, back-up materials, and glazing requires more careful investigation, because the critical components, such as connections and framing, often will be concealed. In some cases it will be necessary to remove materials in order to conduct the evaluation. In addition, some calculations may be necessary to establish capacity to accommodate estimated seismic forces.

Several different types of deficiencies may be identified by the design professional in the Tier 1 Evaluation. Some of these, such as the nonexistence of anchorage or bracing, are clearly in non-compliance, and any further evaluation is not necessary. In other cases, where some bracing or anchorage is provided, or material is deteriorated or corroded, further evaluation and judgment is necessary to ascertain the extent of the deficiency and the consequences of the failure. Some simple calculations of weights, dimensional ratios, and forces are used in this tier of evaluation. A few critical components, such as heavy cladding, may justify a complete analysis (a Tier 3 Evaluation) for ability to withstand forces and drifts and achievement of the desired performance level.

> **Hazards**
>
> Nonstructural elements can pose significant hazards to life safety under certain circumstances. In addition, certain types of building contents can pose hazards (e.g., toxic chemicals) and should be given attention during the evaluation. Special consideration also is warranted for nonstructural elements in essential facilities (e.g., hospitals and police and fire stations) and other facilities that must remain operational after an earthquake.
>
> **Unintended Structural Effects**
>
> Any element with rigidity will be a part of the lateral-force-resisting system until it fails. All walls have some rigidity, and they will participate in resisting lateral forces in proportion to their relative rigidity. Walls of gypsum board or plaster have considerable rigidity. If connected at top and bottom, they can take a significant portion of the lateral load at low force levels; at some higher level they crack and lose strength, and the main system then takes all of the lateral load.

4.8.1 Partitions

4.8.1.1 UNREINFORCED MASONRY: Unreinforced masonry or hollow clay tile partitions shall be braced at a spacing equal to or less than 10 feet in levels of low or moderate seismicity and 6 feet in levels of high seismicity.

Tier 2 Evaluation Procedure: The adequacy of the bracing to resist seismic forces calculated in accordance with Section 4.2.7 shall be evaluated.

> **C4.8.1.1 Unreinforced Masonry**
>
> Hollow clay tile units are brittle and subject to shattering. Unreinforced masonry units may have cracks, loose blocks, or weak mortar. Bracing is needed to prevent portions of the unreinforced masonry from dislodging due to out-of-plane seismic forces, especially at corridors, elevator shafts, and stairs. Door openings often create localized weaknesses due to inadequate support for the block masonry or clay tile at the head and at the sides of the opening.
>
> If bracing is nonexistent, mitigation with elements or connections needed to brace the partitions is necessary to achieve the selected performance level.

4.8.1.2 DRIFT: Rigid cementititous partitions shall be detailed to accommodate a drift ratio of 0.02 in steel moment frame, concrete moment frame, and wood frame buildings. Rigid cementitious partitions shall be detailed to accommodate a drift ratio of 0.005 in other buildings.

Tier 2 Evaluation Procedure: The adequacy of details and lateral bracing for rigid cementitious partitions to resist expected levels of drift calculated in accordance with Section 4.2.7 shall be evaluated.

C4.8.1.2 Drift

Full-height partitions may fail due to lack of provision for building drift. Rigid cementitious partitions should be detailed to provide adequate space for the structure drift without racking the walls, while retaining out-of-plane support. In addition, if not separated from the structure at the top and sides, these walls may alter the response of the building.

4.8.1.3 STRUCTURAL SEPARATIONS: Partitions at structural separations shall have seismic or control joints.

Tier 2 Evaluation Procedure: No Tier 2 Evaluation procedure is available for partitions at structural separations without seismic or control joints. A Tier 3 Evaluation is permitted for evaluation of partitions at structural separations.

C4.8.1.3 Structural Separations

Seismic and control joints are necessary to permit differential structure movement at building separations without causing damage. However, if localized cracking of the partition will not lead to out-of-plane failure of the wall, the costs of a difficult rehabilitation process may not be justified.

4.8.1.4 TOPS: The tops of framed or panelized partitions that extend only to the ceiling line shall have lateral bracing to the building structure at a spacing equal to or less than 6 feet.

Tier 2 Evaluation Procedure: The adequacy of the lateral bracing to resist seismic forces calculated in accordance with Section 4.2.7 shall be evaluated.

C4.8.1.4 Tops

Partitions extending only to suspended ceilings may fall out-of-plane due to lack of bracing. Movement of the partition may damage the ceiling. Cross walls that may frame into the wall will have a beneficial impact on preventing excessive out-of-plane movement and should be considered in the evaluation process.

If lateral bracing is nonexistent, mitigation with elements or connections needed to brace the partitions is necessary to achieve the selected performance level.

4.8.2 Ceiling Systems

4.8.2.1 SUPPORT: The integrated suspended ceiling system shall not be used to laterally support the tops of gypsum board, masonry, or hollow clay tile partitions. Gypsum board partitions need not be evaluated when only the Basic Nonstructural Component Checklist is required by Table 3-2.

Tier 2 Evaluation Procedure: The adequacy of integrated ceiling systems used to laterally support the tops of gypsum board, masonry, or hollow clay tile partitions to resist seismic forces calculated in accordance with Section 4.2.7 shall be evaluated.

Evaluation Phase (Tier 2)

> **C4.8.2.1 Support**
>
> Integrated suspended ceilings braced with diagonal wires will move laterally when subjected to seismic forces. The ability of the gypsum board, masonry, or hollow clay tile partitions to accommodate such ceiling movement without collapse of the partitions should be considered by the design professional.

4.8.2.2 LAY-IN TILES: Lay-in tiles used in ceiling panels located at exits and corridors shall be secured with clips.

Tier 2 Evaluation Procedure: The consequence of non-compliant lay-in tiles shall be evaluated.

> **C4.8.2.2 Lay-In Tiles**
>
> Lay-in board or tile ceilings may drop out of the grid and, depending on their location and weight, could cause injury. In egress areas, falling tiles represent a hazard because they may impede evacuations. Clips can reduce the likelihood of tiles falling, but, depending on the type of ceiling, the likelihood of failure may vary; the design professional should use judgment in assessing the risk.

4.8.2.3 INTEGRATED CEILINGS: Integrated suspended ceilings at exits and corridors or weighing more than 2 pounds per square foot shall be laterally restrained with a minimum of four diagonal wires or rigid members attached to the structure above at a spacing equal to or less than 12 feet.

Tier 2 Evaluation Procedure: The adequacy of the bracing to resist seismic forces calculated in accordance with Section 4.2.7 shall be evaluated.

> **C4.8.2.3 Integrated Ceilings**
>
> Without bracing, integrated ceiling systems are susceptible to vertical and lateral movement which can damage fire sprinkler piping and other elements that penetrate the ceiling grid. Lightweight suspended ceilings may not pose a life safety hazard unless special conditions apply in the judgment of the design professional, such as a large area of ceiling, poor-quality construction, vulnerable occupancy, or egress route.
>
> If bracing is inadequate or nonexistent, mitigation with elements or connections needed to brace the ceilings is necessary to achieve the selected performance level.

4.8.2.4 SUSPENDED LATH AND PLASTER: Ceilings consisting of suspended lath and plaster shall be anchored to resist seismic forces for every 12 square feet of area.

Tier 2 Evaluation Procedure: The adequacy of the anchorage to resist seismic forces calculated in accordance with Section 4.2.7 shall be evaluated.

C4.8.2.4 Suspended Lath and Plaster

Suspended plaster ceilings may behave like structural diaphragms and resist in-plane seismic forces. If the strength of the plaster is exceeded, cracking and spalling of portions of the ceiling are possible. Large areas of suspended plaster may separate from the suspension system and fall if not properly fastened. The interconnection of the plaster to the lath and of the lath to the support framing should also be specifically assessed.

If anchorage is nonexistent, mitigation with elements or connections needed to brace the ceilings is necessary to achieve the selected performance level.

4.8.2.5 EDGES: The edges of integrated suspended ceilings shall be separated from enclosing walls by a minimum of 1/2 inch.

Tier 2 Evaluation Procedure: The adequacy of integrated suspended ceilings to resist expected levels of drift calculated in accordance with Section 4.2.7 shall be evaluated.

C4.8.2.5 Edges

This provision relates especially to large suspended grid ceilings but also may apply to other forms of hung ceilings. The intent is to ensure that the ceiling is sufficiently detached from the surrounding structural walls, such that it can tolerate out-of-plane drift without suffering distortion and damage.

4.8.2.6 SEISMIC JOINT: The ceiling system shall not extend continuously across any seismic joint.

Tier 2 Evaluation Procedure: No Tier 2 Evaluation procedure is available for ceiling systems that extend continuously across any seismic joint. A Tier 3 Evaluation is permitted for evaluation of ceiling systems that extend continuously across seismic joints.

C4.8.2.6 Seismic Joint

Localized damage to ceilings is expected where seismic separations are not provided in the ceiling framing. Seismic or control joints should be provided based on a consideration of the consequences of local ceiling damage. If the damage is unlikely to create a falling hazard or prevent safe egress, the costs of a difficult rehabilitation process may not be justified.

4.8.3 Light Fixtures

4.8.3.1 EMERGENCY LIGHTING: Emergency lighting equipment and signs shall be anchored or braced to prevent falling during an earthquake.

Tier 2 Evaluation Procedure: No Tier 2 Evaluation procedure is available for emergency lighting equipment and signs that are not braced or anchored.

C4.8.3.1 Emergency Lighting

Emergency lighting equipment and signs should be provided with positive anchorage and/or bracing to prevent falling hazards and to enhance the reliability of post-earthquake performance.

If bracing or anchorage is nonexistent, mitigation is necessary to achieve the selected performance level.

4.8.3.2 INDEPENDENT SUPPORT: Light fixtures in suspended grid ceilings shall be supported independently of the ceiling suspension system by a minimum of two wires at diagonally opposite corners of the fixtures.

Tier 2 Evaluation Procedure: No Tier 2 Evaluation procedure is available for light fixtures not independently supported. A Tier 3 Evaluation is permitted for evaluation for light fixtures not independently supported.

C4.8.3.2 Independent Support

With lay-in fluorescent lighting systems, ceiling movement can cause fixtures to separate and fall from suspension systems. These fixtures perform satisfactorily when they are supported separately from the ceiling system or have back-up support that is independent of the ceiling system. If the fixtures are independently supported by methods other than that described, the design professional should exercise judgment as to their adequacy.

If independent support is nonexistent, mitigation is necessary to achieve the selected performance level.

4.8.3.3 PENDANT SUPPORTS: Light fixtures on pendant supports shall be attached at a spacing equal to or less than 6 feet and, if rigidly supported, shall be free to move with the structure to which they are attached without damaging adjoining materials.

Tier 2 Evaluation Procedure: The adequacy of the support to resist seismic forces calculated in accordance with Section 4.2.7 shall be evaluated.

C4.8.3.3 Pendant Supports

With stem-hung incandescent or fluorescent fixtures, the fixtures are usually suspended from stems or chains that allow them to sway. This swaying may cause the light and/or fixture to break after encountering other building components. The stem or chain connection may fail. Long rows of fluorescent fixtures placed end to end have sometimes fallen due to poor support, and their weight makes them hazardous. Long-stem fixtures, which may swing considerably, tend to suffer more damage than short-stem items.

If anchorage is inadequate or nonexistent, mitigation is necessary to achieve the selected performance level.

4.8.3.4 LENS COVERS: Lens covers on light fixtures shall be attached or shall be supplied with safety devices.

Tier 2 Evaluation Procedure: The adequacy of lens covers on light fixtures to resist seismic forces calculated in accordance with Section 4.2.7 shall be evaluated.

C4.8.3.4 Lens Covers

Devices or detailing to prevent lens covers from falling from the fixture are necessary to prevent damage to the lens and items below and may be a safety feature.

4.8.4 Cladding and Glazing

4.8.4.1 CLADDING ANCHORS: Cladding components weighing more than 10 psf shall be mechanically anchored to the exterior wall framing at a spacing equal to or less than 4 feet. A spacing of up to 6 feet is permitted when only the Basic Nonstructural Component Checklist is required by Table 3-2.

Tier 2 Evaluation Procedure: The adequacy of the anchorage to resist seismic forces calculated in accordance with Section 4.2.7 shall be evaluated. The adequacy of cladding components to resist expected levels of drift calculated in accordance with Section 4.2.7 shall be evaluated.

C4.8.4.1 Cladding Anchors

Exterior cladding components, which are often heavy, can fail if their connections to the building frames have insufficient strength and/or ductility. The design professional should assess the consequences of failure, in particular the location of the panels in relation to building occupants and passersby. Adhesive anchorage of heavy exterior cladding components is unacceptable; as such, anchorages typically fail at lower drift ratios than are necessary to ensure life safety performance.

If anchorage is nonexistent, mitigation is necessary to achieve the selected performance level.

Evaluation Phase (Tier 2)

4.8.4.2 DETERIORATION: There shall be no evidence of deterioration, damage, or corrosion in any of the connection elements.

Tier 2 Evaluation Procedure: The cause and extent of damage shall be identified. The consequences of the damage shall be determined. The adequacy of the remaining undeteriorated or undamaged connections to resist seismic forces calculated in accordance with Section 4.2.7 shall be evaluated.

C4.8.4.2 Deterioration

Corrosion can reduce the strength of connections and lead to deterioration of the adjoining materials. The extent of corrosion and its impact on the wall cladding and structure should be considered in the evaluation.

Water leakage into and through exterior walls is a common building problem. Damage due to corrosion, rotting, freezing, or erosion can be concealed in wall spaces. Substantial deterioration can lead to loss of cladding elements or panels.

Exterior walls should be checked for deterioration. Probe into wall spaces if necessary and look for signs of water leakage at vulnerable locations (e.g., at windows and at floor areas). Pay particular attention to elements that tie cladding to the back-up structure and that tie the back-up structure to the floor and roof slabs.

Extremes of temperature can cause substantial structural damage to exterior walls. The resulting weakness may be brought out in a seismic event. Check exterior walls for cracking due to thermal movements.

4.8.4.3 CLADDING ISOLATION: For moment frame buildings of steel or concrete, panel connections shall be detailed to accommodate a story drift ratio of 0.02. Panel connection detailing for a story drift ratio of 0.01 is permitted when only the Basic Nonstructural Component Checklist is required by Table 3-2.

Tier 2 Evaluation Procedure: The adequacy of panel connections to resist expected levels of drift calculated in accordance with Section 4.2.7 shall be evaluated.

C4.8.4.3 Cladding Isolation

High levels of drift and deformation may occur in moment frames. If cladding connections are not detailed to accommodate the drift, failure of connections can result and panels can become dislodged.

4.8.4.4 MULTI-STORY PANELS: For multi-story panels attached at each floor level, panel connections shall be detailed to accommodate a story drift ratio of 0.02. Panel connection detailing for a story drift ratio of 0.01 is permitted when only the Basic Nonstructural Component Checklist is required by Table 3-2.

Tier 2 Evaluation Procedure: The adequacy of the panels and connections to resist expected levels of drift calculated in accordance with Section 4.2.7 shall be evaluated.

C4.8.4.4 Multi-Story Panels

The design professional should determine whether the panels themselves and/or their connections to the structure will deform to accommodate the story drift. If the connectors are expected to deform, they should be capable of doing so without loss of structural support for the panel. If the panels are expected to rack, they should be capable of deforming without becoming unstable and without loss of support for other interconnected systems, such as glazing.

4.8.4.5 BEARING CONNECTIONS: Where bearing connections are required, there shall be a minimum of two bearing connections for each wall panel.

Tier 2 Evaluation Procedure: The adequacy of the connection to resist seismic forces calculated in accordance with Section 4.2.7 shall be evaluated.

4.8.4.5 Bearing Connections

A single bearing connection can result in a dangerous lack of redundancy. The adequacy of single point bearing connections should be evaluated for resistance to in-plane overturning forces including all eccentricities. Small panels, such as some column covers, may have a single bearing connection and still provide adequate safety against failure.

If connections are nonexistent, mitigation is necessary to achieve the selected performance level.

4.8.4.6 INSERTS: Where inserts are used in concrete connections, the inserts shall be anchored to reinforcing steel or other positive anchorage.

Tier 2 Evaluation Procedure: The adequacy of inserts used in concrete connections to resist seismic forces calculated in accordance with Section 4.2.7 shall be evaluated.

C4.8.4.6 Inserts

Out-of-plane panel connections that do not engage panel reinforcement are susceptible to pulling out when subjected to seismic forces.

4.8.4.7 PANEL CONNECTIONS: Exterior cladding panels shall be anchored out-of-plane with a minimum of four connections for each wall panel. Two connections per wall panel are permitted when only the Basic Nonstructural Component Checklist is required by Table 3-2.

Tier 2 Evaluation Procedure: The adequacy of the connections to resist seismic forces calculated in accordance with Section 4.2.7 shall be evaluated.

C4.8.4.7 Panel Connections

A minimum of two connections, usually one at the top and bottom of the panel, are generally required for stability in resisting out-of-plane earthquake forces. Evaluation of connection adequacy should include consideration of all connection eccentricities.

Evaluation Phase (Tier 2)

> If connections are nonexistent, mitigation is necessary to achieve the selected performance level.

4.8.4.8 GLAZING: Glazing in curtain walls and individual panes over 16 square feet in area, located up to a height of 10 feet above an exterior walking surface, shall have safety glazing. Such glazing located over 10 feet above an exterior walking surface shall be laminated annealed or laminated heat-strengthened safety glass or other glazing system that will remain in the frame when glass is cracked.

Tier 2 Evaluation Procedure: Glazing in curtain walls and individual panes over 16 square feet in area shall be shown by analysis or dynamic racking testing to be detailed to accommodate expected levels of drift calculated in accordance with Section 4.2.7.

> **C4.8.4.8 Glazing**
>
> Glazing may shatter and fall due to lack of provision for building drift or racking. If it is safety glazing, it may shatter or crack in a manner that is unlikely to cause injury (life safety). If, in addition, it has racking capability, it may shatter or crack and remain in the frame to provide a temporary weather barrier (Immediate Occupancy). Glass generally fails in earthquakes because of deformation of the frame and lack of space between the glass and frame to allow for independent movement. Special attention should be given to glazing over or close to entrances and exits.

4.8.4.9 GLAZING: All exterior glazing shall be laminated annealed or laminated heat-strengthened safety glass or other glazing system that will remain in the frame when glass is cracked.

Tier 2 Evaluation Procedure: All exterior glazing shall be shown by analysis or dynamic racking testing to be detailed to accommodate expected levels of drift calculated in accordance with Section 4.2.7.

> **C4.8.4.9 Glazing**
>
> Laminated glass will remain in the frame after cracking or shattering, providing a temporary weather barrier and allowing for immediate occupancy following an earthquake.

4.8.5 Masonry Veneer

4.8.5.1 SHELF ANGLES: Masonry veneer shall be supported by shelf angles or other elements at each floor 30 feet or more above ground for Life Safety and at each floor above the first floor for Immediate Occupancy.

Tier 2 Evaluation Procedure: The adequacy of masonry veneer anchors to resist seismic forces calculated in accordance with Section 4.2.7 shall be evaluated.

> **C4.8.5.1 Shelf Angles**
>
> Inadequately fastened masonry veneer can pose a falling hazard if it peels away from its backing. Judgment may be needed to assess the adequacy of various attachments that may be used.

Evaluation Phase (Tier 2)

> If anchorage is nonexistent, mitigation is necessary to achieve the selected performance level.

4.8.5.2 TIES: Masonry veneer shall be connected to the back-up with corrosion-resistant ties. The ties shall have a spacing equal to or less than 24 inches with a minimum of one tie for every 2 2/3 square feet. A spacing of up to 36 inches is permitted when only the Basic Nonstructural Component Checklist is required by Table 3-2.

Tier 2 Evaluation Procedure: The adequacy of the masonry veneer ties to resist seismic forces calculated in accordance with Section 4.2.7 shall be evaluated.

> **C4.8.5.2 Ties**
>
> Inadequately fastened masonry veneer can pose a falling hazard if it peels away from its backing. Judgment may be needed to assess the adequacy of various attachments that may be used. For levels of lower seismicity, it may be easier to show compliance for a larger tie spacing and larger tie area.
>
> Ordinary shop-galvanized wire ties are not very corrosion resistant and are likely to become heavily corroded within 15 years, if the environment is marine or causes continued wetting and drying cycles to the ties, such as at a windward or southern exposure. To be corrosion resistant, the ties should be stainless steel.
>
> If anchorage is nonexistent, mitigation is necessary to achieve the selected performance level.

4.8.5.3 WEAKENED PLANES: Masonry veneer shall be anchored to the back-up adjacent to weakened planes, such as at the locations of flashing.

Tier 2 Evaluation Procedure: The adequacy of masonry veneer and its anchors adjacent to weakened planes created by flashing or other discontinuities shall be evaluated. Anchors shall be evaluated for resistance to seismic forces calculated in accordance with Section 4.2.7.

> **C4.8.5.3 Weakened Planes**
>
> Inadequate attachment at locations of wall discontinuities is a potential source of weakness. Such discontinuities can be created by base flashing or architectural reveals. In areas of moderate and high seismicity, masonry veneer should be anchored to the back-up system immediately above the weakened plane.
>
> If anchorage is nonexistent, mitigation is necessary to achieve the selected performance level.

4.8.5.4 DETERIORATION: There shall be no evidence of deterioration, damage, or corrosion in any of the connection elements.

Tier 2 Evaluation Procedure: The cause and extent of damage shall be identified. The consequences of the damage shall be determined. The adequacy of the remaining undeteriorated or undamaged connections to resist seismic forces calculated in accordance with Section 4.2.7 shall be evaluated. The calculated stress in unreinforced brick veneer shall not exceed the allowable stresses as defined by ACI 530.

Evaluation Phase (Tier 2)

> **C4.8.5.4 Deterioration**
>
> Corrosion can reduce the strength of connections and lead to deterioration of the adjoining materials. The extent of corrosion and its impact on the wall cladding and structure should be considered in the evaluation.
>
> Water leakage into and through exterior walls is a common building problem. Damage due to corrosion, rotting, freezing, or erosion can be concealed in wall spaces. Substantial deterioration can lead to loss of cladding elements or panels.
>
> Exterior walls should be checked for deterioration. Probe into wall spaces if necessary and look for signs of water leakage at vulnerable locations (e.g., at windows and at floor areas). Pay particular attention to elements that tie cladding to the back-up structure and that tie the back-up structure to the floor and roof slabs.
>
> Extremes of temperature can cause substantial structural damage to exterior walls. The resulting weakness may be brought out in a seismic event. Check exterior walls for cracking due to thermal movements.

4.8.5.5 MORTAR: The mortar in masonry veneer shall not be easily scraped away from the joints by hand with a metal tool, and there shall not be significant areas of eroded mortar.

Tier 2 Evaluation Procedure: No Tier 2 Evaluation procedure is available for non-compliant mortar.

> **C4.8.5.5 Mortar**
>
> Inadequate mortar will affect the veneer's ability to withstand seismic motions and maintain attachment to the back-up system.
>
> If mortar is non-compliant, mitigation is necessary to achieve the selected performance level.

4.8.5.6 WEEP HOLES: In veneer braced by stud walls, functioning weep holes and base flashing shall be present.

Tier 2 Evaluation Procedure: No Tier 2 Evaluation procedure is available for non-compliant weep holes.

> **4.8.5.6 Weep Holes**
>
> Absence of weep holes and flashing indicates an inadequately detailed veneer. Water intrusion can lead to deterioration of the veneer and/or substrate. Destructive investigation may be needed to evaluate whether deterioration has taken place and mitigation is necessary.
>
> If weep holes are non-compliant, mitigation is necessary to achieve the selected performance level.

4.8.5.7 STONE CRACKS: There shall be no visible cracks or signs of visible distortion in the stone.

Tier 2 Evaluation Procedure: The extent and consequences of visible cracking shall be evaluated.

C4.8.5.7 Stone Cracks

Cracking in the panel, depending on the material, may be due to weathering or to stresses imposed by movement of the structure or connection system. Severely cracked panels will probably require replacement.

Veins in the stone can create weak points and potential for future cracking and deterioration.

4.8.6 Metal Stud Back-up Systems

4.8.6.1 STUD TRACKS: Stud tracks shall be fastened to the structural framing at a spacing equal to or less than 24 inches on center.

Tier 2 Evaluation Procedure: The adequacy of stud track fasteners to resist seismic forces calculated in accordance with Section 4.2.7 shall be evaluated.

C4.8.6.1 Stud Tracks

Without proper anchorage at top and bottom tracks, metal stud back-up systems are susceptible to excessive movement during an earthquake.

4.8.6.2 OPENINGS: Steel studs shall frame window and door openings.

Tier 2 Evaluation Procedure: The adequacy of window and door framing shall be evaluated.

C4.8.6.2 Openings

This issue is primarily one of the general framing system of the building. Absence of adequate framing around openings indicates a possible out-of-plane weakness in the framing system.

4.8.7 Concrete Block and Masonry Back-up Systems

4.8.7.1 ANCHORAGE: Back-up shall have a positive anchorage to the structural framing at a spacing equal to or less than 4 feet along the floors and roof.

Tier 2 Evaluation Procedure: The adequacy of the concrete block back-up to resist seismic forces calculated in accordance with Section 4.2.7 shall be evaluated.

C4.8.7.1 Anchorage

Back-up is the system that supports veneer for out-of-plane forces. Inadequate anchorage of the back-up wall may affect the whole assembly's ability to withstand seismic motions and maintain attachment to back-up.

Evaluation Phase (Tier 2)

4.8.7.2 URM BACK-UP: There shall be no unreinforced masonry back-up.

Tier 2 Evaluation Procedure: The adequacy of unreinforced masonry to resist seismic forces calculated in accordance with Section 4.2.7 shall be evaluated.

C4.8.7.2 URM Back-up

Unreinforced masonry back-up is common in early steel-framed buildings with cut stone exteriors. The design professional should use judgment in evaluating the condition and integrity of the back-up and necessary remedial measures. Testing may be necessary to determine the strength of the URM back-up.

Complete replacement of back-up is extremely expensive; depending on the state of the installation and the facing materials, alternative methods may be possible.

To qualify as reinforced masonry, the reinforcing steel shall be greater than 0.002 times the gross area of the wall with a minimum of 0.0007 in either of the two directions; the spacing of reinforcing steel shall be less than 48 inches; and all vertical bars shall extend to the top of the back-up walls.

Judgment by the design professional must be used to evaluate the adequacy of concrete block walls not classified as "reinforced." Concrete block walls lacking the minimum reinforcement may be susceptible to in-plane cracking under seismic loads, and portions of the wall may become dislodged.

4.8.8 Parapets, Cornices, Ornamentation, and Appendages

4.8.8.1 URM PARAPETS: There shall be no laterally unsupported unreinforced masonry parapets or cornices with height-to-thickness ratios greater than 1.5. A height-to-thickness ratio of up to 2.5 is permitted when only the Basic Nonstructural Component Checklist is required by Table 3-2.

Tier 2 Evaluation Procedure: The adequacy of the anchorage to resist seismic forces calculated in accordance with Section 4.2.7 shall be evaluated.

C4.8.8.1 URM Parapets

URM parapets present a major falling hazard and potential life safety threat. For sloped roofs, the highest anchorage level should not be taken at the ridge but should vary with roof slope when checking height-to-thickness ratios.

If anchorage is nonexistent, mitigation is necessary to achieve the selected performance level.

4.8.8.2 CANOPIES: Canopies located at building exits shall be anchored to the structural framing at a spacing of 6 feet or less. An anchorage spacing of up to 10 feet is permitted when only the Basic Nonstructural Component Checklist is required by Table 3-2.

Tier 2 Evaluation Procedure: The adequacy of the anchorage to resist seismic forces calculated in accordance with Section 4.2.7 shall be evaluated.

Evaluation Phase (Tier 2)

> **C4.8.8.2 Canopies**
>
> Inadequately supported canopies present a life safety hazard. A common form of failure is pull-out of shallow anchors from building walls.
>
> If anchorage is nonexistent, mitigation is necessary to achieve the selected performance level.

4.8.8.3 CONCRETE PARAPETS: Concrete parapets with height-to-thickness ratios greater than 2.5 shall have vertical reinforcement.

Tier 2 Evaluation Procedure: The adequacy of the anchorage to resist seismic forces calculated in accordance with Section 4.2.7 shall be evaluated.

> **C4.8.8.3 Concrete Parapets**
>
> Inadequately reinforced parapets can be severely damaged during an earthquake.
>
> If anchorage is nonexistent, mitigation is necessary to achieve the selected performance level.

4.8.8.4 APPENDAGES: Cornices, parapets, signs, and other appendages that extend above the highest point of anchorage to the structure or cantilever from exterior wall faces and other exterior wall ornamentation shall be reinforced and anchored to the structural system at a spacing equal to or less than 10 feet for Life Safety and 6 feet for Immediate Occupancy. This requirement need not apply to parapets or cornices compliant with Section 4.8.8.1 or 4.8.8.3.

Tier 2 Evaluation Procedure: The adequacy of the anchorages to resist seismic forces calculated in accordance with Section 4.2.7 shall be evaluated.

> **C4.8.8.4 Appendages**
>
> The above components may vary greatly in size, location, and attachment; the design professional should use judgment in their assessment. If any of these items is of insufficient strength and/or is not securely attached to the structural elements, it may break off and fall onto storefronts, streets, sidewalks, or adjacent property and become a significant life safety hazard.
>
> If anchorages are nonexistent, mitigation is necessary to achieve the selected performance level.

4.8.9 Masonry Chimneys

4.8.9.1 URM CHIMNEYS: No unreinforced masonry chimney shall extend above the roof surface more than twice the least dimension of the chimney. A height above the roof surface of up to three times the least dimension of the chimney is permitted when only the Basic Nonstructural Component Checklist is required by Table 3-2.

Tier 2 Evaluation Procedure: The adequacy of the chimney anchorage to resist seismic forces calculated in accordance with Section 4.2.7 shall be evaluated.

C4.8.9.1 URM Chimneys

Unreinforced masonry chimneys are highly vulnerable to damage in earthquakes. Typically, chimneys extending above the roof more than twice the least dimension of the chimney crack just above the roof line and become dislodged. Chimneys may fall through the roof or onto a public or private walkway, creating a life safety hazard. Experience has shown that the costs of rehabilitating masonry chimneys can sometimes exceed the costs of damage repair.

4.8.9.2 ANCHORAGE: Masonry chimneys shall be anchored at each floor level and the roof.

Tier 2 Evaluation Procedure: The adequacy of the anchorage to resist seismic forces calculated in accordance with Section 4.2.7 shall be evaluated.

C4.8.9.2 Anchorage

Anchorage of chimneys has proven to be problematic at best, ineffective at worst in reducing chimney losses because anchorage alone does not typically account for incompatibility of deformations between the main structure and the chimney. Other rehabilitation strategies – such as the presence of plywood above the ceiling or on the roof to keep the falling masonry from penetrating or relocating occupant activities within a falling radius – may be more effective than anchoring chimneys.

4.8.10 Stairs

4.8.10.1 URM WALLS: Walls around stair enclosures shall not consist of unbraced hollow clay tile or unreinforced masonry with a height-to-thickness ratio greater than 12-to-1. A height-to-thickness ratio of up to 15-to-1 is permitted when only the Basic Nonstructural Component Checklist is required by Table 3-2.

Tier 2 Evaluation Procedure: No Tier 2 Evaluation procedure is available for unbraced hollow clay tile or unreinforced masonry around stair enclosures. A Tier 3 Evaluation is permitted for the evaluation of unbraced hollow clay tile or unreinforced masonry around stair enclosures. Unless otherwise required, a Tier 3 Evaluation of the entire structure need not be performed.

C4.8.10.1 URM Walls

Hollow tile or unreinforced masonry walls may fail and block stairs and corridors. Post-earthquake evacuation efforts can be severely hampered as a result.

There are no Tier 2 Evaluation procedures for this existing condition. FEMA 356 is recommended for analysis of the walls for both in-plane and out-of-plane forces. If bracing is nonexistent, mitigation may be necessary to achieve the selected performance level.

4.8.10.2 STAIR DETAILS: In moment frame structures, the connection between the stairs and the structure shall not rely on shallow anchors in concrete. Alternatively, the stair details shall be capable of accommodating the drift calculated using the Quick Check procedure of Section 3.5.3.1 without inducing tension in the anchors.

Tier 2 Evaluation Procedure: The adequacy of stair connections shall be evaluated when subjected to story drifts calculated in accordance with Section 4.2.

> **C4.8.10.2 Stair Details**
>
> If stairs are not specially detailed to accommodate story drift, they can modify structural response by acting as struts attracting seismic force. Shallow anchors, such as expansion and sleeve anchors, rigidly connect the stairs to the structure. The connection of the stair to the structure must be capable of resisting the imposed forces without loss of gravity support for the stair.

4.8.11 Building Contents and Furnishing

4.8.11.1 TALL NARROW CONTENTS: Contents over 4 feet in height with a height-to-depth or height-to-width ratio greater than 3-to-1 shall be anchored to the floor slab or adjacent structural walls. A height-to-depth or height-to-width ratio of up to 4-to-1 is permitted when only the Basic Nonstructural Component Checklist is required by Table 3-2.

Tier 2 Evaluation Procedure: The adequacy of tall, narrow contents to resist overturning due to seismic forces calculated in accordance with Section 4.2.7 shall be evaluated.

> **C4.8.11.1 Tall Narrow Contents**
>
> Tall, narrow storage or file cabinets or racks can tip over if they are not anchored to resist overturning forces. Commercial kitchen equipment, such as freezer boxes, refrigerators, ovens, and storage racks, can be overturned if not properly fastened to adjacent structural walls and floors.

4.8.11.2 FILE CABINETS: File cabinets arranged in groups shall be attached to one another.

Tier 2 Evaluation Procedure: The adequacy of file cabinets to resist overturning due to seismic forces calculated in accordance with Section 4.2.7 shall be evaluated.

> **C4.8.11.2 File Cabinets**
>
> File cabinets that are grouped together and attached can virtually eliminate the possibility of overturning; the attachment of these file cabinets to the floor then may not be necessary.

4.8.11.3 CABINET DOORS AND DRAWERS: Cabinet doors and drawers shall have latches to keep them closed during an earthquake.

Tier 2 Evaluation Procedure: No Tier 2 Evaluation procedure is available for non-compliant cabinet doors and drawers.

> **C4.8.11.3 Cabinet Doors and Drawers**
>
> Breakable items stored in cabinets should be restrained from falling out by latches, shelf lips, wires, or other methods. It may not be necessary for every cabinet door and drawer to have a latch.

4.8.11.4 ACCESS FLOORS: Access floors over 9 inches in height shall be braced.

Tier 2 Evaluation Procedure: No Tier 2 Evaluation procedure is available for unbraced computer access floors.

> **C4.8.11.4 Access Floors**
>
> Unbraced access floors can collapse onto the structural slab. Small areas of unbraced floors "captured" on all sides within full-height walls may be acceptable; however, the impact of ramps and/or other access openings should be considered in evaluating the adequacy of such unbraced access floors.
>
> If bracing is nonexistent, mitigation is necessary to achieve the selected performance level.

4.8.11.5 EQUIPMENT ON ACCESS FLOORS: Equipment and computers supported on access floor systems shall be either attached to the structure or fastened to a laterally braced floor system.

Tier 2 Evaluation Procedure: No Tier 2 Evaluation procedure is available for unattached equipment supported on access floor systems.

> **C4.8.11.5 Access Floors**
>
> Tall, narrow computers and communications equipment can overturn if not properly anchored. Where overturning is not a concern due to the aspect ratio of the equipment, and it is desirable to provide some isolation between the equipment and the structure, it may be acceptable to support the equipment on a raised floor without positive restraint. In this case, the consequences of equipment movement should be considered. Tethering or some other form of restraint may be appropriate for limiting the range of movement.
>
> If anchorage is nonexistent, mitigation is necessary to achieve the selected performance level.

Evaluation Phase (Tier 2)

4.8.12 Mechanical and Electrical Equipment

4.8.12.1 EMERGENCY POWER: Equipment used as part of an emergency power system shall be anchored to maintain continued operation following an earthquake.

Tier 2 Evaluation Procedure: No Tier 2 Evaluation procedure is available for unanchored equipment used as part of an emergency power system.

> **C4.8.12.1 Emergency Power**
>
> Protection of the emergency power system is critical to post-earthquake recovery, and proper mounting of the components of the system is needed for reliable performance.
>
> Non-emergency equipment located close to or above emergency equipment can be dislodged and fall onto, or cause piping to fail and flood out of, the emergency system.
>
> If anchorage is nonexistent, mitigation is necessary to achieve the selected performance level.

4.8.12.2 HAZARDOUS MATERIAL EQUIPMENT: HVAC or other equipment containing hazardous material shall not have damaged supply lines or unbraced isolation supports.

Tier 2 Evaluation Procedure: No Tier 2 Evaluation procedure is available for HVAC or other equipment containing hazardous material that has damaged supply lines or unbraced isolation supports.

> **C4.8.12.2 Hazardous Material Equipment**
>
> HVAC or other equipment containing hazardous material with damaged supply lines or unbraced isolation supports may release their contents during an earthquake.

4.8.12.3 DETERIORATION: There shall be no evidence of deterioration, damage, or corrosion in any of the anchorage or supports of mechanical or electrical equipment.

Tier 2 Evaluation Procedure: The cause and extent of damage shall be identified. The consequences of the damage shall be determined. The adequacy of the remaining undeteriorated or undamaged anchors or supports to resist seismic forces calculated in accordance with Section 4.2.7 shall be evaluated.

> **C4.8.12.3 Deterioration**
>
> Damaged or corroded anchorage or supports of equipment may not have adequate capacity to resist seismic demands. Suspended or wall-mounted equipment is of more concern than floor- or roof-mounted equipment, as failure of supports would create a falling hazard.

Evaluation Phase (Tier 2)

4.8.12.4 ATTACHED EQUIPMENT: Equipment weighing over 20 pounds that is attached to ceiling, wall, or other support more than 4 feet above the floor shall be braced.

Tier 2 Evaluation Procedure: No Tier 2 Evaluation procedure is available for unbraced equipment weighing over 20 pounds.

C4.8.12.4 Attached Equipment

Equipment located more than 4 feet above the floor poses a falling hazard unless properly anchored and braced. Suspended equipment is more susceptible to damage than floor-, roof-, or wall-mounted equipment. Unbraced suspended equipment can sway during an earthquake, causing damage on impact with other adjacent items.

If bracing is nonexistent, mitigation is necessary to achieve the selected performance level.

4.8.12.5 VIBRATION ISOLATORS: Equipment mounted on vibration isolators shall be equipped with restraints or snubbers.

Tier 2 Evaluation Procedure: No Tier 2 Evaluation procedure is available for non-compliant equipment mounted on vibration isolators.

C4.8.12.5 Vibration Isolators

Many isolation devices for vibration isolated equipment (e.g., fans or pumps,) offer no restraint against lateral movement. As a result, earthquake forces can cause the equipment to fall off its isolators, usually damaging interconnected piping. Snubbers or other restraining devices are needed to prevent horizontal movement in all directions.

Seismic restraints or snubbers must have proper anchors to prevent pull-out. The contact surfaces on the snubbers should be resilient to prevent impact amplification.

If restraints and snubbers are nonexistent, mitigation is necessary to achieve the selected performance level.

4.8.12.6 HEAVY EQUIPMENT: Equipment weighing over 100 pounds shall be anchored to the structure or foundation.

Tier 2 Evaluation Procedure: No Tier 2 Evaluation procedure is available for unbraced equipment weighing over 100 pounds.

C4.8.12.6 Heavy Equipment

For rigidly mounted large equipment (e.g., boilers, chillers, tanks, or generators), inadequate anchorage can lead to horizontal movement. Unanchored equipment, particularly equipment with high aspect ratios such as all tanks, may overturn and/or move and damage utility connections. Performance generally is good when positive attachment to the structure is provided.

If bracing is nonexistent, mitigation is necessary to achieve the selected performance level.

4.8.12.7 ELECTRICAL EQUIPMENT: Electrical equipment and associated wiring shall be laterally braced to the structural system.

Tier 2 Evaluation Procedure: No Tier 2 Evaluation procedure is available for unattached electrical equipment.

C4.8.12.7 Electrical Equipment

Without proper connection to the structure, electrical equipment can move horizontally and/or overturn. The movement can damage the equipment and may create a hazardous condition. Equipment may be mounted to the primary structural system or on walls or ceilings that are capable of resisting the applied loads. Distribution lines that cross structural separations should be investigated. If relative movement of two adjacent buildings can be accommodated by "slack" in the distribution lines, the condition may be acceptable.

If attachment is nonexistent, mitigation is necessary to achieve the selected performance level.

4.8.12.8 DOORS: Mechanically operated doors shall be detailed to operate at a story drift ratio of 0.01.

Tier 2 Evaluation Procedure: The adequacy of the door operation at expected levels of drift calculated in accordance with Section 4.2.7 shall be evaluated.

C4.8.12.8 Doors

Doors that are stuck open or closed, such as fire house garage doors, can greatly impact essential services. Most large doors are not designed to accommodate earthquake-induced transient or permanent drifts in flexible buildings. Fire trucks and ambulances can be delayed in exiting. Critical minutes of emergency response time have been lost in past earthquakes when such doors have been rendered inoperable. Energy conservation measures and vandalism concerns have resulted in an evolution in modern door system designs. Most common door designs are drift intolerant and can result in egress difficulties in flexible buildings, requiring contingency planning and in many cases retrofits. Simple visual evaluations of drift incompatibility between doors that are critical to essential services, their frames, and supporting structures can quickly identify vulnerabilities.

4.8.13 Piping

4.8.13.1 FIRE SUPPRESSION PIPING: Fire suppression piping shall be anchored and braced in accordance with NFPA-13 (NFPA, 1996).

Tier 2 Evaluation Procedure: No Tier 2 Evaluation procedure is available for unbraced and unanchored fire suppression piping.

Evaluation Phase (Tier 2)

C4.8.13.1 Fire Suppression Piping

Fire sprinkler piping has performed poorly in past earthquakes, rendering systems unusable when most needed. Causes of fire sprinkler piping failure included inadequate lateral bracing of sprinkler mains and cross-mains, inadequate flexibility and clearance around sprinkler piping, and impact between sprinkler pipes and other unbraced nonstructural elements. Proper pipe bracing is needed for reliable performance of the system. Note that NFPA-13 is intended to provide an Immediate Occupancy level of performance.

If anchorage and bracing are nonexistent, mitigation is necessary to achieve the selected performance level.

4.8.13.2 FLEXIBLE COUPLINGS: Fluid, gas, and fire suppression piping shall have flexible couplings to allow for building movement at seismic separations.

Tier 2 Evaluation Procedure: No Tier 2 Evaluation procedure is available for fluid, gas, and fire suppression piping without flexible couplings.

C4.8.13.2 Flexible Couplings

Failures may occur in pipes that cross seismic joints due to differential movement of the two adjacent structures. Special detailing is required to accommodate the movement. Flexibility can be provided by a variety of means, including special couplings and pipe bends. Flexible couplings should be evaluated for their ability to accommodate expected seismic movements in all directions.

If flexible couplings are nonexistent, mitigation is necessary to achieve the selected performance level.

4.8.13.3 FLUID AND GAS PIPING: Fluid and gas piping shall be anchored and braced to the building structure to prevent breakage of piping.

Tier 2 Evaluation Procedure: No Tier 2 Evaluation procedure is available for unbraced and unanchored fluid and gas piping.

C4.8.13.3 Fluid and Gas Piping

Piping can fail at elbows, tees, and connections to supported equipment. The potential for failure is dependent on the rigidity, ductility, and expansion or movement capability of the piping system. Joints may separate and hangers may fail. Hanger failures can cause progressive failure of other hangers or supports. Smaller diameter pipes, which generally have greater flexibility, often perform better than larger diameter pipes but they are still subject to damage at the joints. Piping in vertical runs typically performs better than in horizontal runs if it is regularly connected to a vertical shaft.

When using flexible couplings, the following limitations should be considered:

Elastomeric flexible couplings can resist compression, tension, torsion, and bending.

Metal flexible couplings can resist bending only.

Ball joints can resist bending and torsion.

Grooved couplings can resist only minimum bending and torsion.

It should be noted that some building codes permit certain configurations and size of piping without bracing or anchorage. It may be possible to demonstrate compliance by showing that the piping meets current code requirements.

> If anchorage and bracing are nonexistent, mitigation is necessary to achieve the selected performance level.

4.8.13.4   SHUT-OFF VALVES: Shut-off devices shall be present at building utility interfaces to shut off the flow of gas and high-temperature energy in the event of earthquake-induced failure.

Tier 2 Evaluation Procedure: No Tier 2 Evaluation procedure is available for non-compliant shut-off devices.

> **C4.8.13.4   Shut-Off Valves**
>
> Post earthquake recovery efforts have been severely hampered in cases where damaged utility lines could not be expediently isolated from main distribution systems. Shut-off valves are needed to allow for isolation of a building or portions of a building. The valves should be easily accessible and training should be provided for reliable post-earthquake response.
>
> Shut-off valves can be either manually operated or automatic. Automatic shut-off valves should conform to ASCE 25-97, *Earthquake-Actuated Automatic Gas Shutoff Devices*. Manually operated valves should conform to ASME B16.33, *Manually Operated Metallic Gas Valves for Use in Gas Piping Systems up to 125 psig*, or ANSI Z21.15, *Manually Operated Gas Valves for Appliances, Appliance Connector Valves and Hose End Valves*.
>
> If shut-off devices are nonexistent, mitigation is necessary to achieve the selected performance level. The need for and location of shut-off devices should be established in cooperation with local utility companies. Utility companies vary in their policies regarding the installation of shut-off devices.

4.8.13.5   C-CLAMPS: One-sided C-clamps that support piping greater than 2.5 inches in diameter shall be restrained.

Tier 2 Evaluation Procedure: No Tier 2 Evaluation procedure is available for non-compliant C-clamps.

> **C4.8.13.5   C-Clamps**
>
> Unrestrained C-clamps (such as those connected to the bottom flange of structural steel beams) have proven to be unreliable during an earthquake. Pipe movement can cause the C-clamp to work itself off its support, causing local loss of gravity support for the pipe. The loss of a single C-clamp can lead to progressive collapse of other supports.
>
> If C-clamps are non-compliant, mitigation is necessary to achieve the selected performance level.

4.8.14   Ducts

4.8.14.1   STAIR AND SMOKE DUCTS: Stair pressurization and smoke control ducts shall be braced and shall have flexible connections at seismic joints.

Tier 2 Evaluation Procedure: No Tier 2 Evaluation procedure is available for stair pressurization and smoke control ducts without bracing or flexible connection at seismic joints.

C4.8.14.1 Stair and Smoke Ducts

Since these ducts are part of the fire protection system, they are more critical than normal air conditioning ducts. Depending on the duct layout and function of the building, however, the hazard may vary greatly and judgment should be exercised during the evaluation.

If bracing or flexible connections are nonexistent, mitigation is necessary to achieve the selected performance level.

4.8.14.2 DUCT BRACING: Rectangular ductwork exceeding 6 square feet in cross-sectional area, and round ducts exceeding 28 inches in diameter, shall be braced. Maximum spacing of transverse bracing shall not exceed 30 feet. Maximum spacing of longitudinal bracing shall not exceed 60 feet. Intermediate supports shall not be considered part of the lateral-force-resisting system.

Tier 2 Evaluation Procedure: The adequacy of the existing bracing at ductwork to resist seismic forces calculated in accordance with Section 4.2.7 shall be evaluated.

C4.8.14.2 Duct Bracing

Large duct installations are heavy and can cause damage to other materials and may pose a hazard to occupants. Failures may occur in long runs due to large amplitude swaying. Failure usually consists of leakage rather than collapse.

When evaluating the ductwork, the function of the duct system, proximity to occupants, and other materials likely to be damaged should be considered.

If bracing is nonexistent, mitigation is necessary to achieve the selected performance level.

4.8.14.3 DUCT SUPPORT: Ducts shall not be supported by piping or electrical conduit.

Tier 2 Evaluation Procedure: The adequacy of piping or electrical conduit to resist seismic forces calculated in accordance with Section 4.2.7 and gravity forces shall be evaluated.

C4.8.14.3 Duct Support

Though generally undesirable, this condition is only serious when large ducts are supported by other elements that are poorly supported and braced.

4.8.15 Hazardous Materials

4.8.15.1 TOXIC SUBSTANCES: Toxic and hazardous substances stored in breakable containers shall be restrained from falling by latched doors, shelf lips, wires, or other methods.

Tier 2 Evaluation Procedure: No Tier 2 Evaluation procedure is available for toxic and hazardous substances stored in unrestrained breakable containers.

> **C4.8.15.1 Toxic Substances**
>
> Unrestrained containers are susceptible to overturning and falling, resulting in release of materials. Storage conditions should be evaluated in relation to the proximity to occupants, the nature of the substances involved, and the possibility of a toxic condition.
>
> If restraints are nonexistent, mitigation is necessary to achieve the selected performance level.

4.8.15.2 GAS CYLINDERS: Compressed gas cylinders shall be restrained.

Tier 2 Evaluation Procedure: No Tier 2 Evaluation procedure is available for unrestrained compressed gas cylinders.

> **C4.8.15.2 Gas Cylinders**
>
> Unrestrained gas cylinders are highly susceptible to overturning. Release and/or ignition of gas may result. Cylinders should be prevented from overturning by positive means, with restraints above the center of gravity being most effective.
>
> If restraints are nonexistent, mitigation is necessary to achieve the selected performance level.

4.8.15.3 HAZARDOUS MATERIALS: Piping containing hazardous materials shall have shut-off valves or other devices to prevent major spills or leaks.

Tier 2 Evaluation Procedure: No Tier 2 Evaluation procedure is available for non-compliant shut-off devices.

> **C4.8.15.3 Hazardous Materials**
>
> Post-earthquake recovery efforts will be hampered if toxic releases cannot be promptly stopped. Shut-off valves should be accessible, and training should be provided to enhance the reliability of post-earthquake recovery efforts. The specifics of the materials and systems vary greatly. Federal, state, and local codes will govern regarding the installation of shut-off devices.
>
> Large spills of some non-hazardous materials, such as liquid soap or some food products, also can be environmentally damaging and can create a nuisance. Proper shut-off valves and containment structures can help to avert these problems.
>
> If shut-off devices are nonexistent, mitigation is necessary to achieve the selected performance level. The need for and location of shut-off devices should be established in cooperation with local utility companies. Utility companies vary in their policies regarding the installation of shut-off devices.

4.8.16 Elevators

Tier 2 Evaluation Procedure: To evaluate all the items specified below, the elevator installation shall be reviewed by the design professional and an elevator consultant or representative of the elevator manufacturer familiar with elevator seismic requirements. Seismic forces and expected levels of story drift shall be calculated in accordance with Section 4.2.7.

C4.8.16 Elevators

Elevator components are typically not dealt with by design professionals. If necessary, a design professional with experience in elevator design should be consulted.

4.8.16.1 SUPPORT SYSTEM: All elements of the elevator system shall be anchored.

C4.8.16.1 Support System

The successful performance of an elevator system requires that the various elements of the system remain in place, undamaged and capable of operating after inspection. As a minimum, all equipment, including hoistway doors, brackets, controllers, and motors, must be anchored.

4.8.16.2 SEISMIC SWITCH: All elevators shall be equipped with seismic switches that will terminate operations when the ground motion exceeds $0.10g$.

C4.8.16.2 Seismic Switch

Traction elevators, unless carefully designed and constructed, are highly vulnerable to damage during strong shaking. It is very common for the counter-weights to swing out of their rails and collide with the car. Current industry practice and most elevator regulations ensure that the elevator occupants will remain safe by installing seismic switches that sense when strong shaking has begun and automatically shut down the system. Seismic switches are generally located in the elevator machine room and connected directly to the controller. The design professional should verify that the switch is operational, as they are often disabled due to malfunctioning.

4.8.16.3 SHAFT WALLS: All elevator shaft walls shall be anchored and reinforced to prevent toppling into the shaft during strong shaking.

C4.8.16.3 Shaft Walls

Elevator shaft walls are often unreinforced masonry construction using hollow clay tile or concrete masonry block. In the event of strong shaking, these walls may experience significant damage due to in-plane and out-of-plane forces and fall into the shaft.

Evaluation Phase (Tier 2)

4.8.16.4 RETAINER GUARDS: Cable retainer guards on sheaves and drums shall be present to inhibit the displacement of cables.

> **C4.8.16.4 Retainer Guards**
>
> Strong earthquake motions causes the elevator hoist-way cables to whip around and often misalign on the sheaves and drums. Retainer guards are effective at reducing the number of misalignments and improving the possibility that the elevator can continue in service after inspection.

4.8.16.5 RETAINER PLATE: A retainer plate shall be present at the top and bottom of both car and counterweight.

> **C4.8.16.5 Retainer Plate**
>
> Retainer plates are installed just above or below all roller guides and serve to prevent derailment. They are U-shaped, firmly attached to the roller guides, and run not more than 3/4 inch from the rail.

4.8.16.6 COUNTERWEIGHT RAILS: All counterweight rails and divider beams shall be sized in accordance with ASME A17.1.

> **C4.8.16.6 Counterweight Rails**
>
> The typically poor performance of counterweights is due to the size of the rails and the spacing of the rail brackets. Eight-pound rails have routinely shown to be insufficient and are best replaced by 15-pound rails as a minimum.

4.8.16.7 BRACKETS: The brackets that tie the car rails and the counterweight rail to the building structure shall be sized in accordance with ASME A17.1.

> **C4.8.16.7 Brackets**
>
> The brackets that support the rails must be properly spaced and designed to be effective. It is common for brackets to be properly spaced but improperly designed. The design professional should be particularly aware of the eccentricities that often occur within the standard bracket systems most commonly used.

Evaluation Phase (Tier 2)

4.8.16.8   SPREADER BRACKET: Spreader brackets shall not be used to resist seismic forces.

> **C4.8.16.8   Spreader Bracket**
>
> Spreader brackets are a useful element to maintain alignment of counterweight rails between supporting brackets. They have worked very successfully under normal daily operating loads. However, they do not offer any protection to the rails under seismic loading because of the large eccentricities inherent in their shape.

4.8.16.9   GO-SLOW ELEVATORS: The building shall have a go-slow elevator system.

> **C4.8.16.9   Go-Slow Elevators**
>
> The functionality of a building following an earthquake depends on the ability to move through it. However, elevators that are compliant with the code shut down following an earthquake. Therefore, even if the building has the ability to provide immediate occupancy following an earthquake, movement through the building is impeded until the elevators are reactivated. Go-slow elevators alleviate this problem by providing one elevator that functions at a lower speed following an earthquake.

Chapter 5: Detailed Evaluation Phase (Tier 3)

5.1 General

For buildings requiring further investigation, a Tier 3 Evaluation shall be completed in accordance with this chapter. A Tier 3 Evaluation shall be performed either for the entire building after the requirements of Chapter 2 have been met or for those elements identified to be deficient in a Tier 1 and/or Tier 2 Evaluation.

C5.1 General

Tier 1 and Tier 2 Evaluations have the potential for being conservative because of the simplifying assumptions involved in their application. More detailed and presumably more accurate evaluations may employ less conservatism; therefore, they may reveal that buildings or building components identified by Tier 1 and/or Tier 2 Evaluations as having seismic deficiencies are satisfactory to resist seismic forces.

The decision as to whether to employ a Tier 3 Evaluation requires judgment regarding the likelihood of finding that Tier 1 and/or Tier 2 Evaluations are too conservative and regarding whether there would be a significant economic or other advantage to a more detailed evaluation.

No general building evaluation procedures for buildings are currently available that are more detailed than the procedures for Tier 1 and Tier 2. Therefore, in order to make more detailed evaluations, it is necessary to adapt procedures intended for design.

Provisions intended for design may be used for evaluation by inserting existing conditions in the analysis procedures intended for design. Expected performance of existing components can be evaluated by comparing calculated demands on the components with their capacities.

Because Tier 3 uses other provisions not intended for seismic evaluation, the design professional must use caution when adapting them. Topics in which judgment is important include analysis procedure, force levels, and material strengths, properties, and testing.

5.2 Available Procedures

A Tier 3 Evaluation shall be performed using one of the two procedures described in the following sections.

5.2.1 Provisions for Seismic Rehabilitation Design

A component-based evaluation procedure developed for seismic rehabilitation of existing buildings shall be used for a Tier 3 Evaluation. Acceptable analysis procedures for such a detailed evaluation include linear and nonlinear methods for static or dynamic analysis of buildings. Acceptance criteria for such detailed evaluations for various performance levels are based on stiffness, strength, and ductility characteristics of elements and components derived from laboratory tests and analytical studies. The more accurate analysis method and more realistic acceptance criteria developed specifically for rehabilitation of existing buildings shall constitute the detailed evaluation phase. Such a component-based detailed evaluation procedure shall be used in accordance with the authority having jurisdiction.

Demand levels used for analysis in provisions for seismic rehabilitation of existing buildings shall be multiplied by 0.75 when used in a Tier 3 Evaluation. If a linear analysis method is selected, the analysis shall implicitly or explicitly recognize nonlinear response. A building meeting all provisions for the seismic rehabilitation of existing buildings shall be deemed compliant with this section.

> **C5.2.1 Provisions for Seismic Rehabilitation Design**
>
> The only nationally applicable provisions for seismic rehabilitation of existing buildings are FEMA 356, *Prestandard and Commentary for the Seismic Rehabilitation of Buildings* (FEMA, 2000c). Regionally applicable provisions may be available. For example, ATC-40, *Seismic Evaluation and Retrofit of Concrete Buildings, Guidelines for the Seismic Retrofit of Existing Buildings* (ICBO, 2001), the *Tri-Services Manual*, and *Division 95* of the City of Los Angeles Code all were developed specifically for use with buildings in California. Several procedures for nonlinear static analysis and nonlinear dynamic analysis have been developed which also could be used for Tier 3 Evaluations, with the approval of the authority having jurisdiction.
>
> FEMA 356 is the recommended design procedure for adaptation to evaluation. All analysis procedures described in FEMA 356, except for the Simplified Procedure, may be used as permitted by FEMA 356.
>
> The 0.75 reduction factor can be applied to seismic forces because the force levels in these documents are intended for design. Until recently, the 10-percent-in-50-year earthquake was used by most design codes, but the spectra was based on the "mean-plus-one standard deviation." The extra standard deviation provided a factor of safety of design against a stronger ground motion for the same return period. The standard deviation in the earthquake motion is deemed too conservative for an existing building. The 0.75 reduction factor is intended to reduce the earthquake motion from the conservative level used in design to one that is believed to be more appropriate for evaluating existing buildings. The factors that can be used to justify a reduced force level for an existing building are: (1) the actual strength of the components will be greater than that used in the evaluation, and (2) an existing building does not need to have the same level of factor of safety as a new building since the remaining useful life of an existing building may be less than that of a new building. For Tier 1 and 2 Evaluations, this 0.75 factor is taken into account in the methodology in various factors, including material strengths and *m*-factors.

5.2.2 Provisions for Design of New Buildings

Well-established provisions for the design of new buildings approved by the authority having jurisdiction shall be used to perform a Tier 3 Evaluation of an existing building. Acceptable provisions for such a detailed evaluation include Section 9, Earthquake Loads, in ASCE 7-02, *Minimum Design Loads for Buildings and Other Structures,* and the *International Building Code* (ICC, 2000). Such a detailed evaluation shall be performed in accordance with the authority having jurisdiction.

Demand levels used for analysis in provisions for seismic design of new buildings shall be multiplied by 0.75 when used in a Tier 3 Evaluation. If a linear analysis method is selected, the analysis shall implicitly or explicitly recognize nonlinear response. A building meeting all provisions for the design of new buildings shall be deemed compliant with this section.

> **C5.2.2 Provisions for Design of New Buildings**
>
> Provisions for design of new buildings may not be well suited for evaluation of existing buildings because they are based on construction details and building configurations that meet specific standards. These standards may not describe the construction details and configurations or the archaic materials of construction frequently found in existing buildings.

> The 0.75 reduction factor can be applied to seismic forces because the force levels in these documents are intended for design. Until recently, the 10 percent in 50-year earthquake was used by most design codes, but the spectra was based on the "mean-plus-one standard deviation." The extra standard deviation provided a factor of safety of design against a stronger ground motion for the same return period. The standard deviation in the earthquake motion is deemed too conservative for an existing building. The 0.75 reduction factor is intended to reduce the earthquake motion from the conservative level used in design to one that is believed to be more appropriate for evaluating existing buildings. The factors that can be used to justify a reduced force level for an existing building are: 1) the actual strength of the components will be greater than that used in the evaluation, and 2) an existing building does not need to have the same level of factor of safety as a new building since the remaining useful life of an existing building may be less than that of a new building. For Tiers 1 and 2, this 0.75 factor is taken into account in the methodology in various factors, including material strengths and *m*-factors.

5.3 Selection of Detailed Procedures

Buildings with one or more of the following characteristics shall be evaluated using linear dynamic or nonlinear static or dynamic analysis methods (see Section 4.2):

- The fundamental period of the building, T, is greater than or equal to 3.5 times S_{D1}/S_{DS}.
- The ratio of the horizontal dimension at any story to the corresponding dimension at an adjacent story exceeds 1.4 (excluding penthouses and mezzanines).
- The building has a torsional stiffness irregularity in any story. A torsional stiffness irregularity exists in a story if the diaphragm above the story under consideration is not flexible and the results of the analysis indicate that the drift along any side of the structure is more than 150 percent of the average story drift.
- The building has a vertical mass or stiffness irregularity. A vertical mass or stiffness irregularity exists when the average drift in any story (except penthouses) exceeds that of the story above or below by more than 150 percent.
- The building has a non-orthogonal lateral-force-resisting system.

> **C5.3 Selection of Detailed Procedures**
>
> The procedure selected should be based on the judgment as to which procedure is most applicable to the building being evaluated and is likely to yield the most useful data.
>
> Because procedures that explicitly recognize the nonlinear response of building components in earthquakes are likely to yield the most accurate results, nonlinear analysis methods should be selected for complex or irregular buildings and for higher performance levels.

Appendix A: Examples

Introduction

The following examples have been developed to illustrate the evaluation procedures of this standard:

1. Wood frame school building (Building Type W1)
2. Steel moment frame building (Building Type S1A)
3. Concrete frame building with unreinforced masonry infill (Building Type C3)
4. Reinforced masonry bearing wall building with stiff diaphragms (Building Type RM2)
5. Wood frame office building (Building Type W2)
6. Steel braced frame (Building Type S2)
7. Unreinforced masonry bearing wall building with flexible diaphragms (Building Type URM)

The buildings in Examples 1, 2, 3, 5, 6, and 7 are evaluated to the Life Safety Performance Level, while the building in Example 4 is evaluated to the Immediate Occupancy Performance Level.

A Deficiency-Only Tier 2 Evaluation is conducted for Examples 2 and 5. A Full-Building Tier 2 Evaluation is conducted for Example 6. The URM Special Procedure is conducted for Example 7.

Note that the examples are for illustration of the use of this standard only. Judgments, decisions, and conclusions illustrated in these examples may or may not apply to other evaluations. Thus, the design professional must always use judgment when applying the provisions of this standard.

A1 Example 1 – Building Type W1: Wood Light Frames

A1.1 Introduction

This example is based on an actual wood frame school building.

The purpose of this example is to demonstrate the use of the Tier 1 Checklists only. While this structure meets all of the requirements set forth in Tier 1, it should be noted that this is an ideal case; in most evaluations, there will be statements that are non-compliant, requiring that the deficiencies be reported or that a Deficiency-Only Tier 2 Evaluation be conducted.

A1.2 Evaluation Requirements

A1.2.1 Building Description

The building is a one-story wood frame classroom building located in the Los Angeles, California. It was designed and constructed in the 1950s. As-built plans for the building are available.

The total floor area of the building is approximately 4,100 square feet. The foundation is of concrete. The vertical lateral-force-resisting system consists of walls constructed with 2x studs and 1x diagonal sheathing. The roof diaphragm, which consists of 2x rafters and 1x diagonal sheathing and plywood, forms the horizontal lateral-force-resisting system.

A floor plan and elevations of the building are shown in Figures A1-1 and A1-2, respectively.

A1.2.2 Level of Investigation

As-built plans are available for this building (Section 2.2). A site visit was conducted to verify plan information and to assess the condition of the building.

A1.2.3 Level of Performance

The building is evaluated to the Life Safety Performance Level (Section 2.4).

A1.2.4 Level of Seismicity

The building is located in a level of high seismicity (Section 2.5).

$S_S = 1.5$; $S_1 = 0.8$ (ASCE 7-02)
Site Class = D (Section 3.5.2.3.1)
$F_v = 1.5$; $F_a = 1.0$ (Tables 3-5 & 3-6)
$S_{D1} = (2/3)F_vS_1 = 0.8$ (Equation 3-5)
$S_{DS} = (2/3)F_aS_S = 1.00$ (Equation 3-6)

Appendix A: Examples

A1.2.5 Building Type

The building is classified as Building Type W1, Wood Light Frame (Section 2.6). A description of this type of building is included in Table 2-2. Although this building is used as a classroom, it is not classified as Building Type W2, Commercial and Industrial, since it is a one-story building.

A1.3 Screening Phase (Tier 1)

A1.3.1 Introduction

The purpose of a Tier 1 Evaluation is to quickly identify buildings that comply with the provisions of this standard. The data collected is sufficient to conduct a Tier 1 Evaluation.

A1.3.2 Benchmark Buildings

This building was not designed or evaluated in accordance with one of the benchmark documents listed in Table 3-1 (Section 3.2).

A1.3.3 Selection of Checklists

This building is classified as Building Type W1 (Table 2-2). Thus, the Structural Checklist(s) associated with Building Type W1 are used.

A Tier 1 Evaluation of this building involves completing the following checklists (Section 3.3):

- Basic Structural Checklist for Building Type W1 (Section 3.7.1)
- Supplemental Structural Checklist for Building Type W1 (Section 3.7.1S)
- Geologic Site Hazards and Foundations Checklist (Section 3.8)
- Basic Nonstructural Component Checklist (Section 3.9.1)
- Intermediate Nonstructural Component Checklist (Section 3.9.2)

The Supplemental Nonstructural Checklist does not need to be completed for this example since the building is being evaluated to the Life Safety Performance Level.

Each of the checklists listed above has been completed for this example and is included in the following sections.

A1.3.4 Further Evaluation Requirements

While neither a Full-Building Tier 2 Evaluation nor a Tier 3 Evaluation is required for this building based on Section 3.4, the design professional may decide to further evaluate the building if deficiencies are identified by the Tier 1 Evaluation.

Appendix A: Examples

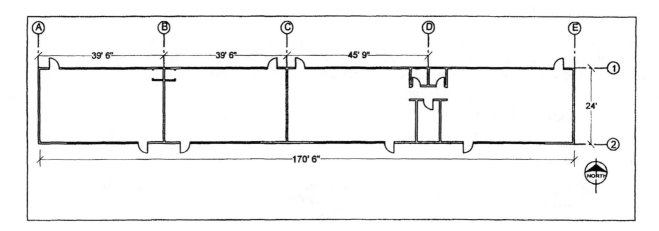

Figure A1-1. Example 1 – Floor Plan

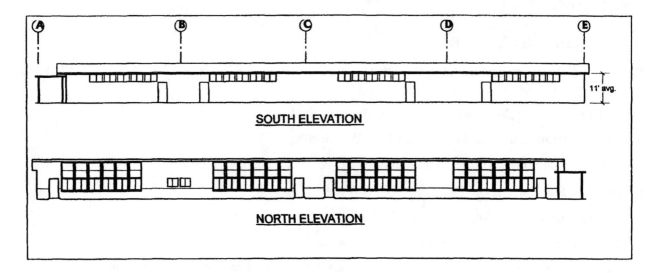

Figure A1-2. Example 1 – Elevations

Appendix A: Examples

A1.3.5 Basic Structural Checklist for Building Type W1: Wood Light Frames

The following is a completed Basic Structural Checklist for Example 1. Each of the evaluation statements has been marked Compliant (C), Non-Compliant (NC) or Not Applicable (N/A). Compliant statements identify issues that are acceptable according to the criteria of this standard, while non-compliant statements identify issues that require further investigation. In addition, an explanation of the evaluation process is included in italics after each evaluation statement. The section numbers in parentheses following each evaluation statement correspond to Tier 2 Evaluation procedures.

Building System

(C) NC N/A LOAD PATH: The structure shall contain a minimum of one complete load path for Life Safety and Immediate Occupancy for seismic force effects from any horizontal direction that serves to transfer the inertial forces from the mass to the foundation. (Tier 2: Sec. 4.3.1.1)

The building contains a complete load path.

(C) NC N/A VERTICAL DISCONTINUITIES: All vertical elements in the lateral-force-resisting system shall be continuous to the foundation. (Tier 2: Sec. 4.3.2.4)

All shear walls are continuous to the foundation.

(C) NC N/A DETERIORATION OF WOOD: There shall be no signs of decay, shrinkage, splitting, fire damage, or sagging in any of the wood members, and none of the metal connection hardware shall be deteriorated, broken, or loose. (Tier 2: Sec. 4.3.3.1)

No deterioration was noted during the site visit.

C NC **(N/A)** WOOD STRUCTURAL PANEL SHEAR WALL FASTENERS: There shall be no more than 15 percent of inadequate fastening such as overdriven fasteners, omitted blocking, excessive fastening spacing, or inadequate edge distance. This statement shall apply to the Immediate Occupancy Performance Level only. (Tier 2: Sec. 4.3.3.2)

Life Safety Performance Level.

Lateral-Force-Resisting System

(C) NC N/A REDUNDANCY: The number of lines of shear walls in each principal direction shall be greater than or equal to 2 for Life Safety and Immediate Occupancy. (Tier 2: Sec. 4.4.2.1.1)

There are two exterior shear walls in the longitudinal direction and five in the transverse direction.

Appendix A: Examples

(C) NC N/A SHEAR STRESS CHECK: The shear stress in the shear walls, calculated using the Quick Check procedure of Section 3.5.3.3, shall be less than the following values for Life Safety and Immediate Occupancy (Tier 2: Sec. 4.4.2.7.1):

Structural panel sheathing	1,000 plf
Diagonal sheathing	700 plf
Straight sheathing	100 plf
All other conditions	100 plf

The shear stress is calculated as follows. The maximum shear stress is 520 plf, which is less than the allowable value of 700 plf.

Pseudo Lateral Force:

$V = CS_aW$ — Equation (3-1)
$C = 1.3$ — Table 3-4
$S_{DI} = 0.8$; $S_{DS} = 1.00$ — Section A1.2.4
$T = C_t h_n^{3/4} = 0.06(11)^{3/4} = 0.36$ sec — Equation (3-8)
$S_a = 0.8/0.362 = 2.3 > S_{DS} = 1.00 = S_a$ — Equation (3-4)
$V = (1.3)(1.00) = 1.3W$
$W = (36 \text{ psf})(24 \text{ ft})(170 \text{ ft}) = 147$ kips
$V = (1.3)(147 \text{ kips}) = 191$ kips

	Shear	L_w^1	$v_1^{avg\ 2}$
Line 1	96 kips	46 ft	520 plf
Line 2	96 kips	69 ft	350 plf
North-South Direction	191 kips	92 ft	520 plf

$^1 L_w$ = length of wall between windows (piers).
^{2} Equation (3-12); ($m=4$).

C NC **(N/A)** STUCCO (EXTERIOR PLASTER) SHEAR WALLS: Multi-story buildings shall not rely on exterior stucco walls as the primary lateral-force-resisting system. (Tier 2: Sec. 4.4.2.7.2)

One-story building.

C NC **(N/A)** GYPSUM WALLBOARD OR PLASTER SHEAR WALLS: Interior plaster or gypsum wallboard shall not be used as shear walls on buildings over one story in height with the exception of the uppermost level of a multi-story building. (Tier 2: Sec. 4.4.2.7.3)

One-story building.

(C) NC N/A NARROW WOOD SHEAR WALLS: Narrow wood shear walls with an aspect ratio greater than 2-to-1 for Life Safety and 1.5-to-1 for Immediate Occupancy shall not be used to resist lateral forces developed in the building in levels of moderate and high seismicity. Narrow wood shear walls with an aspect ratio greater than 2-to-1 for Immediate Occupancy shall not be used to resist lateral forces developed in the building in levels of low seismicity. (Tier 2: Sec. 4.4.2.7.4)

Critical aspect ratio = 11 ft / 12 ft = 0.9 < 2.0.
Length of shear wall taken as shortest distance between windows (shortest pier).

C NC **(N/A)** WALLS CONNECTED THROUGH FLOORS: Shear walls shall have interconnection between stories to transfer overturning and shear forces through the floor. (Tier 2: Sec. 4.4.2.7.5)

One-story building.

C NC **(N/A)** HILLSIDE SITE: For structures that are taller on at least one side by more than half of a story due to a sloping site, all shear walls on the downhill slope shall have an aspect ratio less than 1-to-1 for Life Safety and 1 to 2 for Immediate Occupancy. (Tier 2: Sec. 4.4.2.7.6)

Building is on a level lot.

Appendix A: Examples

C   NC   (N/A)   CRIPPLE WALLS: Cripple walls below first floor level shear walls shall be braced to the foundation with wood structural panels. (Tier 2: Sec. 4.4.2.7.7)

There are no cripple walls.

C   NC   (N/A)   OPENINGS: Walls with openings greater than 80 percent of the length shall be braced with wood structural panel shear walls with aspect ratios of not more than 1.5-to-1 or shall be supported by adjacent construction through positive ties capable of transferring the lateral forces. This statement shall apply to the Immediate Occupancy Performance Level only. (Tier 2: Sec. 4.4.2.7.8)

There are no walls with large openings.

Connections

C   NC   (N/A)   WOOD POSTS: There shall be a positive connection of wood posts to the foundation. (Tier 2: Sec. 4.6.3.3)

Shear wall system.

(C)   NC   N/A   WOOD SILLS: All wood sills shall be bolted to the foundation. (Tier 2: Sec. 4.6.3.4)

There are anchor bolts at 4 feet on center along all shear walls.

C   NC   (N/A)   GIRDER/COLUMN CONNECTION: There shall be a positive connection utilizing plates, connection hardware, or straps between the girder and the column support. (Tier 2: Sec. 4.6.4.1)

Shear wall system.

A1.3.6 Supplemental Structural Checklist for Building Type W1: Wood Light Frames

The following is a completed Supplemental Structural Checklist for Example 1. Each of the evaluation statements has been marked Compliant (C), Non-Compliant (NC) or Not Applicable (N/A). Compliant statements identify issues that are acceptable according to the criteria of this standard, while non-compliant statements identify issues that require further investigation. In addition, an explanation of the evaluation process is included in italics after each evaluation statement. The section numbers in parentheses following each evaluation statement correspond to Tier 2 Evaluation procedures.

Lateral-Force-Resisting System

C   NC   (N/A)   HOLD-DOWN ANCHORS: All shear walls shall have hold-down anchors constructed per acceptable construction practices, attached to the end studs. This statement shall apply to the Immediate Occupancy Performance Level only. (Tier 2: Sec. 4.4.2.7.9)

Life Safety Performance Level.

Diaphragms

(C)   NC   N/A   DIAPHRAGM CONTINUITY: The diaphragms shall not be composed of split-level floors and shall not have expansion joints. (Tier 2: Sec. 4.5.1.1)

No split-level floors or expansion joints.

(C)   NC   N/A   ROOF CHORD CONTINUITY: All chord elements shall be continuous, regardless of changes in roof elevation. (Tier 2: Sec. 4.5.1.3)

All roof chords are continuous.

Appendix A: Examples

C NC (N/A) PLAN IRREGULARITIES: There shall be tensile capacity to develop the strength of the diaphragm at re-entrant corners or other locations of plan irregularities. This statement shall apply to the Immediate Occupancy Performance Level only. (Tier 2: Sec. 4.5.1.7)

Life Safety Performance Level.

C NC (N/A) DIAPHRAGM REINFORCEMENT AT OPENINGS: There shall be reinforcing around all diaphragms openings larger than 50 percent of the building width in either major plan dimension. This statement shall apply to the Immediate Occupancy Performance Level only. (Tier 2: Sec. 4.5.1.8)

Life Safety Performance Level.

C NC (N/A) STRAIGHT SHEATHING: All straight sheathed diaphragms shall have aspect ratios less than 2-to-1 for Life Safety and 1-to-1 for Immediate Occupancy in the direction being considered. (Tier 2: Sec. 4.5.2.1)

No straight sheathed diaphragms.

(C) NC N/A SPANS: All wood diaphragms with spans greater than 24 feet for Life Safety and 12 feet for Immediate Occupancy shall consist of wood structural panels or diagonal sheathing. (Tier 2: Sec. 4.5.2.2)

All diaphragms consist of diagonal sheathing.

(C) NC N/A UNBLOCKED DIAPHRAGMS: All diagonally sheathed or unblocked wood structural panel diaphragms shall have horizontal spans less than 40 feet for Life Safety and 30 feet for Immediate Occupancy and shall have aspect ratios less than or equal to 4-to-1 for Life Safety and 3-to-1 for Immediate Occupancy. (Tier 2: Sec. 4.5.2.3)

Diaphragm span is 40 feet and the aspect ratio is 2-to-1.

(C) NC N/A OTHER DIAPHRAGMS: The diaphragm shall not consist of a system other than wood, metal deck, concrete, or horizontal bracing. (Tier 2: Sec. 4.5.7.1)

All diaphragms are diagonally sheathed.

Connections

(C) NC N/A WOOD SILL BOLTS: Sill bolts shall be spaced at 6 feet or less for Life Safety and 4 feet or less for Immediate Occupancy, with proper edge and end distance provided for wood and concrete. (Tier 2: Sec. 4.6.3.9)

Sill bolts are spaced at 4 feet.

A1.3.7 Geologic Site Hazards and Foundations Checklist

The following is a completed Geologic Site Hazards and Foundations Checklist for Example 1. Each of the evaluation statements has been marked Compliant (C), Non-Compliant (NC) or Not Applicable (N/A). Compliant statements identify issues that are acceptable according to the criteria of this standard, while non-compliant statements identify issues that require further investigation. In addition, an explanation of the evaluation process is included in italics after each evaluation statement. The section numbers in parentheses following each evaluation statement correspond to Tier 2 Evaluation procedures.

Appendix A: Examples

Geologic Site Hazards

The following statements shall be completed for buildings in levels of high or moderate seismicity.

(C) NC N/A   LIQUEFACTION: Liquefaction susceptible, saturated, loose granular soils that could jeopardize the building's seismic performance shall not exist in the foundation soils at depths within 50 feet under the building for Life Safety and Immediate Occupancy. (Tier 2: Sec. 4.7.1.1)

Liquefaction susceptible soils do not exist in the foundation soils at depths within 50 feet under the building as described in project geotechnical investigation.

(C) NC N/A   SLOPE FAILURE: The building site shall be sufficiently remote from potential earthquake-induced slope failures or rockfalls to be unaffected by such failures or shall be capable of accommodating small predicted movements without failure. (Tier 2: Sec. 4.7.1.2)

The building is on a relatively flat site.

(C) NC N/A   SURFACE FAULT RUPTURE: Surface fault rupture and surface displacement at the building site is not anticipated. (Tier 2: Sec. 4.7.1.3)

Surface fault rupture and surface displacements are not anticipated.

Condition of Foundations

The following statements shall be completed for all Tier 1 building evaluations.

(C) NC N/A   FOUNDATION PERFORMANCE: There shall not be evidence of excessive foundation movement such as settlement or heave that would affect the integrity or strength of the structure. (Tier 2: Sec. 4.7.2.1)

The structure does not show evidence of excessive foundation movement.

The following statement shall be completed for buildings in levels of high or moderate seismicity being evaluated to the Immediate Occupancy Performance Level.

C NC (N/A)   DETERIORATION: There shall not be evidence that foundation elements have deteriorated due to corrosion, sulfate attack, material breakdown, or other reasons in a manner that would affect the integrity or strength of the structure. (Tier 2: Sec. 4.7.2.2)

Life Safety Performance Level.

Capacity of Foundations

The following statement shall be completed for all Tier 1 building evaluations.

C NC (N/A)   POLE FOUNDATIONS: Pole foundations shall have a minimum embedment depth of 4 feet for Life Safety and Immediate Occupancy. (Tier 2: Sec. 4.7.3.1)

Pole foundations not used.

The following statements shall be completed for buildings in levels of moderate seismicity being evaluated to the Immediate Occupancy Performance Level and for buildings in levels of high seismicity.

(C) NC N/A   OVERTURNING: The ratio of the horizontal dimension of the lateral-force-resisting system at the foundation level to the building height (base/height) shall be greater than $0.6S_a$. (Tier 2: Sec. 4.7.3.2)

Ratio of base to height = 12 ft / 11 ft = 1.1 > 0.6 S_a = 0.6(1.0) = 0.6.

Appendix A: Examples

C NC (N/A) TIES BETWEEN FOUNDATION ELEMENTS: The foundation shall have ties adequate to resist seismic forces where footings, piles, and piers are not restrained by beams, slabs, or soils classified as Class A, B, or C. (Section 3.5.2.3.1, Tier 2: Sec. 4.7.3.3)

Footings are restrained by slabs.

C NC (N/A) DEEP FOUNDATIONS: Piles and piers shall be capable of transferring the lateral forces between the structure and the soil. This statement shall apply to the Immediate Occupancy Performance Level only. (Tier 2: Sec. 4.7.3.4)

Life Safety Performance Level.

C NC (N/A) SLOPING SITES: The difference in foundation embedment depth from one side of the building to another shall not exceed one story in height. This statement shall apply to the Immediate Occupancy Performance Level only. (Tier 2: Sec. 4.7.3.5)

Life Safety Performance Level.

A1.3.8 Basic Nonstructural Component Checklist

The following is a completed Basic Nonstructural Component Checklist for Example 1. Each of the evaluation statements has been marked Compliant (C), Non-Compliant (NC) or Not Applicable (N/A). Compliant statements identify issues that are acceptable according to the criteria of this standard, while non-compliant statements identify issues that require further investigation. In addition, an explanation of the evaluation process is included in italics after each evaluation statement. The section numbers in parentheses following each evaluation statement correspond to Tier 2 Evaluation procedures.

Partitions

C NC (N/A) UNREINFORCED MASONRY: Unreinforced masonry or hollow clay tile partitions shall be braced at a spacing equal to or less than 10 feet in levels of low or moderate seismicity and 6 feet in levels of high seismicity. (Tier 2: Sec. 4.8.1.1)

No interior hollow clay tile partitions.

Ceiling Systems

C NC (N/A) SUPPORT: The integrated suspended ceiling system shall not be used to laterally support the tops of gypsum board, masonry, or hollow clay tile partitions. Gypsum board partitions need not be evaluated where only the Basic Nonstructural Component Checklist is required by Table 3-2. (Tier 2: Sec. 4.8.2.1)

No gypsum board, masonry, or hollow clay tile partitions.

Light Fixtures

(C) NC N/A EMERGENCY LIGHTING: Emergency lighting shall be anchored or braced to prevent falling during an earthquake. (Tier 2: Sec. 4.8.3.1)

Emergency lighting is anchored at a spacing of less than 6 feet.

Cladding and Glazing

No exterior cladding; statements in this section are Not Applicable (N/A).

Appendix A: Examples

Masonry Veneer

No masonry veneer; statements in this section are Not Applicable (N/A).

Parapets, Cornices, Ornamentation, and Appendages

No parapets, cornices, ornamentation, or appendages; statements in this section are Not Applicable (N/A).

Masonry Chimneys

No masonry chimneys; statements in this section are Not Applicable (N/A).

Stairs

No stairs; statements in this section are Not Applicable (N/A).

Building Contents and Furnishing

(C) NC N/A   TALL NARROW CONTENTS: Contents over 4 feet in height with a height-to-depth or height-to-width ratio greater than 3-to-1 shall be anchored to the floor slab or adjacent structural walls. A height-to-depth or height-to-width ratio of up to 4-to-1 is permitted where only the Basic Nonstructural Component Checklist is required by Table 3-2. (Tier 2: Sec. 4.8.11.1)

Storage racks, bookcases, and file cabinets are anchored to the floor slab.

Mechanical and Electrical Equipment

(C) NC N/A   EMERGENCY POWER: Equipment used as part of an emergency power system shall be mounted to maintain continued operation after an earthquake. (Tier 2: Sec. 4.8.12.1)

Emergency power equipment is anchored.

(C) NC N/A   HAZARDOUS MATERIAL EQUIPMENT: HVAC or other equipment containing hazardous material shall not have damaged supply lines or unbraced isolation supports. (Tier 2: Sec. 4.8.12.2)

HVAC equipment and supply lines are anchored.

(C) NC N/A   DETERIORATION: There shall be no evidence of deterioration, damage, or corrosion in any of the anchorage or supports of mechanical or electrical equipment. (Tier 2: Sec. 4.8.12.3)

No deterioration observed.

(C) NC N/A   ATTACHED EQUIPMENT: Equipment weighing over 20 pounds that is attached to ceilings, walls, or other supports 4 feet above the floor level shall be braced. (Tier 2: Sec. 4.8.12.4)

Equipment is braced.

Piping

(C) NC N/A   FIRE SUPPRESSION PIPING: Fire suppression piping shall be anchored and braced in accordance with NFPA-13 (NFPA, 1996). (Tier 2: Sec. 4.8.13.1)

Fire suppression piping is braced in accordance with NFPA-13.

Appendix A: Examples

(C) NC N/A   **FLEXIBLE COUPLINGS:** Fluid, gas and fire suppression piping shall have flexible couplings. (Tier 2: Sec. 4.8.13.2)

Fluid, gas, and fire suppression piping have flexible couplings.

Hazardous Materials Storage and Distribution

C NC (N/A)   **TOXIC SUBSTANCES:** Toxic and hazardous substances stored in breakable containers shall be restrained from falling by latched doors, shelf lips, wires, or other methods. (Tier 2: Sec. 4.8.15.1)

No toxic or hazardous substances are stored in the building.

A1.3.9 Intermediate Nonstructural Component Checklist

The following is a completed Intermediate Nonstructural Checklist for Example 1. Each of the evaluation statements has been marked Compliant (C), Non-Compliant (NC) or Not Applicable (N/A). Compliant statements identify issues that are acceptable according to the criteria of this standard, while non-compliant statements identify issues that require further investigation. In addition, an explanation of the evaluation process is included in italics after each evaluation statement. The section numbers in parentheses following each evaluation statement correspond to Tier 2 Evaluation procedures.

Ceiling Systems

(C) NC N/A   **LAY-IN TILES:** Lay-in tiles used in ceiling panels located at exits and corridors shall be secured with clips. (Tier 2: Sec. 4.8.2.2)

Lay-in tiles are secured with clips.

(C) NC N/A   **INTEGRATED CEILINGS:** Integrated suspended ceilings at exits and corridors or weighing more than 2 lb/ft^2 shall be laterally restrained with a minimum of 4 diagonal wires or rigid members attached to the structure above at a spacing equal to or less than 12 feet. (Tier 2: Sec. 4.8.2.3)

Integrated suspended ceilings are braced.

(C) NC N/A   **SUSPENDED LATH AND PLASTER:** Ceilings consisting of suspended lath and plaster or gypsum board shall be attached to resist seismic forces for every 12 square feet of area. (Tier 2: Sec. 4.8.2.4)

Each 10 square feet of suspended lath and plaster ceilings is attached.

Light Fixtures

(C) NC N/A   **INDEPENDENT SUPPORT:** Light fixtures in suspended grid ceilings shall be supported independently of the ceiling suspension system by a minimum of two wires at diagonally opposite corners of the fixtures. (Tier 2: Sec. 4.8.3.2)

Light fixtures are independently supported.

Cladding and Glazing

No exterior cladding; statements in this section are Not Applicable (N/A).

Parapets, Cornices, Ornamentation, and Appendages

No parapets, cornices, ornamentation, or appendages; statements in this section are Not Applicable (N/A).

Appendix A: Examples

Masonry Chimneys

No masonry chimneys; statements in this section are Not Applicable (N/A).

Mechanical and Electrical Equipment

C   NC   (**N/A**)   VIBRATION ISOLATORS: Equipment mounted on vibration isolators shall be equipped with restraints or snubbers. (Tier 2: Sec. 4.8.12.5)

No equipment on vibration isolators.

Ducts

C   NC   (**N/A**)   STAIR AND SMOKE DUCTS: Stair pressurization and smoke control ducts shall be braced and shall have flexible connections at seismic joints. (Tier 2: Sec. 4.8.14.1)

No stair pressurization or smoke control ducts.

A1.4 Final Evaluation and Report

The building was found to be either compliant with each of the evaluation statements on the Tier 1 Checklists or the evaluation statement was not applicable; thus, the building complies with the provisions of this standard. No potential deficiencies were identified by the Tier 1 Checklists. Further evaluation is not required. A final report shall be prepared that outlines the results of the Tier 1 Evaluation.

Appendix A: Examples

A2 Example 2 – Building Type S1A: Steel Moment Frame with Flexible Diaphragms

A2.1 Introduction

This example is based on an actual steel moment frame building with a flexible diaphragm.

The purpose of this example is to demonstrate the use of the Tier 1 Evaluation and Deficiency-Only Tier 2 Evaluation.

A2.2 Evaluation Requirements

A2.2.1 Building Description

The building is a three-story steel moment frame building constructed in 1986. It is located in southern California and is used as an office building.

The building is rectangular in plan, approximately 250 feet by 110 feet. The building height is approximately 41 feet. The building area is approximately 27,000 square feet per floor. The floor plan of the building is provided in Figure A2-1.

Grade beams and spread footings are used for the foundation. Roof and floor framing systems consist of truss joists and steel beams with plywood diaphragms. There is lightweight concrete on both floors. The exterior of the building has a brick veneer.

The lateral-force-resisting system consists of single bay steel moment frames with grade beams. Three different frames are used: 1, 1A, and 1B. Frame elevations with member sizes are shown in Figure A2-2. There are a total of four frames in each direction: frames 1 and 1B in the North-South direction and frame 1A in the East-West direction (see plan in Figure A2-1).

A2.2.2 Level of Investigation

As-built plans are available for this building (Section 2.2). A site visit was conducted to verify plan information and check the condition of the building.

A2.2.3 Level of Performance

The building is evaluated to the Life Safety Performance Level (Section 2.4).

Appendix A: Examples

A2.2.4 Level of Seismicity

The building is located in a level of high seismicity (Section 2.5).

$S_S = 1.5$; $S_I = 0.6$	(ASCE 7-02)
Site Class = D	(Section 3.5.2.3.1)
$F_v = 1.5$; $F_a = 1.0$	(Tables 3-5 & 3-6)
$S_{DI} = (2/3)F_vS_I = 0.6$	(Equation 3-5)
$S_{DS} = (2/3)F_aS_S = 1.00$	(Equation 3-6)

A2.2.5 Building Type

The building is classified as Building Type S1A, Steel Moment Frame with Flexible Diaphragms (Section 2.6). A description of this type of building is included in Table 2-2.

A2.3 Screening Phase (Tier 1)

A2.3.1 Introduction

The purpose of a Tier 1 Evaluation is to quickly identify buildings that comply with the provisions of this standard. The data collected is sufficient to conduct a Tier 1 Evaluation.

A2.3.2 Benchmark Buildings

This building was not designed in accordance with one of the benchmark documents listed in Table 3-1 (Section 3.2).

A2.3.3 Selection of Checklists

This building is classified as Building Type S1A (Table 2-2). Thus, the Structural Checklist(s) associated with Building Type S1A are used.

A Tier 1 Evaluation of this building involves completing the following checklists (Section 3.3):

- Basic Structural Checklist for Building Type S1A (Section 3.7.3A)
- Supplemental Structural Checklist for Building Type S1A (Section 3.7.2AS)
- Geologic Site Hazards and Foundations Checklist (Section 3.8)
- Basic Nonstructural Component Checklist (Section 3.9.1)
- Intermediate Nonstructural Component Checklist (Section 3.9.2)

The Supplemental Nonstructural Checklist does not need to be completed for this example since the building is being evaluated to the Life Safety Performance Level.

Each of the checklists listed above has been completed for this example and is included in the following sections.

Appendix A: Examples

A2.3.4 Further Evaluation Requirements

While neither a Full-Building Tier 2 Evaluation nor a Tier 3 Evaluation is required for this building based on Section 3.4, the design professional may decide to further evaluate the building if deficiencies are identified by the Tier 1 Evaluation.

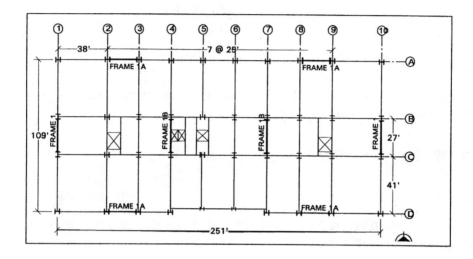

Figure A2-1. Example 2 – Typical Floor Plan

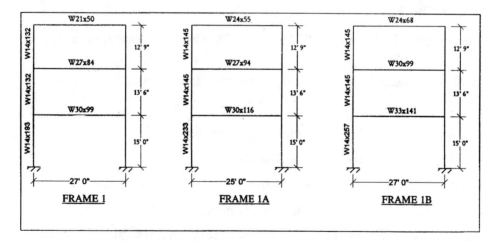

Figure A2-2. Example 2 – Frame Elevations

Appendix A: Examples

A2.3.5 Basic Structural Checklist for Building Type S1A: Steel Moment Frames with Flexible Diaphragms

The following is a completed Basic Structural Checklist for Example 2. Each of the evaluation statements has been marked Compliant (C), Non-Compliant (NC) or Not Applicable (N/A). Compliant statements identify issues that are acceptable according to the criteria of this standard, while non-compliant statements identify issues that require further investigation. In addition, an explanation of the evaluation process is included in italics after each evaluation statement. The section numbers in parentheses following each evaluation statement correspond to Tier 2 Evaluation procedures.

Building System

(C) NC N/A LOAD PATH: The structure shall contain a minimum of one complete load path for Life Safety and Immediate Occupancy for seismic force effects from any horizontal direction that serves to transfer the inertial forces from the mass to the foundation. (Tier 2: Sec. 4.3.1.1)

The building contains a complete load path. The load path consists of plywood diaphragms spanning to moment-resisting frames to grade beams and spread footings.

(C) NC N/A ADJACENT BUILDINGS: The clear distance between the building being evaluated and any adjacent building shall be greater than 4 percent of the height of the shorter building for Life Safety and Immediate Occupancy. (Tier 2: Sec. 4.3.1.2)

No buildings within a distance equal to 4 percent of the building height (approximately 20 inches).

C NC (N/A) MEZZANINES: Interior mezzanine levels shall be braced independently from the main structure, or shall be anchored to the lateral-force-resisting elements of the main structure. (Tier 2: Sec. 4.3.1.3)

No mezzanines.

(C) NC N/A WEAK STORY: The strength of the lateral-force-resisting system in any story shall not be less than 80 percent of the strength in an adjacent story above or below for Life Safety and Immediate Occupancy. (Tier 2: Sec. 4.3.2.1)

There is no strength discontinuity in any of the vertical elements in the lateral-force-resisting system.

(C) NC N/A SOFT STORY: The stiffness of the lateral-force-resisting-system in any story shall not be less than 70 percent of the lateral-force-resisting system stiffness in an adjacent story above or below, or less than 80 percent of the average lateral-force-resisting system stiffness of the three stories above or below for Life Safety and Immediate Occupancy. (Tier 2: Sec. 4.3.2.2)

There is no stiffness discontinuity in any of the vertical elements in the lateral-force-resisting system.

(C) NC N/A GEOMETRY: There shall be no changes in horizontal dimension of the lateral-force-resisting system of more than 30 percent in a story relative to adjacent stories for Life Safety and Immediate Occupancy, excluding one-story penthouses and mezzanines. (Tier 2: Sec. 4.3.2.3)

The horizontal dimension of the moment-resisting frames is constant.

(C) NC N/A VERTICAL DISCONTINUITIES: All vertical elements in the lateral-force-resisting system shall be continuous to the foundation. (Tier 2: Sec. 4.3.2.4)

Frames are continuous to the foundation.

Appendix A: Examples

(C) NC N/A   MASS: There shall be no change in effective mass more than 50 percent from one story to the next for Life Safety and Immediate Occupancy. Light roofs, penthouses, and mezzanines need not be considered. (Tier 2: Sec. 4.3.2.5)

Change in effective mass is less than 50 percent.

Story	Weight	Percent Change
Roof	662 kips	41
3rd	1,122 kips	1
2nd	1,135 kips	

(C) NC N/A   DETERIORATION OF WOOD: There shall be no signs of decay, shrinkage, splitting, fire damage, or sagging in any of the wood members, and none of the metal connection hardware shall be deteriorated, broken, or loose. (Tier 2: Sec. 4.3.3.1)

There is no sign of deterioration of wood members.

(C) NC N/A   DETERIORATION OF STEEL: There shall be no visible rusting, corrosion, cracking, or other deterioration in any of the steel elements or connections in the vertical- or lateral-force-resisting systems. (Tier 2: Sec. 4.3.3.3)

There is no significant rusting, corrosion, cracking, or other deterioration of steel members.

Lateral-Force-Resisting System

C **(NC)** N/A   REDUNDANCY: The number of lines of moment frames in each principal direction shall be greater than or equal to 2 for Life Safety and Immediate Occupancy. The number of bays of moment frames in each line shall be greater than or equal to 2 for Life Safety and 3 for Immediate Occupancy. (Tier 2: Sec. 4.4.1.1.1)

Four frames in each direction. However, in the transverse direction they are all single bay frames.

C NC **(N/A)**   INTERFERING WALLS: All concrete and masonry infill walls placed in moment frames shall be isolated from structural elements. (Tier 2: Sec. 4.4.1.2.1)

No infill walls.

C **(NC)** N/A   DRIFT CHECK: The drift ratio of the steel moment frames, calculated using the Quick Check procedure of Section 3.5.3.1, shall be less than 0.025 for Life Safety and 0.015 for Immediate Occupancy. (Tier 2: Sec. 4.4.1.3.1)

The maximum drift ratio is 0.030, which is greater than the allowable value of 0.025. Drift ratio calculations are provided below.

Pseudo Lateral Force:

$V = CS_aW$ Equation (3-1)
$C = 1.0$ Table 3-4
$S_{D1} = 0.6$; $S_{DS} = 1.00$ Section A2.2.4
$T = C_t h_n^{3/4} = 0.035(41.25)^{3/4} = 0.57$ sec   Equation (3-8)
$S_a = 0.60/0.57 = 1.05 > S_{DS} = 1.00$ Equation (3-4)
$V = (1.0)(1.00)W = 1.00W$
$W = 2,920$ kips
$V = 1.00(2,920$ kips$) = 2,920$ kips

Appendix A: Examples

| | | Frame 1 | | Frame 1B | | Frame 1A | |
Story	Story Shear[1]	Column Shear[2]	Drift Ratio[3]	Column Shear[2]	Drift Ratio[3]	Column Shear[2]	Drift Ratio[3]
3rd	1,051 kips	92 kips	0.018	170 kips	0.02	131 kips	0.019
2nd	227 kips	199 kips	0.020	370 kips	0.031	285 kips	0.027
1st	2,920 kips	256 kips	0.021	474 kips	0.021	365 kips	0.023

[1] Equation (3-3).
[2] Based on tributary area.
[3] Equation (3-10).

(C) NC N/A **AXIAL STRESS CHECK:** The axial stress due to gravity loads in columns subjected to overturning forces shall be less than $0.10F_y$ for Life Safety and Immediate Occupancy. Alternatively, the axial stresses due to overturning forces alone, calculated using the Quick Check procedure of Section 3.5.3.6, shall be less than $0.30F_y$ for Life Safety and Immediate Occupancy. (Tier 2: Sec. 4.4.1.4.2)

Allowable axial stress = $0.1F_y$ = 0.1(36 ksi) = 3.6 ksi; axial stresses are presented in the table below.

Frame	Column	Axial Load[1]	Column Area	Axial Stress
1	W14x132	71 kips	38.8 in.^{2}	1.8 ksi
	W14x193	120 kips	56.8 in.^{2}	2.1 ksi
1A	W14x145	71 kips	42.7 in.^{2}	1.7 ksi
	W14x233	120 kips	68.5 in.^{2}	1.8 ksi
1B	W14x145	94 kips	42.7 in.^{2}	2.2 ksi
	W14x257	159 kips	75.6 in.^{2}	2.1 ksi

[1] Axial Load = (Dead Load + Live Load)(Tributary Area).
Roof Dead Load = 22 psf; Roof Live Load = 12 psf.
Floor Dead Load = 26 psf + 20 psf (partition load); Floor Live Load = 0.6(50 psf).

Connections

(C) NC N/A **TRANSFER TO STEEL FRAMES:** Diaphragms shall be connected for transfer of loads to the steel frames for Life Safety, and the connections shall be able to develop the lesser of the strength of the frames or the diaphragms for Immediate Occupancy. (Tier 2: Sec. 4.6.2.2)

Diaphragms are connected for transfer of loads to the steel frames.

(C) NC N/A **STEEL COLUMNS:** The columns in lateral-force-resisting frames shall be anchored to the building foundation for Life Safety, and the anchorage shall be able to develop the lesser of the tensile capacity of the column, the tensile capacity of the lowest level column splice (if any), or the uplift capacity of the foundation, for Immediate Occupancy. (Tier 2: Sec. 4.6.3.1)

Moment-resisting frame columns are anchored to the foundation.

Appendix A: Examples

A2.3.6 Supplemental Structural Checklist for Building Type S1A: Steel Moment Frames with Flexible Diaphragms

The following is a completed Supplemental Structural Checklist for Example 2. Each of the evaluation statements has been marked Compliant (C), Non-Compliant (NC) or Not Applicable (N/A). Compliant statements identify issues that are acceptable according to the criteria of this standard, while non-compliant statements identify issues that require further investigation. In addition, an explanation of the evaluation process is included in italics after each evaluation statement. The section numbers in parentheses following each evaluation statement correspond to Tier 2 Evaluation procedures.

Lateral-Force-Resisting System

C (NC) N/A MOMENT-RESISTING CONNECTIONS: All moment connections shall be able to develop the strength of the adjoining members or panel zones. (Tier 2: Sec. 4.4.1.3.3)

Pre-Northridge connections (see Figure C4-14)

(C) NC N/A PANEL ZONES: All panel zones shall have the shear capacity to resist the shear demand required to develop 0.8 times the sum of the flexural strengths of the girders framing in at the face of the column. (Tier 2: Sec. 4.4.1.3.4)

Frame	Column	d_z (in.)	w_z (in.)	t_z (in.)	Beam	$0.8M_p$[1] (Beam)	Shear Demand[2]	Shear Capacity[3]	Comply?
1	W14x132	25.59	12.66	0.63	W27x84	7,027 k-in.	274 k	326 k	yes
	W14x193	28.31	12.96	0.88	W30x99	8,986 k-in.	317 k	502 k	yes
1A	W14x145	25.43	12.60	0.68	W27x94	8,006 k-in.	315 k	359 k	yes
	W14x233	28.49	12.60	1.07	W30x116	10,886 k-in.	382 k	659 k	yes
1B	W14x145	28.31	12.60	0.68	W30x99	8,986 k-in.	317 k	354 k	yes
	W14x257	31.38	12.60	1.18	W33x141	14,803 k-in.	472 k	727 k	yes

[1] $M_p = ZF_{ye}$; $F_{ye} = 54$ ksi.
[2] Shear Demand = $0.8M_p/d_z$.
[3] Shear Capacity = AISC Equation 9-1, $\phi = 1$ (AISC, 1997).

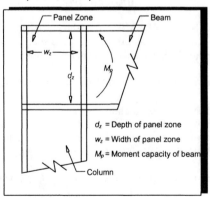

(C) NC N/A COLUMN SPLICES: All column splice details located in moment resisting frames shall include connection of both flanges and the web for Life Safety, and the splice shall develop the strength of the column for Immediate Occupancy. (Tier 2: Sec. 4.4.1.3.5)

Webs have partial penetration welds and flanges have bolted shear panels.

C (NC) N/A STRONG COLUMN/WEAK BEAM: The percentage of strong column/weak beam joints in each story of each line of moment resisting frames shall be greater than 50 percent for Life Safety and Immediate Occupancy. (Tier 2: Sec. 4.4.1.3.6)

Appendix A: Examples

Second-story frames do not have strong column/weak beam joints based on a comparison of column and beam moment capacities considering axial load effects according to the following equation:

$$\frac{\Sigma Z_c (F_{yc} - f_a)}{\Sigma Z_b F_{yb}} > 1.0 \qquad \text{(Equation 8-3, 1997 UBC, Chapter 22, Div. IV)}$$

$F_{yb} = F_{yc} = 36 \; ksi$

Frame	Column	Beam	Z_c	Z_b	Axial Stress (f_a)	Ratio	Comply
1	W14x132	W27x84	234 in.³	244 in.³	1.8 ksi	0.900	no
	W14x193	W30x99	355 in.³	312 in.³	2.1 ksi	1.100	yes
1A	W14x145	W27x94	260 in.³	278 in.³	1.7 ksi	0.900	no
	W14x233	W30x116	436 in.³	378 in.³	1.8 ksi	1.100	yes
1B	W14x145	W30x99	260 in.³	312 in.³	2.2 ksi	0.800	no
	W14x257	W33x141	487 in.³	514 in.³	2.1 ksi	0.900	no

(**C**) NC N/A **COMPACT MEMBERS:** All frame elements shall meet section requirements set forth by *Seismic Provisions for Structural Steel Buildings* Table I-9-1 (AISC, 1997). (Tier 2: Sec. 4.4.1.3.7)

Frame elements are compact sections.

C NC (**N/A**) **BEAM PENETRATIONS:** All openings in frame-beam webs shall be less than 1/4 of the beam depth and shall be located in the center half of the beams. This statement shall apply to the Immediate Occupancy Performance Level only. (Tier 2: Sec. 4.4.1.3.8)

Life Safety Performance Level.

C NC (**N/A**) **GIRDER FLANGE CONTINUITY PLATES:** There shall be girder flange continuity plates at all moment-resisting frame joints. This statement shall apply to the Immediate Occupancy Performance Level only. (Tier 2: Sec. 4.4.1.3.9)

Life Safety Performance Level.

C NC (**N/A**) **OUT-OF-PLANE BRACING:** Beam-column joints shall be braced out-of-plane. This statement shall apply to the Immediate Occupancy Performance Level only. (Tier 2: Sec. 4.4.1.3.10)

Life Safety Performance Level.

C NC (**N/A**) **BOTTOM FLANGE BRACING:** The bottom flanges of beams shall be braced out-of-plane. This statement shall apply to the Immediate Occupancy Performance Level only. (Tier 2: Sec. 4.4.1.3.11)

Life Safety Performance Level.

Diaphragms

(**C**) NC N/A **CROSS TIES:** There shall be continuous cross ties between diaphragm chords. (Tier 2: Sec. 4.5.1.2)

Continuous cross ties are present.

C NC (**N/A**) **PLAN IRREGULARITIES:** There shall be tensile capacity to develop the strength of the diaphragm at re-entrant corners or other locations of plan irregularities. This statement shall apply to the Immediate Occupancy Performance Level only. (Tier 2: Sec. 4.5.1.7)

Life Safety Performance Level.

Appendix A: Examples

C NC (N/A) DIAPHRAGM REINFORCEMENT AT OPENINGS: There shall be reinforcing around all diaphragm openings larger than 50 percent of the building width in either major plan dimension. This statement shall apply to the Immediate Occupancy Performance Level only. (Tier 2: Sec. 4.5.1.8)

Life Safety Performance Level.

C NC (N/A) STRAIGHT SHEATHING: All straight sheathed diaphragms shall have aspect ratios less than 2-to-1 for Life Safety and 1-to-1 for Immediate Occupancy in the direction being considered. (Tier 2: Sec. 4.5.2.1)

No straight board sheathing.

(C) NC N/A SPANS: All wood diaphragms with spans greater than 24 feet for Life Safety and 12 feet for Immediate Occupancy shall consist of wood structural panels or diagonal sheathing. (Tier 2: Sec. 4.5.2.2)

All wood diaphragms consist of wood structural panels.

C NC (N/A) UNBLOCKED DIAPHRAGMS: All diagonally sheathed or unblocked wood structural panel diaphragms shall have horizontal spans less than 40 feet for Life Safety and 30 feet for Immediate Occupancy and shall have aspect ratios less than or equal to 4-to-1 for Life Safety and 3-to-1 for Immediate Occupancy. (Tier 2: Sec. 4.5.2.3)

All diaphragms are blocked.

C NC (N/A) NON-CONCRETE-FILLED DIAPHRAGMS: Untopped metal deck diaphragms or metal deck diaphragms with fill other than concrete shall consist of horizontal spans of less than 40 feet and shall have span/depth ratios less than 4-to-1. This statement shall apply to the Immediate Occupancy Performance Level only. (Tier 2: Sec. 4.5.3.1)

Life Safety Performance Level.

(C) NC N/A OTHER DIAPHRAGMS: The diaphragm shall not consist of a system other than wood, metal deck, concrete, or horizontal bracing. (Tier 2: Sec. 4.5.7.1)

All diaphragms sheathed with plywood.

Connections

C NC (N/A) UPLIFT AT PILE CAPS: Pile caps shall have top reinforcement and piles shall be anchored to the pile caps for Life Safety, and the pile cap reinforcement and pile anchorage shall be able to develop the tensile capacity of the piles for Immediate Occupancy. (Tier 2: Sec. 4.6.3.10)

Pile caps not used.

Appendix A: Examples

A2.3.7 Geologic Site Hazards and Foundations Checklist

The following is a completed Geologic Site Hazards and Foundations Checklist for Example 2. Each of the evaluation statements has been marked Compliant (C), Non-Compliant (NC) or Not Applicable (N/A). Compliant statements identify issues that are acceptable according to the criteria of this standard, while non-compliant statements identify issues that require further investigation. In addition, an explanation of the evaluation process is included in italics after each evaluation statement. The section numbers in parentheses following each evaluation statement correspond to Tier 2 Evaluation procedures.

Geologic Site Hazards

The following statements shall be completed for buildings in levels of high or moderate seismicity.

(C) NC N/A LIQUEFACTION: Liquefaction susceptible, saturated, loose granular soils that could jeopardize the building's seismic performance shall not exist in the foundation soils at depths within 50 feet under the building for Life Safety and Immediate Occupancy. (Tier 2: Sec. 4.7.1.1)

Liquefaction susceptible soils do not exist in the foundation soils at depths within 50 feet of the building as described in project geotechnical investigation.

(C) NC N/A SLOPE FAILURE: The building site shall be sufficiently remote from potential earthquake-induced slope failures or rockfalls to be unaffected by such failures or shall be capable of accommodating small predicted movements without failure. (Tier 2: Sec. 4.7.1.2)

The building is on a relatively flat site.

(C) NC N/A SURFACE FAULT RUPTURE: Surface fault rupture and surface displacement at the building site is not anticipated. (Tier 2: Sec. 4.7.1.3)

Surface fault rupture and surface displacements are not anticipated.

Condition of Foundations

The following statements shall be completed for all Tier 1 building evaluations.

(C) NC N/A FOUNDATION PERFORMANCE: There shall not be evidence of excessive foundation movement such as settlement or heave that would affect the integrity or strength of the structure. (Tier 2: Sec. 4.7.2.1)

The structure does not show evidence of excessive foundation movement.

The following statement shall be completed for buildings in levels of high or moderate seismicity being evaluated to the Immediate Occupancy Performance Level.

C NC (N/A) DETERIORATION: There shall not be evidence that foundation elements have deteriorated due to corrosion, sulfate attack, material breakdown, or other reasons in a manner that would affect the integrity or strength of the structure. (Tier 2: Sec. 4.7.2.2)

Life Safety Performance Level.

Capacity of Foundations

The following statement shall be completed for all Tier 1 building evaluations.

C NC (N/A) POLE FOUNDATIONS: Pole foundations shall have a minimum embedment depth of 4 feet for Life Safety and Immediate Occupancy. (Tier 2: Sec. 4.7.3.1)

Pole foundations not used.

Appendix A: Examples

The following statements shall be completed for buildings in levels of moderate seismicity being evaluated to the Immediate Occupancy Performance Level and for buildings in levels of high seismicity.

(C) NC N/A OVERTURNING: The ratio of the horizontal dimension of the lateral-force-resisting system at the foundation level to the building height (base/height) shall be greater than $0.6S_a$. (Tier 2: Sec. 4.7.3.2)

Ratio of base to height = 27 ft / 41 ft = 0.66 > 0.6 S_a = 0.6(1.05) = 0.63.

C NC (N/A) TIES BETWEEN FOUNDATION ELEMENTS: The foundation shall have ties adequate to resist seismic forces where footings, piles, and piers are not restrained by beams, slabs, or soils classified as Class A, B, or C. (Section 3.5.2.3.1, Tier 2: Sec. 4.7.3.3)

Footings are restrained by grade beams.

C NC (N/A) DEEP FOUNDATIONS: Piles and piers shall be capable of transferring the lateral forces between the structure and the soil. This statement shall apply to the Immediate Occupancy Performance Level only. (Tier 2: Sec. 4.7.3.4)

Life Safety Performance Level.

C NC (N/A) SLOPING SITES: The difference in foundation embedment depth from one side of the building to another shall not exceed one story in height. This statement shall apply to the Immediate Occupancy Performance Level only. (Tier 2: Sec. 4.7.3.5)

Life Safety Performance Level.

A2.3.8 Basic Nonstructural Component Checklist

The following is a completed Basic Nonstructural Component Checklist for Example 2. Each of the evaluation statements has been marked Compliant (C), Non-Compliant (NC) or Not Applicable (N/A). Compliant statements identify issues that are acceptable according to the criteria of this standard, while non-compliant statements identify issues that require further investigation. In addition, an explanation of the evaluation process is included in italics after each evaluation statement. The section numbers in parentheses following each evaluation statement correspond to Tier 2 Evaluation procedures.

Partitions

C NC (N/A) UNREINFORCED MASONRY: Unreinforced masonry or hollow clay tile partitions shall be braced at a spacing equal to or less than 10 feet in levels of low or moderate seismicity and 6 feet in levels of high seismicity. (Tier 2: Sec. 4.8.1.1)

No interior hollow clay tile partitions.

Ceiling Systems

C NC (N/A) SUPPORT: The integrated suspended ceiling system shall not be used to laterally support the tops of gypsum board, masonry, or hollow clay tile partitions. Gypsum board partitions need not be evaluated where only the Basic Nonstructural Component Checklist is required by Table 3-2. (Tier 2: Sec. 4.8.2.1)

No gypsum board, masonry, or hollow clay tile partitions.

Light Fixtures

(C) NC N/A EMERGENCY LIGHTING: Emergency lighting shall be anchored or braced to prevent falling during an earthquake. (Tier 2: Sec. 4.8.3.1)

Appendix A: Examples

Emergency lighting is anchored at a spacing of less than 6 feet.

Cladding and Glazing

Masonry veneer (see statements below); statements in this section are Not Applicable (N/A).

Masonry Veneer

(C) NC N/A   **SHELF ANGLES:** Masonry veneer shall be supported by shelf angles or other elements at each floor 30 feet or more above ground for Life Safety and at each floor above the first floor for Immediate Occupancy. (Tier 2: Sec. 4.8.5.1)

Masonry veneer is supported by shelf angles.

(C) NC N/A   **TIES:** Masonry veneer shall be connected to the back-up with corrosion-resistant ties. The ties shall have a spacing equal to or less than 24 inches with a minimum of one tie for every 2? square feet. A spacing of up to 36 inches is permitted where only the Basic Nonstructural Component Checklist is required by Table 3-2. (Tier 2: Sec. 4.8.5.2)

Masonry veneer connected to the back-up with corrosion-resistant ties at a spacing of less than 24 inches.

(C) NC N/A   **WEAKENED PLANES:** Masonry veneer shall be anchored to the back-up adjacent to weakened planes, such as at the locations of flashing. (Tier 2: Sec. 4.8.5.3)

Masonry veneer is anchored to the back-up at locations of through-wall flashing.

(C) NC N/A   **DETERIORATION:** There shall be no evidence of deterioration, damage, or corrosion in any of the connection elements. (Tier 2: Sec. 4.8.5.4)

No deterioration observed.

Parapets, Cornices, Ornamentation, and Appendages

No parapets, cornices, ornamentation, or appendages; statements in this section are Not Applicable (N/A).

Masonry Chimneys

No masonry chimneys; statements in this section are Not Applicable (N/A).

Stairs

C NC (N/A)   **URM WALLS:** Walls around stair enclosures shall not consist of unbraced hollow clay tile or unreinforced masonry with a height-to-thickness ratio greater than 12-to-1. A height-to-thickness ratio of up to 15-to-1 is permitted where only the Basic Nonstructural Component Checklist is required by Table 3-2. (Tier 2: Sec. 4.8.10.1)

No hollow clay tile or URM walls.

(C) NC N/A   **STAIR DETAILS:** In moment frame structures, the connection between the stairs and the structure shall not rely on shallow anchors in concrete. Alternatively, the stair details shall be capable of accommodating the drift calculated using the Quick Check procedure of Section 3.5.3.1 without including tension in the anchors. (Tier 2: Sec. 4.8.10.2)

Stairs detailed to accommodate drift.

Appendix A: Examples

Building Contents and Furnishing

(C) NC N/A TALL NARROW CONTENTS: Contents over 4 feet in height with a height-to-depth or height-to-width ratio greater than 3-to-1 shall be anchored to the floor slab or adjacent structural walls. A height-to-depth or height-to-width ratio of up to 4-to-1 is permitted where only the Basic Nonstructural Component Checklist is required by Table 3-2. (Tier 2: Sec. 4.8.11.1)

Storage racks, bookcases, and file cabinets are anchored to the floor slab.

Mechanical and Electrical Equipment

(C) NC N/A EMERGENCY POWER: Equipment used as part of an emergency power system shall be mounted to maintain continued operation after an earthquake. (Tier 2: Sec. 4.8.12.1)

Emergency power systems are anchored.

(C) NC N/A HAZARDOUS MATERIAL EQUIPMENT: HVAC or other equipment containing hazardous material shall not have damaged supply lines or unbraced isolation supports. (Tier 2: Sec. 4.8.12.2)

HVAC equipment and supply lines are anchored.

(C) NC N/A DETERIORATION: There shall be no evidence of deterioration, damage, or corrosion in any of the anchorage or supports of mechanical or electrical equipment. (Tier 2: Sec. 4.8.12.3)

No deterioration observed.

(C) NC N/A ATTACHED EQUIPMENT: Equipment weighing over 20 pounds that is attached to ceilings, walls, or other supports 4 feet above the floor level shall be braced. (Tier 2: Sec. 4.8.12.4)

Equipment is braced.

Piping

(C) NC N/A FIRE SUPPRESSION PIPING: Fire suppression piping shall be anchored and braced in accordance with NFPA-13 (NFPA, 1996). (Tier 2: Sec. 4.8.13.1)

Fire suppression piping is braced in accordance with NFPA-13.

(C) NC N/A FLEXIBLE COUPLINGS: Fluid, gas and fire suppression piping shall have flexible couplings. (Tier 2: Sec. 4.8.13.2)

Fluid, gas, and fire suppression piping have flexible couplings.

Hazardous Materials Storage and Distribution

C NC **(N/A)** TOXIC SUBSTANCES: Toxic and hazardous substances stored in breakable containers shall be restrained from falling by latched doors, shelf lips, wires, or other methods. (Tier 2: Sec. 4.8.15.1)

No toxic or hazardous substances are stored in the building.

Appendix A: Examples

A2.3.9 Intermediate Nonstructural Component Checklist

The following is a completed Intermediate Nonstructural Checklist for Example 2. Each of the evaluation statements has been marked Compliant (C), Non-Compliant (NC) or Not Applicable (N/A). Compliant statements identify issues that are acceptable according to the criteria of this standard, while non-compliant statements identify issues that require further investigation. In addition, an explanation of the evaluation process is included in italics after each evaluation statement. The section numbers in parentheses following each evaluation statement correspond to Tier 2 Evaluation procedures.

Ceiling Systems

(C) NC N/A   **LAY-IN TILES:** Lay-in tiles used in ceiling panels located at exits and corridors shall be secured with clips. (Tier 2: Sec. 4.8.2.2)

Lay-in tiles are secured with clips.

(C) NC N/A   **INTEGRATED CEILINGS:** Integrated suspended ceilings at exits and corridors or weighing more than 2 lb/ft² shall be laterally restrained with a minimum of 4 diagonal wires or rigid members attached to the structure above at a spacing equal to or less than 12 feet. (Tier 2: Sec. 4.8.2.3)

Integrated suspended ceilings are braced.

(C) NC N/A   **SUSPENDED LATH AND PLASTER:** Ceilings consisting of suspended lath and plaster or gypsum board shall be attached to resist seismic forces for every 12 square feet of area. (Tier 2: Sec. 4.8.2.4)

Each 10 square feet of suspended lath and plaster ceilings is attached.

Light Fixtures

(C) NC N/A   **INDEPENDENT SUPPORT:** Light fixtures in suspended grid ceilings shall be supported independently of the ceiling suspension system by a minimum of two wires at diagonally opposite corners of the fixtures. (Tier 2: Sec. 4.8.3.2)

Light fixtures are independently supported.

Cladding and Glazing

Masonry veneer; statements in this section are Not Applicable (N/A).

Parapets, Cornices, Ornamentation, and Appendages

No parapets, cornices, ornamentation, or appendages; statements in this section are Not Applicable (N/A).

Masonry Chimneys

No masonry chimneys; statements in this section are Not Applicable (N/A).

Mechanical and Electrical Equipment

C NC (N/A)   **VIBRATION ISOLATORS:** Equipment mounted on vibration isolators shall be equipped with restraints or snubbers. (Tier 2: Sec. 4.8.12.5)

No equipment on vibration isolators.

Appendix A: Examples

Ducts

C   NC   (N/A)   **STAIR AND SMOKE DUCTS:** Stair pressurization and smoke control ducts shall be braced and shall have flexible connections at seismic joints. (Tier 2: Sec. 4.8.14.1)

No stair pressurization or smoke control ducts.

A2.4 Tier 2 Evaluation

The following potential deficiencies were identified by the Tier 1 Evaluation:

1. Redundancy
2. Drift
3. Strong Column/Weak Beam Connections
4. Moment-Resisting Connections
5. Panel Zones

A Deficiency-Only Tier 2 Evaluation for this example is completed by following the Tier 2 Procedures associated with the potential deficiencies listed above. Tier 2 Procedures are located in the following sections of Chapter 4: Section 4.4.1.1.1 (Redundancy), Section 4.4.1.3.1 (Drift Check), Section 4.4.1.3.6 (Strong Column/Weak Beam), Section 4.4.1.3.3 (Moment-Resisting Connections), and Section 4.4.1.3.4 (Panel Zones).

A2.4.1 Seismic Forces

The first step of the Tier 2 Procedure for the evaluation of each of the potential deficiencies listed above is to compute the frame shear forces in accordance with Section 4.2. Frame shear forces are computed as follows:

- Pseudo Lateral Force = 2,920 kips (Computed in Tier 1)

- Vertical distribution of forces:

    $T = 0.57$ sec (Computed in Tier 1)

Floor	h_x (ft)	w_x (kips)	$w_i h_i^k$ (k=1.035)	C_{vx} (Eqn. 4-3)	F_x (kips) (Eqn. 4-2)
Roof	41.25	662	31,104	0.36	1051.2
3rd	28.5	1,122	35,955	0.42	1226.4
2nd	15.0	1,135	18,718	0.22	642.4

- Frame shear forces:

    Compute frame shear forces based on tributary area.

Floor	Frame 1 – $0.175F_x$[1]	Frame 1A – $0.25F_x$[2]	Frame 1B – $0.325F_x$[3]
Roof	184.0 kips	262.8 kips	341.6 kips
3rd	214.6 kips	306.5 kips	398.6 kips
2nd	112.4 kips	160.6 kips	208.8 kips

[1] 0.175 = Area tributary to frame 1/total area.
         = 44 ft/251 ft.
[2] 0.250 = Load distributed evenly to four frames.
[3] 0.325 = (75 ft/2 + 44 ft)/251 ft.

Appendix A: Examples

A2.4.2 Tier 2 Evaluation of Redundancy and Drift (Tier 2 Procedure: Sections 4.4.1.1.1 and 4.4.1.3.1)

A drift ratio of 0.03 was computed for the second-story columns of Frame 1B; this drift ratio exceeds the allowable Tier 1 value of 0.0025. Thus, a Tier 2 Evaluation is conducted to further evaluate the drift in the building.

Computer analysis of each frame (Frame 1, Frame 1A, and Frame 1B) was performed using the frame shear forces computed in Section A2.4.1; P-Δ effects were included in the analyses. The adequacy of the beams and columns was checked using the m-factors in Table 4-5. The results for the Tier 2 Evaluation of drift are presented in the following tables.

- Check adequacy of columns (assuming negligible axial load):

Frame	Floor	Section	Moment[1]	Q_{UD}/m [2]	Q_{CE} [3]	Comply?[4]
1	Roof	W14x132	6,850 k-in.	856 k-in.	10,530 k-in.	yes
1	3rd	W14x132	14,624 k-in.	1,828 k-in.	10,530 k-in.	yes
1	2nd	W14x193	24,698 k-in.	3,087 k-in.	15,975 k-in.	yes
1A	Roof	W14x145	7,222 k-in.	903 k-in.	11,700 k-in.	yes
1A	3rd	W14x145	14,800 k-in.	1,850 k-in.	11,700 k-in.	yes
1A	2nd	W14x233	24,870 k-in.	3,109 k-in.	19,620 k-in.	yes
1B	Roof	W14x145	13,495 k-in.	1,687 k-in.	11,700 k-in.	yes
1B	3rd	W14x145	27,991 k-in.	3,499 k-in.	11,700 k-in.	yes
1B	2nd	W14x257	45,935 k-in.	5,742 k-in.	21,915 k-in.	yes

[1] Based on SAP2000 Analyses of the frames.
[2] m = 8 (Table 4-5; compact members); Q_{UD} = Moment due to earthquake loading.
[3] Expected Strength = $Q_{CE} = 1.25 M_p$; $M_p = Z F_y$. Alternatively, $Q_{CE} = M_{allowable} (1.25)(1.7)$.   Section 4.2.4.4
[4] $Q_{UD}/m < Q_{CE}$?   Equation (4-11)

- Check adequacy of beams:

Frame	Floor	Section	Moment[1]	Q_{UD}/m [2]	Q_{CE} [3]	Comply?
1	Roof	W21x50	6,880 k-in.	860 k-in.	4,950 k-in.	yes
1	3rd	W27x84	20,465 k-in.	2,558 k-in.	10,980 k-in.	yes
1	2nd	W30x99	26,825 k-in.	3,353 k-in.	14,040 k-in.	yes
1A	Roof	W24x55	7,222 k-in.	903 k-in.	6,030 k-in.	yes
1A	3rd	W27x94	19,983 k-in.	2,498 k-in.	12,510 k-in.	yes
1A	2nd	W30x116	26,862 k-in.	3,358 k-in.	17,010 k-in.	yes
1B	Roof	W24x68	13,495 k-in.	1,687 k-in.	7,965 k-in.	yes
1B	3rd	W30x99	36,468 k-in.	4,559 k-in.	14,040 k-in.	yes
1B	2nd	W33x141	52,582 k-in.	6,573 k-in.	23,130 k-in.	yes

[1] Based on SAP2000 Analyses of the frames.
[2] m = 8 (Table 4-5; compact members); Q_{UD} = Moment due to earthquake loading.
[3] Expected Strength = $Q_{CE} = 1.25 M_p$; $M_p = Z F_y$. Alternatively, $Q_{CE} = M_{allowable} (1.25)(1.7)$.   Section 4.2.4.4
[4] $Q_{UD}/m < Q_{CE}$?   Equation (4-11)

Appendix A: Examples

A2.4.3 Tier 2 Evaluation of Strong Column/Weak Beam (Tier 2 Procedure: Section 4.4.1.3.6)

Based on the Tier 1 Evaluation, the percentage of strong column/weak beam joints in each story of each line of moment-resisting frames was less than the required 50 percent. Thus, a Tier 2 Evaluation is conducted to further assess the adequacy of the joints.

An Computer analysis of each frame (Frame 1, Frame 1A, and Frame 1B) was performed using the frame shear forces computed above in accordance with Section 4.2. The adequacy of the columns was checked using an m-factor of 2.5. The results for the Tier 2 Evaluation of strong column/weak beam are presented in the following table.

- Check adequacy of columns:

Frame	Floor	Section	Moment[1]	Q_{UD}/m [2]	Q_{CE} [3]	Comply?[4]
1	Roof	W14x132	6,850 k-in.	2,740 k-in.	10,530 k-in.	yes
	3rd	W14x132	14,624 k-in.	5,850 k-in.	10,530 k-in.	yes
	2nd	W14x193	24,698 k-in.	9,879 k-in.	15,975 k-in.	yes
1A	Roof	W14x145	7,222 k-in.	2,889 k-in.	11,700 k-in.	yes
	3rd	W14x145	14,800 k-in.	5,920 k-in.	11,700 k-in.	yes
	2nd	W14x233	24,870 k-in.	9,948 k-in.	19,620 k-in.	yes
1B	Roof	W14x145	13,495 k-in.	5,398 k-in.	11,700 k-in.	yes
	3rd	W14x145	2,7991 k-in.	11,196 k-in.	11,700 k-in.	yes
	2nd	W14x257	45,935 k-in.	18,374 k-in.	21,915 k-in.	yes

[1] Based on SAP2000 Analyses of the frames.
[2] $m = 8$ (Table 4-5; compact members); Q_{UD} = Moment due to earthquake loading.
[3] Expected Strength = $Q_{CE} = 1.25 M_p$; $M_p = ZF_y$. Alternatively, $Q_{CE} = M_{allowable}(1.25)(1.7)$.   Section 4.2.4.4
[4] $Q_{UD}/m < Q_{CE}$?   Equation (4-11)

A2.4.4 Tier 2 Evaluation of Moment-Resisting Connections (Tier 2 Procedure: Section 4.4.1.3.3)

The moment-resisting connections in this building were designed prior to the Northridge Earthquake. Thus, a Tier 2 Evaluation is conducted to assess the adequacy of these joints.

An Computer analysis of each frame (Frame 1, Frame 1A, and Frame 1B) was performed using the frame shear forces computed above in accordance with Section 4.2. The adequacy of the connections was checked using an m-factor of 2. The results for the Tier 2 Evaluation of moment-resisting connections are presented in the following table.

Appendix A: Examples

- Check adequacy of connections:

Frame	Floor	Section	Moment[1]	Q_{UD}/m [2]	Q_{CE} [3]	Comply?
1	Roof	W21x50	6,880 k-in.	3,440 k-in.	4,950 k-in.	yes
	3rd	W27x84	20,465 k-in.	10,233 k-in.	10,980 k-in.	yes
	2nd	W30x99	26,825 k-in.	13,413 k-in.	14,040 k-in.	yes
1A	Roof	W24x55	7,222 k-in.	3,611 k-in.	6,030 k-in.	yes
	3rd	W27x94	19,983 k-in.	9,992 k-in.	12,510 k-in.	yes
	2nd	W30x116	26,862 k-in.	13,431 k-in.	17,010 k-in.	yes
1B	Roof	W24x68	13,495 k-in.	6,748 k-in.	7,965 k-in.	yes
	3rd	W30x99	36,468 k-in.	18,234 k-in.	14,040 k-in.	no
	2nd	W33x141	52,582 k-in.	26,291 k-in.	23,130 k-in.	no

[1] Based on SAP2000 Analyses of the frames.
[2] m = 8 (Table 4-5; compact members); Q_{UD} = Moment due to earthquake loading.
[3] Expected Strength = Q_{CE} = 1.25M_p; M_p = ZF_y. Alternatively, Q_{CE} = $M_{allowable}$ (1.25)(1.7).   Section 4.2.4.4
[4] $Q_{UD}/m < Q_{CE}$?   Equation (4-11)

A2.4.5 Tier 2 Evaluation of Panel Zones (Tier 2 Procedure: Section 4.4.1.3.4)

Based on the Tier 1 Evaluation, the panel zones were approximately 3 percent of complying with the requirements. While the panel zones may have been judged adequate, a Tier 2 Evaluation is conducted to verify that they are adequate.

An Computer analysis of each frame (Frame 1, Frame 1A, and Frame 1B) was performed using the frame shear forces computed above in accordance with Section 4.2. The adequacy of the panel zones to resist the shear demand was evaluated using an m-factor of 2 (Table 4-5). The results for the Tier 2 Evaluation of panel zones are presented in the following table.

- Check adequacy of panel zones:

Frame	Column	d_z (in.)	w_z (in.)	t_z (in.)	Beam	Beam Moment[1]	Shear Demand[2]	Shear Capacity[3]	Comply?
1	W14x132	25.59	12.66	0.63	W27x84	20,465 k-in.	80 k	287 k	yes
	W14x193	28.31	12.96	0.88	W30x99	26,825 k-in.	95 k	411 k	yes
1A	W14x145	25.43	12.60	0.68	W27x94	19,983 k-in.	79 k	308 k	yes
	W14x233	28.49	12.60	1.07	W30x116	26,862 k-in.	94 k	485 k	yes
1B	W14x145	28.31	12.60	0.68	W30x99	36,468 k-in.	129 k	308 k	yes
	W14x257	31.38	12.60	1.18	W33x141	52,582 k-in.	168 k	535 k	yes

[1] Based on SAP2000 Analyses.
[2] Shear Demand = Beam Moment/(md_z); m = 10.
[3] Shear Capacity = (Shear Area)(F_y) = $w_z t_z F_y$.
w_z = width of panel zone; d_z = depth of panel zone; t_z = thickness of panel zone.

A2.5 Final Evaluation and Report

Based on the Tier 2 Evaluation, the only potential deficiency in this building is the moment-resisting connections. A Tier 3 Evaluation may be conducted to further evaluate the adequacy of the connections or the connections may be rehabilitated. A final report shall be prepared outlining the results of the Tier 1 and Tier 2 Evaluations.

A3 Example 3 – Building Type C3: Concrete Frames with Infill Masonry Shear Walls and Stiff Diaphragms

A3.1 Introduction

This example is based on an actual concrete frame building with infill unreinforced masonry shear walls. Data for this example was taken from FEMA 276: *Example Applications of the NEHRP Guidelines for the Seismic Rehabilitation of Buildings*.

The purpose of this example is to demonstrate the use of the Tier 1 Evaluation and conclusions drawn from the evaluation. Based on the number and significance of the non-compliant statements, it was judged that a Tier 2 and/or Tier 3 Evaluation would not change the conclusion drawn from the Tier 1 Evaluation. Thus, it would be inefficient and uneconomical to continue, and the evaluation was terminated after Tier 1.

A3.2 Evaluation Requirements

A3.2.1 Building Description

The building is an historic three-story retail, office and apartment structure located in the Midwest United States. It was constructed in 1908; no seismic design code was used. As-built plans for the building are available.

The total floor area of the building is approximately 14,200 square feet. Concrete caissons are used for the foundation. The floor framing consists of reinforced concrete slabs and beams. The vertical lateral-force-resisting system consists of reinforced frames and infill, while the horizontal lateral-force-resisting system consists of a reinforced concrete slab. The reinforcing steel consists of smooth bars. The columns have closely spaced spiral transverse reinforcement.

The front of the building has curtain walls while the other three sides are infilled with unreinforced masonry. The parapet is constructed of reinforced concrete. The concrete columns have flared capitals. A floor plan and elevations of the building are shown in Figure A3-1 and Figure A3-2.

A3.2.2 Level of Investigation

As-built plans are available for this building (Section 2.2). A site visit was conducted to verify the plan information and to assess the condition of the building.

A3.2.3 Level of Performance

The building is evaluated to the Life Safety Performance Level (Section 2.4).

A3.2.4 Level of Seismicity

The building is located in a region of moderate seismicity (Section 2.5).

$S_S = 0.6$; $S_1 = 0.15$ (ASCE 7-02)
Site Class = C (Section 3.5.2.3.1)
$F_v = 1.65$; $F_a = 1.15$ (Tables 3-5 & 3-6)
$S_{D1} = (2/3)F_v S_1 = 0.17$ (Equation 3-5)
$S_{DS} = (2/3)F_a S_S = 0.46$ (Equation 3-6)

A3.2.5 Building Type

The building is classified as Building Type C3, Concrete Frames with Infill Masonry Shear Walls and Stiff Diaphragms (Section 2.6). A description of this type of building is included in Table 2-2.

A3.3 Screening Phase (Tier 1)

A3.3.1 Introduction

The purpose of a Tier 1 Evaluation is to quickly identify buildings that comply with the provisions of this standard. The data collected is sufficient to conduct a Tier 1 Evaluation.

A3.3.2 Benchmark Buildings

This building was not designed in accordance with one of the benchmark documents listed in Table 3-1 (Section 3.2).

A3.3.3 Selection of Checklists

This building is classified as Building Type C3 (Table 2-2). Thus, the Structural Checklist(s) associated with Building Type C3 are used.

A Tier 1 Evaluation of this building involves completing the following checklists (Section 3.3):

- Basic Structural Checklist for Building Type C3 (Section 3.7.10)
- Geologic Site Hazards and Foundations Checklist (Section 3.8)
- Basic Nonstructural Component Checklist (Section 3.9.1)

The Supplemental Structural and Intermediate Checklists and Supplemental Nonstructural Checklist do not need to be completed for this example since the building is in a level of moderate seismicity and is being evaluated to the Life Safety Performance Level.

Each of the checklists listed above has been completed for this example and is included in the following sections.

Appendix A: Examples

A3.3.4 Further Evaluation Requirements

While neither a Full-Building Tier 2 Evaluation nor a Tier 3 Evaluation is required for this building based on Section 3.4, the design professional may decide to further evaluate the building if deficiencies are identified by the Tier 1 Evaluation.

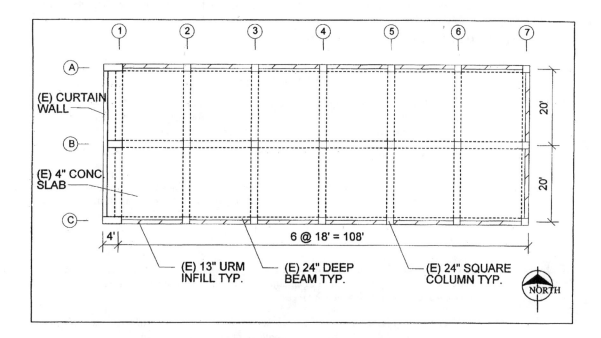

Figure A3-1. Example 3 – Typical Floor Plan

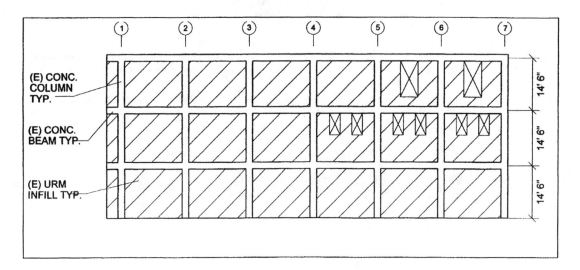

Figure A3-2. Example 3 – North and South Elevation

Appendix A: Examples

A3.3.5 Basic Structural Checklist for Building Type C3: Concrete Frames with Infill Masonry Shear Walls with Stiff Diaphragms

The following is a completed Basic Structural Checklist for Example 3. Each of the evaluation statements has been marked Compliant (C), Non-Compliant (NC) or Not Applicable (N/A). Compliant statements identify issues that are acceptable according to the criteria of this standard, while non-compliant statements identify issues that require further investigation. In addition, an explanation of the evaluation process is included in italics after each evaluation statement. The section numbers in parentheses following each evaluation statement correspond to Tier 2 Evaluation procedures.

Building System

(C) NC N/A LOAD PATH: The structure shall contain a minimum of one complete load path for Life Safety and Immediate Occupancy for seismic force effects from any horizontal direction that serves to transfer the inertial forces from the mass to the foundation. (Tier 2: Sec. 4.3.1.1)

The structure contains a complete load path.

C NC (N/A) MEZZANINES: Interior mezzanine levels shall be braced independently from the main structure, or shall be anchored to the lateral-force-resisting elements of the main structure. (Tier 2: Sec. 4.3.1.3)

No mezzanines.

(C) NC N/A WEAK STORY: The strength of the lateral-force-resisting system in any story shall not be less than 80 percent of the strength in an adjacent story above or below for Life Safety and Immediate Occupancy. (Tier 2: Sec. 4.3.2.1)

There are no significant strength discontinuities in any of the vertical elements of the lateral-force-resisting system; all of the stories are identical in height and member sizes.

(C) NC N/A SOFT STORY: The stiffness of the lateral-force-resisting system in any story shall not be less than 70 percent of the stiffness in an adjacent story above or below or less than 80 percent of the average stiffness of the three stories above or below for Life Safety and Immediate Occupancy. (Tier 2: Sec. 4.3.2.2)

There are no significant stiffness discontinuities in any of the vertical elements of the lateral-force-resisting system; all of the stories are identical in height and member sizes.

(C) NC N/A GEOMETRY: There shall be no changes in horizontal dimension of the lateral-force-resisting system of more than 30 percent in a story relative to adjacent stories for Life Safety and Immediate Occupancy, excluding one-story penthouses and mezzanines. (Tier 2: Sec. 4.3.2.3)

The lateral-force-resisting system has constant horizontal dimension.

(C) NC N/A VERTICAL DISCONTINUITIES: All vertical elements in the lateral-force-resisting system shall be continuous to the foundation. (Tier 2: Sec. 4.3.2.4)

All infilled walls and frames are continuous to the foundation.

Appendix A: Examples

(C) NC N/A   MASS: There shall be no change in effective mass more than 50 percent from one story to the next for Life Safety and Immediate Occupancy. Light roofs, penthouses, and mezzanines need not be considered. (Tier 2: Sec. 4.3.2.5)

The change in effective mass between any story is equal to or less than 31 percent, as shown in the table below.

Story	Weight	Mass	Percent Change
Roof	688 kips	1.78 k-sec^2/in.	31
3rd	996 kips	2.58 k-sec^2/in.	0
2nd	996 kips	2.58 k-sec^2/in.	

C **(NC)** N/A   TORSION: The estimated distance between the story center of mass and the story center of rigidity shall be less than 20 percent of the building width in either plan dimension for Life Safety and Immediate Occupancy. (Tier 2: Sec. 4.3.2.6)

The maximum distance between the story center of rigidity and the center of mass is approximately 45 feet, which is 42 percent of the plan dimension of the building in the East-West direction.

(C) NC N/A   DETERIORATION OF CONCRETE: There shall be no visible deterioration of concrete or reinforcing steel in any of the vertical- or lateral-force-resisting elements. (Tier 2: Sec. 4.3.3.4)

There is no visible deterioration of concrete or reinforcing steel in any of the vertical- or lateral-force-resisting elements.

(C) NC N/A   MASONRY UNITS: There shall be no visible deterioration of masonry units. (Tier 2: Sec. 4.3.3.7)

There is no visible deterioration of large areas of masonry units.

(C) NC N/A   MASONRY JOINTS: The mortar shall not be easily scraped away from the joints by hand with a metal tool, and there shall be no areas of eroded mortar. (Tier 2: Sec.4.3.3.8)

The mortar is not easily scraped away from the joints by hand or with a metal tool, and there are no significant areas of eroded mortar.

C **(NC)** N/A   CRACKS IN INFILL WALLS: There shall be no existing diagonal cracks in the infilled walls that extend throughout a panel greater than 1/8 inch for Life Safety and 1/16 inch for Immediate Occupancy, or out-of-plane offsets in the bed joint greater than 1/8 inch for Life Safety and 1/16 inch for Immediate Occupancy. (Tier 2: Sec. 4.3.3.12)

There are approximately 80 linear feet of cracks in the infilled unreinforced masonry walls that extend through a panel or are greater than 1/8 inch wide.

(C) NC N/A   CRACKS IN BOUNDARY COLUMNS: There shall be no existing diagonal cracks wider than 1/8 inch for Life Safety and 1/16 inch for Immediate Occupancy in concrete columns that encase masonry infills. (Tier 2: Sec. 4.3.3.13)

There are no diagonal cracks in the concrete columns wider than 1/8 inch.

Lateral-Force-Resisting System

C **(NC)** N/A   REDUNDANCY: The number of lines of shear walls in each principal direction shall be greater than or equal to 2 for Life Safety and Immediate Occupancy. (Tier 2: Sec. 4.4.2.1.1)

There is only one line of shear walls in the North-South direction.

Appendix A: Examples

C   NC   (N/A)   **SHEAR STRESS CHECK:** The shear stress in the reinforced masonry shear walls, calculated using the Quick Check procedure of Section 3.5.3.3, shall be less than 70 psi for Life Safety and Immediate Occupancy. (Tier 2: Sec. 4.4.2.4.1)

No reinforced masonry shear walls.

C   (NC)   N/A   **SHEAR STRESS CHECK:** The shear stress in the unreinforced masonry shear walls, calculated using the Quick Check procedure of Section 3.5.3.3, shall be less than 30 psi for clay units and 70 psi for concrete units for Life Safety and Immediate Occupancy. (Tier 2: Sec. 4.4.2.5.1)

The shear stress in the unreinforced masonry walls exceeds the allowable value of 30 psi. The shear stress is computed in accordance with the Quick Check procedure of Section 3.5.3.3 as shown below.

Pseudo Lateral Force:

$V = CS_aW$		Equation (3-1)
$C = 1.1$		Table 3-4
$S_{D1} = 0.17$; $S_{DS} = 0.46$		Section A3.2.4
$T = C_t h_n^{3/4} = 0.02(43.5)^{3/4} = 0.34$ sec		Equation (3-8)
$S_a = 0.17/0.34 = 0.50 > S_{DS} = 0.46$		Equation (3-4)
$V = (1.1)(0.46)W = 0.51W$		
$V = 0.51(2,680$ kips$) = 1,356$ kips		

Story	Story Shear	East-West Direction		North-South Direction	
		A_w^1	$V_j^{avg\ 2}$	A_w^1	$V_j^{avg\ 2}$
3rd	554 kips	130 ft²	20 psi	30 ft²	86 psi
2nd	1,089 kips	117 ft²	43 psi	30 ft²	168 psi
1st	1,356 kips	145 ft²	43 psi	30 ft²	209 psi

$^1 A_w = 0.7$ (length of wall) (13 in.); 0.7 used to convert from gross area to net area; width of wall = 13 in.
^{2} Equation (3-12); $m = 1.5$.

C   (NC)   N/A   **WALL CONNECTIONS:** Masonry shall be in full contact with frame for Life Safety and Immediate Occupancy. (Tier 2: Sec. 4.4.2.6.1)

There are gaps between infill walls and the frame.

Connections

(C)   NC   N/A   **TRANSFER TO SHEAR WALLS:** Diaphragms shall be connected for transfer of loads to the shear walls for Life Safety, and the connections shall be able to develop the lesser of the shear strength of the walls or diaphragms for Immediate Occupancy. (Tier 2: Sec. 4.6.2.1)

Diaphragms are connected for transfer of loads to the shear walls.

(C)   NC   N/A   **CONCRETE COLUMNS:** All concrete columns shall be doweled into the foundation for Life Safety, and the dowels shall be able to develop the tensile capacity of reinforcement in columns of lateral-force-resisting system for Immediate Occupancy. (Tier 2: Sec. 4.6.3.2)

Longitudinal steel is doweled into the foundation.

Appendix A: Examples

A3.3.6 Geologic Site Hazards and Foundations Checklist

The following is a completed Geologic Site Hazards and Foundations Checklist for Example 3. Each of the evaluation statements has been marked Compliant (C), Non-Compliant (NC) or Not Applicable (N/A). Compliant statements identify issues that are acceptable according to the criteria of this standard, while non-compliant statements identify issues that require further investigation. In addition, an explanation of the evaluation process is included in italics after each evaluation statement. The section numbers in parentheses following each evaluation statement correspond to Tier 2 Evaluation procedures.

Geologic Site Hazards

The following statements shall be completed for buildings in levels of high or moderate seismicity.

(C) NC N/A LIQUEFACTION: Liquefaction susceptible, saturated, loose granular soils that could jeopardize the building's seismic performance shall not exist in the foundation soils at depths within 50 feet under the building for Life Safety and Immediate Occupancy. (Tier 2: Sec. 4.7.1.1)

Liquefaction susceptible soils do not exist in the foundation soils at depths within 50 feet of the building as described in project geotechnical investigation.

(C) NC N/A SLOPE FAILURE: The building site shall be sufficiently remote from potential earthquake-induced slope failures or rockfalls to be unaffected by such failures or shall be capable of accommodating small predicted movements without failure. (Tier 2: Sec. 4.7.1.2)

The building is on a relatively flat site.

(C) NC N/A SURFACE FAULT RUPTURE: Surface fault rupture and surface displacement at the building site is not anticipated. (Tier 2: Sec. 4.7.1.3)

Surface fault rupture and surface displacements are not anticipated.

Condition of Foundations

The following statements shall be completed for all Tier 1 building evaluations.

(C) NC N/A FOUNDATION PERFORMANCE: There shall not be evidence of excessive foundation movement such as settlement or heave that would affect the integrity or strength of the structure. (Tier 2: Sec. 4.7.2.1)

The structure does not show evidence of excessive foundation movement.

The following statement shall be completed for buildings in levels of high or moderate seismicity being evaluated to the Immediate Occupancy Performance Level.

C NC (N/A) DETERIORATION: There shall not be evidence that foundation elements have deteriorated due to corrosion, sulfate attack, material breakdown, or other reasons in a manner that would affect the integrity or strength of the structure. (Tier 2: Sec. 4.7.2.2)

Life Safety Performance Level.

Appendix A: Examples

Capacity of Foundations

The following statement shall be completed for all Tier 1 building evaluations.

C   NC   (N/A)   POLE FOUNDATIONS: Pole foundations shall have a minimum embedment depth of 4 feet for Life Safety and Immediate Occupancy. (Tier 2: Sec. 4.7.3.1)

Pole foundations not used.

The remaining statements on the Geologic Site Hazards and Foundations Checklist do not need to be completed for this example since the building is in a level of moderate seismicity and is being evaluated to the Life Safety Performance Level.

A3.3.7 Basic Nonstructural Component Checklist

The following is a completed Basic Nonstructural Component Checklist for Example 3. Each of the evaluation statements has been marked Compliant (C), Non-Compliant (NC) or Not Applicable (N/A). Compliant statements identify issues that are acceptable according to the criteria of this standard, while non-compliant statements identify issues that require further investigation. In addition, an explanation of the evaluation process is included in italics after each evaluation statement. The section numbers in parentheses following each evaluation statement correspond to Tier 2 Evaluation procedures.

Partitions

C   (NC)   N/A   UNREINFORCED MASONRY: Unreinforced masonry or hollow clay tile partitions shall be braced at a spacing equal to or less than 10 feet in levels of low or moderate seismicity and 6 feet in levels of high seismicity. (Tier 2: Sec. 4.8.1.1)

Unreinforced masonry partitions are not braced.

Ceiling Systems

(C)   NC   N/A   SUPPORT: The integrated suspended ceiling system shall not be used to laterally support the tops of gypsum board, masonry, or hollow clay tile partitions. Gypsum board partitions need not be evaluated where only the Basic Nonstructural Component Checklist is required by Table 3-2. (Tier 2: Sec. 4.8.2.1)

Ceiling system does not laterally brace unreinforced masonry partitions.

Light Fixtures

(C)   NC   N/A   EMERGENCY LIGHTING: Emergency lighting shall be anchored or braced to prevent falling during an earthquake. (Tier 2: Sec. 4.8.3.1)

Emergency lighting is anchored at a spacing of less than 6 feet.

Cladding and Glazing

No exterior cladding; statements in this section are Not Applicable (N/A).

Masonry Veneer

No masonry veneer; statements in this section are Not Applicable (N/A).

Appendix A: Examples

Parapets, Cornices, Ornamentation, and Appendages

No parapets, cornices, ornamentation, or appendages; statements in this section are Not Applicable (N/A).

Masonry Chimneys

No masonry chimneys; statements in this section are Not Applicable (N/A).

Stairs

C   NC   (N/A)   URM WALLS: Walls around stair enclosures shall not consist of unbraced hollow clay tile or unreinforced masonry with a height-to-thickness ratio greater than 12-to-1. A height-to-thickness ratio of up to 15-to-1 is permitted where only the Basic Nonstructural Component Checklist is required by Table 3-2. (Tier 2: Sec. 4.8.10.1)

No hollow clay tile or URM walls around stairs.

C   NC   (N/A)   STAIR DETAILS: In moment frame structures, the connection between the stairs and the structure shall not rely on shallow anchors in concrete. Alternatively, the stair details shall be capable of accommodating the drift calculated using the Quick Check procedure of Section 3.5.3.1 without including tension in the anchors. (Tier 2: Sec. 4.8.10.2)

Not a moment frame structure.

Building Contents and Furnishing

(C)   NC   N/A   TALL NARROW CONTENTS: Contents over 4 feet in height with a height-to-depth or height-to-width ratio greater than 3-to-1 shall be anchored to the floor slab or adjacent structural walls. A height-to-depth or height-to-width ratio of up to 4-to-1 is permitted where only the Basic Nonstructural Component Checklist is required by Table 3-2. (Tier 2: Sec. 4.8.11.1)

Storage racks, bookcases, and file cabinets are anchored to the floor slab.

Mechanical and Electrical Equipment

(C)   NC   N/A   EMERGENCY POWER: Equipment used as part of an emergency power system shall be mounted to maintain continued operation after an earthquake. (Tier 2: Sec. 4.8.12.1)

Emergency power systems are anchored.

(C)   NC   N/A   HAZARDOUS MATERIAL EQUIPMENT: HVAC or other equipment containing hazardous material shall not have damaged supply lines or unbraced isolation supports. (Tier 2: Sec. 4.8.12.2)

HVAC equipment and supply lines are anchored.

(C)   NC   N/A   DETERIORATION: There shall be no evidence of deterioration, damage, or corrosion in any of the anchorage or supports of mechanical or electrical equipment. (Tier 2: Sec. 4.8.12.3)

No deterioration observed.

(C)   NC   N/A   ATTACHED EQUIPMENT: Equipment weighing over 20 pounds that is attached to ceilings, walls, or other supports 4 feet above the floor level shall be braced. (Tier 2: Sec. 4.8.12.4)

Equipment is braced.

Appendix A: Examples

Piping

(C) NC N/A   FIRE SUPPRESSION PIPING: Fire suppression piping shall be anchored and braced in accordance with NFPA-13 (NFPA, 1996). (Tier 2: Sec. 4.8.13.1)

Fire suppression piping is braced in accordance with NFPA-13.

(C) NC N/A   FLEXIBLE COUPLINGS: Fluid, gas and fire suppression piping shall have flexible couplings. (Tier 2: Sec. 4.8.13.2)

Fluid, gas, and fire suppression piping have flexible couplings.

Hazardous Materials Storage and Distribution

(C) NC N/A   TOXIC SUBSTANCES: Toxic and hazardous substances stored in breakable containers shall be restrained from falling by latched doors, shelf lips, wires, or other methods. (Tier 2: Sec. 4.8.15.1)

No toxic or hazardous substances are stored in the building.

A3.4 Further Evaluation Phase (Tier 2)

The potential deficiencies identified by the Tier 1 Evaluation presented in Section A3.3 would likely also be identified as potential deficiencies by a Tier 2 Evaluation. Further evaluation, therefore, is judged to be unnecessary. The final report shall recommend that the potential deficiencies be mitigated.

A3.5 Final Evaluation and Report

The following is a list of the potential deficiencies identified by a Tier 1 Evaluation:

- Torsion
- Cracks in Infill Walls
- Redundancy
- Masonry Wall Shear Stress
- Infill Wall Connections
- Unreinforced Masonry Partitions

The deficiencies listed above should be mitigated. Possible resources for information regarding mitigation measures are listed in Section 1.3. A final report that outlines the results of the Tier 1 Evaluation shall be prepared.

Appendix A: Examples

A4 Example 4 – Building Type RM2: Reinforced Masonry Bearing Walls with Stiff Diaphragms

A4.1 Introduction

This example is based on an actual reinforced masonry bearing wall building with stiff diaphragms.

The purpose of this example is to demonstrate that a building can meet the Immediate Occupancy criteria set forth in this standard through a Tier 1 Evaluation only. It should be noted that this example is an ideal case; in most evaluations, there will be statements that are non-compliant, requiring that the deficiencies be reported or that a Deficiency-Only Tier 2 Evaluation be conducted.

A4.2 Evaluation Requirements

A4.2.1 Building Description

The building is a one-story concrete block masonry building located in Los Angeles, California. It was designed in accordance with the 1988 *Uniform Building Code* as a retail facility. It is currently being altered, however, to be used as a police station.

The concrete block masonry walls are 8 inches thick and fully grouted. The height of the walls is approximately 17 feet. The walls do not have any openings. There is a storefront and canopy at the front of the building.

A floor plan of the building is shown in Figure A4-1.

A4.2.2 Level of Investigation

As-built plans are available for this building (Section 2.2). A site visit was conducted to verify the plan information and to assess the condition of the building. In addition, testing was performed to quantify the strength of the masonry and location of reinforcement.

A4.2.3 Level of Performance

The building is evaluated to the Immediate Occupancy Performance Level since it is going to be used as a police station (Section 2.4).

A4.2.4 Level of Seismicity

The building is located in a level of high seismicity (Section 2.5).

$S_S = 1.5$; $S_1 = 0.6$ (ASCE 7-02)
Site Class = C (Section 3.5.2.3.1)
$F_v = 1.3$; $F_a = 1.0$ (Tables 3-5 & 3-6)
$S_{D1} = (2/3)F_v S_1 = 0.52$ (Equation 3-5)
$S_{DS} = (2/3)F_a S_S = 1.0$ (Equation 3-6)

A4.2.5 Building Type

The building is classified as Building Type RM2, Reinforced Masonry Bearing Walls with Stiff Diaphragms (Section 2.6). A description of this type of building is included in Table 2-2.

A4.3 Screening Phase (Tier 1)

A4.3.1 Introduction

The purpose of a Tier 1 Evaluation is to quickly identify buildings that comply with the provisions of this standard. The data collected is sufficient to conduct a Tier 1 Evaluation.

A4.3.2 Benchmark Buildings

The building is being evaluated to the Immediate Occupancy Performance Level; thus, benchmark documents do not apply (Section 3.2).

A4.3.3 Selection of Checklists

This building is classified as Building Type RM2 (Table 2-2). Thus, the Structural Checklist(s) associated with Building Type RM2 are used.

A Tier 1 Evaluation of this building involves completing the following checklists (Section 3.3):

- Basic Structural Checklist for Building Type RM2 (Section 3.7.14)
- Supplemental Structural Checklist for Building Type RM2 (Section 3.7.14S)
- Geologic Site Hazards and Foundations Checklist (Section 3.8)
- Basic Nonstructural Component Checklist (Section 3.9.1)
- Intermediate Nonstructural Component Checklist (Section 3.9.2
- Supplemental Nonstructural Checklist (Section 3.9.3)

Each of the checklists listed above has been completed for this example and is included in the following sections.

Appendix A: Examples

A4.3.4 Further Evaluation Requirements

While neither a Full-Building Tier 2 Evaluation nor a Tier 3 Evaluation is required for this building based on Section 3.4, the design professional may decide to further evaluate the building if deficiencies are identified by the Tier 1 Evaluation.

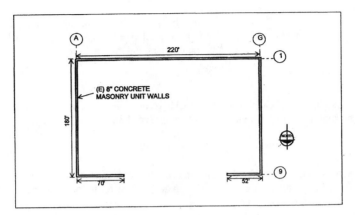

Figure A4-1. Example 4 – Floor Plan

A4.3.5 Basic Structural Checklist for Building Type RM2: Reinforced Masonry Bearing Walls with Stiff Diaphragms

The following is a completed Basic Structural Checklist for Example 4. Each of the evaluation statements has been marked Compliant (C), Non-Compliant (NC) or Not Applicable (N/A). Compliant statements identify issues that are acceptable according to the criteria of this standard, while non-compliant statements identify issues that require further investigation. In addition, an explanation of the evaluation process is included in italics after each evaluation statement. The section numbers in parentheses following each evaluation statement correspond to Tier 2 Evaluation procedures.

Building System

(C) NC N/A LOAD PATH: The structure shall contain a minimum of one complete load path for Life Safety and Immediate Occupancy for seismic force effects from any horizontal direction that serves to transfer the inertial forces from the mass to the foundation. (Tier 2: Sec. 4.3.1.1)

The building contains a complete load path.

C NC (N/A) MEZZANINES: Interior mezzanine levels shall be braced independently from the main structure, or shall be anchored to the lateral-force-resisting elements of the main structure. (Tier 2: Sec. 4.3.1.3)

No mezzanine.

C NC (N/A) WEAK STORY: The strength of the lateral-force-resisting system in any story shall not be less than 80 percent of the strength in an adjacent story above or below for Life Safety and Immediate Occupancy. (Tier 2: Sec. 4.3.2.1)

One-story building.

Appendix A: Examples

C   NC   (N/A)   SOFT STORY: The stiffness of the lateral-force-resisting-system in any story shall not be less than 70 percent of the lateral-force-resisting system stiffness in an adjacent story above or below, or less than 80 percent of the average lateral-force-resisting system stiffness of the three stories above or below for Life Safety and Immediate Occupancy. (Tier 2: Sec. 4.3.2.2)

One-story building.

(C)   NC   N/A   GEOMETRY: There shall be no changes in horizontal dimension of the lateral-force-resisting system of more than 30 percent in a story relative to adjacent stories for Life Safety and Immediate Occupancy, excluding one-story penthouses and mezzanines. (Tier 2: Sec. 4.3.2.3)

One-story building.

(C)   NC   N/A   VERTICAL DISCONTINUITIES: All vertical elements in the lateral-force-resisting system shall be continuous to the foundation. (Tier 2: Sec. 4.3.2.4)

All shear walls are continuous to the foundation.

C   NC   (N/A)   MASS: There shall be no change in effective mass more than 50 percent from one story to the next for Life Safety and Immediate Occupancy. Light roofs, penthouses, and mezzanines need not be considered. (Tier 2: Sec. 4.3.2.5)

One-story building.

(C)   NC   N/A   TORSION: The estimated distance between the story center of mass and the story center of rigidity shall be less than 20 percent of the building width in either plan dimension for Life Safety and Immediate Occupancy. (Tier 2: Sec. 4.3.2.6)

The distance between the story center of mass and the story center of rigidity is less than 20 percent.

(C)   NC   N/A   DETERIORATION OF CONCRETE: There shall be no visible deterioration of concrete or reinforcing steel in any of the vertical- or lateral-force-resisting elements. (Tier 2: Sec. 4.3.3.4)

No deterioration noted.

(C)   NC   N/A   MASONRY UNITS: There shall be no visible deterioration of masonry units. (Tier 2: Sec. 4.3.3.7)

No deterioration of masonry units.

(C)   NC   N/A   MASONRY JOINTS: The mortar shall not be easily scraped away from the joints by hand with a metal tool, and there shall be no areas of eroded mortar. (Tier 2: Sec.4.3.3.8)

Mortar cannot be easily scraped away from the joints.

(C)   NC   N/A   REINFORCED MASONRY WALL CRACKS: All existing diagonal cracks in wall elements shall be less than 1/8 inch for Life Safety and 1/16 inch for Immediate Occupancy, shall not be concentrated in one location, and shall not form an X pattern. (Tier 2: Sec. 4.3.3.10)

No visible cracks in the concrete block walls.

Lateral-Force-Resisting System

(C)   NC   N/A   REDUNDANCY: The number of lines of shear walls in each principal direction shall be greater than or equal to 2 for Life Safety and Immediate Occupancy. (Tier 2: Sec. 4.4.2.1.1)

Two lines of shear walls in each direction.

Appendix A: Examples

(C) NC N/A   **SHEAR STRESS CHECK:** The shear stress in the reinforced masonry shear walls, calculated using the Quick Check procedure of Section 3.5.3.3, shall be less than 70 psi for Life Safety and Immediate Occupancy. (Tier 2: Sec. 4.4.2.4.1)

Pseudo Lateral Force:

$V = CS_aW$	Equation (3-1)
$C = 1.4$	Table 3-4
$S_{DI} = 0.52; S_{DS} = 1.0$	Section A4.2.4
$T = C_t h_n^{3/4} = 0.02(17)^{3/4} = 0.17$ sec	Equation (3-8)
$S_a = 0.52/0.17 = 3.11 > S_{DS} = 1.0$	Equation (3-4)
$V = (1.4)(1.0)W = 1.4\ W$	
$W = 1{,}270$ kips	
$V = 1.4\ (1{,}270$ kips$) = 1{,}778$ kips	

Gridline	Shear[1]	A_w[2]	V_1^{avg} [3]	Comply?
1.0	1,160 kips	20,130 in.²	29 psi	yes
9.0	618 kips	11,163 in.²	28 psi	yes
A	889 kips	16,470 in.²	27 psi	yes
G	889 kips	16,470 in.²	27 psi	yes

[1] Based on stiffness.
[2] A_w = (length of wall)(7.625 in.); 7.625 in. = wall thickness.
[3] Equation (3-12); $m = 2$ for Immediate Occupancy (Table 3-7).

(C) NC N/A   **REINFORCING STEEL:** The total vertical and horizontal reinforcing steel ratio in reinforced masonry walls shall be greater than 0.002 for Life Safety and Immediate Occupancy of the wall with the minimum of 0.0007 for Life Safety and Immediate Occupancy in either of the two directions; the spacing of reinforcing steel shall be less than 48 inches for Life Safety and Immediate Occupancy; and all vertical bars shall extend to the top of the walls. (Tier 2: Sec. 4.4.2.4.2)

Vertical reinforcement: #6 @ 16 inches on center; reinforcing ratio = 0.004 > 0.0007;
Horizontal reinforcement: #5 @ 24 inches on center; reinforcing ratio = 0.002 > 0.0007;
Total reinforcing ratio = 0.006 > 0.002.

(C) NC N/A   **TOPPING SLAB:** Precast concrete diaphragms shall be interconnected by a reinforced concrete topping slab. (Tier 2: Sec. 4.5.5.1)

Precast concrete diaphragm is interconnected by a reinforced concrete topping slab.

Connections

(C) NC N/A   **WALL ANCHORAGE:** Exterior concrete or masonry walls, which are dependent on the diaphragm for lateral support, shall be anchored for out-of-plane forces at each diaphragm level with steel anchors, reinforcing dowels, or straps that are developed into the diaphragm. Connections shall have adequate strength to resist the connection force calculated in the Quick Check procedure of Section 3.5.3.7. (Tier 2: Sec. 4.6.1.1)

Walls are anchored at a spacing of 4 feet with steel straps.

(C) NC N/A   **TRANSFER TO SHEAR WALLS:** Diaphragms shall be connected for transfer of loads to the shear walls for Life Safety, and the connections shall be able to develop the lesser of the shear strength of the walls or diaphragms for Immediate Occupancy. (Tier 2: Sec. 4.6.2.1)

Diaphragms are connected for transfer of loads to the shear walls and can develop strength of diaphragms.

Appendix A: Examples

(C) NC N/A   TOPPING SLAB TO WALLS OR FRAMES: Reinforced concrete topping slabs that interconnect the precast concrete diaphragm elements shall be doweled for transfer of forces into the shear wall or frame elements for Life Safety, and the dowels shall be able to develop the lesser of the shear strength of the walls, frames, or slabs for Immediate Occupancy. (Tier 2: Sec. 4.6.2.3)

Topping slabs are doweled into the walls and able to develop the shear strength of the walls.

(C) NC N/A   FOUNDATION DOWELS: Wall reinforcement shall be doweled into the foundation for Life Safety, and the dowels shall be able to develop the lesser of the strength of the walls or the uplift capacity of the foundation for Immediate Occupancy. (Tier 2: Sec. 4.6.3.5)

Vertical wall reinforcing is doweled into the foundation.

(C) NC N/A   GIRDER/COLUMN CONNECTION: There shall be a positive connection utilizing plates, connection hardware, or straps between the girder and the column support. (Tier 2: Sec. 4.6.4.1)

There is a positive connection between the girder and column support utilizing steel plates.

A4.3.6 Supplemental Structural Checklist for Building Type RM2: Reinforced Masonry Bearing Walls with Stiff Diaphragms

The following is a completed Supplemental Structural Checklist for Example 4. Each of the evaluation statements has been marked Compliant (C), Non-Compliant (NC) or Not Applicable (N/A). Compliant statements identify issues that are acceptable according to the criteria of this standard, while non-compliant statements identify issues that require further investigation. In addition, an explanation of the evaluation process is included in italics after each evaluation statement. The section numbers in parentheses following each evaluation statement correspond to Tier 2 Evaluation procedures.

Lateral-Force-Resisting System

(C) NC N/A   REINFORCING AT OPENINGS: All wall openings that interrupt rebar shall have trim reinforcing on all sides. This statement shall apply to the Immediate Occupancy Performance Level only. (Tier 2: Sec. 4.4.2.4.3)

Trim reinforcing present at all openings.

(C) NC N/A   PROPORTIONS: The height-to-thickness ratio of the shear walls at each story shall be less than 30. This statement shall apply to the Immediate Occupancy Performance Level only. (Tier 2: Sec. 4.4.2.4.4)

Height-to-thickness ratio = (17 ft)(12 in./ft)/(7.625 in.) = 26.8 < 30

Diaphragms

(C) NC N/A   OPENINGS AT SHEAR WALLS: Diaphragm openings immediately adjacent to the shear walls shall be less than 25 percent of the wall length for Life Safety and 15 percent of the wall length for Immediate Occupancy. (Tier 2: Sec. 4.5.1.4)

Diaphragm openings adjacent to shear walls are less than 15 percent of the wall length.

(C) NC N/A   OPENINGS AT EXTERIOR MASONRY SHEAR WALLS: Diaphragm openings immediately adjacent to exterior masonry shear walls shall not be greater than 8 feet long for Life Safety and 4 feet long for Immediate Occupancy. (Tier 2: Sec. 4.5.1.6)

Diaphragm openings adjacent to shear walls are less than 4 feet long.

Appendix A: Examples

C NC (N/A)   **PLAN IRREGULARITIES:** There shall be tensile capacity to develop the strength of the diaphragm at re-entrant corners or other locations of plan irregularities. This statement shall apply to the Immediate Occupancy Performance Level only. (Tier 2: Sec. 4.5.1.7)

No re-entrant corners.

(C) NC N/A   **DIAPHRAGM REINFORCEMENT AT OPENINGS:** There shall be reinforcing around all diaphragms openings larger than 50 percent of the building width in either major plan dimension. This statement shall apply to the Immediate Occupancy Performance Level only. (Tier 2: Sec. 4.5.1.8)

No openings larger than 50 percent of the building width.

A4.3.7 Geologic Site Hazards and Foundations Checklist

The following is a completed Geologic Site Hazards and Foundations Checklist for Example 4. Each of the evaluation statements has been marked Compliant (C), Non-Compliant (NC) or Not Applicable (N/A). Compliant statements identify issues that are acceptable according to the criteria of this standard, while non-compliant statements identify issues that require further investigation. In addition, an explanation of the evaluation process is included in italics after each evaluation statement. The section numbers in parentheses following each evaluation statement correspond to Tier 2 Evaluation procedures.

Geologic Site Hazards

The following statements shall be completed for buildings in levels of high or moderate seismicity.

(C) NC N/A   **LIQUEFACTION:** Liquefaction susceptible, saturated, loose granular soils that could jeopardize the building's seismic performance shall not exist in the foundation soils at depths within 50 feet under the building for Life Safety and Immediate Occupancy. (Tier 2: Sec. 4.7.1.1)

Liquefaction susceptible soils do not exist in the foundation soils at depths within 50 feet under the building as described in project geotechnical investigation.

(C) NC N/A   **SLOPE FAILURE:** The building site shall be sufficiently remote from potential earthquake-induced slope failures or rockfalls to be unaffected by such failures or shall be capable of accommodating small predicted movements without failure. (Tier 2: Sec. 4.7.1.2)

The building is on a relatively flat site.

(C) NC N/A   **SURFACE FAULT RUPTURE:** Surface fault rupture and surface displacement at the building site is not anticipated. (Tier 2: Sec. 4.7.1.3)

Surface fault rupture and surface displacements are not anticipated.

Condition of Foundations

The following statements shall be completed for all Tier 1 building evaluations.

(C) NC N/A   **FOUNDATION PERFORMANCE:** There shall not be evidence of excessive foundation movement such as settlement or heave that would affect the integrity or strength of the structure. (Tier 2: Sec. 4.7.2.1)

The structure does not show evidence of excessive foundation movement.

Appendix A: Examples

The following statement shall be completed for buildings in levels of high or moderate seismicity being evaluated to the Immediate Occupancy Performance Level.

(C) NC N/A   DETERIORATION: There shall not be evidence that foundation elements have deteriorated due to corrosion, sulfate attack, material breakdown, or other reasons in a manner that would affect the integrity or strength of the structure. (Tier 2: Sec. 4.7.2.2)

No evidence of deterioration.

Capacity of Foundations

The following statement shall be completed for all Tier 1 building evaluations.

C NC (N/A)   POLE FOUNDATIONS: Pole foundations shall have a minimum embedment depth of 4 feet for Life Safety and Immediate Occupancy. (Tier 2: Sec. 4.7.3.1)

Pole foundations not used.

The following statements shall be completed for buildings in levels of moderate seismicity being evaluated to the Immediate Occupancy Performance Level and for buildings in levels of high seismicity.

(C) NC N/A   OVERTURNING: The ratio of the horizontal dimension of the lateral-force-resisting system at the foundation level to the building height (base/height) shall be greater than $0.6S_a$. (Tier 2: Sec. 4.7.3.2)

Ratio of base to height = 52 ft / 17 ft = 3.1 > $0.6 S_a$ = 0.6(1.05) = 0.63.

C NC (N/A)   TIES BETWEEN FOUNDATION ELEMENTS: The foundation shall have ties adequate to resist seismic forces where footings, piles, and piers are not restrained by beams, slabs, or soils classified as Class A, B, or C. (Section 3.5.2.3.1, Tier 2: Sec. 4.7.3.3)

Footings are restrained by slabs.

C NC (N/A)   DEEP FOUNDATIONS: Piles and piers shall be capable of transferring the lateral forces between the structure and the soil. This statement shall apply to the Immediate Occupancy Performance Level only. (Tier 2: Sec. 4.7.3.4)

No piles and piers.

(C) NC N/A   SLOPING SITES: The difference in foundation embedment depth from one side of the building to another shall not exceed one story in height. This statement shall apply to the Immediate Occupancy Performance Level only. (Tier 2: Sec. 4.7.3.5)

The site is relatively level.

A4.3.8 Basic Nonstructural Component Checklist

The following is a completed Basic Nonstructural Component Checklist for Example 4. Each of the evaluation statements has been marked Compliant (C), Non-Compliant (NC) or Not Applicable (N/A). Compliant statements identify issues that are acceptable according to the criteria of this standard, while non-compliant statements identify issues that require further investigation. In addition, an explanation of the evaluation process is included in italics after each evaluation statement. The section numbers in parentheses following each evaluation statement correspond to Tier 2 Evaluation procedures.

Appendix A: Examples

Partitions

C NC (N/A) UNREINFORCED MASONRY: Unreinforced masonry or hollow clay tile partitions shall be braced at a spacing equal to or less than 10 feet in levels of low or moderate seismicity and 6 feet in levels of high seismicity. (Tier 2: Sec. 4.8.1.1)

No interior hollow clay tile partitions.

Ceiling Systems

C NC (N/A) SUPPORT: The integrated suspended ceiling system shall not be used to laterally support the tops of gypsum board, masonry, or hollow clay tile partitions. Gypsum board partitions need not be evaluated where only the Basic Nonstructural Component Checklist is required by Table 3-2. (Tier 2: Sec. 4.8.2.1)

No gypsum board, masonry, or hollow clay tile partitions.

Light Fixtures

(C) NC N/A EMERGENCY LIGHTING: Emergency lighting shall be anchored or braced to prevent falling during an earthquake. (Tier 2: Sec. 4.8.3.1)

Emergency lighting is anchored at a spacing of less than 6 feet.

Cladding and Glazing

No exterior cladding; statements in this section are Not Applicable (N/A).

Masonry Veneer

No masonry veneer; statements in this section are Not Applicable (N/A).

Parapets, Cornices, Ornamentation, and Appendages

C NC (N/A) URM PARAPETS: There shall be no laterally unsupported unreinforced masonry parapets or cornices with height-to-thickness ratios greater than 1.5. A height-to-thickness ratio of up to 2.5 is permitted where only the Basic Nonstructural Component Checklist is required by Table 3-2. (Tier 2: Sec. 4.8.8.1)

No URM parapets.

(C) NC N/A CANOPIES: Canopies located at building exits shall be anchored to the structural framing at a spacing of 6 feet or less. An anchorage spacing of up to 10 feet is permitted where only the Basic Nonstructural Component Checklist is required by Table 3-2. (Tier 2: Sec. 4.8.8.2)

Canopy at storefront is anchored every 6 feet.

Masonry Chimneys

No masonry chimneys; statements in this section are Not Applicable (N/A).

Stairs

No stairs in one-story building; statements in this section are Not Applicable (N/A).

Appendix A: Examples

Building Contents and Furnishing

(C) NC N/A **TALL NARROW CONTENTS:** Contents over 4 feet in height with a height-to-depth or height-to-width ratio greater than 3-to-1 shall be anchored to the floor slab or adjacent structural walls. A height-to-depth or height-to-width ratio of up to 4-to-1 is permitted where only the Basic Nonstructural Component Checklist is required by Table 3-2. (Tier 2: Sec. 4.8.11.1)

Storage racks, bookcases, and file cabinets are anchored to the floor slab.

Mechanical and Electrical Equipment

(C) NC N/A **EMERGENCY POWER:** Equipment used as part of an emergency power system shall be mounted to maintain continued operation after an earthquake. (Tier 2: Sec. 4.8.12.1)

Emergency power systems are anchored.

(C) NC N/A **HAZARDOUS MATERIAL EQUIPMENT:** HVAC or other equipment containing hazardous material shall not have damaged supply lines or unbraced isolation supports. (Tier 2: Sec. 4.8.12.2)

HVAC equipment and supply lines are anchored.

(C) NC N/A **DETERIORATION:** There shall be no evidence of deterioration, damage, or corrosion in any of the anchorage or supports of mechanical or electrical equipment. (Tier 2: Sec. 4.8.12.3)

No deterioration observed.

(C) NC N/A **ATTACHED EQUIPMENT:** Equipment weighing over 20 pounds that is attached to ceilings, walls, or other supports 4 feet above the floor level shall be braced. (Tier 2: Sec. 4.8.12.4)

Equipment is braced.

Piping

(C) NC N/A **FIRE SUPPRESSION PIPING:** Fire suppression piping shall be anchored and braced in accordance with NFPA-13 (NFPA, 1996). (Tier 2: Sec. 4.8.13.1)

Fire suppression piping is braced in accordance with NFPA-13.

(C) NC N/A **FLEXIBLE COUPLINGS:** Fluid, gas and fire suppression piping shall have flexible couplings. (Tier 2: Sec. 4.8.13.2)

Fluid, gas, and fire suppression piping have flexible couplings.

Hazardous Materials Storage and Distribution

C NC (N/A) **TOXIC SUBSTANCES:** Toxic and hazardous substances stored in breakable containers shall be restrained from falling by latched doors, shelf lips, wires, or other methods. (Tier 2: Sec. 4.8.15.1)

No toxic or hazardous substances are stored in the building.

Appendix A: Examples

A4.3.9 Intermediate Nonstructural Component Checklist

The following is a completed Intermediate Nonstructural Checklist for Example 4. Each of the evaluation statements has been marked Compliant (C), Non-Compliant (NC) or Not Applicable (N/A). Compliant statements identify issues that are acceptable according to the criteria of this standard, while non-compliant statements identify issues that require further investigation. In addition, an explanation of the evaluation process is included in italics after each evaluation statement. The section numbers in parentheses following each evaluation statement correspond to Tier 2 Evaluation procedures.

Ceiling Systems

(C) NC N/A   **LAY-IN TILES:** Lay-in tiles used in ceiling panels located at exits and corridors shall be secured with clips. (Tier 2: Sec. 4.8.2.2)

Lay-in tiles are secured with clips.

(C) NC N/A   **INTEGRATED CEILINGS:** Integrated suspended ceilings at exits and corridors or weighing more than 2 lb/ft^2 shall be laterally restrained with a minimum of 4 diagonal wires or rigid members attached to the structure above at a spacing equal to or less than 12 feet (Tier 2: Sec. 4.8.2.3)

Integrated suspended ceilings are braced.

(C) NC N/A   **SUSPENDED LATH AND PLASTER:** Ceilings consisting of suspended lath and plaster or gypsum board shall be attached to resist seismic forces for every 12 square feet of area. (Tier 2: Sec. 4.8.2.4)

Each 10 square feet of suspended lath and plaster ceilings is attached.

Light Fixtures

(C) NC N/A   **INDEPENDENT SUPPORT:** Light fixtures in suspended grid ceilings shall be supported independently of the ceiling suspension system by a minimum of two wires at diagonally opposite corners of the fixtures. (Tier 2: Sec. 4.8.3.2)

Light fixtures are independently supported.

Cladding and Glazing

(C) NC N/A   **GLAZING:** Glazing in curtain walls and individual panes over 16 square feet in area, located up to a height of 10 feet above an exterior walking surface, shall be safety glazing. Such glazing located over 10 feet above an exterior walking surface shall be laminated annealed or laminated heat-strengthened safety glass or other glazing system that will remain in the frame where glass is cracked. (Tier 2: Sec. 4.8.4.8)

Glazing is safety glass.

Parapets, Cornices, Ornamentation, and Appendages

C   NC   **(N/A)**   **CONCRETE PARAPETS:** Concrete parapets with height-to-thickness ratios greater than 2.5 shall have vertical reinforcement. (Tier 2: Sec. 4.8.8.3)

No concrete parapets.

(C) NC N/A   **APPENDAGES:** Cornices, parapets, signs, and other appendages that extend above the highest point of anchorage to the structure or cantilever from exterior wall faces and other exterior wall ornamentation shall be reinforced and anchored to the structural system at a spacing equal to or less than 10 feet for Life Safety and 6 feet for Immediate Occupancy. This requirement need not apply to parapets or cornices compliant with Section 4.8.8.1 or 4.8.8.3. (Tier 2: Sec. 4.8.8.4)

Appendix A: Examples

Masonry parapets are reinforced at a spacing of 5 feet.

Masonry Chimneys

No masonry chimneys; statements in this section are Not Applicable (N/A).

Mechanical and Electrical Equipment

C NC (N/A) VIBRATION ISOLATORS: Equipment mounted on vibration isolators shall be equipped with restraints or snubbers. (Tier 2: Sec. 4.8.12.5)

No equipment mounted on vibration isolators.

Ducts

C NC (N/A) STAIR AND SMOKE DUCTS: Stair pressurization and smoke control ducts shall be braced and shall have flexible connections at seismic joints. (Tier 2: Sec. 4.8.14.1)

No stair pressurization or smoke control ducts.

A4.3.10 Supplemental Nonstructural Component Checklist

The following is a completed Supplemental Nonstructural Checklist for Example 4. Each of the evaluation statements has been marked Compliant (C), Non-Compliant (NC) or Not Applicable (N/A). Compliant statements identify issues that are acceptable according to the criteria of this standard, while non-compliant statements identify issues that require further investigation. In addition, an explanation of the evaluation process is included in italics after each evaluation statement. The section numbers in parentheses following each evaluation statement correspond to Tier 2 Evaluation procedures.

Partitions

C NC (N/A) DRIFT: Rigid cementititous partitions shall be detailed to accommodate a drift ratio of 0.02 in steel moment frame, concrete moment frame, and wood frame buildings. Rigid cementititous partitions shall be detailed to accommodate a drift ratio of 0.005 in other buildings. (Tier 2: Sec. 4.8.1.2)

No rigid cementitious partitions.

C NC (N/A) STRUCTURAL SEPARATIONS: Partitions at structural separations shall have seismic or control joints. (Tier 2: Sec. 4.8.1.3)

No structural separations.

C NC (N/A) TOPS: The tops of framed or panelized partitions that only extend to the ceiling line shall have lateral bracing to the building structure at a spacing equal to or less than 6 feet. (Tier 2: Sec. 4.8.1.4)

No partitions extend only to the ceiling line.

Ceiling Systems

(C) NC N/A EDGES: The edges of integrated suspended ceilings shall be separated from enclosing walls by a minimum of 1/2 inch. (Tier 2: Sec. 4.8.2.5)

The edges of integrated ceilings are separated from enclosing walls by at least 1/2 inch.

Appendix A: Examples

C NC **(N/A)** SEISMIC JOINT: The ceiling system shall not extend continuously across any seismic joint. (Tier 2: Sec. 4.8.2.6)

No seismic joints.

Light Fixtures

(C) NC N/A PENDANT SUPPORTS: Light fixtures on pendant supports shall be attached at a spacing equal to or less than 6 feet and, if rigidly supported, shall be free to move with the structure to which they are attached without damaging adjoining materials. (Tier 2: Sec. 4.8.3.3)

Light fixtures attached every 6 feet.

(C) NC N/A LENS COVERS: Lens covers on light fixtures shall be attached or supplied with safety devices. (Tier 2: Sec. 4.8.3.4)

Lens covers on light fixtures are positively attached.

Cladding and Glazing

(C) NC N/A GLAZING: All exterior glazing shall be laminated, annealed or laminated heat-strengthened safety glass or other glazing system that will remain in the frame when glass is cracked. (Tier 2: Sec. 4.8.4.9)

Glazing is safety glass.

Masonry Veneer

No masonry veneer; statements in this section are Not Applicable (N/A).

Metal Stud Back-Up Systems

No metal stud back-up system; statements in this section are Not Applicable (N/A).

Concrete Block and Masonry Back-Up Systems

No concrete block or masonry back-up system; statements in this section are Not Applicable (N/A).

Building Contents and Furnishing

(C) NC N/A FILE CABINETS: File cabinets arranged in groups shall be attached to one another. (Tier 2: Sec. 4.8.11.2)

File cabinets are attached to one another.

(C) NC N/A CABINET DOORS AND DRAWERS: Cabinet doors and drawers shall have latches to keep them closed during an earthquake. (Tier 2: Sec. 4.8.11.3)

Cabinet drawers have latches.

C NC **(N/A)** ACCESS FLOORS: Access floors over 9 inches in height shall be braced. (Tier 2: Sec. 4.8.11.4)

No access floor systems.

C NC **(N/A)** EQUIPMENT ON ACCESS FLOORS: Equipment and computers supported on access floor systems shall be either attached to the structure or fastened to a laterally braced floor system. (Tier 2: Sec. 4.8.11.5)

Appendix A: Examples

No access floor systems.

Mechanical and Electrical Equipment

(C) NC N/A   HEAVY EQUIPMENT: Equipment weighing over 100 pounds shall be anchored to the structure or foundation. (Tier 2: Sec. 4.8.12.6)

Equipment is anchored.

(C) NC N/A   ELECTRICAL EQUIPMENT: Electrical equipment and associated wiring shall be laterally braced to the structural system. (Tier 2: Sec. 4.8.12.7)

All electrical equipment laterally braced to the structure.

C NC **(N/A)**   DOORS: Mechanically operated doors shall be detailed to operate at a story drift ratio of 0.01. (Tier 2: Sec. 4.8.12.8)

No mechanically operated doors.

Piping

(C) NC N/A   FLUID AND GAS PIPING: Fluid and gas piping shall be anchored and braced to the structure to prevent breakage in piping. (Tier 2: Sec. 4.8.13.3)

Fluid and gas piping is anchored and braced.

(C) NC N/A   SHUT-OFF VALVES: Shut-off devices shall be present at building utility interfaces to shut off the flow of gas and high-temperature energy in the event of earthquake-induced failure. (Tier 2: Sec. 4.8.13.4)

Shut-off valves are present.

C NC **(N/A)**   C-CLAMPS: One-sided C-clamps that support piping greater than 2.5 inches in diameter shall be restrained. (Tier 2: Sec. 4.8.13.5)

One-sided C-clamps are not used to support major piping.

Ducts

(C) NC N/A   DUCT BRACING: Rectangular ductwork exceeding 6 square feet in cross-sectional area, and round ducts exceeding 28 inches in diameter, shall be braced. Maximum spacing of transverse bracing shall not exceed 30 feet. Maximum spacing of longitudinal bracing shall not exceed 60 feet. Intermediate supports shall not be considered part of the lateral-force-resisting system. (Tier 2: Sec. 4.8.14.2)

Ductwork is adequately braced.

C NC **(N/A)**   DUCT SUPPORT: Ducts shall not be supported by piping or electrical conduit. (Tier 2: Sec. 4.8.14.3)

No ducts are supported by piping or other nonstructural elements.

Hazardous Materials Storage and Distribution

C NC **(N/A)**   GAS CYLINDERS: Compressed gas cylinders shall be restrained. (Tier 2: Sec. 4.8.15.2)

No compressed gas cylinders in the building.

Appendix A: Examples

C   NC   HAZARDOUS MATERIALS: Piping containing hazardous materials shall have shut-off valves or other devices to prevent major spills or leaks. (Tier 2: Sec. 4.8.15.3)

No toxic or hazardous substances are stored in the building.

Elevators

No elevators; statements in this section are Not Applicable (N/A).

A4.4 Final Evaluation and Report

This building complies with the provisions of this standard. No deficiencies were identified by the Tier 1 Checklists; thus, further evaluation is not required. A final report that outlines the findings of the Tier 1 Evaluation shall be prepared.

A5 Example 5: Building Type W2: Wood Frames, Commercial and Industrial

A5.1 Introduction

This example is based on a Case Study - Three Story Wood Building presented at the November 6-7, 1996 BSSC FEMA 273 Ballot Symposium in Denver, Colorado.

The purpose of this example is to demonstrate the use of the Tier 1 Evaluation and Deficiency-Only Tier 2 Evaluation.

A5.2 Evaluation Requirements

A5.2.1 Building Description

The building is a three-story wood frame office building located in Los Angeles, California. The dimensions of the building are 120 feet by 75 feet in plan. Each story is 12 feet in height.

The lateral-force-resisting system consists of the following:
- Plywood floor and roof diaphragms
- Interior and exterior plywood shear panels
- Concrete footing and grade beams under shear walls

A floor plan of the building is provided in Figure A5-1.

A5.2.2 Level of Investigation

As-built plans are available for this building (Section 2.2). A site visit was conducted to verify plan information and to assess the condition of the building.

A5.2.3 Level of Performance

The building is evaluated to the Life Safety Performance Level (Section 2.4).

A5.2.4 Level of Seismicity

The building is located in a level of high seismicity (Section 2.5).

$S_S = 1.5$; $S_I = 0.8$ (ASCE 7-02)
Site Class = D (Section 3.5.2.3.1)
$F_v = 1.5$; $F_a = 1.0$ (Tables 3-5 & 3-6)
$S_{DI} = (2/3)F_vS_I = 0.8$ (Equation 3-5)
$S_{DS} = (2/3)F_aS_S = 1.00$ (Equation 3-6)

A5.2.5 Building Type

The building is classified as Building Type W2: Wood Frames, Commercial and Industrial (Section 2.6). A description of this type of building is included in Table 2-2.

Appendix A: Examples

A5.3 Screening Phase (Tier 1)

A5.3.1 Introduction

The purpose of a Tier 1 Evaluation is to quickly identify buildings that comply with the provisions of this standard. The data collected is sufficient to conduct a Tier 1 Evaluation.

A5.3.2 Benchmark Buildings

This building was not designed or evaluated in accordance with one of the benchmark documents listed in Table 3-1 (Section 3.2).

A5.3.3 Selection of Checklists

This building is classified as Building Type W2 (Table 2-2). Thus, the Structural Checklist(s) associated with Building Type W2 are used.

A Tier 1 Evaluation of this building involves completing the following checklists (Section 3.3):

- Basic Structural Checklist for Building Type W2 (Section 3.7.2)
- Supplemental Structural Checklist for Building Type W2 (Section 3.7.2S)
- Geologic Site Hazards and Foundations Checklist (Section 3.8)
- Basic Nonstructural Component Checklist (Section 3.9.1)
- Intermediate Nonstructural Component Checklist (Section 3.9.2)

The Supplemental Nonstructural Checklist does not need to be completed for this example since the building is being evaluated to the Life Safety Performance Level.

Each of the checklists listed above has been completed for this example and is included in the following sections.

A5.3.4 Further Evaluation Requirements

While neither a Full-Building Tier 2 Evaluation nor a Tier 3 Evaluation is required for this building based on Section 3.4, the design professional may decide to perform a Deficiency-Only Tier 2 Evaluation if deficiencies are identified by the Tier 1 Evaluation.

Appendix A: Examples

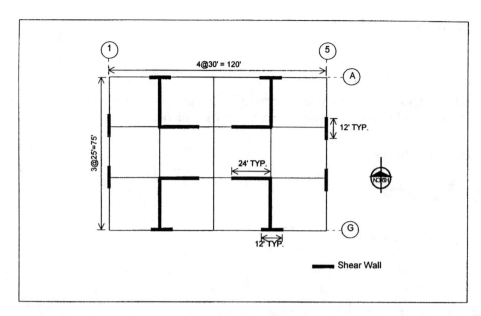

Figure A5-1. Example 5 – Typical Floor Plan

A5.3.5 Basic Structural Checklist for Building Type W2: Wood Frames, Commercial and Industrial

The following is a completed Basic Structural Checklist for Example 1. Each of the evaluation statements has been marked Compliant (C), Non-Compliant (NC) or Not Applicable (N/A). Compliant statements identify issues that are acceptable according to the criteria of this standard, while non-compliant statements identify issues that require further investigation. In addition, an explanation of the evaluation process is included in italics after each evaluation statement. The section numbers in parentheses following each evaluation statement correspond to Tier 2 Evaluation procedures.

Building System

(C) NC N/A LOAD PATH: The structure shall contain a minimum of one complete load path for Life Safety and Immediate Occupancy for seismic force effects from any horizontal direction that serves to transfer the inertial forces from the mass to the foundation. (Tier 2: Sec. 4.3.1.1)

The building contains a complete load path.

C NC **(N/A)** MEZZANINES: Interior mezzanine levels shall be braced independently from the main structure, or shall be anchored to the lateral-force-resisting elements of the main structure. (Tier 2: Sec. 4.3.1.3)

No mezzanine.

(C) NC N/A WEAK STORY: The strength of the lateral-force-resisting system in any story shall not be less than 80 percent of the strength in an adjacent story above or below for Life Safety and Immediate Occupancy. (Tier 2: Sec. 4.3.2.1)

Shear walls are the same in each story.

(C) NC N/A SOFT STORY: The stiffness of the lateral-force-resisting-system in any story shall not be less than 70 percent of the lateral-force-resisting system stiffness in an adjacent story above or below, or less than 80 percent of the average lateral-force-resisting system stiffness of the three stories above or below for Life Safety and Immediate Occupancy. (Tier 2: Sec. 4.3.2.2)

Shear walls are the same in each story.

Appendix A: Examples

(C) NC N/A GEOMETRY: There shall be no changes in horizontal dimension of the lateral-force-resisting system of more than 30 percent in a story relative to adjacent stories for Life Safety and Immediate Occupancy, excluding one-story penthouses and mezzanines. (Tier 2: Sec. 4.3.2.3)

The lateral-force-resisting system has constant horizontal dimension.

(C) NC N/A VERTICAL DISCONTINUITIES: All vertical elements in the lateral-force-resisting system shall be continuous to the foundation. (Tier 2: Sec. 4.3.2.4)

All shear walls are continuous to the foundation.

(C) NC N/A MASS: There shall be no change in effective mass more than 50 percent from one story to the next for Life Safety and Immediate Occupancy. Light roofs, penthouses, and mezzanines need not be considered. (Tier 2: Sec. 4.3.2.5)

The change in effective mass between the first, second, and third stories is negligible.

(C) NC N/A DETERIORATION OF WOOD: There shall be no signs of decay, shrinkage, splitting, fire damage, or sagging in any of the wood members, and none of the metal connection hardware shall be deteriorated, broken, or loose. (Tier 2: Sec. 4.3.3.1)

No deterioration noted.

C NC **(N/A)** WOOD STRUCTURAL PANEL SHEAR WALL FASTENERS: There shall be no more than 15 percent of inadequate fastening such as overdriven fasteners, omitted blocking, excessive fastening spacing, or inadequate edge distance. This statement shall apply to the Immediate Occupancy Performance Level only. (Tier 2: Sec. 4.3.3.2)

Life Safety Performance Level.

Lateral-Force-Resisting System

(C) NC N/A REDUNDANCY: The number of lines of shear walls in each principal direction shall be greater than or equal to 2 for Life Safety and Immediate Occupancy. (Tier 2: Sec. 4.4.2.1.1)

There are multiple lines of shear walls in each direction.

C **(NC)** N/A SHEAR STRESS CHECK: The shear stress in the shear walls, calculated using the Quick Check procedure of Section 3.5.3.3, shall be less than the following values for Life Safety and Immediate Occupancy (Tier 2: Sec. 4.4.2.7.1):

Structural panel sheathing	1,000 plf
Diagonal sheathing	700 plf
Straight sheathing	100 plf
All other conditions	100 plf

The shear stress is calculated as follows. The maximum shear stress is 1,365 plf, which is greater than the allowable value of 1,000 plf.

Pseudo Lateral Force:

$V = CS_aW$ — Equation (3-1)

$V = 0.75W$ does not control by inspection ($m>2$)

$C = 1.0$ — Table 3-4

$S_{D1} = 0.8$; $S_{DS} = 1.00$ — Section A5.2.4

$T = C_t h_n^{3/4} = 0.06(36)^{3/4} = 0.88$ sec — Equation (3-8)

$S_a = 0.8/0.88 = 0.91 < S_{DS} = 1.00$ — Equation (3-4)

$V = (1.0)(0.91) = 0.91W$

$W = 864$ kips

$V = (0.91)(864 \text{ kips}) = 786$ kips

Appendix A: Examples

Direction	Shear (V_1)	L_w^1	$V_1^{avg\ 2}$
North-South	786 kips	148 ft	1,328 plf
East-West	786 kips	144 ft	1,365 plf

$^1 L_w$ = Length of wall.
^{2}Equation (3-12); (m = 4)

C NC **(N/A)** STUCCO (EXTERIOR PLASTER) SHEAR WALLS: Multi-story buildings shall not rely on exterior stucco walls as the primary lateral-force-resisting system. (Tier 2: Sec. 4.4.2.7.2)

No stucco walls.

(C) NC N/A GYPSUM WALLBOARD OR PLASTER SHEAR WALLS: Interior plaster or gypsum wallboard shall not be used as shear walls on buildings over one story in height with the exception of the uppermost level of a multi-story building. (Tier 2: Sec. 4.4.2.7.3)

Gypsum or plaster wallboard not used as shear walls.

(C) NC N/A NARROW WOOD SHEAR WALLS: Narrow wood shear walls with an aspect ratio greater than 2-to-1 for Life Safety and 1.5-to-1 for Immediate Occupancy shall not be used to resist lateral forces developed in the building in levels of moderate and high seismicity. Narrow wood shear walls with an aspect ratio greater than 2-to-1 for Immediate Occupancy shall not be used to resist lateral forces developed in the building in levels of low seismicity. (Tier 2: Sec. 4.4.2.7.4)

Critical aspect ratio = 12 ft / 12 ft = 1.0 < 2.0.

(C) NC N/A WALLS CONNECTED THROUGH FLOORS: Shear walls shall have interconnection between stories to transfer overturning and shear forces through the floor. (Tier 2: Sec. 4.4.2.7.5)

Shear walls are continuous through floors.

C NC **(N/A)** HILLSIDE SITE: For structures that are taller on at least one side by more than half of a story due to a sloping site, all shear walls on the downhill slope shall have an aspect ratio less than 1-to-1 for Life Safety and 1 to 2 for Immediate Occupancy. (Tier 2: Sec. 4.4.2.7.6)

Building is on a level lot.

C NC **(N/A)** CRIPPLE WALLS: Cripple walls below first floor level shear walls shall be braced to the foundation with wood structural panels. (Tier 2: Sec. 4.4.2.7.7)

There are no cripple walls.

C NC **(N/A)** OPENINGS: Walls with openings greater than 80 percent of the length shall be braced with wood structural panel shear walls with aspect ratios of not more than 1.5-to-1 or shall be supported by adjacent construction through positive ties capable of transferring the lateral forces. (Tier 2: Sec. 4.4.2.7.8)

There are no walls with large openings.

Connections

C NC **(N/A)** WOOD POSTS: There shall be a positive connection of wood posts to the foundation. (Tier 2: Sec. 4.6.3.3)

Shear wall system.

(C) NC N/A WOOD SILLS: All wood sills shall be bolted to the foundation. (Tier 2: Sec. 4.6.3.4)

There are anchor bolts along all shear walls.

Appendix A: Examples

C NC (N/A) GIRDER/COLUMN CONNECTION: There shall be a positive connection utilizing plates, connection hardware, or straps between the girder and the column support. (Tier 2: Sec. 4.6.4.1)

Shear wall system.

A5.3.6 Supplemental Structural Checklist for Building Type W2: Wood Frames, Commercial and Industrial

The following is a completed Supplemental Structural Checklist for Example 5. Each of the evaluation statements has been marked Compliant (C), Non-Compliant (NC) or Not Applicable (N/A). Compliant statements identify issues that are acceptable according to the criteria of this standard, while non-compliant statements identify issues that require further investigation. In addition, an explanation of the evaluation process is included in italics after each evaluation statement. The section numbers in parentheses following each evaluation statement correspond to Tier 2 Evaluation procedures.

Lateral-Force-Resisting System

C NC (N/A) HOLD-DOWN ANCHORS: All shear walls shall have hold-down anchors constructed per acceptable construction practices, attached to the end studs. This statement shall apply to the Immediate Occupancy Performance Level only. (Tier 2: Sec. 4.4.2.7.9)

Life Safety Performance Level.

Diaphragms

(C) NC N/A DIAPHRAGM CONTINUITY: The diaphragms shall not be composed of split-level floors and shall not have expansion joints. (Tier 2: Sec. 4.5.1.1)

The diaphragm does not consist of split-level framing or have expansion joints.

(C) NC N/A ROOF CHORD CONTINUITY: All chord elements shall be continuous, regardless of changes in roof elevation. (Tier 2: Sec. 4.5.1.3)

All roof chords are continuous.

(C) NC N/A PLAN IRREGULARITIES: There shall be tensile capacity to develop the strength of the diaphragm at re-entrant corners or other locations of plan irregularities. This statement shall apply to the Immediate Occupancy Performance Level only. (Tier 2: Sec. 4.5.1.7)

Building is rectangular.

C NC (N/A) DIAPHRAGM REINFORCEMENT AT OPENINGS: There shall be reinforcing around all diaphragms openings larger than 50 percent of the building width in either major plan dimension. This statement shall apply to the Immediate Occupancy Performance Level only. (Tier 2: Sec. 4.5.1.8)

Life Safety Performance Level.

C NC (N/A) STRAIGHT SHEATHING: All straight sheathed diaphragms shall have aspect ratios less than 2-to-1 for Life Safety and 1-to-1 for Immediate Occupancy in the direction being considered. (Tier 2: Sec. 4.5.2.1)

There is no straight-board sheathing in the building.

(C) NC N/A SPANS: All wood diaphragms with spans greater than 24 feet for Life Safety and 12 feet for Immediate Occupancy shall consist of wood structural panels or diagonal sheathing. (Tier 2: Sec. 4.5.2.2)

Appendix A: Examples

Wood structural panels used in diaphragms.

C   NC   **(N/A)**   UNBLOCKED DIAPHRAGMS: All diagonally sheathed or unblocked wood structural panel diaphragms shall have horizontal spans less than 40 feet for Life Safety and 30 feet for Immediate Occupancy and shall have aspect ratios less than or equal to 4-to-1 for Life Safety and 3-to-1 for Immediate Occupancy. (Tier 2: Sec. 4.5.2.3)

All diaphragms are blocked and sheathed with wood structural panels.

(C)   NC   N/A   OTHER DIAPHRAGMS: The diaphragm shall not consist of a system other than wood, metal deck, concrete, or horizontal bracing. (Tier 2: Sec. 4.5.7.1)

All diaphragms are sheathed with wood structural panels.

Connections

(C)   NC   N/A   WOOD SILL BOLTS: Sill bolts shall be spaced at 6 feet or less for Life Safety and 4 feet or less for Immediate Occupancy, with proper edge and end distance provided for wood and concrete. (Tier 2: Sec. 4.6.3.9)

Sill bolts are spaced at 6 feet.

A5.3.7 Geologic Site Hazards and Foundations Checklist

The following is a completed Geologic Site Hazards and Foundations Checklist for Example 5. Each of the evaluation statements has been marked Compliant (C), Non-Compliant (NC) or Not Applicable (N/A). Compliant statements identify issues that are acceptable according to the criteria of this standard, while non-compliant statements identify issues that require further investigation. In addition, an explanation of the evaluation process is included in italics after each evaluation statement. The section numbers in parentheses following each evaluation statement correspond to Tier 2 Evaluation procedures.

Geologic Site Hazards

The following statements shall be completed for buildings in levels of high or moderate seismicity.

(C)   NC   N/A   LIQUEFACTION: Liquefaction susceptible, saturated, loose granular soils that could jeopardize the building's seismic performance shall not exist in the foundation soils at depths within 50 feet under the building for Life Safety and Immediate Occupancy. (Tier 2: Sec. 4.7.1.1)

Liquefaction susceptible soils do not exist in the foundation soils as described in project geotechnical investigation.

(C)   NC   N/A   SLOPE FAILURE: The building site shall be sufficiently remote from potential earthquake-induced slope failures or rockfalls to be unaffected by such failures or shall be capable of accommodating small predicted movements without failure. (Tier 2: Sec. 4.7.1.2)

The building is on a relatively flat site.

(C)   NC   N/A   SURFACE FAULT RUPTURE: Surface fault rupture and surface displacement at the building site is not anticipated. (Tier 2: Sec. 4.7.1.3)

Surface fault rupture and surface displacements are not anticipated.

Appendix A: Examples

Condition of Foundations

The following statements shall be completed for all Tier 1 building evaluations.

(C) NC N/A   FOUNDATION PERFORMANCE: There shall not be evidence of excessive foundation movement such as settlement or heave that would affect the integrity or strength of the structure. (Tier 2: Sec. 4.7.2.1)

The structure does not show evidence of excessive foundation movement.

The following statement shall be completed for buildings in levels of high or moderate seismicity being evaluated to the Immediate Occupancy Performance Level.

C NC (N/A)   DETERIORATION: There shall not be evidence that foundation elements have deteriorated due to corrosion, sulfate attack, material breakdown, or other reasons in a manner that would affect the integrity or strength of the structure. (Tier 2: Sec. 4.7.2.2)

Life Safety Performance Level.

Capacity of Foundations

The following statement shall be completed for all Tier 1 building evaluations.

C NC (N/A)   POLE FOUNDATIONS: Pole foundations shall have a minimum embedment depth of 4 feet for Life Safety and Immediate Occupancy. (Tier 2: Sec. 4.7.3.1)

Pole foundations not used.

The following statements shall be completed for buildings in levels of moderate seismicity being evaluated to the Immediate Occupancy Performance Level and for buildings in levels of high seismicity.

C (NC) N/A   OVERTURNING: The ratio of the horizontal dimension of the lateral-force-resisting system at the foundation level to the building height (base/height) shall be greater than $0.6S_a$. (Tier 2: Sec. 4.7.3.2)

Ratio of base to height = 12 ft / 36 ft = 0.33 < 0.6 S_a = 0.6(0.91) = 0.55.

(C) NC N/A   TIES BETWEEN FOUNDATION ELEMENTS: The foundation shall have ties adequate to resist seismic forces where footings, piles, and piers are not restrained by beams, slabs, or soils classified as Class A, B, or C. (Section 3.5.2.3.1, Tier 2: Sec. 4.7.3.3)

Footings are restrained by slabs.

C NC (N/A)   DEEP FOUNDATIONS: Piles and piers shall be capable of transferring the lateral forces between the structure and the soil. This statement shall apply to the Immediate Occupancy Performance Level only. (Tier 2: Sec. 4.7.3.4)

Life Safety Performance Level.

C NC (N/A)   SLOPING SITES: The difference in foundation embedment depth from one side of the building to another shall not exceed one story in height. This statement shall apply to the Immediate Occupancy Performance Level only. (Tier 2: Sec. 4.7.3.5)

Life Safety Performance Level.

Appendix A: Examples

A5.3.8 Basic Nonstructural Component Checklist

The following is a completed Basic Nonstructural Component Checklist for Example 5. Each of the evaluation statements has been marked Compliant (C), Non-Compliant (NC) or Not Applicable (N/A). Compliant statements identify issues that are acceptable according to the criteria of this standard, while non-compliant statements identify issues that require further investigation. In addition, an explanation of the evaluation process is included in italics after each evaluation statement. The section numbers in parentheses following each evaluation statement correspond to Tier 2 Evaluation procedures.

Partitions

C NC (N/A)   UNREINFORCED MASONRY: Unreinforced masonry or hollow clay tile partitions shall be braced at a spacing equal to or less than 10 feet in levels of low or moderate seismicity and 6 feet in levels of high seismicity. (Tier 2: Sec. 4.8.1.1)

No URM or hollow clay tile partitions.

Ceiling Systems

C NC (N/A)   SUPPORT: The integrated suspended ceiling system shall not be used to laterally support the tops of gypsum board, masonry, or hollow clay tile partitions. Gypsum board partitions need not be evaluated where only the Basic Nonstructural Component Checklist is required by Table 3-2. (Tier 2: Sec. 4.8.2.1)

No gypsum board, masonry, or hollow clay tile partitions.

Light Fixtures

(C) NC N/A   EMERGENCY LIGHTING: Emergency lighting shall be anchored or braced to prevent falling during an earthquake. (Tier 2: Sec. 4.8.3.1)

Emergency lighting is anchored at a spacing of less than 6 feet.

Cladding and Glazing

No exterior cladding; statements in this section are Not Applicable (N/A).

Masonry Veneer

No masonry veneer; statements in this section are Not Applicable (N/A).

Parapets, Cornices, Ornamentation, and Appendages

No parapets, cornices, ornamentation, or appendages; statements in this section are Not Applicable (N/A).

Masonry Chimneys

No masonry chimneys; statements in this section are Not Applicable (N/A).

Appendix A: Examples

Stairs

C   NC   (N/A)   **URM WALLS:** Walls around stair enclosures shall not consist of unbraced hollow clay tile or unreinforced masonry with a height-to-thickness ratio greater than 12-to-1. A height-to-thickness ratio of up to 15-to-1 is permitted where only the Basic Nonstructural Component Checklist is required by Table 3-2. (Tier 2: Sec. 4.8.10.1)

No hollow clay tile or URM walls.

C   NC   (N/A)   **STAIR DETAILS:** In moment frame structures, the connection between the stairs and the structure shall not rely on shallow anchors in concrete. Alternatively, the stair details shall be capable of accommodating the drift calculated using the Quick Check procedure of Section 3.5.3.1 without including tension in the anchors. (Tier 2: Sec. 4.8.10.2)

Not a moment frame structure.

Building Contents and Furnishing

(C)   NC   N/A   **TALL NARROW CONTENTS:** Contents over 4 feet in height with a height-to-depth or height-to-width ratio greater than 3-to-1 shall be anchored to the floor slab or adjacent structural walls. A height-to-depth or height-to-width ratio of up to 4-to-1 is permitted where only the Basic Nonstructural Component Checklist is required by Table 3-2. (Tier 2: Sec. 4.8.11.1)

Storage racks, bookcases, and file cabinets are anchored to the floor slab.

Mechanical and Electrical Equipment

(C)   NC   N/A   **EMERGENCY POWER:** Equipment used as part of an emergency power system shall be mounted to maintain continued operation after an earthquake. (Tier 2: Sec. 4.8.12.1)

Emergency power systems are anchored.

(C)   NC   N/A   **HAZARDOUS MATERIAL EQUIPMENT:** HVAC or other equipment containing hazardous material shall not have damaged supply lines or unbraced isolation supports. (Tier 2: Sec. 4.8.12.2)

HVAC equipment and supply lines are anchored.

(C)   NC   N/A   **DETERIORATION:** There shall be no evidence of deterioration, damage, or corrosion in any of the anchorage or supports of mechanical or electrical equipment. (Tier 2: Sec. 4.8.12.3)

No deterioration observed.

(C)   NC   N/A   **ATTACHED EQUIPMENT:** Equipment weighing over 20 lb that is attached to ceilings, walls, or other supports 4 feet above the floor level shall be braced. (Tier 2: Sec. 4.8.12.4)

Equipment is braced.

Piping

(C)   NC   N/A   **FIRE SUPPRESSION PIPING:** Fire suppression piping shall be anchored and braced in accordance with NFPA-13 (NFPA, 1996). (Tier 2: Sec. 4.8.13.1)

Fire suppression piping is braced in accordance with NFPA-13.

(C)   NC   N/A   **FLEXIBLE COUPLINGS:** Fluid, gas and fire suppression piping shall have flexible couplings. (Tier 2: Sec. 4.8.13.2)

Appendix A: Examples

Fluid, gas, and fire suppression piping have flexible couplings.

Hazardous Materials Storage and Distribution

C   NC   (N/A)   TOXIC SUBSTANCES: Toxic and hazardous substances stored in breakable containers shall be restrained from falling by latched doors, shelf lips, wires, or other methods. (Tier 2: Sec. 4.8.15.1)

No toxic or hazardous substances are stored in the building.

A5.3.9 Intermediate Nonstructural Component Checklist

The following is a completed Intermediate Nonstructural Checklist for Example 5. Each of the evaluation statements has been marked Compliant (C), Non-Compliant (NC) or Not Applicable (N/A). Compliant statements identify issues that are acceptable according to the criteria of this standard, while non-compliant statements identify issues that require further investigation. In addition, an explanation of the evaluation process is included in italics after each evaluation statement. The section numbers in parentheses following each evaluation statement correspond to Tier 2 Evaluation procedures.

Ceiling Systems

(C)   NC   N/A   LAY-IN TILES: Lay-in tiles used in ceiling panels located at exits and corridors shall be secured with clips. (Tier 2: Sec. 4.8.2.2)

Lay-in tiles are secured with clips.

(C)   NC   N/A   INTEGRATED CEILINGS: Integrated suspended ceilings at exits and corridors or weighing more than 2 lb/ft^2 shall be laterally restrained with a minimum of 4 diagonal wires or rigid members attached to the structure above at a spacing equal to or less than 12 feet. (Tier 2: Sec. 4.8.2.3)

Integrated suspended ceilings are braced.

(C)   NC   N/A   SUSPENDED LATH AND PLASTER: Ceilings consisting of suspended lath and plaster or gypsum board shall be attached to resist seismic forces for every 12 square feet of area. (Tier 2: Sec. 4.8.2.4)

Each 10 square feet of suspended lath and plaster ceilings is attached.

Light Fixtures

(C)   NC   N/A   INDEPENDENT SUPPORT: Light fixtures in suspended grid ceilings shall be supported independently of the ceiling suspension system by a minimum of two wires at diagonally opposite corners of the fixtures. (Tier 2: Sec. 4.8.3.2)

Light fixtures are independently supported.

Cladding and Glazing

(C)   NC   N/A   GLAZING: Glazing in curtain walls and individual panes over 16 square feet in area, located up to a height of 10 feet above an exterior walking surface, shall be safety glazing. Such glazing located over 10 feet above an exterior walking surface shall be laminated annealed or laminated heat-strengthened safety glass or other glazing system that will remain in the frame where glass is cracked. (Tier 2: Sec. 4.8.4.8)

Glazing is safety glass.

Appendix A: Examples

Parapets, Cornices, Ornamentation, and Appendages

No parapets, cornices, ornamentation, or appendages; statements in this section are Not Applicable (N/A).

Masonry Chimneys

No masonry chimneys; statements in this section are Not Applicable (N/A).

Mechanical and Electrical Equipment

C   NC   (N/A)   VIBRATION ISOLATORS: Equipment mounted on vibration isolators shall be equipped with restraints or snubbers. (Tier 2: Sec. 4.8.12.5)

No equipment mounted on vibration isolators.

Ducts

C   NC   (N/A)   STAIR AND SMOKE DUCTS: Stair pressurization and smoke control ducts shall be braced and shall have flexible connections at seismic joints. (Tier 2: Sec. 4.8.14.1)

No stair pressurization or smoke control ducts.

A5.4 Further Evaluation (Tier 2)

The following potential deficiencies were identified by the Tier 1 Checklists:
1. Shear Stress Check
2. Overturning

A Deficiency-Only Tier 2 Evaluation is completed for this building by following the Tier 2 Procedure associated with each potential deficiency. Tier 2 Procedures for this example are located in the following sections: Section 4.4.2.7.1 (Shear Stress), and Section 4.7.3.2 (Overturning).

A5.4.1 Seismic Forces

The first step of the Tier 2 Evaluation is to compute the shear forces in accordance with Section 4.2. Shear forces are computed as follows:

- Pseudo Lateral Force = 786 kips   (Computed in Tier 1)
- Vertical Distribution of Forces:

    $T = 0.88$ sec   (Computed in Tier 1)

Floor	h_x (ft)	w_x (kips)	$w_i h_i^k$ ($k = 1.19$)	C_{vx} Eqn. (4-3)	F_x (kips) Eqn. (4-2)
Roof	36	198	14,082	0.40	315
3rd	24	333	14,618	0.42	328
2nd	12	333	6,407	0.18	143

A5.4.2 Tier 2 Evaluation of Shear Stress in Walls (Tier 2 Procedure: Section 4.4.2.7.1)

In the Tier 1 Evaluation, a shear stress of approximately 1,350 pounds per linear foot was computed; this exceeds the allowable shear stress value of 1,000 pounds per linear foot. Thus, a Tier 2 Evaluation is conducted to further evaluate the shear stress in the walls.

Appendix A: Examples

The Tier 2 Procedure for evaluating shear stress states that an analysis shall be performed in accordance with Section 4.2 and the adequacy of the shear walls shall be evaluated using the *m*-factors in Table 4-8.

Seismic forces are computed in Section A5.4.1. Wall shear stresses are computed based on these computed and compared with capacities as presented in the following table:

Direction	Length of Wall (ft)	Trib. Area / Total Area[1]	Shear[2] (kips)	h/L[3]	m (Table 4-8)	Q_{UD}/m[4]	Comply?[5]
East-West	12.0	0.083	65.2	3.0	3.5	1.55 klf	within 3 percent
	24.0	0.167	131.3	1.5	4.0	1.39 klf	yes
North-South	12.0	0.063	49.1	3.0	3.5	1.17 klf	yes
	25.0	0.188	147.4	1.4	4.1	1.44 klf	yes

[1] Tributary area of wall / total area.
[2] Based on tributary area; shear = pseudo lateral force × tributary area / total area.
[3] Wall height / wall length; used to compute *m*-factor.
[4] Q_{UD} = Shear demand/length of wall.
[5] $Q_{CE} > Q_{UD}/m$ [Equation (4-11)].
Q_{CE} = Expected strength = (1.25)(0.78 k/ft)/(0.65) = 1.50 k/ft.
$F Q_n$ = 0.78 k/ft (based on the *2000 NEHRP Provisions*: 15/32-inch Str. I plywood with 10d @ 3 in.; F = 0.65).

A5.4.3 Tier 2 Evaluation of Overturning (Tier 2 Procedure: Section 4.7.3.2)

In the Tier 1 Evaluation, the ratio of the base to the height of the 12-foot-long shear walls was computed as 0.33, which is less than the required value of 0.55 for this building. Thus, a Tier 2 Evaluation is conducted to further assess the adequacy of these shear walls to resist overturning.

The Tier 2 Procedure for evaluating overturning states that an analysis shall be conducted in accordance with Section 4.2 and the adequacy of the foundation including all gravity and seismic overturning forces shall be evaluated. For this example, the adequacy of the foundation to resist overturning is evaluated using the R_{OT} factor in accordance with Section 4.2.4.3.4.

Seismic forces are computed in Section A5.4.1 and used to compute overturning moments. Overturning moments for the 12-foot-long walls are presented in the following table:

	Story Force[1] (kips)	Story Height (ft)	East-West Direction		North-South Direction	
			Wall Force (k)[2]	M_o (k-ft)[3]	Wall Force (k)[2]	M_o (k-ft)[3]
Roof	315.0	36.0	26.2	945.0	19.7	708.8
3rd	328.0	24.0	27.3	656.0	20.5	492.4
2nd	143.0	12.0	11.9	143.0	8.9	107.3
				Σ = 1,744.0		Σ = 1,308.1

[1] Computed in Section A5.4.1.
[2] Story force on 12-foot-long shear walls; based on tributary area.
[3] Overturning moment = story force × story height.

Appendix A: Examples

The adequacy of the building to resist overturning is evaluated by checking for uplift based on the overturning moments computed above using Equation (4-9) as follows:

Direction	M_o^1 (k-ft)	Q_E^2 (kips)	Q_G^3 (kips)	Q_{UF}^4 (kips)	Comply?[5]
East-West	1,744.0	145.3	21.3	3.1	yes
North-South	1,308.1	109.0	21.3	7.7	yes

[1] M_o = Overturning moment = summation (story forces × height).
[2] $Q_E = M_o$ / length of wall.
[3] Q_G computed in Section A5.4.2.
[4] $Q_{UF} = Q_G - Q_E / R_{OT}$; $R_{OT} = 8$.
[5] $Q_{UF} > 0$; no uplift; Equation (4-13).

A5.5 Final Evaluation and Report

Based on the Tier 2 Evaluation, this building complies with the provisions of this standard. No potential deficiencies were identified by the Tier 2 Evaluation; thus, further evaluation is not required. A final report that outlines the findings of the Tier 1 and Tier 2 Evaluations shall be prepared.

A6 Example 6 – Building Type S2: Steel Braced Frames with Stiff Diaphragms

A6.1 Introduction

The purpose of this example is to demonstrate the use of the Tier 1 Evaluation and the Full-Building Tier 2 Evaluation.

A6.2 Evaluation Requirements

A6.2.1 Building Description

The building is a two-story steel braced frame office building located in the Los Angeles area.

The total floor area of the building is approximately 37,500 square feet.

The building was designed according to the 1976 *Uniform Building Code*.

A floor plan and elevation of the building are shown in Figures A6-1 and A6-2, respectively.

A6.2.2 Level of Investigation

As-built plans are available for this building (Section 2.2). A site visit was conducted to verify the plan information and to assess the condition of the building.

A6.2.3 Level of Performance

The building is evaluated to the Life Safety Performance Level (Section 2.4).

A6.2.4 Level of Seismicity

The building is located in a level of high seismicity (Section 2.5).

$S_S = 1.5$; $S_1 = 0.8$ (ASCE 7-02)
Site Class = D (Section 3.5.2.3.1)
$F_v = 1.5$; $F_a = 1.0$ (Tables 3-5 & 3-6)
$S_{D1} = (2/3)F_v S_1 = 0.8$ (Equation 3-5)
$S_{DS} = (2/3)F_a S_S = 1.00$ (Equation 3-6)

A6.2.5 Building Type

The building is classified as Building Type S2, Steel Braced Frames with Stiff Diaphragms (Section 2.6). A description of this type of building is included in Table 2-2.

Appendix A: Examples

A6.3 Screening Phase (Tier 1)

A6.3.1 Introduction

The purpose of a Tier 1 Evaluation is to quickly identify buildings that comply with the provisions of this standard. The data collected is sufficient to conduct a Tier 1 Evaluation.

A6.3.2 Benchmark Buildings

This building was not designed or evaluated in accordance with one of the benchmark documents listed in Table 3-1 (Section 3.2).

A6.3.3 Selection of Checklists

This building is classified as Building Type S2 (Table 2-2). Thus, the Structural Checklist(s) associated with Building Type S2 are used.

A Tier 1 Evaluation of this building involves completing the following checklists (Section 3.3):

- Basic Structural Checklist for Building Type S2 (Section 3.7.4)
- Supplemental Structural Checklist for Building Type S2 (Section 3.7.4S)
- Geologic Site Hazards and Foundations Checklist (Section 3.8)
- Basic Nonstructural Component Checklist (Section 3.9.1)
- Intermediate Nonstructural Component Checklist (Section 3.9.2)

The Supplemental Nonstructural Checklist does not need to be completed for this example since the building is being evaluated to the Life Safety Performance Level.

Each of the checklists listed above has been completed for this example and is included in the following sections.

A6.3.4 Further Evaluation Requirements

While neither a Full-Building Tier 2 Evaluation nor a Tier 3 Evaluation is required for this building based on Section 3.4, the design professional may decide to further evaluate the building if deficiencies are identified by the Tier 1 Evaluation.

Appendix A: Examples

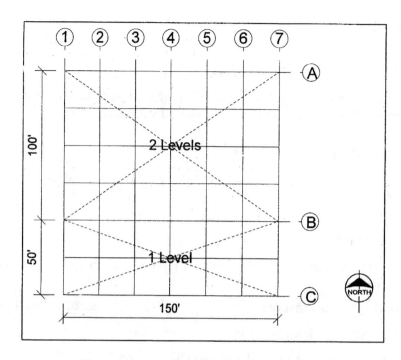

Figure A6-1. Example 6 – Floor Plan

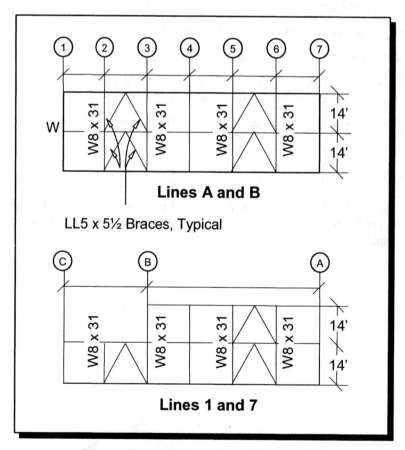

Figure A6-2. Example 6 – Elevations

Appendix A: Examples

A6.3.5 Basic Structural Checklist for Building Type S2: Steel Braced Frames with Stiff Diaphragms

The following is a completed Basic Structural Checklist for Example 6. Each of the evaluation statements has been marked Compliant (C), Non-Compliant (NC) or Not Applicable (N/A). Compliant statements identify issues that are acceptable according to the criteria of this standard, while non-compliant statements identify issues that require further investigation. In addition, an explanation of the evaluation process is included in italics after each evaluation statement. The section numbers in parentheses following each evaluation statement correspond to Tier 2 Evaluation procedures.

Building System

(C) NC N/A LOAD PATH: The structure shall contain a minimum of one complete load path for Life Safety and Immediate Occupancy for seismic force effects from any horizontal direction that serves to transfer the inertial forces from the mass to the foundation. (Tier 2: Sec. 4.3.1.1)

The building contains a complete load path.

C NC **(N/A)** MEZZANINES: Interior mezzanine levels shall be braced independently from the main structure, or shall be anchored to the lateral-force-resisting elements of the main structure. (Tier 2: Sec. 4.3.1.3)

There is no mezzanine in this building.

(C) NC N/A WEAK STORY: The strength of the lateral-force-resisting system in any story shall not be less than 80 percent of the strength in an adjacent story above or below for Life Safety and Immediate Occupancy. (Tier 2: Sec. 4.3.2.1)

The strength of the second story is not less than 80 percent of the strength of the first story.

(C) NC N/A SOFT STORY: The stiffness of the lateral-force-resisting-system in any story shall not be less than 70 percent of the lateral-force-resisting system stiffness in an adjacent story above or below, or less than 80 percent of the average lateral-force-resisting system stiffness of the three stories above or below for Life Safety and Immediate Occupancy. (Tier 2: Sec. 4.3.2.2)

The stiffness of the second story is not less than 70 percent of the stiffness of the first story.

(C) NC N/A GEOMETRY: There shall be no changes in horizontal dimension of the lateral-force-resisting system of more than 30 percent in a story relative to adjacent stories for Life Safety and Immediate Occupancy, excluding one-story penthouses and mezzanines. (Tier 2: Sec. 4.3.2.3)

There is no change in horizontal dimension of the lateral-force-resisting system of more than 30 percent relative to adjacent stories.

(C) NC N/A VERTICAL DISCONTINUITIES: All vertical elements in the lateral-force-resisting system shall be continuous to the foundation. (Tier 2: Sec. 4.3.2.4)

All braced frames are continuous to the foundation.

(C) NC N/A MASS: There shall be no change in effective mass more than 50 percent from one story to the next for Life Safety and Immediate Occupancy. Light roofs, penthouses, and mezzanines need not be considered. (Tier 2: Sec. 4.3.2.5)

The mass of the building varies by approximately 30 percent from the first to the second story.
  Roof: (100 ft)(150 ft)(100 psf) = 1,500 kips
  Second: (150 ft)(150 ft)(100 psf) = 2,250 kips

Appendix A: Examples

(C) NC N/A   TORSION: The estimated distance between the story center of mass and the story center of rigidity shall be less than 20 percent of the building width in either plan dimension for Life Safety and Immediate Occupancy. (Tier 2: Sec. 4.3.2.6)

The distance between the story center of mass and the story center of rigidity is less than 20 percent of the building width.

(C) NC N/A   DETERIORATION OF STEEL: There shall be no visible rusting, corrosion, cracking, or other deterioration in any of the steel elements or connections in the vertical- or lateral-force-resisting systems. (Tier 2: Sec. 4.3.3.3)

No deterioration noted.

(C) NC N/A   DETERIORATION OF CONCRETE: There shall be no visible deterioration of concrete or reinforcing steel in any of the vertical- or lateral-force-resisting elements. (Tier 2: Sec. 4.3.3.4)

No deterioration noted.

Lateral-Force-Resisting System

C (NC) N/A   AXIAL STRESS CHECK: The axial stress due to gravity loads in columns subjected to overturning forces shall be less than $0.10F_y$ for Life Safety and Immediate Occupancy. Alternatively, the axial stress due to overturning forces alone, calculated using the Quick Check procedure of Section 3.5.3.6, shall be less than $0.30F_y$ for Life Safety and Immediate Occupancy. (Tier 2: Sec. 4.4.1.3.2)

Axial stress due to gravity loads (typical column):

	Tributary Area (ft²)	Dead Load (kips)	Live Load (kips)	Gravity Load (kips)
Roof	(25)(25) = 625	(A_{trib})(0.1 ksf) = 62.5	(A_{trib})(0.02 ksf) = 12.5	75.0
2nd Floor	(25)(25) = 625	(A_{trib})(0.1 ksf) = 62.5	(A_{trib})(0.04 ksf) = 25.0	87.5

Column area (W8 × 31) = 9.13 in².

Axial stress = (75 kips + 87.5 kips) / 9.13 in.² = 17.8 ksi > 3.6 ksi = 0.10 F_y.

Alternatively, axial stress due to overturning forces alone (Quick Check: Section 3.5.3.6):

$$\text{Overturning force} = p_{ot} = \frac{1}{m}\left(\frac{2}{3}\right)\left(\frac{Vh_n}{Ln_f}\right) \qquad \text{Equation (3-15)}$$

Check axial stress in column @ line B (Axial Stress = P_{ot}):

m = 2.0   Section 3.5.3.6
Case 1: $V = CS_aW$ Equation (3-1)
     $C = 1.2$ Table 3-4
     $S_a = S_{D1}/T$ Equation (3-4)
     $S_{D1} = 0.8$ ($S_{DS} = 1.0$)   Equation (3-5) & (3-6)
     $T = C_t h_n^{3/4}$ Equation (3-8)
         = (0.020)(28 ft)³ᐟ⁴ = 0.243 sec.,
     $S_a = 0.8/0.243 = 3.3 > S_{D1} = 1.0$,
     $S_a = 1.0$,
     W = 1,500 kips + 2,250 kips = 3,750 kips,
     V (Case 1) = (1.2)(1.0)(3750 kips) = 4,500 kips,

Appendix A: Examples

Case 2: $V = 0.75W = 2{,}813$ *kips; by inspection, Case 2 does not control since J and m are greater than the ratio of the pseudo lateral forces;* $J = 2.5$; $4{,}500\ k / 2{,}813\ k = 1.6$.

$h_n = 28\ ft$
$L = 25\ ft = 300\ in.$
$p_{ot} = (1/2.0)(2/3)((4{,}500\ kips \times 28\ ft)/(4 \times 25\ ft)) = 420\ kips$ Equation (3-15)
Axial stress $= 420\ kips/9.13\ in.^2 = 46\ ksi > 0.3F_y = 10.8\ ksi$

(C) NC N/A **REDUNDANCY:** The number of lines of braced frames in each principal direction shall be greater than or equal to 2 for Life Safety and Immediate Occupancy. The number of braced bays in each line shall be greater than 2 for Life Safety and 3 for Immediate Occupancy. (Tier 2: Sec. 4.4.3.1.1)

There are 2 lines of braced frames in each principal direction and 2 braced bays in each line.

(C) NC N/A **AXIAL STRESS CHECK:** The axial stress in the diagonals, calculated using the Quick Check procedure of Section 3.5.3.4, shall be less than $0.50F_y$ for Life Safety and for Immediate Occupancy. (Tier 2: Sec. 4.4.3.1.2)

$$f_j^{avg} = \frac{1}{m}\left(\frac{V_j}{sN_{br}}\right)\left(\frac{L_{br}}{A_{br}}\right)$$ Equation (3-13)

$m = 6.0$ (double angle) Table 3-8
$s = 25\ ft$
$A_{br} = 9.5\ in.^2$
$N_{br} = 8$
f_1^{avg} (Case 1) $= (1/6.0)(4{,}500/(12.5 \times 8))(18.77\ ft/9.5\ in.^2) = 14.8\ ksi < 18\ ksi$

C NC (N/A) **COLUMN SPLICES:** All column splice details located in braced frames shall develop the tensile strength of the column. This statement shall apply to the Immediate Occupancy Performance Level only. (Tier 2: Sec. 4.4.3.1.5)

Life Safety Performance Level.

Connections

(C) NC N/A **TRANSFER TO STEEL FRAMES:** Diaphragms shall be connected for transfer of loads to the steel frames for Life Safety, and the connections shall be able to develop the lesser of the strength of the frames or the diaphragms for Immediate Occupancy. (Tier 2: Sec. 4.6.2.2)

Diaphragms are connected for transfer of loads to the steel frames.

(C) NC N/A **STEEL COLUMNS:** The columns in lateral-force-resisting frames shall be anchored to the building foundation for Life Safety, and the anchorage shall be able to develop the lesser of the tensile capacity of the column, the tensile capacity of the lowest level column splice (if any), or the uplift capacity of the foundation, for Immediate Occupancy. (Tier 2: Sec. 4.6.3.1)

The columns in lateral-force-resisting frames are anchored to the foundation.

Appendix A: Examples

A6.3.6 Supplemental Structural Checklist for Building Type S2: Steel Braced Frames with Stiff Diaphragms

The following is a completed Supplemental Structural Checklist for Example 6. Each of the evaluation statements has been marked Compliant (C), Non-Compliant (NC) or Not Applicable (N/A). Compliant statements identify issues that are acceptable according to the criteria of this standard, while non-compliant statements identify issues that require further investigation. In addition, an explanation of the evaluation process is included in italics after each evaluation statement. The section numbers in parentheses following each evaluation statement correspond to Tier 2 Evaluation procedures.

Lateral-Force-Resisting System

C   NC   (N/A)   MOMENT-RESISTING CONNECTIONS: All moment connections shall be able to develop the strength of the adjoining members or panel zones. (Tier 2: Sec. 4.4.1.3.3)

No moment-resisting connections.

(C)   NC   N/A   COMPACT MEMBERS: All frame elements shall meet section requirements set forth by *Seismic Provisions for Structural Steel Buildings* Table I-9-1 (AISC, 1997). (Tier 2: Sec. 4.4.1.3.7)

All frame elements compact.

(C)   NC   N/A   SLENDERNESS OF DIAGONALS: All diagonal elements required to carry compression shall have Kl/r ratios less than 120. (Tier 2: Sec. 4.4.3.1.3)

All braces have Kl/r < 120.

(C)   NC   N/A   CONNECTION STRENGTH: All the brace connections shall develop the yield capacity of the diagonals. (Tier 2: Sec. 4.4.3.1.4)

Connection can develop capacity of braces.

C   NC   (N/A)   OUT-OF-PLANE BRACING: Braced frame connections attached to beam bottom flanges located away from beam-column joints shall be braced out-of-plane at the bottom flange of the beams. This statement shall apply to the Immediate Occupancy Performance Level only. (Tier 2: Sec. 4.4.3.1.6)

Life Safety Performance Level.

(C)   NC   N/A   K-BRACING: The bracing system shall not include K-braced bays. (Tier 2: Sec. 4.4.3.2.1)

K-bracing not used; V-bracing is used.

C   NC   (N/A)   TENSION-ONLY BRACES: Tension-only braces shall not comprise more than 70 percent of the total lateral-force-resisting capacity in structures over two stories in height. This statement shall apply to the Immediate Occupancy Performance Level only. (Tier 2: Sec. 4.4.3.2.2)

Life Safety Performance Level.

C   NC   (N/A)   CHEVRON BRACING: The bracing system shall not include chevron, or V-braced, bays. This statement shall apply to the Immediate Occupancy Performance Level only. (Tier 2: Sec. 4.4.3.2.3)

Life Safety Performance Level.

C   NC   (N/A)   CONCENTRICALLY BRACED FRAME JOINTS: All the diagonal braces shall frame into the beam-column joints concentrically. This statement shall apply to the Immediate Occupancy Performance Level only. (Tier 2: Sec. 4.4.3.2.4)

Life Safety Performance Level.

Appendix A: Examples

Diaphragms

(C) NC N/A   OPENINGS AT BRACED FRAMES: Diaphragm openings immediately adjacent to the braced frames shall extend less than 25 percent of the frame length for Life Safety and 15 percent of the frame length for Immediate Occupancy. (Tier 2: Sec. 4.5.1.5)

Diaphragm openings extend less than 25 percent of the frame length.

C NC **(N/A)**   PLAN IRREGULARITIES: There shall be tensile capacity to develop the strength of the diaphragm at re-entrant corners or other locations of plan irregularities. This statement shall apply to the Immediate Occupancy Performance Level only. (Tier 2: Sec. 4.5.1.7)

Life Safety Performance Level.

C NC **(N/A)**   DIAPHRAGM REINFORCEMENT AT OPENINGS: There shall be reinforcing around all diaphragms openings larger than 50 percent of the building width in either major plan dimension. This statement shall apply to the Immediate Occupancy Performance Level only. (Tier 2: Sec. 4.5.1.8)

Life Safety Performance Level.

Connections

C NC **(N/A)**   UPLIFT AT PILE CAPS: Pile caps shall have top reinforcement and piles shall be anchored to the pile caps for Life Safety, and the pile cap reinforcement and pile anchorage shall be able to develop the tensile capacity of the piles for Immediate Occupancy. (Tier 2: Sec. 4.6.3.10)

No pile caps.

A6.3.7 Geologic Site Hazards and Foundations Checklist

The following is a completed Geologic Site Hazards and Foundations Checklist for Example 6. Each of the evaluation statements has been marked Compliant (C), Non-Compliant (NC) or Not Applicable (N/A). Compliant statements identify issues that are acceptable according to the criteria of this standard, while non-compliant statements identify issues that require further investigation. In addition, an explanation of the evaluation process is included in italics after each evaluation statement. The section numbers in parentheses following each evaluation statement correspond to Tier 2 Evaluation procedures.

Geologic Site Hazards

The following statements shall be completed for buildings in levels of high or moderate seismicity.

(C) NC N/A   LIQUEFACTION: Liquefaction susceptible, saturated, loose granular soils that could jeopardize the building's seismic performance shall not exist in the foundation soils at depths within 50 feet under the building for Life Safety and Immediate Occupancy. (Tier 2: Sec. 4.7.1.1)

Liquefaction susceptible soils do not exist in the foundation soils at depths within 50 feet under the building as described in project geotechnical investigation.

(C) NC N/A   SLOPE FAILURE: The building site shall be sufficiently remote from potential earthquake-induced slope failures or rockfalls to be unaffected by such failures or shall be capable of accommodating small predicted movements without failure. (Tier 2: Sec. 4.7.1.2)

The building is on a relatively flat site.

(C) NC N/A   SURFACE FAULT RUPTURE: Surface fault rupture and surface displacement at the building site is not anticipated. (Tier 2: Sec. 4.7.1.3)

Appendix A: Examples

Surface fault rupture and surface displacements are not anticipated.

Condition of Foundations

The following statements shall be completed for all Tier 1 building evaluations.

 NC   N/A   FOUNDATION PERFORMANCE: There shall not be evidence of excessive foundation movement such as settlement or heave that would affect the integrity or strength of the structure. (Tier 2: Sec. 4.7.2.1)

The structure does not show evidence of excessive foundation movement.

The following statement shall be completed for buildings in levels of high or moderate seismicity being evaluated to the Immediate Occupancy Performance Level.

C   NC   (N/A)   DETERIORATION: There shall not be evidence that foundation elements have deteriorated due to corrosion, sulfate attack, material breakdown, or other reasons in a manner that would affect the integrity or strength of the structure. (Tier 2: Sec. 4.7.2.2)

Life Safety Performance Level.

Capacity of Foundations

The following statement shall be completed for all Tier 1 building evaluations.

C   NC   (N/A)   POLE FOUNDATIONS: Pole foundations shall have a minimum embedment depth of 4 feet for Life Safety and Immediate Occupancy. (Tier 2: Sec. 4.7.3.1)

Pole foundations not used.

The following statements shall be completed for buildings in levels of moderate seismicity being evaluated to the Immediate Occupancy Performance Level and for buildings in levels of high seismicity.

 NC   N/A   OVERTURNING: The ratio of the horizontal dimension of the lateral-force-resisting system at the foundation level to the building height (base/height) shall be greater than $0.6S_a$. (Tier 2: Sec. 4.7.3.2)

Ratio of base to height = 25 ft / 28 ft = 0.89 > 0.6 S_a = 0.6(1.00) = 0.6.

C   NC   (N/A)   TIES BETWEEN FOUNDATION ELEMENTS: The foundation shall have ties adequate to resist seismic forces where footings, piles, and piers are not restrained by beams, slabs, or soils classified as Class A, B, or C. (Section 3.5.2.3.1, Tier 2: Sec. 4.7.3.3)

Footings are restrained by slabs.

C   NC   (N/A)   DEEP FOUNDATIONS: Piles and piers shall be capable of transferring the lateral forces between the structure and the soil. This statement shall apply to the Immediate Occupancy Performance Level only. (Tier 2: Sec. 4.7.3.4)

Life Safety Performance Level.

C   NC   (N/A)   SLOPING SITES: The difference in foundation embedment depth from one side of the building to another shall not exceed one story in height. This statement shall apply to the Immediate Occupancy Performance Level only. (Tier 2: Sec. 4.7.3.5)

Life Safety Performance Level.

Appendix A: Examples

A6.3.8 Basic Nonstructural Component Checklist

The following is a completed Basic Nonstructural Component Checklist for Example 6. Each of the evaluation statements has been marked Compliant (C), Non-Compliant (NC) or Not Applicable (N/A). Compliant statements identify issues that are acceptable according to the criteria of this standard, while non-compliant statements identify issues that require further investigation. In addition, an explanation of the evaluation process is included in italics after each evaluation statement. The section numbers in parentheses following each evaluation statement correspond to Tier 2 Evaluation procedures.

Partitions

C   NC   (N/A)   UNREINFORCED MASONRY: Unreinforced masonry or hollow clay tile partitions shall be braced at a spacing equal to or less than 10 feet in levels of low or moderate seismicity and 6 feet in levels of high seismicity. (Tier 2: Sec. 4.8.1.1)

No URM or hollow clay tile partitions.

Ceiling Systems

C   NC   (N/A)   SUPPORT: The integrated suspended ceiling system shall not be used to laterally support the tops of gypsum board, masonry, or hollow clay tile partitions. Gypsum board partitions need not be evaluated where only the Basic Nonstructural Component Checklist is required by Table 3-2. (Tier 2: Sec. 4.8.2.1)

No gypsum board, masonry, or hollow clay tile partitions.

Light Fixtures

(C)   NC   N/A   EMERGENCY LIGHTING: Emergency lighting shall be anchored or braced to prevent falling during an earthquake. (Tier 2: Sec. 4.8.3.1)

Emergency lighting is anchored at a spacing of less than 6 feet.

Cladding and Glazing

No exterior cladding; statements in this section are Not Applicable (N/A).

Masonry Veneer

No masonry veneer; statements in this section are Not Applicable (N/A).

Parapets, Cornices, Ornamentation, and Appendages

No parapets, cornices, ornamentation, or appendages; statements in this section are Not Applicable (N/A).

Masonry Chimneys

No masonry chimneys; statements in this section are Not Applicable (N/A).

Appendix A: Examples

Stairs

C NC (N/A) URM WALLS: Walls around stair enclosures shall not consist of unbraced hollow clay tile or unreinforced masonry with a height-to-thickness ratio greater than 12-to-1. A height-to-thickness ratio of up to 15-to-1 is permitted where only the Basic Nonstructural Component Checklist is required by Table 3-2. (Tier 2: Sec. 4.8.10.1)

No hollow clay tile or URM walls.

C NC (N/A) STAIR DETAILS: In moment frame structures, the connection between the stairs and the structure shall not rely on shallow anchors in concrete. Alternatively, the stair details shall be capable of accommodating the drift calculated using the Quick Check procedure of Section 3.5.3.1 without including tension in the anchors. (Tier 2: Sec. 4.8.10.2)

Not a moment frame structure.

Building Contents and Furnishing

(C) NC N/A TALL NARROW CONTENTS: Contents over 4 feet in height with a height-to-depth or height-to-width ratio greater than 3-to-1 shall be anchored to the floor slab or adjacent structural walls. A height-to-depth or height-to-width ratio of up to 4-to-1 is permitted where only the Basic Nonstructural Component Checklist is required by Table 3-2. (Tier 2: Sec. 4.8.11.1)

Storage racks, bookcases, and file cabinets are anchored to the floor slab.

Mechanical and Electrical Equipment

(C) NC N/A EMERGENCY POWER: Equipment used as part of an emergency power system shall be mounted to maintain continued operation after an earthquake. (Tier 2: Sec. 4.8.12.1)

Emergency power systems are anchored.

(C) NC N/A HAZARDOUS MATERIAL EQUIPMENT: HVAC or other equipment containing hazardous material shall not have damaged supply lines or unbraced isolation supports. (Tier 2: Sec. 4.8.12.2)

HVAC equipment and supply lines are anchored.

(C) NC N/A DETERIORATION: There shall be no evidence of deterioration, damage, or corrosion in any of the anchorage or supports of mechanical or electrical equipment. (Tier 2: Sec. 4.8.12.3)

No deterioration observed.

(C) NC N/A ATTACHED EQUIPMENT: Equipment weighing over 20 lb that is attached to ceilings, walls, or other supports 4 feet above the floor level shall be braced. (Tier 2: Sec. 4.8.12.4)

Equipment is braced.

Piping

(C) NC N/A FIRE SUPPRESSION PIPING: Fire suppression piping shall be anchored and braced in accordance with NFPA-13 (NFPA, 1996). (Tier 2: Sec. 4.8.13.1)

Fire suppression piping is braced in accordance with NFPA-13.

(C) NC N/A FLEXIBLE COUPLINGS: Fluid, gas and fire suppression piping shall have flexible couplings. (Tier 2: Sec. 4.8.13.2)

Appendix A: Examples

Fluid, gas, and fire suppression piping have flexible couplings.

Hazardous Materials Storage and Distribution

C   NC   **(N/A)**   TOXIC SUBSTANCES: Toxic and hazardous substances stored in breakable containers shall be restrained from falling by latched doors, shelf lips, wires, or other methods. (Tier 2: Sec. 4.8.15.1)

No toxic or hazardous substances are stored in the building.

A6.3.9 Intermediate Nonstructural Component Checklist

The following is a completed Intermediate Nonstructural Checklist for Example 6. Each of the evaluation statements has been marked Compliant (C), Non-Compliant (NC) or Not Applicable (N/A). Compliant statements identify issues that are acceptable according to the criteria of this standard, while non-compliant statements identify issues that require further investigation. In addition, an explanation of the evaluation process is included in italics after each evaluation statement. The section numbers in parentheses following each evaluation statement correspond to Tier 2 Evaluation procedures.

Ceiling Systems

(C)   NC   N/A   LAY-IN TILES: Lay-in tiles used in ceiling panels located at exits and corridors shall be secured with clips. (Tier 2: Sec. 4.8.2.2)

Lay-in tiles are secured with clips.

(C)   NC   N/A   INTEGRATED CEILINGS: Integrated suspended ceilings at exits and corridors or weighing more than 2 lb/ft^2 shall be laterally restrained with a minimum of 4 diagonal wires or rigid members attached to the structure above at a spacing equal to or less than 12 feet. (Tier 2: Sec. 4.8.2.3)

Integrated suspended ceilings are braced.

(C)   NC   N/A   SUSPENDED LATH AND PLASTER: Ceilings consisting of suspended lath and plaster or gypsum board shall be attached to resist seismic forces for every 12 square feet of area. (Tier 2: Sec. 4.8.2.4)

Each 10 square feet of suspended lath and plaster ceilings is attached.

Light Fixtures

(C)   NC   N/A   INDEPENDENT SUPPORT: Light fixtures in suspended grid ceilings shall be supported independently of the ceiling suspension system by a minimum of two wires at diagonally opposite corners of the fixtures. (Tier 2: Sec. 4.8.3.2)

Light fixtures are independently supported.

Cladding and Glazing

(C)   NC   N/A   GLAZING: Glazing in curtain walls and individual panes over 16 square feet in area, located up to a height of 10 feet above an exterior walking surface, shall be safety glazing. Such glazing located over 10 feet above an exterior walking surface shall be laminated annealed or laminated heat-strengthened safety glass or other glazing system that will remain in the frame where glass is cracked. (Tier 2: Sec. 4.8.4.8)

Glazing is safety glass.

Appendix A: Examples

Parapets, Cornices, Ornamentation, and Appendages

No parapets, cornices, ornamentation, or appendages; statements in this section are Not Applicable (N/A).

Masonry Chimneys

No masonry chimneys; statements in this section are Not Applicable (N/A).

Mechanical and Electrical Equipment

C NC (N/A) VIBRATION ISOLATORS: Equipment mounted on vibration isolators shall be equipped with restraints or snubbers. (Tier 2: Sec. 4.8.12.5)

No equipment mounted on vibration isolators.

Ducts

C NC (N/A) STAIR AND SMOKE DUCTS: Stair pressurization and smoke control ducts shall be braced and shall have flexible connections at seismic joints. (Tier 2: Sec. 4.8.14.1)

No stair pressurization or smoke control ducts.

A6.4 Further Evaluation Phase (Tier 2)

The only potential deficiency identified by the Tier 1 Evaluation is that the axial stress in columns due to overturning exceeded the allowable value.

Although it is not required, a Full-Building Tier 2 Evaluation is conducted for this building to illustrate the evaluation process. A Full-Building Tier 2 Evaluation is conducted by (1) computing seismic forces in accordance with Section 4.2, and (2) checking the adequacy of the braces, connections, and columns. The braces are considered deformation-controlled components, while the columns and connections are considered force-controlled components. For simplicity, the effect of eccentricity is neglected.

A6.4.1 Seismic Forces

The first step of the Full-Building Tier 2 Evaluation is to compute the seismic forces in accordance with Section 4.2. Seismic forces are computed as follows:

- Pseudo lateral force: $V = CS_aW = 4,500$ kips   Case 1 Equation (3-1)

  $V = 0.75W = 2,812.5$ kips   Case 2 Equation (3-2)

  By inspection, Case 2 does not govern since J and m are greater than the ratio of the pseudo lateral forces; 4,500k / 2,813k = 1.6.

- Vertical distribution of seismic forces: Section 4.2.2.1.3

  $T = 0.24$ sec   (Computed in Tier 1)

Floor	h_x (ft)	w_x (kips)	$w_ih_i^k$ (k = 1.0)	C_{vx} (Eqn. 4-3)	F_x (kips) (Eqn. 4-2)
Roof	28	1,500	42,000	0.571	2,570
2nd	14	2,250	31,500	0.429	1,930

Appendix A: Examples

A6.4.2 Axial Force in the Braces

The adequacy of the braces to resist axial force is assessed. The braces are considered deformation-controlled components. The axial force in the braces due to seismic forces as computed in Section 4.2 is divided by an appropriate m-factor and compared with the expected strength of the braces.

Floor	V_x (kips)[1] (Eqn. 3-3b)	$Q_E = V_x(18.8 / 12.5) / 8$[2]	Q_{UD}/m[3]	Comply?[4]
Roof	2,570	483 kips	81 kips	yes
2nd	4,500	846 kips	141 kips	yes

[1] Computed in Section A6.4.1.
[2] Q_E = Axial force in each brace; number of braces = 8; 18.8 / 12.5 = angle of brace.
[3] $Q_{UD} = Q_E$; $m = 6$.   Table 4-5
[4] $Q_{CE} > Q_{UD} / m$.   Equation (4-12)
$Q_{CE} = (1.25)(1.7)(P_{ASD})$.   Section 4.2.4.4
P_{ASD} = Brace capacity based on allowable stress design tables = 70 kips.
$Q_{CE} = (1.25)(1.7)(70 \text{ kips}) = 149 \text{ kips}$.

A6.4.3 Brace Connection at the Roof

The adequacy of the brace connection at the roof is assessed. The brace connection is considered force controlled.

Using Method 2 of 4.2.4.3.2, the shear demand below the roof is computed in accordance with Section 4.2 (see Section A6.4.2). The demand on the brace connection is as follows:

$Q_E = 483$ kips   Section A6.4.2

The force-controlled action, Q_{UF}, is computed by dividing the shear demand, Q_E, by C and J as follows:

$Q_{UF} = Q_E / CJ = 161$ kips   Equation (4-9); $C = 1.2$; $J = 2.5$

The expected strength of the connection then is computed as follows:

Assumed connection strength (working stress) = 97 kips

$Q_{CN} = (1.7)(97 \text{ kips}) = 165$ kips   Section 4.2.4.4

Finally, the expected strength of the connection is compared with the action at the connection:

$Q_{CN} = 165$ kips $> Q_{UF} = 161$ kips   Equation (4-13)

Since the expected strength of the connection is greater than the demand, the brace connection at the roof is found to be adequate.

A6.4.4 Axial Force in Columns

The adequacy of the columns to resist axial forces due to gravity loads as well as overturning forces then is determined. The columns are considered force controlled.

In this case, the action due to earthquake forces is based on the brace capacity (or the amount of force that the brace can deliver to the columns). The action due to earthquake forces is calculated as follows:

Brace capacity = $(1.25)(1.7)(70 \text{ kips}) = 149$ kips;

P_{ASD} = Brace capacity based on allowable stress design tables = 70 kips;

$Q_E = (149 \text{ kips})(14 \text{ ft}/18.8 \text{ ft} = 111$ kips.

The gravity load is computed based on tributary area:

$Q_G = 1.1(Q_D + Q_L) = 1.1[(31.3 + 15.6) + 0.25(3.1 + 12.5)] = 56$ kips.   Equation (4-6)

Appendix A: Examples

The total axial force in the columns is computed as follows:

$Q_{UF} = Q_G + Q_E = 56 + 111 = 167$ kips.

Finally, the expected strength of the column is compared with the action in the column:

Allowable stress design capacity of column = 137 kips;

$Q_{CN} = (1.7)(137 \text{ kips}) = 233 \text{ kips} > Q_{UF}$ Equation (4-13)

Since the expected strength of the column is greater than the axial force, the column is found to be adequate.

A6.4.5 Drag Force at the Roof

The adequacy of the connections to resist drag forces is determined. The drag force is considered a force-controlled action and is checked using two methods.

Assumed drag capacity (working stress) = 93 kips

$Q_{CN} = (1.7)(93 \text{ kips}) = 158$ kips

Method 1:

Roof load on lines 1 and 7 associated with the brace capacity of $(1.25)(1.7)(70 \text{ kips}) = 149$ kips;

Diaphragm shear force = 149 kips(12.5/18.8)(4 braces) = 396 kips;

Shear force = 396 kips/(4x25 ft) = 3.96 k/ft;

Maximum drag length = 1.5x25 ft = 37.5 ft;

$Q_{UF} = 3.96 \text{ k/f} \times 37.5 \text{ ft} = 148.4 \text{ kips} < Q_{CN} = 158$ kips. Equation (4-13)

Method 2:

$Q_E = 2{,}570$ kips/2 sides = 1,285 kips/side;

Diaphragm shear force = 1,285 kips /((1.2)(2.5)) = 428 kips. Equation (4-9)

Since the diaphragm force in Method 2 is larger than in Method 1, use the results of Method 1.

A6.5 Final Evaluation and Report

This building complies with the provisions of this standard. No deficiencies were identified by the Tier 2 Evaluation; thus, further evaluation is not required. A final report that outlines the findings of the Tier 1 and Tier 2 Evaluations shall be prepared.

Appendix A: Examples

A7 Example 7 – Building Type URM: Unreinforced Masonry Bearing Walls with Flexible Diaphragms

A7.1 Introduction

This example is based on an unreinforced masonry bearing wall building with flexible diaphragms.

The purpose of this example is to demonstrate the use of a Tier 1 Evaluation and the Tier 2 URM Special Procedure.

A7.2 Evaluation Requirements

A7.2.1 Building Description

The building is a two-story URM building constructed prior to 1940. The first level has six stores. The second level consists of apartments with interior partitions. The building is located in Los Angeles, California.

The building is rectangular in plan, 30 feet × 100 feet. The building height is 25 feet. The building area is 3,000 square feet per floor. Floor plans and elevations of the building are shown in Figures A7-1 and A7-2.

The roof has diagonal sheathing. The weight of the roof is 27 psf. The floor has finished wood flooring over diagonal sheathing. The weight of the floor is 35 psf.

Unreinforced masonry walls are on the perimeter of the building at both story levels. The walls are of unreinforced brick: a 1.5-foot-high parapet and an 11-foot-high second-level wall 9 inches thick (with a weight of 90 psf) and a 14-foot-high first level wall 13 inches thick (130 psf). The second-level walls are 75 percent solid (25 percent of the area is windows); the first level walls are 80 percent solid. Walls are attached to the diaphragms by 3/4-in.-diameter government anchors at 4 feet on center typical at the perimeter.

A7.2.2 Level of Investigation

As-built plans are available for this building (Section 2.2). A site visit was conducted to verify the plan information and to assess the condition of the building. Destructive testing of the walls and anchors was performed in accordance with Section 4.2.6.2.2.

A7.2.3 Level of Performance

The building is evaluated to the Life Safety Performance Level (Section 2.4).

A7.2.4 Level of Seismicity

The building is located in a level of high seismicity (Section 2.5).

$S_S = 1.5$; $S_1 = 0.6$ (ASCE 7-02)
Site Class = C   (Section 3.5.2.3.1)
$F_v = 1.3$; $F_a = 1.0$ (Tables 3-5 & 3-6)
$S_{D1} = (2/3)F_v S_1 = 0.52$ (Equation 3-5)
$S_{DS} = (2/3)F_a S_S = 1.0$ (Equation 3-6)

Appendix A: Examples

A7.2.5 Building Type

The building is classified as Building Type URM, Unreinforced Masonry Bearing Walls with Flexible Diaphragms (Section 2.6). A description of this type of building is included in Table 2-2.

A7.3 Screening Phase (Tier 1)

A7.3.1 Introduction

The purpose of a Tier 1 Evaluation is to quickly identify buildings that comply with the provisions of this standard. The data collected is sufficient to conduct a Tier 1 Evaluation.

A7.3.2 Benchmark Buildings

The building was constructed prior to 1970; thus, this building was not designed in accordance with the benchmark documents listed in Table 3-1 (Section 3.2).

A7.3.3 Selection of Checklists

This building is classified as Building Type URM (Table 2-2). Thus, the Structural Checklist(s) associated with Building Type URM are used.

A Tier 1 Evaluation of this building involves completing the following checklists (Section 3.3):

- Basic Structural Checklist for Building Type URM (Section 3.7.15)
- Supplemental Structural Checklist for Building Type URM (Section 3.7.15S)
- Geologic Site Hazards and Foundations Checklist (Section 3.8)
- Basic Nonstructural Component Checklist (Section 3.9.1)
- Intermediate Nonstructural Component Checklist (Section 3.9.2)

The Supplemental Nonstructural Checklist does not need to be completed for this example since the building is being evaluated to the Life Safety Performance Level.

Each of the checklists listed above has been completed for this example and is included in the following sections.

A7.3.4 Further Evaluation Requirements

While neither a Full-Building Tier 2 Evaluation nor a Tier 3 Evaluation is required for this building based on Section 3.4, the design professional may decide to perform a Tier 2 URM Special Procedure if deficiencies are identified by the Tier 1 Evaluation.

Appendix A: Examples

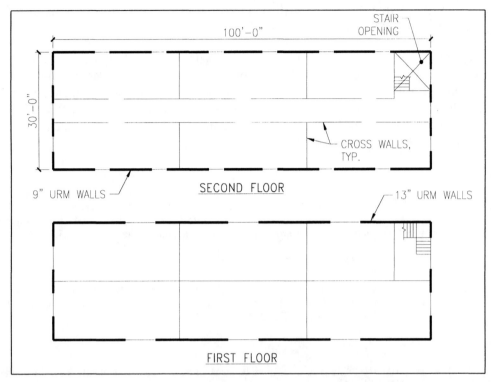

Figure A7-1. Example 7 – Floor Plans

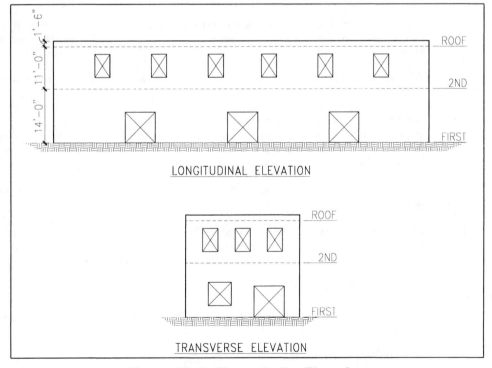

Figure A7-2. Example 7 – Elevations

Appendix A: Examples

A7.3.5 Basic Structural Checklist for Building Type URM: Unreinforced Masonry Bearing Walls with Flexible Diaphragms

The following is a completed Basic Structural Checklist for Example 7. Each of the evaluation statements has been marked Compliant (C), Non-Compliant (NC) or Not Applicable (N/A). Compliant statements identify issues that are acceptable according to the criteria of this standard, while non-compliant statements identify issues that require further investigation. In addition, an explanation of the evaluation process is included in italics after each evaluation statement. The section numbers in parentheses following each evaluation statement correspond to Tier 2 Evaluation procedures.

Building System

(C) NC N/A LOAD PATH: The structure shall contain a minimum of one complete load path for Life Safety and Immediate Occupancy for seismic force effects from any horizontal direction that serves to transfer the inertial forces from the mass to the foundation. (Tier 2: Sec. 4.3.1.1)

The structure contains a complete load path.

(C) NC N/A ADJACENT BUILDINGS: The clear distance between the building being evaluated and any adjacent building shall be greater than 4 percent of the height of the shorter building for Life Safety and Immediate Occupancy. (Tier 2: Sec. 4.3.1.2)

No adjacent buildings.

C NC (N/A) MEZZANINES: Interior mezzanine levels shall be braced independently from the main structure, or shall be anchored to the lateral-force-resisting elements of the main structure. (Tier 2: Sec. 4.3.1.3)

No mezzanine.

(C) NC N/A WEAK STORY: The strength of the lateral-force-resisting system in any story shall not be less than 80 percent of the strength in an adjacent story above or below for Life Safety and Immediate Occupancy. (Tier 2: Sec. 4.3.2.1)

There are no significant strength discontinuities in any of the vertical elements of the lateral-force-resisting system.

(C) NC N/A SOFT STORY: The stiffness of the lateral-force-resisting-system in any story shall not be less than 70 percent of the lateral-force-resisting system stiffness in an adjacent story above or below, or less than 80 percent of the average lateral-force-resisting system stiffness of the three stories above or below for Life Safety and Immediate Occupancy. (Tier 2: Sec. 4.3.2.2)

There are no significant stiffness discontinuities in any of the vertical elements of the lateral-force-resisting system.

(C) NC N/A GEOMETRY: There shall be no changes in horizontal dimension of the lateral-force-resisting system of more than 30 percent in a story relative to adjacent stories for Life Safety and Immediate Occupancy, excluding one-story penthouses and mezzanines. (Tier 2: Sec. 4.3.2.3)

No building setbacks.

(C) NC N/A VERTICAL DISCONTINUITIES: All vertical elements in the lateral-force-resisting system shall be continuous to the foundation. (Tier 2: Sec. 4.3.2.4)

All walls are continuous to the foundation.

Appendix A: Examples

(**C**) NC N/A MASS: There shall be no change in effective mass more than 50 percent from one story to the next for Life Safety and Immediate Occupancy. Light roofs, penthouses, and mezzanines need not be considered. (Tier 2: Sec. 4.3.2.5)

Change in effective mass is less than 50 percent.

Story	Weight	Percent Change
Roof	213 kips	46
2nd	391 kips	-

(**C**) NC N/A DETERIORATION OF WOOD: There shall be no signs of decay, shrinkage, splitting, fire damage, or sagging in any of the wood members, and none of the metal connection hardware shall be deteriorated, broken, or loose. (Tier 2: Sec. 4.3.3.1)

No deterioration observed.

(**C**) NC N/A MASONRY UNITS: There shall be no visible deterioration of masonry units. (Tier 2: Sec. 4.3.3.7)

No deterioration observed.

(**C**) NC N/A MASONRY JOINTS: The mortar shall not be easily scraped away from the joints by hand with a metal tool, and there shall be no areas of eroded mortar. (Tier 2: Sec. 4.3.3.8)

Mortar cannot be easily scraped away from the joints.

(**C**) NC N/A UNREINFORCED MASONRY WALL CRACKS: There shall be no existing diagonal cracks in the wall elements greater than 1/8 inch for Life Safety and 1/16 inch for Immediate Occupancy, or out-of-plane offsets in the bed joint greater than 1/8 inch for Life Safety and 1/16 inch for Immediate Occupancy, and shall not form an X pattern. (Tier 2: Sec. 4.3.3.11)

No cracking observed.

Lateral-Force-Resisting System

(**C**) NC N/A REDUNDANCY: The number of lines of shear walls in each principal direction shall be greater than or equal to 2.0 for Life Safety and Immediate Occupancy. (Tier 2: Sec. 4.4.2.1.1)

There are two lines of shear walls in each principal direction.

C (**NC**) N/A SHEAR STRESS CHECK: The shear stress in the unreinforced masonry shear walls, calculated using the Quick Check procedure of Section 3.5.3.3, shall be less than 30 psi for clay units and 70 psi for concrete units for Life Safety and Immediate Occupancy. (Tier 2: Sec. 4.4.2.5.1)

Pseudo Lateral Force:

$V = CS_aW$ — Equation (3-1)
$C = 1.0$ — Table 3-4
$S_{D1} = 0.52; S_{DS} = 1.0$ — Section A4.2.4
$T = C_t h_n^{3/4} = 0.02(22)^{3/4} = 0.203 \text{ sec}$ — Equation (3-8)
$S_a = 0.52/0.203 = 2.56 > S_{DS} = 1.0$ — Equation (3-4)
$V = (1.0)(1.0)W = 1.0\ W$
$W = 604 \text{ kips}$
$V = 1.0\ (604 \text{ kips}) = 604 \text{ kips}$

Appendix A: Examples

Story	Shear	Longitudinal Direction		Transverse Direction		Compy?[3]
		A_w^1	$V_1^{avg\,2}$	A_w^1	$V_1^{avg\,2}$	
1st	604 kips	24,960 in.²	16 psi	7,488 in.²	54 psi	no

[1] A_w = (length of wall)(thickness); thickness = 13 inches.
[2] Equation (3-12); m = 1.5 for Life Safety (Table 3-7).
[3] Clay masonry units: the allowable shear stress is 30 psi.

Connections

C (NC) N/A **WALL ANCHORAGE:** Exterior concrete or masonry walls, which are dependent on the diaphragm for lateral support, shall be anchored for out-of-plane forces at each diaphragm level with steel anchors, reinforcing dowels, or straps that are developed into the diaphragm. Connections shall have adequate strength to resist the connection force calculated in the Quick Check procedure of Section 3.5.3.7. (Tier 2: Sec. 4.6.1.1)

Second Level $T_c = \Psi S_{DS} w_p A_p$ *Equation (3-16)*
$\Psi = 0.9$
$S_{DS} = 1.0$ *Section A7.2.4*
$T_c = 0.9(1.0)[0.09(5.5')(4') + 0.13(7.0')(4')] = 3.9$ kips
$Q_{CE} = 1.5$ kips   *Tested capacity*
$T_c > Q_{CE} \Rightarrow$ *Non-compliant*

(C) NC N/A **WOOD LEDGERS:** The connection between the wall panels and the diaphragm shall not induce cross-grain bending or tension in the wood ledgers. (Tier 2: Sec. 4.6.1.2)

Connections are 3/4-inch diameter government anchors that are attached from the wood joists and embedded into the masonry wall.

(C) NC N/A **TRANSFER TO SHEAR WALLS:** Diaphragms shall be connected for transfer of loads to the shear walls for Life Safety, and the connections shall be able to develop the lesser of the shear strength of the walls or diaphragms for Immediate Occupancy. (Tier 2: Sec. 4.6.2.1)

Diaphragms are connected to masonry walls.

C NC (N/A) **GIRDER/COLUMN CONNECTION:** There shall be a positive connection utilizing plates, connection hardware, or straps between the girder and the column support. (Tier 2: Sec. 4.6.4.1)

Joists span to interior bearing cross walls. There are no columns in the building.

Appendix A: Examples

A7.3.6 Supplemental Structural Checklist for Building Type URM: Unreinforced Masonry Bearing Walls with Flexible Diaphragms

The following is a completed Supplemental Structural Checklist for Example 7. Each of the evaluation statements has been marked Compliant (C), Non-Compliant (NC) or Not Applicable (N/A). Compliant statements identify issues that are acceptable according to the criteria of this standard, while non-compliant statements identify issues that require further investigation. In addition, an explanation of the evaluation process is included in italics after each evaluation statement. The section numbers in parentheses following each evaluation statement correspond to Tier 2 Evaluation procedures.

Lateral-Force-Resisting System

C (NC) N/A   PROPORTIONS: The height-to-thickness ratio of the shear walls at each story shall be less than the following for Life Safety and Immediate Occupancy (Tier 2: Sec. 4.4.2.5.2):

        Top story of multi-story building   9
        First story of multi-story building   15
        All other conditions   13

Second Level   h/t = 11 ft/9 in. = 14.6   Non-compliant
First Level   h/t = 14 ft/13 in. = 12.9   Compliant

(C) NC N/A   MASONRY LAY-UP: Filled collar joints of multiwythe masonry walls shall have negligible voids. (Tier 2: Sec. 4.4.2.5.3)

During testing of the walls, the condition of the collar joints was observed. At all of the locations observed, the collar joints were observed to have negligible voids.

Diaphragms

C (NC) N/A   CROSS TIES: There shall be continuous cross ties between diaphragm chords. (Tier 2: Sec. 4.5.1.2)

Cross ties are not continuous.

C (NC) N/A   OPENINGS AT SHEAR WALLS: Diaphragm openings immediately adjacent to the shear walls shall be less than 25 percent of the wall length for Life Safety and 15 percent of the wall length for Immediate Occupancy. (Tier 2: Sec. 4.5.1.4)

Stair opening at second level is 33 percent of the wall length in the transverse direction.

C (NC) N/A   OPENINGS AT EXTERIOR MASONRY SHEAR WALLS: Diaphragm openings immediately adjacent to exterior masonry shear walls shall not be greater than 8 feet long for Life Safety and 4 feet long for Immediate Occupancy. (Tier 2: Sec. 4.5.1.6)

Stair openings at second level longer than 8 feet.

C NC (N/A)   PLAN IRREGULARITIES: There shall be tensile capacity to develop the strength of the diaphragm at re-entrant corners or other locations of plan irregularities. This statement shall apply to the Immediate Occupancy Performance Level only. (Tier 2: Sec. 4.5.1.7)

Life Safety Performance Level.

C NC (N/A)   DIAPHRAGM REINFORCEMENT AT OPENINGS: There shall be reinforcing around all diaphragm openings larger than 50 percent of the building width in either major plan dimension. This statement shall apply to the Immediate Occupancy Performance Level only. (Tier 2: Sec. 4.5.1.8)

Life Safety Performance Level.

Appendix A: Examples

C   NC   (N/A)   **STRAIGHT SHEATHING:** All straight sheathed diaphragms shall have aspect ratios less than 2-to-1 for Life Safety and 1-to-1 for Immediate Occupancy in the direction being considered. (Tier 2: Sec. 4.5.2.1)

Diagonal sheathing throughout.

(C)   NC   N/A   **SPANS:** All wood diaphragms with spans greater than 24 feet for Life Safety and 12 feet for Immediate Occupancy shall consist of wood structural panels or diagonal sheathing (Tier 2: Sec. 4.5.2.2)

Diagonal sheathing throughout.

C   (NC)   N/A   **UNBLOCKED DIAPHRAGMS:** All diagonally sheathed or unblocked wood structural panel diaphragms shall have horizontal spans less than 40 feet for Life Safety and 30 feet for Immediate Occupancy and shall have aspect ratios less than or equal to 4-to-1 for Life Safety and 3-to-1 for Immediate Occupancy. (Tier 2: Sec. 4.5.2.3)

Diagonal sheathing aspect ratio is 3-to-1, but span is 100 feet.

C   NC   (N/A)   **NON-CONCRETE-FILLED DIAPHRAGMS:** Untopped metal deck diaphragms or metal deck diaphragms with fill other than concrete shall consist of horizontal spans of less than 40 feet and shall have span/depth ratios less than 4-to-1. This statement shall apply to the Immediate Occupancy Performance Level only. (Tier 2: Sec. 4.5.3.1)

No metal deck.

(C)   NC   N/A   **OTHER DIAPHRAGMS:** The diaphragm shall not consist of a system other than wood, metal deck, concrete, or horizontal bracing. (Tier 2: Sec. 4.5.7.1)

Diagonal sheathing throughout.

Connections

(C)   NC   N/A   **STIFFNESS OF WALL ANCHORS:** Anchors of concrete or masonry walls to wood structural elements shall be installed taut and shall be stiff enough to limit the relative movement between the wall and the diaphragm to no greater than 1/8 inch prior to engagement of the anchors. (Tier 2: Sec. 4.6.1.4)

Government anchors are observed to be tight between the wall and diaphragm.

C   NC   (N/A)   **BEAM, GIRDER, AND TRUSS SUPPORTS:** Beams, girders, and trusses supported by unreinforced masonry walls or pilasters shall have independent secondary columns for support of vertical loads. (Tier 2: Sec. 4.6.4.5)

No girders.

Appendix A: Examples

A7.3.7 Geologic Site Hazards and Foundations Checklist

The following is a completed Geologic Site Hazards and Foundations Checklist for Example 7. Each of the evaluation statements has been marked Compliant (C), Non-Compliant (NC) or Not Applicable (N/A). Compliant statements identify issues that are acceptable according to the criteria of this standard, while non-compliant statements identify issues that require further investigation. In addition, an explanation of the evaluation process is included in italics after each evaluation statement. The section numbers in parentheses following each evaluation statement correspond to Tier 2 Evaluation procedures.

Geologic Site Hazards

The following statements shall be completed for buildings in levels of high or moderate seismicity.

(C) NC N/A LIQUEFACTION: Liquefaction susceptible, saturated, loose granular soils that could jeopardize the building's seismic performance shall not exist in the foundation soils at depths within 50 feet under the building for Life Safety and Immediate Occupancy. (Tier 2: Sec. 4.7.1.1)

Liquefaction susceptible soils do not exist in the foundation soils at depths within 50 feet under the building as described in project geotechnical investigation.

(C) NC N/A SLOPE FAILURE: The building site shall be sufficiently remote from potential earthquake-induced slope failures or rockfalls to be unaffected by such failures or shall be capable of accommodating any predicted movements without failure. (Tier 2: Sec. 4.7.1.2)

The building is on a relatively flat site.

(C) NC N/A SURFACE FAULT RUPTURE: Surface fault rupture and surface displacement at the building site is not anticipated. (Tier 2: Sec. 4.7.1.3)

Surface fault rupture and surface displacements are not anticipated.

Condition of Foundations

The following statement shall be completed for all Tier 1 building evaluations.

(C) NC N/A FOUNDATION PERFORMANCE: There shall be no evidence of excessive foundation movement such as settlement or heave that would affect the integrity or strength of the structure. (Tier 2: Sec. 4.7.2.1)

The structure does not show evidence of excessive foundation movement.

The following statement shall be completed for buildings in levels of high or moderate seismicity being evaluated to the Immediate Occupancy Performance Level.

(C) NC N/A DETERIORATION: There shall not be evidence that foundation elements have deteriorated due to corrosion, sulfate attack, material breakdown, or other reasons in a manner that would affect the integrity or strength of the structure. (Tier 2: Sec. 4.7.2.2)

No evidence of deterioration.

Capacity of Foundations

The following statement shall be completed for all Tier 1 building evaluations.

C NC (N/A) POLE FOUNDATIONS: Pole foundations shall have a minimum embedment depth of 4 feet for Life Safety and Immediate Occupancy. (Tier 2: Sec. 4.7.3.1)

Pole foundations not used.

Appendix A: Examples

The following statements shall be completed for buildings in levels of moderate seismicity being evaluated to the Immediate Occupancy Performance Level and for buildings in levels of high seismicity.

(C) NC N/A OVERTURNING: The ratio of the horizontal dimension of the lateral-force-resisting system at the foundation level to the building height (base/height) shall be greater than $0.6S_a$. (Tier 2: Sec. 4.7.3.2)

Ratio of base to height = 30 ft / 25 ft = 1.2 > 0.6 S_a = 0.6(1.0) = 0.6.

C NC **(N/A)** TIES BETWEEN FOUNDATION ELEMENTS: The foundation shall have ties adequate to resist seismic forces where footings, piles, and piers are not restrained by beams, slabs, or soils classified as Class A, B, or C. (Section 3.5.2.3.1, Tier 2: Sec. 4.7.3.3)

Footings are restrained by slabs.

C NC **(N/A)** DEEP FOUNDATIONS: Piles and piers shall be capable of transferring the lateral forces between the structure and the soil. This statement shall apply to the Immediate Occupancy Performance Level only. (Tier 2: Sec. 4.7.3.4)

Life Safety Performance Level.

C NC **(N/A)** SLOPING SITES: The difference in foundation embedment depth from one side of the building to another shall not exceed one story in height. This statement shall apply to the Immediate Occupancy Performance Level only. (Tier 2: Sec. 4.7.3.5)

Life Safety Performance Level.

A7.3.8 Basic Nonstructural Component Checklist

The following is a completed Basic Nonstructural Component Checklist for Example 7. Each of the evaluation statements has been marked Compliant (C), Non-Compliant (NC) or Not Applicable (N/A). Compliant statements identify issues that are acceptable according to the criteria of this standard, while non-compliant statements identify issues that require further investigation. In addition, an explanation of the evaluation process is included in italics after each evaluation statement. The section numbers in parentheses following each evaluation statement correspond to Tier 2 Evaluation procedures.

Partitions

C NC **(N/A)** UNREINFORCED MASONRY: Unreinforced masonry or hollow clay tile partitions shall be braced at a spacing equal to or less than 10 feet in levels of low or moderate seismicity and 6 feet in levels of high seismicity. (Tier 2: Sec. 4.8.1.1)

No URM or hollow clay tile partitions.

Ceiling Systems

C NC **(N/A)** SUPPORT: The integrated suspended ceiling system shall not be used to laterally support the tops of gypsum board, masonry, or hollow clay tile partitions. Gypsum board partitions need not be evaluated where only the Basic Nonstructural Component Checklist is required by Table 3-2. (Tier 2: Sec. 4.8.2.1)

No gypsum board, masonry, or hollow clay tile partitions.

Light Fixtures

(C) NC N/A EMERGENCY LIGHTING: Emergency lighting shall be anchored or braced to prevent falling during an earthquake. (Tier 2: Sec. 4.8.3.1)

Appendix A: Examples

Emergency lighting is anchored at a spacing of less than 6 feet.

Cladding and Glazing

No exterior cladding; seven statements in this section are Not Applicable (N/A).

(C) NC N/A   GLAZING: Glazing in curtain walls and individual panes over 16 square feet in area, located up to a height of 10 feet above an exterior walking surface, shall be safety glazing. Such glazing located over 10 feet above an exterior walking surface shall be laminated annealed or laminated heat-strengthened safety glass or other glazing system that will remain in the frame where glass is cracked. (Tier 2: Sec. 4.8.4.8)

Glazing is safety glass.

Masonry Veneer

C NC **(N/A)**   SHELF ANGLES: Masonry veneer shall be supported by shelf angles or other elements at each floor 30 feet or more above ground for Life Safety and at each floor above the first floor for Immediate Occupancy. (Tier 2: Sec. 4.8.5.1)

Building height is 25 feet.

C **(NC)** N/A   TIES: Masonry veneer shall be connected to the back-up with corrosion-resistant ties. The ties shall have a spacing equal to or less than 24 inches with a minimum of one tie for every 2-2/3 square feet. A spacing of up to 36 inches is permitted where only the Basic Nonstructural Component Checklist is required by Table 3-2. (Tier 2: Sec. 4.8.5.2)

Veneer is anchored with 16d nails at 24 inches on center mortared in. The nails were observed to be corroded.

C NC **(N/A)**   WEAKENED PLANES: Masonry veneer shall be anchored to the back-up adjacent to weakened planes, such as at the locations of flashing. (Tier 2: Sec. 4.8.5.3)

Masonry veneer is anchored to footing at the ground level, and there are no weakened planes.

(C) NC N/A   DETERIORATION: There shall be no evidence of deterioration, damage, or corrosion in any of the connection elements. (Tier 2: Sec. 4.8.5.4)

No deterioration observed.

Parapets, Cornices, Ornamentation, and Appendages

C **(NC)** N/A   URM PARAPETS: There shall be no laterally unsupported unreinforced masonry parapets or cornices with height-to-thickness ratios greater than 1.5. A height-to-thickness ratio of up to 2.5 is permitted where only the Basic Nonstructural Component Checklist is required by Table 3-2. (Tier 2: Sec. 4.8.8.1)

$h/t = 1.5\ ft/9\ in. = 2 > 1.5$.

(C) NC N/A   CANOPIES: Canopies located at building exits shall be anchored to the structural framing at a spacing of 6 feet or less. An anchorage spacing of up to 10 feet is permitted where only the Basic Nonstructural Component Checklist is required by Table 3-2. (Tier 2: Sec. 4.8.8.2)

Canopy at storefront is anchored every 6 feet.

Masonry Chimneys

No masonry chimneys; statements in this section are Not Applicable (N/A).

Appendix A: Examples

Stairs

C (NC) N/A URM WALLS: Walls around stair enclosures shall not consist of unbraced hollow clay tile or unreinforced masonry with a height-to-thickness ratio greater than 12-to-1. A height-to-thickness ratio of up to 15-to-1 is permitted where only the Basic Nonstructural Component Checklist is required by Table 3-2. (Tier 2: Sec. 4.8.10.1)

See Proportions statement, Section A7.3.6.

C NC (N/A) STAIR DETAILS: In moment frame structures, the connection between the stairs and the structure shall not rely on shallow anchors in concrete. Alternatively, the stair details shall be capable of accommodating the drift calculated using the Quick Check procedure of Section 3.5.3.1 without including tension in the anchors. (Tier 2: Sec. 4.8.10.2)

Not a moment frame structure.

Building Contents and Furnishing

(C) NC N/A TALL NARROW CONTENTS: Contents over 4 feet in height with a height-to-depth or height-to-width ratio greater than 3-to-1 shall be anchored to the floor slab or adjacent structural walls. A height-to-depth or height-to-width ratio of up to 4-to-1 is permitted where only the Basic Nonstructural Component Checklist is required by Table 3-2. (Tier 2: Sec. 4.8.11.1)

Storage racks, bookcases, and file cabinets are anchored to the floor slab.

Mechanical and Electrical Equipment

(C) NC N/A EMERGENCY POWER: Equipment used as part of an emergency power system shall be mounted to maintain continued operation after an earthquake. (Tier 2: Sec. 4.8.12.1)

Emergency power systems are anchored.

(C) NC N/A HAZARDOUS MATERIAL EQUIPMENT: HVAC or other equipment containing hazardous material shall not have damaged supply lines or unbraced isolation supports. (Tier 2: Sec. 4.8.12.2)

HVAC equipment and supply lines are anchored.

(C) NC N/A DETERIORATION: There shall be no evidence of deterioration, damage, or corrosion in any of the anchorage or supports of mechanical or electrical equipment. (Tier 2: Sec. 4.8.12.3)

No deterioration observed.

(C) NC N/A ATTACHED EQUIPMENT: Equipment weighing over 20 lb that is attached to ceilings, walls, or other supports 4 feet above the floor level shall be braced. (Tier 2: Sec. 4.8.12.4)

Equipment is braced.

Piping

(C) NC N/A FIRE SUPPRESSION PIPING: Fire suppression piping shall be anchored and braced in accordance with NFPA-13 (NFPA, 1996). (Tier 2: Sec. 4.8.13.1)

Fire suppression piping is braced in accordance with NFPA-13.

(C) NC N/A FLEXIBLE COUPLINGS: Fluid, gas and fire suppression piping shall have flexible couplings. (Tier 2: Sec. 4.8.13.2)

Appendix A: Examples

Fluid, gas, and fire suppression piping have flexible couplings.

Hazardous Materials Storage and Distribution

C   NC   (N/A)   TOXIC SUBSTANCES: Toxic and hazardous substances stored in breakable containers shall be restrained from falling by latched doors, shelf lips, wires, or other methods. (Tier 2: Sec. 4.8.15.1)

No toxic or hazardous substances are stored in the building.

A7.3.9 Intermediate Nonstructural Component Checklist

The following is a completed Intermediate Nonstructural Checklist for Example 7. Each of the evaluation statements has been marked Compliant (C), Non-Compliant (NC) or Not Applicable (N/A). Compliant statements identify issues that are acceptable according to the criteria of this standard, while non-compliant statements identify issues that require further investigation. In addition, an explanation of the evaluation process is included in italics after each evaluation statement. The section numbers in parentheses following each evaluation statement correspond to Tier 2 Evaluation procedures.

Ceiling Systems

(C)   NC   N/A   LAY-IN TILES: Lay-in tiles used in ceiling panels located at exits and corridors shall be secured with clips. (Tier 2: Sec. 4.8.2.2)

Lay-in tiles are secured with clips.

(C)   NC   N/A   INTEGRATED CEILINGS: Integrated suspended ceilings at exits and corridors or weighing more than 2 lb/ft^2 shall be laterally restrained with a minimum of 4 diagonal wires or rigid members attached to the structure above at a spacing equal to or less than 12 feet. (Tier 2: Sec. 4.8.2.3)

Integrated suspended ceilings are braced.

(C)   NC   N/A   SUSPENDED LATH AND PLASTER: Ceilings consisting of suspended lath and plaster or gypsum board shall be attached to resist seismic forces for every 12 square feet of area. (Tier 2: Sec. 4.8.2.4)

Each 10 square feet of suspended lath and plaster ceilings is attached.

Light Fixtures

(C)   NC   N/A   INDEPENDENT SUPPORT: Light fixtures in suspended grid ceilings shall be supported independently of the ceiling suspension system by a minimum of two wires at diagonally opposite corners of the fixtures. (Tier 2: Sec. 4.8.3.2)

Light fixtures are independently supported.

Cladding and Glazing

(C)   NC   N/A   GLAZING: All exterior glazing shall be laminated, annealed or laminated heat-strengthened safety glass or other glazing system that will remain in the frame when glass is cracked. (Tier 2: Sec. 4.8.4.9)

Glazing is safety glass.

Parapets, Cornices, Ornamentation, and Appendages

C   NC   (N/A)   CONCRETE PARAPETS: Concrete parapets with height-to-thickness ratios greater than 2.5 shall have vertical reinforcement. (Tier 2: Sec. 4.8.8.3)

Appendix A: Examples

No concrete parapets.

C   NC   (N/A)   APPENDAGES: Cornices, parapets, signs, and other appendages that extend above the highest point of anchorage to the structure or cantilever from exterior wall faces and other exterior wall ornamentation shall be reinforced and anchored to the structural system at a spacing equal to or less than 10 feet for Life Safety and 6 feet for Immediate Occupancy. This requirement need not apply to parapets or cornices compliant with Section 4.8.8.1 or 4.8.8.3. (Tier 2: Sec. 4.8.8.4)

No appendages.

Masonry Chimneys

No masonry chimneys; statements in this section are Not Applicable (N/A).

Mechanical and Electrical Equipment

C   NC   (N/A)   VIBRATION ISOLATORS: Equipment mounted on vibration isolators shall be equipped with restraints or snubbers. (Tier 2: Sec. 4.8.12.5)

No equipment mounted on vibration isolators.

Ducts

C   NC   (N/A)   STAIR AND SMOKE DUCTS: Stair pressurization and smoke control ducts shall be braced and shall have flexible connections at seismic joints. (Tier 2: Sec. 4.8.14.1)

No stair pressurization or smoke control ducts.

A7.4 Further Evaluation Phase (Tier 2)

The following potential deficiencies were identified by the Tier 1 Checklists:

1. Shear Stress Check
2. Wall Anchorage
3. Proportions
4. Cross Ties
5. Openings at Shear Walls
6. Openings at Exterior Masonry Shear Walls
7. Diaphragm Span
8. Masonry Veneer Ties
9. URM Parapets
10. URM Walls around Stairs

Since the building is an unreinforced masonry bearing wall structure with flexible diaphragms, a Tier 2 Evaluation is completed by following the Special Procedure for Unreinforced Masonry in accordance with Section 4.2.6. A Tier 3 Evaluation is not required.

Appendix A: Examples

A7.4.1 Evaluation Requirements

A7.4.1.1 Condition of Materials

The building meets the requirements of Section 4.3.3. No significant deterioration of the wood or masonry was observed. No significant brick cracking was observed. In addition, the building meets the requirements of Section 4.4.2.5.3. Collar joints in the walls appear to have negligible voids.

A7.4.1.2 Testing

A7.4.1.2.1 In-Place Mortar Test

In-place shear tests were performed to determine the quality of mortar in the URM walls. In accordance with Section 4.2.6.2.2, two tests were performed per line of wall elements per floor. A total of 16 tests was completed for this building. This meets the minimum requirement of two tests per wall at the top and bottom stories, eight tests per building, and one test per 1,500 square feet of wall surface. At each location, the test strength was reduced by axial stress due to the weight of the wall above in accordance with Equation (4-16). These values ranged from 30 psi to 180 psi.

Since all test specimens were tested to have a mortar shear strength, v_{te}, greater than 30 psi, Section 4.2.6.2.2.1 is satisfied. Refer to the following table of testing values.

Test #	V_{test} (lbs)	A_b (in.^{2})	p_{D+L} (psi)	v_{to} (psi)
1	14,400	64	48	177
2	7,000	67	48	56
3	10,800	72	48	102
4	8,600	69	48	77
5	9,900	70	48	93
6	11,200	68	48	117
7	5,800	66	48	40
8	8,800	67	48	83
9	6,400	72	21	68
10	4,000	69	21	37
11	6,700	66	21	81
12	5,500	70	21	58
13	3,400	64	21	32
14	5,100	80	21	43
15	6,900	74	21	72
16	4,700	65	21	51

A7.4.1.2.2 Wall Anchors

Fourteen wall anchors or 11 percent of the wall anchors were pull tested in accordance with Section 4.2.6.2.2.3. This meets the minimum requirement of four anchors per floor or 10 percent of the total number of tension anchors at each level. The average of tension test values of the anchors was 1.5 kips.

Appendix A: Examples

A7.4.1.2.3 Masonry Strength

A7.4.1.2.3.1 Shear Strength

The expected unreinforced masonry strength, v_{me}, was determined for each masonry class in accordance with Equation (4-19).

The transverse direction is critical because it has greater load on shorter walls.

Weight of parapet and upper half of the wall: $P_D = (1.5 \text{ ft} + 5.5 \text{ ft})(30 \text{ ft})(0.090 \text{ ksf})(0.75) = 14.1$ kips.

Net area through the piers: $A_n = (9 \text{ in.})(30 \text{ ft} - 6(3 \text{ ft})) = 1,296 \text{ in.}^2$.

v_{te} was taken to be the 43 psi since 80 percent of the testing values exceeded this value.

Therefore, $v_{me} = 0.56\, v_{te} + 0.75 P_D/A_n = 0.56(43) + 0.75(14,100)/1,296 = 32$ psi.

A7.4.1.2.3.2 Axial Strength

The allowable compressive stress in unreinforced masonry due to dead plus live loads is taken to be 300 psi in accordance with Section 4.2.6.2.3.2. Tensile stress is not permitted in unreinforced masonry.

A7.4.2 Analysis

A7.4.2.1 Cross Walls

A7.4.2.1.1 General

Cross walls are spaced less than 40 feet on center in each direction and are in each story of the building. The cross walls extend the full story height between diaphragms and have a length-to-height ratio between openings equal to or greater than 1.5.

A7.4.2.1.2 Shear Strength

Cross walls consist of plaster on wood. By Table 4-1, the allowable shear strength of the cross walls is 600 plf. This is equal to one-third of the floor diaphragm strength for a 40-foot width along the span of the diaphragm.

A7.4.2.2 Diaphragms

A7.4.2.2.1 Shear Strength

The materials meet the requirements of Section 4.3.3. Therefore, the diaphragm strengths can be assumed from Table 4-2. The floor diaphragm with diagonal sheathing has an allowable shear strength of 1,800 plf. The roof diaphragm with diagonal sheathing has an allowable shear strength of 750 plf.

A7.4.2.2.2 Demand-Capacity Ratios and Acceptability Criteria

Transverse:

Calculate the diaphragm weight:

- Roof: $(30 \text{ ft})(100 \text{ ft})(0.027 \text{ ksf}) = 81$ kips
- Floor: $(30 \text{ ft})(100 \text{ ft})(0.035 \text{ ksf}) = 105$ kips

The weight of the walls is obtained by multiplying the gross area by a factor of 0.75 at the upper story and of 0.80 at the lower story to account for openings.

- North wall: Parapet: (1.5 ft)(100 ft)(0.090 ksf) = 14 kips

    2nd story wall: (11 ft)(100 ft)(0.090 ksf)(0.75) = 74 kips

    1st story wall: (7 ft)(100 ft)(0.130 ksf)(0.80) = 73 kips
- South wall: Same as above = 14 + 74 + 73 = 161 kips

Thus: For roof diaphragm, W_d = 81 + 2(14+74/2) = 183 kips, and

for roof and second floor diaphragm, ΣW_d = 81 + 105 + 2(161) = 508 kips.

Check the roof diaphragm per Equation (4-25):

$$DCR = 2.1 S_{D1} \Sigma W_d / \Sigma(\Sigma v_u D) = 2.1(0.52)(183 \text{ kips})/2(30 \text{ ft})(0.75 \text{ klf} + 1.8 \text{ klf}) = 1.31.$$

In Figure 4-1, with DCR = 1.31 and L = 100 feet, the point falls below the curve, in region 3. Therefore, the roof diaphragm is compliant.

Check the second-floor diaphragm per Equation (4-24).

$$DCR = 2.1 S_{D1} \Sigma W_d / \Sigma(\Sigma v_u D + V_{cb}) = 2.1(0.52)(508 \text{ kips})/2(30 \text{ ft})(0.75 \text{ klf} + 1.8 \text{ klf} + 4(0.6 \text{ klf})) = 1.87.$$

In Figure 4-1, with DCR = 1.87 and L = 100 feet, the point falls below the curve, in region 3. Therefore, the second-floor diaphragm is compliant.

Longitudinal:

The weight of the walls is obtained by multiplying the gross area by a factor of 0.75 at the upper story and of 0.80 at the lower story to account for openings.

- North wall: Parapet: (1.5 ft)(30 ft)(0.090 ksf) = 4 kips

    2nd story wall: (11 ft)(30 ft)(0.090 ksf)(0.75) = 22 kips

    1st story wall: (7 ft)(30 ft)(0.130 ksf)(0.80) = 22 kips
- South wall: Same as above = 4 + 22 + 22 = 48 kips

Thus, for roof diaphragm, W_d = 81 + 2(4+22/2) = 111 kips, and

for roof and second-floor diaphragm, ΣW_d = 81 + 105 + 2(48) = 282 kips

Check the roof diaphragm per Equation (4-25):

$$DCR = 2.1 S_{D1} \Sigma W_d / (\Sigma v_u D + V_{cb}) = 2.1(0.52)(111 \text{ kips})/2(100 \text{ ft})(0.75 \text{ klf} + 1.8 \text{ klf}) = 0.24.$$

In Figure 4-1, with DCR = 0.24 and L = 30 feet, the point falls below the curve, in region 3. Therefore, the roof diaphragm is compliant.

Check the second-floor diaphragm per Equation (4-24).

$$DCR = 2.1 S_{D1} \Sigma W_d / \Sigma(\Sigma v_u D + V_{cb}) = 2.1(0.52)(282 \text{ kips})/2(100 \text{ ft})(0.75 \text{ klf} + 1.8 \text{ klf} + 2(0.6 \text{ klf})) = 0.41.$$

In Figure 4-1, with DCR = 0.41 and L = 30 feet, the point falls below the curve, in region 3. Therefore, the second-floor diaphragm is compliant.

A7.4.2.2.3 Chords and Collectors

Walls extend the length of the diaphragm, so chords and collectors are not required.

A7.4.2.2.4 Diaphragm Openings

There is a 10-foot × 10-foot stair opening on the second floor. Therefore, the diaphragm must be checked around the opening.

Appendix A: Examples

Transverse:

Check the second-floor diaphragm per Equation (4-24).

$$DCR = 2.1 S_{DI} \Sigma W_d / \Sigma(\Sigma v_u D + V_{cb}) = 2.1(0.52)(508 \text{ kips})/2(20 \text{ ft})(0.75 \text{ klf} + 1.8 \text{ klf} + 4(0.6 \text{ klf})) = 2.81.$$

In Figure 4-1, with $DCR = 2.81$ and $L = 100$ feet, the point falls below the curve, in region 2. Therefore, the second-floor diaphragm is acceptable.

Calculate the collector force needed to develop diaphragm opening.

$$\text{Collector force} = 2.1(0.52)(508 \text{ kips})(10 \text{ ft})/2(30 \text{ ft}) = 92 \text{ kips},$$

but needs not exceed the capacity of the diaphragm:

$$\text{Collector force} = 1.8 \text{ klf}(10 \text{ ft}) = 18 \text{ kips}.$$

There are no metal straps present to develop this force.

Longitudinal:

Check the second-floor diaphragm per Equation (4-24).

$$DCR = 2.1 S_{DI} \Sigma W_d / \Sigma(\Sigma v_u D + V_{cb}) = 2.1(0.52)(282 \text{ kips})/2(90 \text{ ft})(0.75 \text{ klf} + 1.8 \text{ klf} + 2(0.6 \text{ klf})) = 0.46.$$

In Figure 4-1, with $DCR = 0.46$ and $L = 30$ ft, the point falls below the curve, in Region 3. Therefore, the second floor diaphragm is acceptable.

Calculate the collector force needed to develop diaphragm opening:

$$\text{Collector force} = 2.1(0.52)(282 \text{ kips})(10 \text{ ft})/2(100 \text{ ft}) = 15 \text{ kips} < 18 \text{ kips}.$$

There are no metal straps present to develop this force.

A7.4.2.2.5 Diaphragm Shear Transfer

The joists are parallel to the wall. The government anchors at 4 feet on center are hooked into blocking in the first two joist spaces. The response factor, C_p, for a diaphragm with a single layer of boards with applied roofing is 0.5 per Table 4-3. The connection demand is determined using Equations (4-26) and (4-27):

$$V_d = 1.25 \; S_{DI} \; C_p W_d = 1.25(0.52)(0.5)(183/2) = 29.7 \text{ kips},$$

but not to exceed $v_u D = 0.75(30) = 22.5$ kips.

The demand is 22.5 kips or (22.5 kips)(4 ft)/(30 ft) = 3 kips per anchor. The shear capacity of the existing anchors is less than the 1.5 kips tested value for the pull-out tests. Thus, the connections are deficient.

A7.4.2.3 Shear Walls

A7.4.2.3.1 Shear Wall Actions

Transverse:

Calculate the in-plane shear demands in accordance with Equations (4-30), (4-31), (4-32) and (4-33):

Story force at roof (one wall):

$$F_{wr} = 0.75 \; S_{DI} \; (W_{wx} + 0.5 W_d) = 0.75(0.52)(15 \text{ kips} + 0.5(183 \text{ kips})) = 41.5 \text{ kips},$$

but needs not exceed:

$$F_{wr} = 0.75 \; S_{DI} \; W_{wx} + v_u D = 0.75(0.52)(15 \text{ kips}) + (0.75 \text{ klf})(30 \text{ ft}) = 28.4 \text{ kips}.$$

Story force at second floor (one wall):

Appendix A: Examples

$F_{w2} = 0.75\ S_{D1}(W_{wx}+0.5W_d) = 0.75(0.52)(33\text{ kips}+0.5(325\text{ kips})) = 76.2$ kips,

but needs not exceed:

$F_{w2} = 0.75\ S_{D1}(W_{wx}+\Sigma W_d\ v_u D/\Sigma\Sigma\ v_u D) = 0.75(0.52)(33\text{ kips}+(508\text{ kips})(1.8/2.55)) = 152.7$ kips,

and needs not exceed:

$F_{w2} = 0.75\ S_{D1}\ W_{wx}+v_u D = 0.75(0.52)(33\text{ kips})+(1.8\text{ klf})(30\text{ ft}) = 66.9$ kips.

Total shear (one wall):

$V = \Sigma F_{wx} = 28.4\text{ kips} + 66.9\text{ kips} = 95.3$ kips.

Longitudinal:

Calculate the in-plane shear demands in accordance with Equations (4-30), (4-31), (4-32), and (4-33).

Story force at roof (one wall):

$F_{wr} = 0.75\ S_{D1}(W_{wx}+0.5W_d) = 0.75(0.52)(51\text{ kips}+0.5(111\text{ kips})) = 41.5$ kips,

but needs not exceed:

$F_{wr} = 0.75\ S_{D1}\ W_{wx}+v_u D = 0.75(0.52)(51\text{ kips})+(0.75\text{ klf})(100\text{ ft}) = 94.9$ kips.

Story force at second floor (one wall):

$F_{w2} = 0.75\ S_{D1}(W_{wx}+0.5W_d) = 0.75(0.52)(73.5\text{ kips}+0.5(171\text{ kips})) = 62.0$ kips,

but needs not exceed:

$F_{w2} = 0.75\ S_{D1}(W_{wx}+\Sigma W_d\ v_u D/\Sigma\Sigma\ v_u D) = 0.75(0.52)(73.5\text{ kips}+(282\text{ kips})(1.8/2.55)) = 106.3$ kips,

and needs not exceed:

$F_{w2} = 0.75\ S_{D1}\ W_{wx}+v_u D = 0.75(0.52)(73.5\text{ kips})+(1.8\text{ klf})(100\text{ ft}) = 208.7$ kips.

Total shear (one wall):

$V = \Sigma F_{wx} = 41.5\text{ kips} + 62.0\text{ kips} = 103.5$ kips.

A7.4.2.3.2 Shear Wall Strengths

Transverse:

- Second story:

    There are three 4-feet × 6-feet openings (windows) along the length of the wall. The typical pier width is 4.5 feet wide × 6 feet high.

    Calculate the shear wall strength in accordance with Equation (4-34).

    $V_a = 0.67 v_{me} Dt = 0.67(32\text{ psi})(4.5\text{ ft})(9\text{ in.}) = 10.4$ kips.

    Calculate the rocking shear strength in accordance with Equation (4-36).

    $V_r = 0.9 P_D(D/H) = 0.9(4.3\text{ kips})(4.5\text{ ft}/6\text{ ft}) = 2.9$ kips.

- First story:

    There are two 6-feet × 6-feet openings (windows) along the length of the wall. The typical pier width is 6 feet wide × 6 feet high.

    Calculate the shear wall strength in accordance with Equation (4-34).

    $V_a = 0.67 v_{me} Dt = 0.67(32\text{ psi})(6\text{ ft})(13\text{ in.}) = 20.1$ kips.

Calculate the rocking shear strength in accordance with Equation (4-36).

$$V_r = 0.9P_D(D/H) = 0.9(15.8 \text{ kips})(6 \text{ ft}/6 \text{ ft}) = 14.2 \text{ kips.}$$

Longitudinal:

- Second story:

There are six 4-feet × 6-feet openings (windows) along the length of the wall. The typical pier width is 11 feet wide × 6 feet high.

Calculate the shear wall strength in accordance with Equation (4-34).

$$V_a = 0.67v_{me}Dt = 0.67(32 \text{ psi})(11 \text{ ft})(9 \text{ in.}) = 25.5 \text{ kips.}$$

Calculate the rocking shear strength in accordance with Equation (4-36).

$$V_r = 0.9P_D(D/H) = 0.9(7.9 \text{ kips})(11 \text{ ft}/6 \text{ ft}) = 13 \text{ kips.}$$

- First story:

There are three 8-feet × 8-feet openings (doors) along the length of the wall. The typical pier width is 19 feet wide × 8 feet high.

Calculate the shear wall strength in accordance with Equation (4-34).

$$V_a = 0.67v_{me}Dt = 0.67(32 \text{ psi})(19 \text{ ft})(13 \text{ in.}) = 63.5 \text{ kips.}$$

Calculate the rocking shear strength in accordance with Equation (4-36).

$$V_r = 0.9P_D(D/H) = 0.9(39.4 \text{ kips})(19 \text{ ft}/8 \text{ ft}) = 84.2 \text{ kips.}$$

A7.4.2.3.3 Shear Wall Acceptance Criteria

Transverse:

- Second story:

$$V_{w2} = 28.4 \text{ kips.}$$

Since $V_r < V_a$, check acceptability of the walls in accordance with Equation (4-37).

$$0.7V_{wx} = 0.7(28.4 \text{ kips}) = 19.9 \text{ kips,}$$

$$\Sigma V_r = 4(2.9 \text{ kips}) = 11.6 \text{ kips} < 19.9 \text{ kips.}$$

The shear walls at the second floor are non-compliant.

- First story:

$$V_{w2} = 95.3 \text{ kips.}$$

Since $V_r < V_a$, check acceptability of the walls in accordance with Equation (4-37).

$$0.7V_{wx} = 0.7(95.3 \text{ kips}) = 66.7 \text{ kips,}$$

$$\Sigma V_r = 3(14.2 \text{ kips}) = 42.6 \text{ kips} < 66.7 \text{ kips.}$$

The shear walls at the first floor are non-compliant.

Longitudinal:

- Second story:

$$V_{w2} = 41.5 \text{ kips.}$$

Since $V_r < V_a$, check acceptability of the walls accordance with Equation (4-37).

Appendix A: Examples

$$0.7V_{wx} = 0.7(41.5 \text{ kips}) = 29.1 \text{ kips},$$

$$\sum V_r = 7(13 \text{ kips}) = 91 \text{ kips} > 29.1 \text{ kips}.$$

The shear walls at the second floor are compliant.

- First story:

$$V_{wl} = 103.5 \text{ kips}.$$

Since $V_a < V_r$, check acceptability of the walls in accordance with Equations (4-38) and (4-39).

$$V_p = 103.5 \text{ kips}/4 = 25.9 \text{ kips} < 63.5 \text{ kips and } 84.2 \text{ kips}.$$

The shear walls at the first floor are compliant.

A7.4.2.3.4 Out-of-Plane Demands

The demand-capacity ratios for the diaphragms all fell within Region 3. Therefore, the h/t limit from Table 4-4 is 9.0 for the roof. The actual h/t ratio is 12 ft/9 in. = 14.6. Therefore, the second-story walls are non-compliant for out-of-plane demands.

For the first story, the h/t limit is 15.0. The actual h/t ratio is 14 ft/13 in. = 12.9. Therefore, the first-story walls are compliant.

A7.4.2.3.5 Wall Anchorage

The existing anchors are 3/4-inch-diameter government anchors at 4 feet on center. In accordance with Section 4.2.6.3.5, anchors shall be capable of developing the maximum of:

- $2.1(S_{D1})$(weight of wall) = $2.1(0.52)(4 \text{ ft})((5.5 \text{ ft})(0.09 \text{ ksf})+(7 \text{ ft})(0.13 \text{ ksf})) = 6.1$ kips
- 200 lb per lineal foot = (0.2 kips)(4 ft) = 0.8 kips per anchor

Pull-out tests of the anchors limit the anchor tension force to 1.5 kips. Therefore, the connections are deficient.

A7.4.2.3.6 Buildings with Open Fronts

This building is two stories. Therefore, this provision is not applicable.

A7.4.3 Nonstructural Components

For the masonry veneer ties, the nails were observed to be corroded such that they cannot be used to resist seismic forces. Therefore, the masonry veneer ties remain non-compliant.

The h/t ratio for the parapets is non-compliant from the nonstructural checklist. Since there is no anchorage, no check is performed. The h/t ratio of the walls at the stair also is non-compliant.

A7.5 Final Evaluation and Report

The following is a list of the deficiencies remaining following a Tier 2 Evaluation:
- Development of Diaphragm Opening
- Transfer of Diaphragm Shear
- Shear Wall Strength in the Transverse Direction
- Out-of-Plane Wall Forces
- Out-of-Plane Wall Anchorage
- Masonry Veneer Ties
- Parapets

The deficiencies listed above should be mitigated. Possible resources for information regarding mitigation measures are listed in Section 1.3. A final report that outlines the findings of the Tier 1 and Tier 2 Evaluations shall be prepared.

Appendix B: Summary Data Sheet

BUILDING DATA
Building Name: _____ Date: _____
Building Address: _____
Latitude: _____ Longitude: _____ By: _____
Year Built: _____ Year(s) Remodeled: _____ Original Design Code: _____
Area (sf): _____ Length (ft): _____ Width (ft): _____
No. of Stories: _____ Story Height: _____ Total Height: _____

USE ☐ Industrial ☐ Office ☐ Warehouse ☐ Hospital ☐ Residential ☐ Educational ☐ Other: _____

CONSTRUCTION DATA
Gravity Load Structural System: _____
Exterior Transverse Walls: _____ Openings? _____
Exterior Longitudinal Walls: _____ Openings? _____
Roof Materials/Framing: _____
Intermediate Floors/Framing: _____
Ground Floor: _____
Columns: _____ Foundation: _____
General Condition of Structure: _____
Levels Below Grade? _____
Special Features and Comments: _____

LATERAL-FORCE-RESISTING SYSTEM

	Longitudinal	Transverse
System:	_____	_____
Vertical Elements:	_____	_____
Diaphragms:	_____	_____
Connections:	_____	_____

EVALUATION DATA
Spectral Response Accelerations: $S_s=$ _____ $S_1=$ _____
Soil Factors: Class= _____ $F_a=$ _____ $F_v=$ _____
Design Spectral Response Accelerations: $S_{DS}=$ _____ $S_{D1}=$ _____
Level of Seismicity: _____ Performance Level: _____
Building Period: $T=$ _____
Spectral Acceleration: $S_a=$ _____
Modification Factor: $C=$ _____ Building Weight: $W=$ _____
Pseudo Lateral Force: $V=CS_aW=$ _____

BUILDING CLASSIFICATION: _____

REQUIRED TIER 1 CHECKLISTS
	Yes	No
Basic Structural Checklist	☐	☐
Supplemental Structural Checklist	☐	☐
Geologic Site Hazards and Foundations checklist	☐	☐
Basic Nonstructural Component Checklist	☐	☐
Intermediate Nonstructural Checklist	☐	☐
Supplemental Nonstructural Checklist	☐	☐

FURTHER EVALUATION REQUIREMENT: _____

Index

access floors 4-141
actions: components 4-9; deformation-controlled 4-10, 4-12–13; force-controlled 4-10–11, 4-13; shear walls 4-23–24
anchorage: connections 4-106–108; deterioration 4-142; intermediate nonstructural component checklist 3-125; out-of-plane 4-14; Tier 2 evaluation procedures 4-136, 4-139; walls 3-78, 3-82, 3-85, 3-92, 3-96, 3-99, 3-103, 3-110, 4-106–107
anchors: hold-down 3-24, 3-27, 3-30, 3-115, 4-86; post-tensioning 3-63, 3-67, 3-82, 3-85, 3-89, 3-106; stiffness of wall 3-57, 3-76, 3-94, 3-101, 3-117, 4-108; strength 2-1; Tier 2 evaluation procedures 4-45
appendages 3-125, 4-138
axial stress: braced frames 4-88; concrete moment frames 4-60; general basic structural checklist 3-107, 3-109; moment frames 4-54; PC2A buildings 3-89; S1 buildings 3-32; S1A buildings 3-35; S2 buildings 3-39; S2A buildings 3-42; S3 buildings 3-45; Tier 1 analysis 3-18

back-up systems: concrete block and masonry 4-136–137; metal stud 4-136; supplemental nonstructural component checklist 3-128; URM 4-137
beams: bars 3-60, 3-113, 4-62; coupling 3-50, 3-64, 3-68, 3-79, 3-83, 3-86, 3-114, 4-73; penetrations 3-33, 3-36, 3-47, 3-112, 4-57; supports 3-101, 3-104, 3-118, 4-117
benchmark buildings 3-3–4; by building type 3-4
bracing: bottom flange 3-33, 3-36, 3-47, 3-112, 4-58–59; chevron 3-40, 3-43, 4-91; diagonal 3-17; horizontal 4-105; k-bracing 3-40, 3-43, 3-116, 4-90; openings 3-26, 3-29; out-of-plane 3-33, 3-36, 3-40, 3-43, 3-47, 3-112, 3-115, 4-58, 4-89; tension-only 3-43, 4-91

building contents 3-128, 4-140–141
building data 2-2–3
building systems, Tier 2 evaluation procedures 4-33–49
building types 2-5–11; Tier 1 checklist locator 3-21
buildings, adjacent: building systems evaluation 4-34; C1 buildings 3-58; C2A buildings 3-66; C3A buildings 3-73; general basic structural checklist 3-105; PC1 buildings 3-77; PC2A buildings 3-88; RM1 buildings 3-91; S1 buildings 3-31; S1A buildings 3-34; S2 buildings 3-38; S2A buildings 3-41; S5A buildings 3-54; URM buildings 3-98
buildings, benchmark 3-3–4; by building type 3-4

C1 building type: basic structural checklist 3-58–59; described 2-9, 3-58; supplemental structural checklist 3-60–61
C2 building type: basic structural checklist 3-62–63; described 2-9, 3-62; supplemental structural checklist 3-64–65
C2A building type: basic structural checklist 3-66–67; described 2-9, 3-66; supplemental structural checklist 3-68–69
C3 building type: basic structural checklist 3-70–71; described 2-10, 3-70; evaluation example A-33–42; supplemental structural checklist 3-72
C3A building type: basic structural checklist 3-73–74; described 2-10, 3-73; supplemental structural checklist 3-75–76
canopies 3-20, 3-123, 4-137–138
case studies: C3 buildings A-33–42; RM2 buildings A-43–57; S1A building type A-14–32; S2 buildings A-72–86; URM buildings A-87–108; W1 buildings A-2–13; W2 buildings A-58–71
ceiling systems: edges 4-128; integrated ceilings 4-127; intermediate nonstructural component checklist 3-125; lay-in tiles 4-

127; seismic joint 4-128; supplemental structural checklist 3-127; support 4-126–127; suspended lath and plaster 4-127–128; Tier 2 evaluation procedures 4-126–128

checklists: basic nonstructural component 3-122–124; geologic site hazards and foundations 3-119–120; intermediate nonstructural component 3-125–126; by level of seismicity 3-6; low seismicity 3-20; nonstructural 3-121–129; selection 3-5–6; structural 3-21–118; supplemental nonstructural component 3-127–129; for Tier 1 evaluation 3-6

checklists, basic structural: C1 buildings 3-58–59; C2 buildings 3-62–63; C2A buildings 3-66–67; C3 buildings 3-70–71; C3A buildings 3-73–74; general checklist 3-105–111; PC1 buildings 3-77–78; PC1A buildings 3-81–82; PC2 buildings 3-84–85; PC2A buildings 3-88–89; RM1 buildings 3-91–92; RM2 buildings 3-95–96; S1 buildings 3-31–32; S1A buildings 3-34–35; S2 buildings 3-38–39; S2A buildings 3-41–42; S3 buildings 3-45–46; S4 buildings 3-48–49; S5 buildings 3-51–52; S5A buildings 3-54–55; URM buildings 3-98–99; URMA buildings 3-102–103; W1 buildings 3-22–23; W1A buildings 3-25–26; W2 buildings 3-28–29

checklists, supplemental structural: C1 buildings 3-60–61; C2 buildings 3-64–65; C2A buildings 3-68–69; C3 buildings 3-72; C3A buildings 3-75–76; general checklist 3-112–118; PC1 buildings 3-79–80; PC1A buildings 3-83; PC2 buildings 3-86–87; PC2A buildings 3-90; RM1 buildings 3-93–94; RM2 buildings 3-97; S1 buildings 3-33; S1A buildings 3-36–37; S2 buildings 3-40; S2A buildings 3-43–44; S3 buildings 3-47; S4 buildings 3-50; S5 buildings 3-53; S5A buildings 3-56–57; URM buildings 3-100–101; URMA buildings 3-104; W1 buildings 3-24; W1A buildings 3-27; W2 buildings 3-30

chimneys, masonry 3-123, 4-138–139
cladding 3-20, 3-122–123, 4-130–133
columns: captive 3-60, 3-90, 3-112, 4-61; concrete 3-59, 3-71, 3-74, 3-110, 4-111; cracks in boundary columns 3-71, 3-74, 3-106, 4-49; shear-wall-boundary 3-49, 3-110, 4-113; splices 3-109, 3-112, 4-56; steel 3-32, 3-35, 3-39, 3-42, 3-46, 3-52, 3-55, 3-110, 4-110–111; tie spacing 3-60, 3-113

components: actions 4-9; concrete 4-29; demands 4-26–27; gravity loads 4-8; masonry 4-30; primary 4-7; secondary 4-7; steel 4-28; strength 4-11–12; vertical 4-110–115; wood 4-30

components, nonstructural: acceptance criteria 4-27; Tier 2 evaluation procedures 4-26–32, 4-31–32, 4-124–151

concrete 4-44–45

concrete deterioration: C1 buildings 3-59; C2 buildings 3-63; C2A buildings 3-67; C3 buildings 3-71; C3A buildings 3-74; general basic structural checklist 3-106; PC2A buildings 3-85, 3-89; RM2 buildings 3-96; S1 buildings 3-32; S2 buildings 3-39; S4 buildings 3-49; S5 buildings 3-52; URMA buildings 3-103

concrete frames with infill masonry shear walls: *see* C3 or C3A building types

concrete moment frames: *see* C1 building type

concrete shear walls: *see* C2 or C2A building types

connections 4-11; bearing 4-132; deterioration 4-131, 4-134–135; girder/column 3-23, 3-26, 3-29, 3-78, 3-82, 3-85, 3-89, 3-92, 3-96, 3-99, 3-103, 3-111, 4-115; inserts 4-132; moment-resisting 3-33, 3-36, 3-47, 3-112, 4-55; panel-to-panel 3-79, 3-83, 3-114, 4-78; panels 4-117–118, 4-132–133; precast 3-86, 3-107, 3-113, 4-66–67; precast panel 3-80, 3-83, 3-117, 4-108; roof chord 3-27, 3-30; roof panel 4-118; roof panels 3-118; strength 3-40, 3-43, 3-115, 4-89; Tier 2 evaluation procedures 4-106–118; walls

3-50, 3-52, 3-55, 3-108, 3-114, 4-75, 4-81
construction documents 2-1
continuity: diaphragms 3-24, 3-27, 3-30; roof chord 3-24, 4-98
continuity plates, girder flange 3-33, 3-36, 3-112, 4-58
corbels 4-116; general supplemental structural checklist 3-118; PC2A buildings 3-90
cornices 4-137–139
cracks, stone 4-135–136
cross walls 4-19–20

data sheet, summary B-1
deflection compatibility: C2 buildings 3-64; C3 buildings 3-72; general supplemental structural checklist 3-113; PC1A buildings 3-83; PC2 buildings 3-86; PC2A buildings 3-90; Tier 2 evaluation procedures 4-68
deformations 4-5–6
demand-capacity ratios 4-21
destructive testing 2-1–2
deterioration of structural materials 4-43–49
diagonals 3-40, 3-43, 3-115, 4-89
diaphragms: C1 buildings 3-61; C2 buildings 3-64; C2A buildings 3-68–69; C3 buildings 3-72; C3A buildings 3-75–76; concrete 4-104; continuity 3-24, 3-27, 3-30, 3-68, 3-72, 4-96–97; flexible 3-11, 3-18–19; general supplemental structural checklist 3-116–117; linear dynamic procedure 4-5, 4-7; linear static procedure 4-4; metal deck 4-103–104; non-concrete filled 4-104; PC1 buildings 3-79–80; PC1A buildings 3-83; PC2 buildings 3-86; PC2A buildings 3-90; plan irregularities 4-100; precast concrete 4-105; reinforcement 3-27, 4-101; RM1 buildings 3-93; RM2 buildings 3-97; S1 buildings 3-33; S1A buildings 3-36–37; S2 buildings 3-40; S2A buildings 3-43; S3 buildings 3-47; S4 buildings 3-50; S5 buildings 3-53; S5A buildings 3-56; shear transfer 4-23, 4-108–110; spans of wood 4-102–103; Tier 2 evaluation procedures 4-94–106; unblocked 3-24, 4-103; URM buildings 3-100–101, 4-20–23; wood 4-102–103
discontinuities, vertical: building systems evaluation 4-38–39; C1 buildings 3-58; C2 buildings 3-62; C2A buildings 3-66; C3 buildings 3-70; C3A buildings 3-74; general basic structural checklist 3-105; PC1 buildings 3-77; PC1A buildings 3-81; PC2 buildings 3-84; PC2A buildings 3-88; RM1 buildings 3-91; RM2 buildings 3-95; S1 buildings 3-32; S1A buildings 3-35; S2 buildings 3-38; S2A buildings 3-42; S3 buildings 3-45; S4 buildings 3-48; S5 buildings 3-52; S5A buildings 3-55; URM buildings 3-98; URMA buildings 3-102; W1 buildings 3-22; W1A buildings 3-25; W2 buildings 3-28
displacement, surface 3-119, 4-119
doors 3-128, 4-144
drift 4-125–126
drift check 3-35; general basic structural checklist 3-107; moment frames 3-32, 4-53–54
ducts 3-126, 3-129, 4-147

elevators 3-129, 4-149–151
equipment, attached 4-143
equipment, electrical 3-126, 3-128, 4-144
equipment, hazardous material 3-20, 3-124, 4-142
equipment, heavy 4-143
equipment, mechanical 3-126, 3-128, 4-142–144
evaluation phase 1-4–5; Tier 2 requirements 4-1–151
evaluation process: checklists 3-5–6; Tier 1 3-2; Tier 1 analysis 3-9–19; Tier 2 requirements 3-7–8
exemptions 1-2

fasteners 3-22, 3-25, 3-29, 3-106, 4-43–44
foundations 3-119–120; capacity 4-121–123; conditions 4-120–121; deep 4-122–123; deterioration 4-120–121; dowels 3-49, 3-63, 3-67, 3-85, 3-92, 3-96, 3-110, 4-113; foundation/soil interface 4-11;

overturning 4-121–122; performance 3-20, 4-120; pole 4-121; Tier 2 evaluation procedures 4-118–123; ties between elements 4-122

frames: braced 4-87–93; complete 3-49, 3-63, 3-85, 3-107, 4-67; concentrically braced 4-90–91; eccentrically braced 4-92–93; flat slab 3-60, 3-112, 4-60; infill walls 4-81–83; openings 4-99; precast 3-86, 3-113, 4-66; prestressed elements 3-60, 3-90, 4-60–61; steel 3-32, 3-35, 3-39, 3-42, 3-46, 4-109; walls 4-85

furnishings 3-128, 4-140–141

geologic site hazards 4-118–123

geometric irregularities 4-38

girders: general supplemental structural checklist 3-118; PC1 buildings 3-80; PC1A buildings 3-83; PC2A buildings 3-90; supports 3-101, 3-104, 3-118, 4-117; Tier 2 evaluation procedures 4-115

glazing 3-127, 4-133

ground shaking 1-1

historic structures 1-2–3

immediate occupancy performance level 1-11, 2-1, 2-3

investigation requirements 2-1–2

IO: *see* immediate occupancy performance level

joints: concentric 4-91; concentrically braced frame 3-40, 3-43, 3-116; eccentric 3-60, 3-113, 4-65; masonry 3-52, 3-74; reinforcing 3-60, 3-90, 3-113, 4-64; seismic 4-128; strong column/weak beam 3-33, 3-36, 3-60, 3-112–113, 4-56-57, 4-62

LDP: *see* linear dynamic procedure

ledgers, wood 3-92, 3-99, 3-110, 4-107

level of performance 2-3–4

level of seismicity 2-4–5; low level checklist 3-20

life safety performance level 1-11, 2-1, 2-3, 4-15

lighting 3-125, 3-127, 4-129–130; emergency 3-20, 3-122, 4-129; lens covers 4-130; supports 4-129–130

linear dynamic procedure 4-5–6; acceptance criteria 4-8–13, 4-12–13; mathematical model 4-6–8

linear methods 4-1, 4-2–6

linear static procedure 4-2–3; acceptance criteria 4-8–13, 4-12–13; mathematical model 4-6–8

liquefaction 3-119, 4-119

load path: C1 buildings 3-58; C2 buildings 3-62; C2A buildings 3-66; C3 buildings 3-70; C3A buildings 3-73; general basic structural checklist 3-105; low seismicity checklist 3-20; PC1 buildings 3-77; PC1A buildings 3-81; PC2 buildings 3-84; PC2A buildings 3-88; RM1 buildings 3-91; RM2 buildings 3-95; S1 buildings 3-31; S1A buildings 3-34; S2 buildings 3-38; S2A buildings 3-41; S3 buildings 3-45; S4 buildings 3-48; S5 buildings 3-51; S5A buildings 3-54; Tier 2 4-33; URM buildings 3-98; URMA buildings 3-102; W1 buildings 3-22; W1A buildings 3-25; W2 buildings 3-28

load transfer: C2 buildings 3-63; C2A buildings 3-67; C3 buildings 3-71; C3A buildings 3-74; general basic structural checklist 3-110; PC1 buildings 3-78; PC1A buildings 3-82; PC2 buildings 3-85; RM1 buildings 3-92; RM2 buildings 3-96; S2 buildings 3-39; S2A buildings 3-42; S3 buildings 3-46; S4 buildings 3-49; S5 buildings 3-52; S5A buildings 3-55; URM buildings 3-99; URMA buildings 3-103

LS: *see* life safety performance level

LSP: *see* linear static procedure

masonry: deterioration 3-52, 3-71, 3-74, 3-92, 3-96, 3-99, 3-103, 3-106; joints 4-46; lay-up 3-100, 3-104, 3-115, 4-79; units 4-46

masonry veneer: *see* veneer, masonry

mass, effective: C1 buildings 3-59; C2 buildings 3-62; C2A buildings 3-67; C3 buildings 3-71; C3A buildings 3-74; general basic structural checklist 3-105; PC1 buildings 3-78; PC1A buildings 3-81; PC2 buildings 3-84; PC2A buildings 3-88; RM1 buildings 3-92; RM2 buildings 3-95; S1 buildings 3-32; S1A buildings 3-35; S2 buildings 3-39; S2A buildings 3-42; S4 buildings 3-49; S5 buildings 3-52; S5A buildings 3-55; URM buildings 3-99; URMA buildings 3-102; W2 buildings 3-28

mass irregularities 4-40

materials, hazardous 3-129, 4-148–149

maximum considered earthquake 2-3

MCE: *see* maximum considered earthquake

members, compact 3-33, 3-36, 3-40, 3-43, 3-47, 3-112, 4-57

mezzanines: building systems evaluation 4-35; C1 buildings 3-58; C2 buildings 3-62; C2A buildings 3-66; C3 buildings 3-70; C3A buildings 3-73; general basic structural checklist 3-105; PC1 buildings 3-77; PC1A buildings 3-81; PC2 buildings 3-84; PC2A buildings 3-88; RM1 buildings 3-91; RM2 buildings 3-95; S1 buildings 3-31; S1A buildings 3-34; S2 buildings 3-38; S2A buildings 3-41; S3 buildings 3-45; S4 buildings 3-48; S5 buildings 3-51; S5A buildings 3-54; URM buildings 3-98; URMA buildings 3-102; W2 buildings 3-28

mitigation measures 1-1, 1-3

modal responses 4-5

moment frames 3-15–16; axial stress check 4-54; concrete 4-59–65; drift check 4-53–54; infill walls 4-52–53; precast concrete 4-66–67; redundancy 4-51–52; steel 4-53–59; Tier 2 evaluation procedures 4-50–69

mortar shear strength 2-1

openings 4-86; braced frames 4-99; bracing 3-23, 3-26, 3-29; corner 3-79, 3-83, 3-114, 4-77–78; diaphragm reinforcement 3-24, 3-27, 4-101; masonry shear walls 3-104, 4-100; reinforcing 3-56, 3-68, 3-72, 3-75, 3-86, 3-93, 3-97, 4-74, 4-79; shear walls 3-64, 3-104, 4-98–100; steel studs 4-136; walls 3-79, 3-83, 3-114, 4-77

ornamentation 4-137–139

overturning: C2 buildings 3-64; C2A buildings 3-68; general supplemental structural checklist 3-114; geologic site hazards and foundations checklist 3-119; PC2 buildings 3-86; S4 buildings 3-50; Tier 2 evaluation procedures 4-74, 4-121–122

panel zones 3-33, 3-36, 3-112, 4-56

panels: connections 4-132–133; multi-story 4-131–132; precast wall 3-82, 3-111, 4-113; roof 3-46, 3-111, 4-117; roof connections 4-118; walls 3-46, 4-114, 4-117

parapets: basic nonstructural component checklist 3-123; concrete 4-138; intermediate nonstructural component checklist 3-125; low seismicity checklist 3-20; Tier 2 evaluation procedures 4-137–139; URM 4-137

partitions 3-127, 4-125–126, 4-126

PC1 building type: basic structural checklist 3-77–78; described 2-10, 3-77; supplemental structural checklist 3-79–80

PC1A building type: basic structural checklist 3-81–82; described 2-10, 3-81; supplemental structural checklist 3-83

PC2 building type: basic structural checklist 3-84–85; described 2-10, 3-84; supplemental structural checklist 3-86–87

PC2A building type 2-11

PC2A building types: basic structural checklist 3-88–89; described 3-88; supplemental structural checklist 3-90

penthouses 3-28

period 3-14–15, 4-3

pile caps: C1 buildings 3-61; C2 buildings 3-65; C2A buildings 3-69; C3 buildings 3-72; C3A buildings 3-76; general supplemental structural checklist 3-117;

PC1 buildings 3-80; PC1A buildings 3-83; PC2 buildings 3-87; PC2A buildings 3-90; S1 buildings 3-33; S1A buildings 3-37; S2 buildings 3-40; S2A buildings 3-44; S3 buildings 3-47; S4 buildings 3-50; S5 buildings 3-53; S5A buildings 3-57; Tier 2 evaluation procedures 4-114–115
piping 3-124, 3-128–129, 4-144–146
posts, wood 3-23, 3-26, 3-29, 3-110, 4-112
power, emergency 3-20, 3-124, 4-142
precast concrete frames: *see* PC2 or PC2A building types
precast connections 3-17–18
precast/tilt-up concrete shear walls: *see* PC1 or PC1A building types
pseudo lateral forces 3-9–11, 4-3

Quick Checks 3-5, 3-9; for strength and stiffness 3-15–19

redundancy: braced frames 4-87–88; C1 buildings 3-59; C2 buildings 3-63; C2A buildings 3-67; C3 buildings 3-71; C3A buildings 3-74; general basic structural checklist 3-107, 3-109; moment frames 4-51–52; PC1 buildings 3-78; PC1A buildings 3-82; PC2 buildings 3-85; PC2A buildings 3-89; RM1 buildings 3-92; RM2 buildings 3-96; S1 buildings 3-32; S1A buildings 3-35; S2 buildings 3-39; S2A buildings 3-42; S4 buildings 3-49; S5 buildings 3-52; S5A buildings 3-55; shear walls 3-22, 3-26, 4-70–71; URM buildings 3-99; URMA buildings 3-103; W2 buildings 3-29
reinforced masonry bearing walls: *see* RM1 or RM2 building types
reinforcing: confinement 3-50, 3-64, 3-68, 3-86, 3-114, 4-74; joints 3-90; at openings 3-50, 3-53, 3-56, 3-64, 3-68, 3-72, 3-75, 3-86, 3-93, 3-97, 3-114, 4-74, 4-79
RM1 building type 2-11; basic structural checklist 3-91–92; described 3-91; supplemental structural checklist 3-93–94
RM2 building type 2-11; basic structural checklist 3-95–96; described 3-95;
evaluation example A-43–57; supplemental structural checklist 3-97
roofs: chord connections 3-30; panel connections 4-118; panels 4-117

S1 building type 2-7; basic structural checklist 3-31–32; described 3-31; supplemental structural checklist 3-33
S1A building type: basic structural checklist 3-34–35; described 2-8, 3-34; evaluation example A-14–32; supplemental structural checklist 3-36–37
S2 building type: basic structural checklist 3-38–39; described 2-8, 3-38; evaluation example A-72–86; supplemental structural checklist 3-40
S2A building type: basic structural checklist 3-41–42; described 2-8, 3-41; supplemental structural checklist 3-43–44
S3 building type: basic structural checklist 3-45–46; described 2-8, 3-45; supplemental structural checklist 3-47
S4 building type: basic structural checklist 3-48–49; described 2-8, 3-48; supplemental structural checklist 3-50
S5 building type: basic structural checklist 3-51–52; described 2-9, 3-51; supplemental structural checklist 3-53
S5A building type: basic structural checklist 3-54–55; described 2-9, 3-54; supplemental structural checklist 3-56–57
screening phase 1-4; Tier 1 3-1–129
seismic evaluation: evaluation process 1-7; phases 1-3–6; requirements 2-1–11; summary data sheet B-1
seismic forces 3-9–14, 4-3–4
shear capacity 4-61
shear stress: C1 buildings 4-59; concrete frame columns 3-16–17; concrete shear walls 4-72; precast concrete shear walls 4-76; reinforced masonry 4-78–79; unreinforced masonry 4-79; W1 buildings 3-23, 4-83; W1A buildings 3-26, 4-83; W2 buildings 3-29, 4-83
shear walls: acceptance criteria 4-24; concrete 4-71–72; gypsum wallboard 3-

23, 3-26, 3-29, 4-84; load transfer 4-109; openings 4-98–99; plaster 3-23, 3-26, 3-29, 4-84; precast concrete 4-76–78; redundancy 3-22, 3-26, 4-70–71; reinforced masonry 4-78–79; stucco 3-23, 3-26, 3-29, 4-83–84; Tier 2 evaluation procedures 4-23–24, 4-69–86; unreinforced masonry 4-79; wood 3-23, 3-26, 3-29, 4-84

sheathing: straight 3-24, 3-27, 3-30, 3-37, 3-43, 3-56, 3-68, 3-75, 3-79, 3-93; wood diaphragms 4-102

shelf angles 4-133–134

sills, wood: bolts 3-24, 3-27, 3-30, 3-117, 4-114; general basic structural checklist 3-110; general supplemental structural checklist 3-117; Tier 2 evaluation procedures 4-112; W1 buildings 3-23; W1A buildings 3-26; W2 buildings 3-29

site visits 2-2–3

sites, hillside: general basic structural checklist 3-109; Tier 2 evaluation procedures 4-85; W1 buildings 3-23; W1A buildings 3-26; W2 buildings 3-29

sites, sloping 3-120, 4-123

slabs, flat 3-61, 3-64, 3-72, 3-114, 4-68–9

slabs, topping 3-82, 3-85, 3-89, 3-96, 3-110, 4-105, 4-110

slope failure 3-119, 4-119

soil profiles 3-12–13

spans 3-24, 3-27, 3-30, 3-37, 3-56, 3-69

spectral acceleration 3-11–14

splices: beam-bar 3-60, 4-63; column-bar 3-60, 3-113, 4-63; columns 3-33, 3-36, 3-39, 3-42, 3-49, 4-76, 4-88

stairs 3-123, 4-139–140

steel 4-44

steel beams 4-13

steel braced frames: *see* S1A, S2, or S2a building types

steel deterioration: general basic structural checklist 3-106; S1 buildings 3-32; S1A buildings 3-35; S2 buildings 3-39; S2A buildings 3-42; S3 buildings 3-45; S4 buildings 3-49; S5 buildings 3-52; S5A buildings 3-55

steel frames with concrete shear walls: *see* S4 building type

steel frames with infill masonry shear walls: *see* S5 or S5A building types

steel light frames: *see* S3 building type

steel moment frames: *see* S1 building type

steel, reinforcing 4-72–73, 4-77, 4-79

stirrups: spacing 3-60, 3-113, 4-64; tie hooks 3-61, 3-113, 4-65

story drift 3-15–16

story shear forces 3-11

story, soft: building systems evaluation 4-37; C1 buildings 3-58; C2 buildings 3-62; C2A buildings 3-66; C3 buildings 3-70; C3A buildings 3-73; general basic structural checklist 3-105; PC1 buildings 3-77; PC1A buildings 3-81; PC2 buildings 3-84; PC2A buildings 3-88; RM1 buildings 3-91; RM2 buildings 3-95; S1 buildings 3-31; S1A buildings 3-34; S2 buildings 3-38; S2A buildings 3-41; S4 buildings 3-48; S5 buildings 3-51; S5A buildings 3-54; URM buildings 3-98; URMA buildings 3-102; W1A buildings 3-25; W2 buildings 3-28

story, weak: building systems evaluation 4-36–37; C1 buildings 3-58; C2 buildings 3-62; C2A buildings 3-66; C3 buildings 3-70; C3A buildings 3-73; general basic structural checklist 3-105; PC1 buildings 3-77; PC1A buildings 3-81; PC2 buildings 3-84; PC2A buildings 3-88; RM1 buildings 3-91; RM2 buildings 3-95; S1 buildings 3-31; S1A buildings 3-34; S2 buildings 3-38; S2A buildings 3-41; S4 buildings 3-48; S5 buildings 3-51; S5A buildings 3-54; URM buildings 3-98; URMA buildings 3-102; W1A buildings 3-25; W2 buildings 3-28

strength: axial 4-19; masonry 4-18; out-of-plane 4-14; shear 4-18–19, 4-19–20; shear walls 4-24

structural materials 4-43

structural separations 4-126

stud tracks 4-136

supports, truss 3-101, 3-104, 3-118, 4-117

testing: masonry 4-17; mortar 4-16
tie spacing 4-63–64
Tier 1: analysis 3-9–19; checklist locator 3-21; checklists 3-5–6
Tier 2 analysis 4-1–32
ties, cross 3-43, 3-56, 4-97
torsion 4-85; building systems evaluation 4-41–42; C1 buildings 3-59; C2 buildings 3-63; C3 buildings 3-71; general basic structural checklist 3-105; horizontal 4-6–7; PC1A buildings 3-82; PC2 buildings 3-84; PC2A buildings 3-89; RM2 buildings 3-95; S1 buildings 3-32; S2 buildings 3-39; S3 buildings 3-45; S4 buildings 3-49; S5 buildings 3-52; URMA buildings 3-103
truss supports 3-101, 3-104, 3-118, 4-117

unreinforced masonry bearing walls: *see* URM or URMA building types
URM building type: basic nonstructural component checklist 3-122; basic structural checklist 3-98–99; described 2-11, 3-98; evaluation example A-87–108; special evaluation procedures (Tier 2) 4-15–26; supplemental structural checklist 3-100–101; testing (Tier 2) 4-15–16
URMA building type: basic structural checklist 3-102–103; described 2-11, 3-102; supplemental structural checklist 3-104

veneer, masonry: basic nonstructural component checklist 3-123; mortar 4-135; stone cracks 4-135–136; supplemental nonstructural component 3-128; Tier 2 evaluation procedures 4-133–136; ties 4-134; weakened planes 4-134; weep holes 4-135
vibration isolators 4-143

W1 building type: basic structural checklist 3-22–23; described 2-7, 3-22; evaluation example A-2–13; supplemental structural checklist 3-24

W1A building type: basic structural checklist 3-25–26; described 2-7, 3-25; supplemental structural checklist 3-27
W2 building type 4-83–86; basic structural checklist 3-28–29; described 2-7, 3-28; evaluation example A-58–71; supplemental structural checklist 3-30
walls: anchorage 3-20, 3-67, 3-78, 3-82, 3-85, 3-92, 3-96, 3-99, 3-103, 3-110, 4-18, 4-25, 4-106–107; connections 3-50, 3-52, 3-55, 3-71, 3-74, 3-108, 3-114–115, 4-75, 4-81; cracks in concrete 3-49, 3-63, 3-67, 3-85, 3-106, 4-47; cracks in infill 3-52, 3-55, 3-71, 3-74, 3-106, 4-48; cracks in reinforced masonry 3-92, 3-96, 3-106, 4-48; cracks in unreinforced masonry 3-99, 3-106, 4-48; cripple 3-23, 3-26, 3-29, 3-109, 4-85–86; infill 3-32, 3-35, 3-72, 3-75, 3-115, 4-81–83; interfering 3-107; multi-wythe 4-15; openings 3-79, 3-83, 3-114, 4-77; out-of-plane forces 4-13–14; panels 3-78, 4-117; precast concrete 3-78, 3-82, 3-106, 4-46; proportions 3-53, 4-79, 4-82; reinforcement 3-49; solid 3-53, 3-56, 3-72, 3-75, 3-115, 4-82; stucco 3-109; thickness 3-50, 3-64, 3-68, 3-79, 3-83, 3-86, 3-114, 4-75, 4-78; unreinforced masonry 3-123; URM 4-139
weep holes 4-135
wood: ledgers 3-92, 3-99, 3-110, 4-107; posts 3-23, 3-26, 3-29, 3-110, 4-112; sills 3-23, 3-24, 3-26, 3-27, 3-29, 3-30, 3-110, 3-117, 4-112, 4-114
wood deterioration: C2A buildings 3-67; C3A buildings 3-74; general basic structural checklist 3-106; PC1 buildings 3-78; RM1 buildings 3-92; S1A buildings 3-35; S2A buildings 3-42; S5A buildings 3-55; Tier 2 evaluation procedures 4-43; URM buildings 3-99; W1 buildings 3-22; W1A buildings 3-25; W2 buildings 3-29
wood frames, commercial and industrial: *see* W2 building type
wood light frames: *see* W1 or W1A building types